# Mycorrhizal Symbiosis

# Mycorrhizal Symbiosis

J. L. HARLEY, FRS

*Department of Forestry*
*University of Oxford*
*South Parks Road, Oxford*

and

S. E. SMITH

*Department of Agricultural Biochemistry*
*Waite Agricultural Research Institute*
*University of Adelaide*
*Glen Osmond, South Australia*

1983

ACADEMIC PRESS

*A Subsidiary of Harcourt Brace Jovanovich, Publishers*

London · New York

Paris · San Diego · San Francisco · São Paulo

Sydney · Tokyo · Toronto

Academic Press Inc. (London) Ltd
24/28 Oval Road
London NW1

*US Edition published by*
Academic Press Inc.
111 Fifth Avenue
New York, New York 10003

*British Library Cataloguing in Publication Data*

Harley, J. L.
    Mycorrhizal symbiosis.
    1. Mycorrhiza
    I. Title        II. Smith, S. E.
    589.2′045′24        QK604

        ISBN 0-12-325560-0

Filmset by Latimer Trend & Company Ltd, Plymouth
Printed in Great Britain by
St Edmundsbury Press, Bury St Edmunds, Suffolk

# Preface

There has been so great an increase of interest in mycorrhizal symbiosis in the last ten years that it is now almost impossible for one person or even two to keep up with all the experimental work and speculation upon it. This book gives our personal appreciation of the subject and reflects our own interests; it does not pretend to review the subject in all its detail. Only a fraction of the relevant papers have been mentioned especially on matters which have received repeated experimental examination. Part 1 aims to provide the reader with a general account of what has been discovered about the commoner sorts of mycorrhiza, and also to emphasize gaps in our knowledge where new information is required. In Part 2 we have written essays on several subjects which seem to us to be fundamental to the appreciation of mycorrhizal symbiosis as a biological phenomenon. In these essays we have tried to ensure that our speculations are restricted to suggestions that can be experimentally examined by existing methods. We have avoided evolutionary discussions for we feel that these will only be really valuable when more is known of the physiology, genetics, and the ecological implications of mycorrhizal symbiosis. The practical application of knowledge of mycorrhiza to plant production is an extremely important subject and is expanding fast. It is at the moment very empirical and requires a different kind of treatment from those aspects of mycorrhizal symbiosis that we have discussed. Throughout our account we have indicated the kinds of enquiry which are essential before a change in the practical application of mycorrhiza from a "suck it and see" state to one in which reliable predictions can be made. It has been our aim to make each of the essays in the second part of the book capable of being read by themselves. Consequently there is some repetition of information but we have tried to keep this to a minimum. It is our hope that a book organized in this way will prove useful to students, researchers and to teachers by providing information and opinion and above all by stimulating discussion.

Many friends and colleagues have assisted us in countless ways, but especially in discussions and by letting us use unpublished results of their work. It would be tedious to name them all but we must mention a few. We have had valuable discussions with G. D. Bowen, Vivienne Gianinazzi-Pearson, Gösta Lindeberg,

Karen Cooper, André Fortin, David Read, Jane Duddridge, J. Warren Wilson, D. H. Lewis and B. C. Loughman. Professor H. Wallace read and commented upon an early draft of our essay on specificity. Andrew Smith gave us advice on many physiological subjects, especially on the absorption of ions. In addition, our conversations with students, research colleagues, and mycorrhizasts of all kinds have probably contributed much more than we are aware. Many people allowed us to read their papers before publication, or corresponded with us about them. For this we thank especially N. A. Walker, L. K. Abbott, J. H. Silsbury, D. N. Pegler, Kristina Vogt, J. M. Trappe, A. J. Oliver, Randy Molina, Francis Sanders, P. B. Tinker, M. H. Ivory and M. D. Ramsey. We received the greatest kindness from many colleagues who let us use their photographs or figures. We would like to thank, in addition to many already mentioned, Paola Bonfante-Fasolo, C. A. Chambers, M. A. Atkinson, M. F. Brown and P. T. N. Spencer-Phillips and D. Mousain.

We must take the responsibility for the ideas and views expressed, whether good or bad. We hope that any hypotheses we have put forward will stimulate our friends to tear them down or perhaps modify them by further experiment.

F. A., C. L. and H. F. Smith have all given us help in sorting and checking references and preparing diagrams. E. L. Harley, who could read our handwriting, prepared the first typescript for correction, gave us botanical advice and helped make the index. Mrs G. Hogg and Mrs C. Schoenfeldt typed drafts of a number of chapters. Mrs Hilda Pengelly prepared the final typescript. We must mention Tony Fox, Stuart Young, Brian Longworth, David Kerr and Paul Embden for their skilled photography. C. C. McCready, besides cooperating in experimental work over more than 35 years, read proofs and criticized the text. To all these we are extremely grateful.

We also acknowledge the help of those who allowed us to use their copyright published material including especially the publishers, editors or trustees of the following:

| | |
|---|---|
| Academic Press Inc. | Plates 8B and 14 |
| *Acta Oecologica* | Plate 3 |
| *Annales des Sciences naturelles; Botanique* | Figure 23 |
| Cambridge University Press | Table 81 |
| *Canadian Journal of Microbiology* | Plate 4 |
| McGraw Hill Book Company | Table 82 |
| *New Phytologist* | Figures 3, 7 and 9, Plates 11D, 12A–E, and 15A–L |
| *Zeitschrift für Pflanzenphysiologie* | Plate 5 |

Two points need mention concerning the references to literature and the names of plants. Where a primary source of information is quoted it appears in the text with the author's name and date. For secondary sources of information, such as review articles which quote and collate information, the word *see* precedes the author's name. The names of plants given by authors are quoted, but where a nomenclatural change has occurred or a second name is current in mycorrhizal literature, we have also given it, e.g. *Suillus (Boletus) grevillii (elegans)*. This has seemed to us a common-sense way of coping with this problem without loading the text with footnotes.

J. L. HARLEY

S. E. SMITH

*December 1982*

# Contents

## PART 2: ESSAYS ON MYCORRHIZAL SUBJECTS

*Introduction:*

# Mycorrhizal Symbiosis

### Fungal symbiosis

During the last two decades there has been a great increase of interest in symbiosis which has taken two forms. On the one hand speculations about the origin of life and of living cells have led to the conclusion that organelles of eucaryotes were derived by modification of symbiotic procaryotic organisms. On the other hand it has been realized that mutualistic symbiosis at the level of complex organisms is more common and of greater ecological importance than previously believed. These mutually beneficial symbiotic systems of relatively recent origin, such as those between fungus and alga in lichens, plant and fungus in mycorrhiza, alga and coelenterate in corals, and bacteria and angiosperm in nitrogen-fixing nodules, play an exceedingly influential part in natural ecosystems and some may be of great importance also in man-made biological systems. In these symbioses, two or more physiologically different organisms act as component parts of a single super-organism, of which both the whole and each component may be subject to variation, selection and evolution, for each has a separate reproductive process and often a potential for free life. Research upon them is important in itself and has a bearing upon the speculations about the symbiotic origin of cell structure.

The word symbiosis was first used by de Bary (see 1887) to signify the common life of parasite and host. In the course of time the meaning of symbiosis and the meaning of parasite changed. Symbiosis was used more and more, especially by plant biologists, for mutually beneficial associations between dissimilar organisms, and parasite and parasitism came to be almost synonymous with pathogen and pathogenesis. de Bary had pointed out that there was every conceivable gradation between the parasite that quickly destroys its victim and those that "further and support" their hosts, and plant biologists have in recent years come back to this view; it is the basis of much comparative work and thought on symbiotic systems.

As indicated previously (Harley, 1968, 1969a), much of the credit for the unification of symbiotic studies depended upon work with lichens, coelenterates, ectomycorrhiza, and upon uredinalean pathogens. These lines of experimental investigation received comparative comment in an influential article by D. C. Smith, Muscatine and Lewis (1969), especially in respect of carbon metabolism, a process with many common features in diverse biotrophic systems.

de Bary (1887) believed that there was some degree of common life, i.e. of symbiosis, in all or almost all examples of associated growth. Now, we tend to classify associations as either *biotrophic* or *necrotrophic* according to whether both associates remain alive or whether substances released by the death of one are absorbed by the other. There is clearly, as indeed de Bary realized, a great range of behaviour in this regard and a grey penumbra between biotrophy and necrotrophy. There is variation between different kinds of association, within the same kind of association in different environments, and at different times in the existence of a single association. The changes which occur during the normal development of a symbiosis may be very great, but have been given little attention. For instance, behind a root apex, mycorrhizal infection may be initiated and develop into a relationship in which there is an exchange of materials between the partners. Subsequently, *local* death of one partner and the absorption of its contents by the other may also occur. This last phase has been viewed as "parasitism" by some authors, so that Nylund (1981) considered that root cells are parasitized by ectomycorrhizal fungi when hyphae colonize them, and Lewis (1973) views the orchid as parasitic upon its fungus when disintegration of the intracellular hyphae takes place. Both these occurrences are part of the normal development of the mycorrhizal symbioses. Similar problems of nomenclature occur in attempts to classify fungal pathogens; Cooke calls fungi "hemibiotrophs" (see Cooke, 1977), which have both a biotrophic and necrotrophic stage in the normal progress of the diseases that they cause. It is clear that since these kinds of variation occur, much of the discussion based on classification and nomenclature is pedantic unless it is helpful in formulating clear questions which lead to experiments from which answers may be obtained.

Mutualistic bacterial and fungal symbioses with higher plants were first described about a century ago. Since then much has been learned of their structure, distribution, physiology and ecology, but, mainly owing to the lack of suitable techniques of investigation, many wild or speculative hypotheses and many sharp differences of view have been, from time to time, put forward. Owing to advances in technical methods over the last 40 years we now have a sufficient knowledge of the physiological functioning and ecological significance of mycorrhiza not only to make use of it in agricultural and forest production, but also to ask fundamental questions about fungal symbiosis. We should now begin to ask not only questions about the physiological and biochemical functioning of symbiotic organs, but also why so many diverse fungi of different sub-orders, families and genera of Phycomycetes, Ascomycetes and Basidiomycetes play a part in mycorrhizal and lichen symbioses? What have they in common with, and what in difference from, other kinds of fungi? Were they evolved into diverse taxa before becoming symbionts? Was each taxon pre-adapted to mutualisitic symbiosis? What have the autotrophic partners of each kind of symbiosis in common? How do they differ from related organisms? And so on. Other questions demanding attention and having perhaps even wider implications relate to specificity of pairs of organisms one to another and recognition of suitable partners by each participant. Is it not remarkable that mycorrhizal fungi, which set up long-lived mutualistic associations with their hosts, appear to be less specific by far than biotrophic pathogens such as Uredinales? Again, if mycorrhizal fungi are so unspecific and numerous, does the host recognize a suitable fungus or an unsuitable one?

In considering these and similar questions there is one matter which must be emphasized because it is of very great importance and has up till now received too little attention. Species of fungi delimited by reputable taxonomic methods may vary biochemically. Genetical changes involving the activity of one or few enzymes may be of great significance. For instance some strains included in *Boletus subtomentosus* can hydrolyse lignin and cellulose. Some of these have been found to be unable to form mycorrhizas with conifers, whereas strains which are unable to hydrolyse these compounds do so. The particular relevance of such examples, apart from the need to be careful of generalizing about symbiotic activity on the basis of taxonomic similarity, will be described later. Here it is sufficient to point out that the designation of a species of fungus as a "facultative mycorrhiza former" has two possible meanings. It can mean either that the individuals of that species differ genetically from one another in properties which affect their symbiotic activity, or that environmental factors affect genetically similar members of the species and determine whether they behave symbiotically or not.

In subsequent pages, an account will be given of the present state of knowledge of mycorrhizal symbioses. In this it will not be our aim to review the literature but to present an adequately documented readable account. Much that was discussed in detail in "The Biology of Mycorrhiza" will be brought up to date but treated succinctly. Following this, we will present essays on important aspects of mycorrhizal symbiosis, discussing each subject in a comparative way so that they may excite experimentally useful questions.

The aim of much of the research on mycorrhiza must be to provide a corpus of information which will improve the understanding of the interaction of associated organisms. Such information, apart from its scientific interest, would assist the practical users of mycorrhizal symbioses who at the moment rely mainly upon the results of empirical investigations of the effects of mycorrhizal inoculation on plant productivity.

## The nature of mycorrhizal association

Although the term "mycorrhiza" would seem to imply the association of fungi with roots, relationships called "mycorrhizal associations" are found between hyphal fungi and the organs of higher plants of whatever morphological origin, which are concerned with absorption of substances from the soil. Mycorrhizas are indeed the chief organs involved in nutrient uptake of most land plants. The presence of the fungal associate of mycorrhizal systems in the root region, on or in the root tissues and surrounding soil, inevitably ensures that it influences the absorption of soil-derived substances by the host and is influenced by substances exuded or lost by the host. In this respect the mycorrhizal fungus is a specialized member of the root-region or rhizosphere entourage of microorganisms. In particular, mycorrhizal infection usually increases the efficiency of nutrient absorption by its host, from which the fungus directly obtains carbon compounds. The essential difference between mycorrhizal associations and the general association of organisms with the root surface or rhizosphere lies in the closeness of the relationship. In mycorrhizas there is always some penetration of the tissues, or a recognizable structure conforming to one of the many common patterns, or both. The difference of the mycorrhizal condition from disease depends on it being the normal state of both partners in at least some ecological situations. The partners are dependent one upon the other and interchange of material takes place between their living cells, although their association may result in the death of some of their parts, e.g. in the degeneration of hyphae, haustoria or arbuscules, or the death of cortical or epidermal cells or root hairs. For the reason that mycorrhizas possess a recognizable structure, their description by Frank in 1885 preceded by many years the recognition of the universal association of microorganisms and roots—rhizosphere populations— by Hiltner in 1904. The first examples that Frank recognized and named "mykorrhizen" were those of trees of temperate forests, beech and pine, which are morphologically very different from uninfected roots. In 1887 he further categorized them as "ectotrophisch" or ectotrophic, because they possessed a conspicuous fungal tissue, the sheath or mantle, surrounding the host root. In contrast, many other kinds of permanent association of fungi with roots were described soon after his first publication. These lacked a compact external fungal tissue and their hyphae penetrated into the cells of the host. They were called "endotrophisch".

Although it was Frank who gave the name "mycorrhiza" to permanent associations of roots with hyphal fungi, there had been previous work on them. For instance, Reissek (1847) described hyphae in the cells of various angiosperms, especially in the Orchidaceae, and in 1881 Kamienski published an account of *Monotropa*, describing its association with fungal hyphae and showing that a complete fungal layer was formed about the roots. He was the first to point out that whatever was absorbed from the soil must pass through the fungal layer. Frank's paper published a year later was followed in due course by experiments (1894) which showed that ectomycorrhizal plants of pine grew faster than non-mycorrhizal plants. Hence the interest in the association of fungi with root systems began, as might be expected, in the investigation of those kinds of mycorrhiza (the ectomycorrhizas) which were most easily recognized visually. Research on the mycorrhiza of members of the Ericaceae and Orchidaceae from which fungi were fairly readily isolated soon followed. It was only much later that anything more than descriptive work was done on vesicular-arbuscular mycorrhiza, the fungi of which have still not been successfully brought into culture. Although there are hints of it in Rayner's work (1927; Rayner and Neilson-Jones, 1944), only after 1948 when mycorrhiza and soil ecology were reviewed and further elaborated in "The Biology of Mycorrhiza" (Harley, 1948; Harley, 1959) did mycorrhizal associations come to be seen as specialized end-terms, as it were, of the general association of microorganisms with roots.

The presence of active microorganisms on the root surface used to be ascribed to the presence in the rhizosphere of compounds essential for growth. Amino acids, vitamins, growth factors, etc. in particular were believed to be important (see Harley, 1948). It is, however, now apparent that carbon compounds suitable as main substrates for growth are not plentiful in the soil (see Gray and Williams, 1971) and that locations such as the root surface, where exudations and sloughing of tissue provide continuous supplies of readily available carbon substrates, are inevitably occupied by active populations. In contrast, the great mass of the soil, which does not contain such abundant supplies of readily available carbon compounds, contains resting propagules of many fungi which germinate and flare into activity when a carbon source arrives. Both the rhizosphere population and most mycorrhizal fungi have been found to depend upon carbon supplies provided by current or recent photosynthesis by the host plant, rather than carbonaceous detritus from litter fall or dead roots (see Harley, 1971, 1973). These problems will take up a later chapter but here it should be noted that although the mycorrhizal associations of chlorophyll-free plants of various taxa, e.g. of the Orchidaceae, Burmanniaceae, Monotropaceae and Pteridophyta, might appear to be exceptions, not all of them have proved to be so. As will be seen, it is inadvisable to group these higher plants as necrotrophic parasites on the fungi which infect them, although it is true that their carbon compounds are derived in whole or in part from those fungi. Only part of the fungus degenerates when so-called "digestion" occurs, and the efficiency of the remainder as compared with the free-living or parasitic state is uninvestigated.

Mycorrhizal organs may take one of a number of forms depending in particular on the nature of the higher plant and of the fungus, and common kinds have been classified and named. This has aided investigation of their functioning, but it is now becoming clear that there is much similarity in general physiology with some detailed specialization between many of the different kinds, so that the previous bases of their separation have been questioned.

Frank distinguished two main types of mycorrhiza, ectotrophic and endotrophic, the first having, amongst other attributes, a well-defined external sheath of fungal tissue enclosing the root, and the second having no sheath, but inter- and intracellular penetration of the host by the fungus. These names, ectotrophic and endotrophic, have been recently rejected and the contracted forms, ectomycorrhiza and endomycorrhiza, substituted. Many authors have been at pains to indicate that whereas the group ectomycorrhizas have some recognized cohesion, the endomycorrhizas are diverse and include many kinds with different functioning as well as different structure. It is possible, however, to see a continuity, a continuous gradation of kinds from the ectomycorrhizas of pine and beech first described by Frank, into endomycorrhizal forms with septate higher fungal associates on the one hand, and into a plexus of ectendomycorrhizas (and pseudo-mycorrhizas) which have varied development of sheath and intracellular as well as intercellular hyphal penetration on the other. Such advances of knowledge must lead to some modification of the classification of mycorrhizal symbioses, but we would plead that modifications should lead to clearer thinking and should stimulate experimentation and that classification should not be an end in itself.

In addition, one general term needs especial consideration, and that is the term "host". It has been traditionally used in cases of pathogenicity for the partner which is damaged in an association, whether it be a fungus, e.g. in *Mucor-Piptocephalis* associations, or the higher plant in a disease of a crop plant. Since in mycorrhizal associations it is only possible to describe give and take of substances between associates, the use of the term "host" for one of the associates appears to be misleading. Similarly the use of "higher plant" for one associate immediately raises all sorts of irrational discussions of what is meant by "higher". The term "autobiont" has sometimes been used instead and, provided its etymology is not examined too closely and it is used as a *name* for the non-fungal partners of mycorrhizas, etc., it is acceptable. In this book we use both the terms "autobiont" and "host" on occasion. The dilemma of using the term "host" for many achlorophyllous, non-fungal symbionts, especially in associations involving a third truly carbon autotrophic organism, will have to be resolved by a nomenclatural change which will depend upon an increase of knowledge of the interaction between the associates.

## Classification of mycorrhizas

The classification of mycorrhizas to be adopted here is shown in Table 1. It is designed to be descriptive and to emphasize problems in need of solution rather than to gloss over difficulties. Predictions made about the properties of the members of the groups and tested by experiment will reveal the extent of usefulness of the groupings and the probability that further generalizations can be made about them.

The original breakdown into ectotrophic (now ecto-) mycorrhizas and endotrophic (now endo-) mycorrhizas has fallen into disrepute. Yet it had one great merit. It attempted to separate those mycorrhizas where the fungus does not penetrate the cells of the host from those where the fungus enters the cells and forms various fungal organs or structures within them. This difference is not now stressed as much as it used to be on the grounds of the diversity of endomycorrhizas in the kind of fungus, in the kind of host, as well as in structure. Nevertheless the difference is real and there is a danger of throwing the baby away with the bath water. Indeed more and more, the study of the fine structure of endomycorrhizas demonstrates a great similarity of structural relationship between host cell and fungus in all of them; and what is more, between endomycorrhizas, *Rhizobium*, *Frankia* and intracellular parasites in the invagination of plasmalemma, in their interfacial matrices and in their wall structures.

There have been repeated classifications of mycorrhizas made in recent years (for instance, Lewis, 1973, 1975, 1976; Read, 1982) which have been useful in raising discussions, but carry the danger that they may be taken too seriously. This is not said in disparagement of the papers quoted but because any well argued case may carry a sense of authority which is detrimental to the purpose of this kind of classification which differs from a taxonomic classification, for it should emphasize problems. In Table 1 no attempt has been made to link together the different kinds of mycorrhizas beyond entering the presence or absence of various characters as far as they are known. It is impossible to tabulate all the differences, especially details of behaviour, without making the Table unwieldy. When the reader has considered discussions of each kind of mycorrhiza in Part 1 of the book, any further points bearing on classification will become clear. The kinds of mycorrhiza are divided first on the basis of their fungi into those formed with aseptate, phycomycetous endophytes, and those formed with septate endophytes with affinity to Ascomycetes or Basidiomycetes. The autobionts of mycorrhizas are so many and taxonomically so diverse that primary classification on that basis would not be feasible.

In the first group the fungi are aseptate Phycomycetes which belong to the Endogonaceae and are related to other Zygomycetes. However, most are not known in the free-living form, and have not, with minor exceptions, been brought into pure culture. The associated autobionts may belong to all phyla: Bryophyta (especially Hepaticae), almost all groups of Pteridophyta, all groups of Gymnospermae, and the majority of families of the Angiospermae. They may be trees, shrubs or herbs. The fungi have aseptate hyphae which flourish around

## TABLE 1

The characteristics of the important kinds of mycorrhiza. The structural characters given relate to the mature state, not the developing or senescent states. Entries in brackets rare

| | Kinds of mycorrhiza | | | | | | |
|---|---|---|---|---|---|---|---|
| | Vesicular-arbuscular | Ectomycorrhiza | Ectendomycorrhiza | Arbutoid | Monotropoid | Ericoid | Orchid |
| Fungi: septate | – | + | + | + | + | + | + |
| aseptate | + | (+) | – | – | – | – | – |
| Hyphae enter cells | + | – | + | + | + | + | + |
| Fungal sheath present | – | + | + or – | + | + | – | – |
| Hartig net formed | – | + | + | + | + | – | – |
| Hyphal coils in cells | + | – | + | + | – | + | + |
| Haustoria dichotomous | + | – | – | – | – | – | – |
| not dichotomous | – | – | – | – | + | – | + or – |
| Vesicles in cells or tissues | + (or –) | – | – | – | – | – | – |
| Achlorophylly | – (or +) | – | – | – (or +) | + | – | + |
| Fungal taxon | Phyco | Basidio Asco Phyco | Basidio Asco? | Basidio | Basidio | Asco (Basidio) | Basidio |
| Host taxon | Bryo Pterido Gymno Angio | Gymno Angio | Gymno Angio | Ericales | Monotropaceae | Ericales | Orchidaceae |

the root surface and penetrate both between and within the cells of the autobiont. They typically form branched haustoria (arbuscules) in the cells and swollen vesicles inside and outside the host tissues. Many form spores or sporocarps of complex structure. This type of mycorrhiza is called vesicular-arbuscular mycorrhiza.

Aseptate fungi, members of the genus *Endogone* (*sensu stricto*) may also form a kind of ectomycorrhiza with conifer and angiosperm trees. These mycorrhizas usually have a very thin sheath and well-developed intercellular penetration (Hartig net), but the fungus does not invade the cortical cells.

The septate fungi of the remaining kinds of mycorrhiza include members of almost all groups of Basidiomycetes, many hypogeous Ascomycetes and others with large fruit bodies. The autobionts are usually trees, shrubs or dwarf shrubs especially those that form ecto-, ectendo-, and ericoid mycorrhizas. Arbutoid mycorrhizas are also formed by trees and shrubs though some of the hosts, such as species of Pyrolaceae, are herbs, often partially achlorophyllous. The Orchidaceae and Monotropaceae, all partially or wholly achlorophyllous, are also herbaceous. Different members of the same species of host and of fungus are reported to form arbutoid or ectomycorrhizas, ecto- or ectendomycorrhizas, so that there is a plexus of behaviour amongst the species of autobiont and the septate fungi with regard to mycorrhizal structures that they may produce.

In ectomycorrhizas the fungus (whether aseptate or septate) forms a structure called the mantle or sheath which encloses the rootlet. From it hyphae or hyphal strands may radiate outwards into the substrate. Hyphae also penetrate inwards between the cells of the root to form a complex intercellular system which appears as a network of hyphae in section, called the Hartig net. There is little or no intracellular penetration. At times or as a stage in development, the Hartig net may be almost or totally absent. This state has been called "superficial" or "perirhizic" mycorrhiza. A mycorrhizal association of this kind has recently been described in *Pisonia grandis* (Nyctaginaceae) by Ashford and Allaway (1982). In this the external cells of the host in contact with the fungal sheath are modified as transfer cells.

In ectendomycorrhizas the sheath may be much reduced or even absent; the Hartig net is usually well developed but the hyphae also penetrate into the cells of the host. The same fungus which produces ectomycorrhizas may, on different species of autobiont or in different conditions, form ectendomycorrhizas.

Arbutoid mycorrhizas possess sheath, external hyphae and usually a well-developed Hartig net. In addition there is intracellular penetration of the host cells to form extensive fungal infection in the form of hyphal coils.

As will be seen from the Table, ectomycorrhizas, ectendomycorrhizas and arbutoid mycorrhizas have several structural features in common. Indeed the last two are indistinguishable from one another in the characters given except in the nature of their host plant and the fact that the sheath of ectendomycorrhizas may be poorly developed. It might be thought therefore that it is illogical to separate them. They form an external sheath of some sort and a Hartig net, and in both the hyphae penetrate the epidermal or cortical cells to form coils. In this last character they differ from the ectomycorrhizas in which there is as a general rule no intracellular penetration of the cells except in senescent phases. The grounds for keeping these kinds separate are, however, quite clear at the moment. To put them together would simply hide problems yet to be solved. The ectendomycorrhiza of conifers is particularly formed in the seedling phase in some hosts and it is important to examine why these hosts, as seedlings, have intracellular penetration by the fungus when others do not. Arbutoid mycor- rhiza is found as the usual mycorrhizal state of adult trees and shrubs and perhaps of some green and partially achlorophyllous herbs. The self-same strains of fungi form arbutoid mycorrhiza with plants in the Arbutoideae and ectomycorrhizas with members of the Coniferae. In addition, it is important to ascertain in the future what contribution intracellular penetration makes to the physiology of the symbiosis which is not made by the same fungus in ectomycorrhizas.

Two kinds of mycorrhiza have been found especially associated with hosts that are totally achlorophyllous for part or the whole of their lives. They differ very much in structure. The monotropoid mycorrhizal roots that are found on members of the Monotropaceae are somewhat similar to the three kinds of mycorrhiza just considered, as they have a well developed sheath of fungal tissues enclosing them. From this a system of hyphae runs between the cells of the epidermis or cortex forming a network, the Hartig net, between the cells. In addition specialized unbranched haustoria penetrate the epidermal cells and go through a complicated development pattern as the host grows and flowers. The fungus is also associated with the mycorrhizal roots of neighbouring plants.

In the Orchidaceae the autobionts are partially or wholly achlorophyllous for some part of their life. They are associated with Basidiomycetes of various affinity which are able to break down cellulose and sometimes lignin. The fungus penetrates the tissues of the autobiont and after a period of vegetative activity undergoes general or local disintegration. Some of the autobionts may be associated with fungi actively parasitic on other autotrophic plants and indirectly obtain essential nutrients from them.

In many Ericaceae and related families the hair-like roots are enmeshed in an extensive external weft of hyphae which also penetrate the cortical cells of the rootlet where they eventually become disorganized. The septate fungi are apparently usually Ascomycetes and one of the endophytes has formed a perfect stage of *Pezizella* in culture. However the basidiomycete genus, *Clavaria*, may also have members forming ericoid mycorrhizas and there is no doubt that basidiomycetous hyphae with dolipore septa may be found, together with ascomycete hyphae, within the host root cells. In these ericoid mycorrhizas no sheath is formed, although some species of Ericaceae are said to have both ericoid and arbutoid mycorrhizas.

There are descriptions in the literature of other kinds of mycorrhiza about which little is known. One of these, mycorrhiza of the Cistaceae, has fairly recently again been investigated. As a result many of the Cistaceae are now known to form ectomycorrhizas as adults and possibly vesicular-arbuscular mycorrhizas in their early stages of growth. A note on these mycorrhizas is given below in the chapters on ectomycorrhiza. Too little is known about mycorrhizas formed between septate fungi and liverworts, Gentianaceae and Lobeliaceae, to allow further comment.

In the following chapters these types of mycorrhizas will be discussed separately so as to emphasize what is known about each, and the outstanding problems about them. After that, general and comparative aspects of mycorrhizal symbiosis will be discussed in a series of essays in which we consider problems not always widely investigated by experimental means.

# PART 1
# KINDS OF MYCORRHIZA

*Chapter 1*

# The Symbionts

## Introduction

This, the commonest type of mycorrhiza, is formed in the roots of an enormously wide variety of host plants by aseptate fungi belonging to the Endogonaceae. The name is derived from characteristic structures, the arbuscules which occur within the cortical cells, and vesicles which occur within or between them (see Plate 2). A vesicular-arbuscular mycorrhiza has three important components: the root itself, the fungal structures within the cells of the root and an extramatrical mycelium in the soil. The latter may be quite extensive under some conditions, but does not form any vegetative pseudoparenchymatous structures comparable to the fungal sheath typical of ectomycorrhizas and some of the mycorrhizas of Ericaceae. A few of the fungi, however, do form fruiting bodies with limited amounts of sterile mycelium. Little change in root morphology occurs following infection. Root apices continue to grow in an apparently normal manner and root hair production is not suppressed on plants which normally bear them. It is therefore not usually possible to tell if a root system is mycorrhizal without microscopic examination unless there is characteristic yellowing of the infected region of the roots, which occurs in some plants (e.g. onions and some members of the Leguminosae).

Vesicular-arbuscular mycorrhiza was first recognized and described in the last decades of the nineteenth century. Its widespread occurrence and its common presence in plants of many phyla in most parts of the world but especially in the tropics, was realized very soon, but very little of importance was learnt about it until the mid-1950s. The early work is reviewed by Rayner (1927) and very briefly by Harley (1950), but even in the latter article the nature of the fungal symbiont was discussed without a convincing conclusion. Indeed, much of the effort put into researches on this type of mycorrhiza during the twenties and thirties of this century was vitiated by the relative ease with which fungal denizens of the root-surface and of senescing cells could be isolated into culture and the difficulty, still unsurmounted, of isolating the fungal symbionts themselves. The symbionts were placed in the genus *Rhizophagus* by Dangeard (1896, 1900) and that name was current for some isolates until very recently. We may ignore almost all writings about the nature of the fungus until 1953, except for those of Peyronel who, in 1923, showed that the hyphae of the endophyte might be traced to the sporocarps of species of Endogonaceae in the surrounding soil, and of Butler (1939) who, in an influential review, agreed that the fungi called *Rhizophagus* were almost certainly imperfect members of the Endogonaceae. The work of Mosse (1953), which showed convincingly that mycorrhizal strawberry plants were infected by a species of Endogonaceae, may be said to have heralded the modern period. Soon Mosse (1956–63), Gerdemann (1955–65), Nicolson (1959, 1960, 1967), and Daft and Nicolson (1966–69) greatly extended these early observations and demonstrated by inoculation that endogonaceous fungi were symbiotic with many kinds of plants forming the so-called phycomycetous or vesicular-arbuscular endomycorrhizas.

The family Endogonaceae was little studied before the discovery of the association of some of its species with vesicular-arbuscular mycorrhiza, although Thaxter (1922) and Godfrey (1957a,b,c) had written on the subject. The mycorrhizal strains or species were referred at first to the single genus *Endogone*, although the name *Rhizophagus* was still used both generally and for some particular strains or species, e.g. those isolated into culture by Barrett (1947, 1958, 1962). The reappraisal of the taxonomy of the Endogonaceae by Gerdemann and Trappe (1974, 1975), which included the recognition of a number of mycorrhizal genera and species, finally put the study of vesicular-arbuscular mycorrhiza on a firm basis. Much valuable basic work on structure, ecology and physiology was done between 1956 and 1974 during which period the names *Rhizophagus* and *Endogone* were still current. Comparisons between different investigations done at that time therefore require some care.

## Host plants

The wide range of potential hosts of vesicular-arbuscular mycorrhizal fungi has been responsible for the well-worn statement (e.g. Gerdemann, 1968) that it is easier to name families and genera which do not form vesicular-arbuscular mycorrhizas than those which do, and this continues to hold good. Members of some plant families, however, characteristically form mycorrhizas of other types and are rarely found to carry vesicular-arbuscular infections, although recent observations have greatly increased the number of plants known to form more than one kind of mycorrhiza. Species of some genera are reported to form both vesicular-arbuscular and ectomycorrhizas; examples are *Prunus*, *Populus*, *Salix*, *Acacia* and *Casuarina* to name only a few. In other cases, vesicular-arbuscular mycorrhizal infection has been reported to occur on individuals of species usually forming ectomycorrhizas, e.g. *Eucalyptus*. Indeed it is possible that vesicular-arbuscular mycorrhizal fungi might perhaps be made to invade the underground organs of almost all land plants.

Some members of most families of angiosperms and gymnosperms, together with ferns, lycopods and bryophytes, develop vesicular-arbuscular infections. Fossil vesicular-arbuscular mycorrhizas have also been identified. The earliest are in the famous plants from the Rhynie chert, long thought to be the oldest known land plants and now dated at about 370 million years (Chaloner, 1970). They did not possess true roots, but the protostelic rhizomes of *Rhynia* and *Asteroxylon* were clearly infected by fungi (Kidston and Lang, 1921), very similar indeed to modern vesicular-arbuscular mycorrhizas, and similar again to the mycorrhizas of *Psilotum* and *Tmesipteris*, believed to be their closest modern relatives. Later, in carboniferous deposits, many gymnosperm fossils with vesicular-arbuscular mycorrhizas have been identified, of which the best known and preserved is *Amyelon radicans* (Osborn, 1909; Halket, 1930), which again resembled the vesicular-arbuscular mycorrhizas of living gymnosperms (see Nicolson, 1975). These examples show that vesicular-arbuscular mycorrhiza has been part of the equipment of the absorbing organs of many kinds of plant for a very long period indeed and imply that co-evolution of absorbing organs has not resulted in the specialization of the fungi in respect of their host range, for as will be discussed later (Chapter 18) they are very unspecific in modern plants.

We cannot, of course, obtain evidence about the physiology of these fossil mycorrhizas, but it is likely that if they functioned in a manner similar to present-day forms their role in colonization of the land and in subsequent plant evolution may have been considerable, and has been reviewed, in a speculative manner, by Nicolson (1975), Pyrozinski and Malloch (1975) and Raven, S. E. Smith and F. A. Smith (1978). The view has been put forward that the soil available to early land plants might well have been deficient in suitable phosphorous, nitrogenous and other essential compounds, so that the intervention of the fungi in their absorption might well have been important to the success of the hosts invading the terrestrial environment. Indeed, so small a number of modern angiosperm families are characterized by a general deficiency or absence of mycorrhizal infection that special investigation of these possibly specialized plants may well be a profitable line of research.

## Extent of infection

As will be seen later, the magnitude of the effect of mycorrhizal infection upon growth and nutrient absorption is generally correlated with the extent of infection, i.e. the proportion of the root system or absorbing organs colonized by the fungus. It is therefore important not to assume that in any species in which the intensity of infection is small or sporadic the influence of the fungus will be great or even significant under natural conditions. This applies especially to species of the families Cyperaceae, Juncaceae, Urticaceae, Chenopodiaceae, Caryophyllaceae and Cruciferae. It is certainly true that their species are much less commonly infected than those of other families and there may be several reasons for this. The plants may not be susceptible to mycorrhizal infection and this might be very well worth investigating. Other examples of poorly susceptible plants include lupin, an exception in the generally high mycorrhizal Leguminosae. Weak or atypical mycorrhizal infection of the roots of *Lupinus angustifolius*, *L. cosentinii* and *L. luteus* has been observed (Morley and Mosse, 1976; Trinick, 1977). In one investigation (Morley and Mosse, 1976) the presence of lupins also influenced the formation of mycorrhizas on *Trifolium repens* growing in the same pots, so that abnormal appressoria and intercellular hyphae were produced. This was attributed to the effects of root or seed-coat exudates. A similar explanation has been proposed to account for reduction in vesicular-arbuscular mycorrhizal infection in onion roots when grown in association with swede (itself a member of the poorly susceptible Cruciferae) (Hayman, Johnson and Ruddlesdin, 1975). Black and Tinker (1979) observed that field-grown kale repressed spore production in soil and the vesicular-arbuscular infection of a subsequent crop of barley. However, more rapid spread of infection through the roots of a susceptible crop occurred if the mycorrhizal fungus was introduced in transplanted individuals, presumably because of reduced competition (Black and Tinker, 1979; Powell, 1979). The matter is clearly complex, for Trinick (1977) observed increased vesicular-arbuscular mycorrhizal infection in *Lupinus* spp. when these were grown in association with *Trifolium pratense* and similar effects have been observed in some members of the Chenopodiaceae (Hirrell *et al.*, 1978). Ocampo (1980) who studied the influence of cabbage and radish (Cruciferae) on infection in lavender and lettuce, found that these non-hosts had no effect on mycorrhizal infection of the susceptible plants. If diffusible or volatile compounds are responsible for the reduced infection when it occurs, then it is not surprising that conflicting results have been obtained, for soil conditions could very well affect their movement from one plant to another. Not all authors consider that chemical inhibition of fungal infection is important, for Ocampo *et al.* (1980) attribute failure of infection in non-hosts to intrinsic resistance of the epidermis or cortex of the roots.

Soil conditions may influence vesicular-arbuscular mycorrhizal infection in more direct ways. The effect of nutrients, particularly phosphate, will be discussed later. In the present discussion the effects of waterlogging are relevant for the generally non-susceptible Cyperaceae and Juncaceae have many members which are characteristic of wet or waterlogged habitats and even in susceptible plants infection may be much reduced under wet conditions (Boullard, 1956; Mejstřik, 1965). One reason for this may be that numbers of hyphal entry-points on the root epidermis are lower in wet soil for this occurs, at least in *Medicago* sp. (Reid and Bowen, 1978). It is therefore significant to note that some species of Cyperaceae and Juncaceae from drier habitats, such as grassland, scrub and woodland, may be more infected than plants from wet situations (Read *et al.*, 1976). Water-table fluctuations have been shown to be correlated with changes in intensity of mycorrhizal infection of plants in a moorland community dominated by *Cladium mariscus* (Mejstrik, 1965, 1972). In contrast, some water plants are typically mycorrhizal. These include *Lobelia dortmanna*, *Littorella uniflora*, *Cyanotis cristata* and *Eichhornia crassipes* which may be highly infected, and *Phragmites communis*, *Eleocharis palustris*, and *Salvinia cucullata* which have only low levels of infection (Søndergaard and Laegard, 1977; Bagyaraj *et al.*, 1979). The latter showed that *Isoetes lacustris* growing in the same habitat was not infected, nor was *Eichhornia* in another investigation (Khan, 1974). All bottom-rooting aquatic plants have well-developed air passages in both root and shoot tissues so that oxygen can reach the root system. Oxygen supply certainly affects the development of vesicular-arbuscular mycorrhizas as well as ectomycorrhizas (see later). Saif (1981), using *Eupatorium odoratum* and *Glomus mosseae*, observed an increase in infection and in development of vesicles as partial pressure of $O_2$ in the soil was raised from 0 to 21%. Satisfactory pathways for gaseous oxygen transfer through tissues have also been suggested to explain the incidence of mycorrhizas in *Nyssa sylvatica* and *Deschampsia caespitosa* under waterlogged conditions (S. E. Smith, 1974; Armstrong, 1979; Keeley, 1980).

Vesicular-arbuscular mycorrhiza is found in most herbaceous plants that have been studied (see above for exceptions) but it is by no means restricted to herbs, whereas by contrast ectomycorrhiza and ericoid and arbutoid mycorrhizas are almost restricted to woody plants including forest dominants in sub-tropical and temperate forests. As long ago as 1897 Janse examined 46 species of tree in Java and found them all to have vesicular-arbuscular mycorrhiza. Again in a recent paper, Alwis and Abeyanake (1980) found that 58 species of tree from 25 families in Sri Lanka were similarly infected. It is, however, by no means only tropical trees that have vesicular-arbuscular mycorrhiza. Baylis (1961, 1962) states that vesicular-arbuscular mycorrhiza is ecologically the most important mycorrhizal infection of New Zealand forests and it is of first importance in the great coniferous families of the southern hemisphere. Indeed, whereas the Pinaceae are ectomycorrhizal, all other conifer families are dominantly vesicular-arbuscular mycorrhizal, as are most other gymnosperms, all of which are woody. Although vesicular-arbuscular mycorrhiza is often ignored by foresters, it is the characteristic infection of such valuable trees as *Araucaria*, *Podocarpus*, *Agathis*, *Khaya* and *Fraxinus*, as well as all the Cupressaceae, Taxodiaceae, Taxaceae, Cephalotaxaceae and the majority of tropical hardwoods.

While most of the experimental work on vesicular-arbuscular mycorrhiza has been done with herbs, some trees have also been used and include apple, *Citrus*,

*Salix*, *Populus*, avocado, *Araucaria*, *Khaya* and *Liquidambar*. Indeed, one of the first of what we may call "modern" investigations of vesicular-arbuscular mycorrhizal physiology was carried out on apple (Mosse, 1957).

In most cases, experimental work has been carried out on annual or perennial herbaceous species because these are easier to manage under laboratory conditions. However, mycorrhizas may be equally important in nutrient absorption and in nutrient cycling of arborescent species in forest ecosystems, in a similar manner to ectomycorrhizas, so that work with trees and other perennials is very important both from an ecological point of view and from a need to consider forest production. Indeed, although work with herbs allows greater control of conditions in growth rooms etc., the propagation of some woody species from cuttings may have great advantages in providing genetically uniform experimental material which may partly offset the long growth periods necessary for the study of long-lived plants. The work on *Citrus* mycorrhiza by Menge and his colleagues is an example where an arborescent species of economic importance has been used in experiment. In such work, not only can genetically uniform host and fungal strains be used, but after soil sterilization experiments in nursery beds can be carried out in what may be considered semi-natural conditions.

We have already mentioned the capacity of some species to form either ectomycorrhizas or vesicular-arbuscular mycorrhizas. It will be seen later that these two mycorrhizal types have many physiological similarities, so there may be ecological conditions where the two types of infection are equally efficient and the occurrence of a particular kind of mycorrhiza, on a plant capable of forming more than one, may be dictated by availability of inoculum and hence upon the mycorrhizal characteristic of the dominant species in a particular plant community. The wide range of possible host plants and the non-specificity of their fungi might appear to give a little credence to the suggestion that vesicular-arbuscular mycorrhizas are typical of species-rich plant communities, in which several tree species may be co-dominant, while ectomycorrhizas are found on plants which commonly grow in monospecific stands (Malloch *et al.*, 1980). But the mixed deciduous woods of Europe and of America are not monospecific and have several or many species in the dominant storey which are ectomycorrhizal. Neither vesicular-arbuscular nor ectomycorrhizal fungi are closely specific to their hosts, so the suggestion may have no very great physiological or ecological significance.

## Fungi

As has been mentioned, the regular association of spores and sporocarps of members of the Endogonaceae with vesicular-arbuscular mycorrhizal roots was established long ago by Peyronel (1923), but they were not immediately accepted by many as the true mycorrhizal fungi. Only later, after the work of

Butler (1939) and most especially Mosse (1953 and 1956), were they recognized as the chief causal organisms of vesicular-arbuscular mycorrhizas. There followed renewed interest in the taxonomy of the family Endogonaceae, of which the species typically produce very large, globose zygospores, chlamydospores or sporangia. Interest in mycorrhizas has led to the realization that members of this family are among the most common soil fungi and that spores or sporocarps can be collected from almost any soil.

It is now clearly recognized that the Endogonaceae should be placed in the Mucorales (Gerdemann and Trappe, 1974, 1975) because some (but certainly not all) members have been shown to reproduce by zygospores which arise from gametangia. The gametangia and zygospores are often produced within sporocarps, which may indicate that the family Endogonaceae is related to the Mortierellaceae in which a loose hyphal mantle surrounding the zygospores may also be produced.

As increasing numbers of fungi with very large spores were collected and described from roots and soil, the genus *Endogone* grew into an unwieldy and variable assemblage of species about which few generalizations could be made. It became clear that while some species are zygosporic, others produce chlamydospores or azygospores which can be either free or borne in sporocarps. Determination of the relationship between these spore types was (and still is) hindered by the fact that very few of the fungi have been successfully cultured, despite many strenuous efforts (see Harley, 1969). Those that have been cultured belong to the genus *Endogone sensu stricto*, and either seem to be non-mycorrhizal, or like *E. eucalypti* recently isolated by Warcup (1975) form ectomycorrhiza on suitable trees.

Special methods have therefore had to be adopted to maintain pure strains of fungi of vesicular-arbuscular mycorrhiza for experimental or taxonomic purposes. As far as is possible, isolates from single spore types are grown in "pot culture" on the roots of host plants, so that their spore characteristics and mode of infection can be followed by sequential sampling. In many cases spore collections from soil form the only basis for taxonomic study.

Recent revision of the large genus *Endogone* (Gerdemann and Trappe, 1974) has resulted in the establishment of several genera, some of which are known to form mycorrhizas and others have unknown mycorrhizal affinities. *Endogone* (Link ex Fries) now contains only fungi which produce zygospores in sporocarps. Three species (*E. lactiflua*, *E. flammicorona* and *E. eucalypti*) have been shown to form ectomycorrhizas (Fassi 1965, Fassi *et al.*, 1969; Gerdemann and Trappe, 1974; Warcup, 1975) and evidence from the association of spores and sporocarps with ectomycorrhizal hosts suggests that most if not all *Endogone* (*sensu stricto*) are ectomycorrhizal or possibly free-living. Nevertheless, vesicular-arbuscular mycorrhizal fungi were, until about 1974, assigned to the genus *Endogone* (*sensu lato*) and the reader should be aware of this.

Fungi now established as forming vesicular-arbuscular mycorrhizas are included in four separate genera. These are *Gigaspora* (Gerdemann and Trappe), *Acaulospora* (Gerdemann and Trappe), *Glomus* (Tulasne and Tulasne) and *Sclerocystis* (Berkeley and Broome). Two further genera *Glaziella* (Berkeley) and *Modicella* (Kanouse) are of unknown mycorrhizal affinities. Both *Acaulospora* and *Gigaspora* produce azygospores singly in soil. In *Glomus* the chlamydospores are borne either singly or in sporocarps, while in *Sclerocystis* chlamydospores are borne in a single, orderly layer in sporocarps—resulting in a blackberry-like structure (Plate 1). A key to approximately 100 taxa of endogonaceous fungi has recently been produced (Hall and Fish, 1978) which should facilitate the identification of spore types isolated from soil. These spore types were, in the past, frequently assigned numbers or letters, such as the now widely known $E_3$ and YV (yellow vacuolate) strains (see, for example, Mosse and Bowen, 1968a). In many cases these spore types can be assigned to genera and species (see Hall and Fish, 1978) but some care in this is clearly required as we know very little indeed about geographical variation and physiological specialization of the fungi producing the spores.

In some published work, fungi forming vesicular-arbuscular mycorrhizal infections have been assigned to the genus *Rhizophagus* (see Butler, 1939; Baylis, 1962; Greenall, 1963; Johnston, 1977). "*R. populinus*" appears to include many fungi now known from spores, such as *Glomus*, *Acaulospora* and *Gigaspora* which produce "coarse" infections in which intercellular hyphae are $5-10\,\mu m$ in diameter or more. "Fine" infections, by very narrow hyphae ($1-3\,\mu m$ diameter), have been assigned to *Rhizophagus tenuis*. Small spores (diameter $10\,\mu m$) only have been found associated with this fungus so its taxonomic position remains in doubt. It may be a *Glomus* species and has been referred by Hall (1977) to *Glomus tenuis* (Greenall) Hall. The "fine" endophyte is, however, extremely common in many soils and the problems that it poses are not only taxonomic but also ecological and physiological (see below).

Vesicular-arbuscular mycorrhizal fungi are almost certainly ecologically obligate symbionts, although the extent to which they may be intrusive in newly formed plant detritus, as described by Dowding (1959), requires further investigation. Heap and Newman (1980a,b) have demonstrated hyphal connections between roots of different plants and found that decapitation of some plants facilitated movement of $^{32}P$ from these to intact plants nearby—indicating fungal activity in the senescent or dead plants. This view is supported by work of Tommerup and Abbot (1981) who found that regrowth of hyphae of *Glomus monosporus*, *G. fasciculatus* and *Gigaspora calospora* occurred from dead root fragments after these had been stored in dry soil for 6 months or more and subsequently re-wetted. Mycorrhizal infection of clover occurred as a result of this regrowth, which did not depend upon the presence of vesicles within the root fragments (see Table 2). Spores will germinate after surface sterilization and

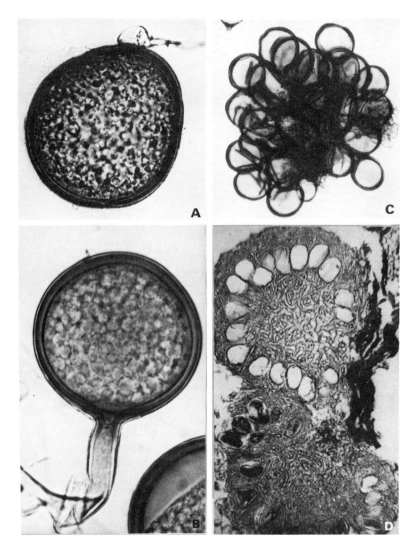

PLATE 1. A. Spore and subtending hypha of *Gigaspora heterogama*. Spore diameter approximately 160 μm.

B. Spore and subtending hypha of *Glomus invermaius*. Spore diameter approximately 75 μm.

C. Sporocarp of *Sclerocystis rubiformis* (slightly squashed). Individual spore diameter approximately 45 μm.

D. Section of a sporocarp of *S. coremioides* (stained). Sporocarp diameter approximately 410 μm.

From Hall and Abbott (1981).

produce a small amount of mycelium (a few centimetres hyphal length) in pure culture. Some further growth is stimulated in the presence of susceptible root systems (Mosse, 1959a; Mosse and Hepper, 1975; Hepper and Mosse, 1975), but an extensive mycelium is not formed unless the roots become infected. The many attempts to grow these fungi in pure culture for long periods and to subculture mycelium, separated from the subtending spore, have so far been unsuccessful, except in one case. Barrett (1947, 1958, 1962) reported the isolation of a "*Rhizophagus*" strain on hemp seed, its growth on malt agar and its back-inoculation with mycorrhiza formation on maize and other plants. The back-inoculation was repeated by Murdoch *et al.* (1967) and by Greenall (1963) using Barrett's isolates without success; but Mosse (1961) obtained infection with apple but not with other plants, using these fungi. This work may well indicate that with persistent efforts the vesicular-arbuscular mycorrhiza formers may be brought into culture as the ectomycorrhiza-forming *Endogone eucalypti* has been (Warcup, 1975a). It must be concluded that the endophytes have little or no saprophytic ability and that hyphal growth in soil is at the expense of either spore reserves or nutrients derived from root systems. It is not surprising, therefore, that vesicular-arbuscular mycorrhizal fungi have never been isolated from soil by traditional methods of studying soil fungi (e.g. soil dilution plates, etc.) and their high frequency in almost all soils has only been recognized after the application of methods particularly appropriate to the separation of large spores. The most widely applied method is that of "wet-sieving and decanting" (Gerdemann, 1955; Gerdemann and Nicolson, 1963). An aqueous soil suspension is mixed thoroughly, allowed to sediment briefly and the supernatant decanted through a series of soil sieves of different mesh sizes. Spores retained on the sieves are then collected. Other methods include "flotation adhesion"

TABLE 2

The ability of mycorrhizal fragments of root to initiate hyphal growth and mycorrhiza formation on clover roots after drying to $-50$ mPa and storage in dry soil for 6 months. Results were recorded after the root fragments had been incubated between membrane filters in pots of steamed moist soil for four weeks (results of Tommerup and Abbott, 1981)

| | | % root fragments | |
| Fungus | Vesicles in root fragments | Initiating hyphal growth | Initiating mycorrhiza |
| --- | --- | --- | --- |
| *Glomus fasciculatus* | yes | $62 \pm 6$ | $41 \pm 4$ |
| *G. monosporus* | — | $58 \pm 5$ | $51 \pm 4$ |
| *G. caledonius* | — | 0 | 0 |
| *Acaulospora laevis* | — | 0 | 0 |
| *Gigaspora calospora* | no | $55 \pm 3$ | $43 \pm 5$ |

(Sutton and Barron, 1972) in which endogonaceous spores both float in aqueous solutions and adhere to glass surfaces; "flotation bubbling" (Furlan and Fortin, 1975); a sucrose centrifugation method, originally developed to extract nematodes from soil (see G. W. Smith and Skipper, 1979) and a method which simply involves the "direct plating" of soil suspensions on filter paper. These methods have been compared for their effectiveness in recovering the total spore population by G. W. Smith and Skipper (1979). Sporocarps of those endogonaceous fungi producing relatively large structures can also be recovered by painstaking search with a truffle fork (Tandy, 1975b).

Both qualitative and quantitative surveys of numbers of spores per unit volume or weight of soil have been attempted using these methods and it has been established that spores of many types are widespread in soil and may be present at very high frequencies (e.g. Mosse and Bowen, 1968b; Hayman, 1970; Powell, 1977, and many others). However, not surprisingly, the number of spores in the rhizosphere is frequently unrelated to the intensity of mycorrhizal infection in roots of associated plants (Hayman and Stovold, 1979; Porter, 1979). Many factors must contribute to this. Extraction techniques may not recover all the spores present in soil. Certainly, the "direct-plating" method appears to yield consistently higher spore counts than either "wet-sieving" or "flotation" (G. W. Smith and Skipper, 1979). The number of spores per plant or per gram of soil, even if accurately determined, may reflect the previous history of a mycorrhizal symbiosis in the soil rather than its potential for immediate infection of roots. Spore production, as well as infection of the root systems, is affected by such factors as plant growth, fertilizer application, light intensity and so on. Daft and Nicolson (1972) showed that the application of increasing quantities of phosphate to tomato plants reduced not only the proportion of the root system colonized by *Endogone* (*Glomus*) *macrocarpa* var. *caledonia* but also the number of spores associated with each plant after 84 days' growth (Table 3). Similar effects of phosphate on spores produced in field soils have been found (e.g. Ross, 1971; Hayman *et al.*, 1975; Porter *et al.*, 1978). Stage of growth of the crop may also influence the number of spores present, so that Saif (1977) observed a general decline in numbers during early growth of several summer vegetables followed by an increase as the plants matured. Baylis (1969) investigated the possibility that a stimulation of sporing might be provided by drought or intermittent root growth, but found no evidence for either in his experiments with "honey coloured" spores and the host plant *Coprosma robusta*. Hayman and Stovold (1979) found a very large variation in the number of spores (between 2 and 1952 spores per 100 g soil) in the soils they sampled. There were higher populations in agricultural soils than in either native grassland or undisturbed bushland, confirming similar results obtained by Mosse and Bowen (1968b). The infectivity of these soils towards clover plants was, however, not correlated with spore numbers per gram of soil. Powell (1977) has also investigated the relationship between spore numbers and infectivity and once again found that the speed with which test clover plants became infected was not closely correlated with numbers of spores that could be extracted from his 37 test soils. Several suggestions have been made to explain this discrepancy.

TABLE 3

The effect of phosphate fertilizer and inoculation with *Endogone (Glomus) macrocarpa* var *caledonia* on growth and mycorrhiza development in *Lycopersicon esculentum* (results of Daft and Nicolson, 1972)

| Phosphate level mEq per plant | Infection | Dry weight foliage g | % infection | Spores per plant |
|---|---|---|---|---|
| 0·6 | + M[a] | 0·78 ± 0·06 | 46·5 ± 14·7 | 147 ± 56 |
|  | − M | 0·14 ± 0·002 | — | — |
| 1·2 | + M | 0·96 ± 0·09 | 33·3 ± 9·9 | 71 ± 22 |
|  | − M | 0·86 ± 0·04 | — | — |
| 2·4 | + M | 1·14 ± 0·14 | 18·5 ± 5·5 | 60 ± 23 |
|  | − M | 1·01 ± 0·11 | — | — |

[a] + M, inoculated with 50 spores of *E. macrocarpa*.
− M, not inoculated.

First, any factors which have different effects on spore production and germination will contribute to the discrepancy between spore populations in soil and extent of infection of plant roots, in those situations where spores are the main source of soil inoculum. Studies of spore germination on agar have provided clues to the factors affecting survival and germination, but it is important to remember that conditions in agar may differ considerably from those in soil. Hepper and G. A. Smith (1976) found that zinc and manganese inhibited spore germination, while McIlveen and Cole (1979) found that trace amounts of zinc increased germination on agar. In soil the infection of soybean plants was increased at 18 mg kg$^{-1}$ of zinc in the soil but reduced when levels were 45 or 135 mg kg$^{-1}$. Spores also may themselves contain an inhibitor, for Watrud *et al.* (1978) found that germination of *Gigaspora margarita* was improved if activated charcoal was incorporated in the medium used for germination studies. Tommerup (quoted in Sward, 1981) also believes that inhibitors in the spore walls may be important in germination. The pH may have a direct or indirect effect upon spore germination; that is, it may affect the solubility of essential substances or toxins in the soil apart from affecting growth of germtubes directly. Mosse and Hepper (1975) found that spore germination and subsequent infection of excised roots of clover by *Glomus mosseae* in agar was prevented at pH 4.9. Fungi may differ in this regard, for Green *et al.* (1976) showed that while *Glomus mosseae* had a pH optimum for germination of 7 and that no germination took place below pH 5, *G. heterogama* and *G. coralloidea* had optima for germination of pH 5 and 6 respectively. These may only be examples of fungal specialization to different habitats, for Porter *et al.* (1978) suggested that endophytes might become adapted to different soil conditions, although their experimental results gave only a small indication of such "adaptation".

Secondly, some spores (e.g. those of *Gigaspora gigantea*) are capable of producing up to 10 germ-tubes successively if the earlier ones are cut off (Koske, 1981a). Moreover, although young spores of *G. margarita* germinate with a single germ-tube, old spores frequently produce several (Sward, 1981). Both these courses of behaviour might result in a discrepancy between the number of spores per gram of soil and intensity of infection of roots. The same might be concluded from the ability of *Glomus epigaeus* to produce secondary sporocarps in long-term storage without any intervening infection of roots (Daniels and Menge, 1980). Such activities are in line with the observation of Hepper (1979) that during spore germination and hyphal growth of *Glomus caledonius* measurable protein and RNA synthesis occur.

A further problem in correlating spore numbers with infectivity of soil arises from the fact that while spores are certainly an important source of inoculum they are not the only one. In some soils spores of vesicular-arbuscular mycorrhizal fungi are rare, even when plants growing in them are extensively mycorrhizal. We must therefore accept the conclusion reached by many workers that other sources of inoculum can be very important and ask what these may be.

Infection can certainly take place from hyphae growing out from living infected roots, and there is evidence that hyphae recently detached from roots can also act as extremely infective sources of inoculum. For instance, Johnson (1977) found that infection of *Coprosma robusta* was more rapid with hyphae of *Acaulospora* picked off *Coprosma* roots than with spores or with infected roots washed free of spores and hyphae. Similarly Hepper (1981) showed that hyphae of *Glomus mosseae* or *G. caledonius*, produced in agar culture in association with a host plant, can infect a second plant introduced into the culture after the first plant together with all its roots had been removed leaving only hyphae behind. Infection can also take place from infected roots severed from their parent plant and indeed this, together with attached spores and hyphae, is a very commonly used source of inoculum for experiments. We do not know how long vegetative hyphae may survive in senescent or dead roots but the results of Dowding (1959) and Tommerup and Abbott (1981) suggest that they may do so. Resting structures, e.g. vesicles, may also persist in dead roots and later germinate in a similar manner to the large external spores. None of these inocula would be separated from soil by the extraction methods which are used for spores.

Recently the "most probable numbers" method (or method of ultimate dilution) has been adapted to the problem of estimating numbers of propagules of vesicular-arbuscular fungi in soil (S. E. Smith and Bowen, 1979; Porter, 1979; Powell, 1980). Numbers of "propagules" determined in this way invariably exceed numbers of "germinable" spores. However, in one case (Porter, 1979) in which infections due to "fine" endophyte were distinguished from "coarse" infections, the numbers of propagules producing the latter were similar to numbers of germinable spores (Table 4). Even though the confidence limits are wide, the most probable numbers method should provide a way of studying the relationship between numbers of propagules and rate of development of infection (which is discussed in a later chapter) and perhaps help to explain the incidence of infection observed under natural and man-made ecological conditions.

TABLE 4

A comparison of the numbers of spores of vesicular-arbuscular endophytes recovered from soils by wet sieving, with the numbers of mycorrhizal propagules determined by the most probable numbers method (MPN) (results of Porter, 1979)

| Soil | Spore numbers[a] per 50 g soil (by wet sieving) | Propagule number[b] per 50 g soil (by MPN) | |
|---|---|---|---|
| | | "Fine" endophyte | "Coarse" endophyte |
| Ulva | 82–108 | 136–430 | 29–85 |
| Merredin | 464–534 | — | 237–1320 |

[a] Range represents two replicate spore counts.
[b] Range represents 95% confidence limits of MPN method.

None of the propagules of vesicular-arbuscular mycorrhizal fungi appear to be adapted for widespread dispersal, except perhaps as the stomach contents of animal or human truffle-hunters or as blown dust. The occurrence of infection when plants are grown in soil devoid of living roots shows that propagules of mycorrhizal fungi must survive in the soil, as do those of many other soil-inhabiting and root-inhabiting fungi. The length of time that propagules survive varies. In some habitats (e.g. the arid regions of Australia), there may be no plant cover for long periods during which vesicular-arbuscular mycorrhizal fungi survive successfully and re-infect annual plants almost as soon as seed germination takes place. The ability of G. gigantea spores to germinate repeatedly (Koske, 1981a) could be advantageous under these conditions, for it is possible following rain to envisage situations where the soil may be moist enough for spore germination to occur, but insufficiently moist for successful seed germination and root production. In other habitats (see for instance, Johnson, 1977), living roots may be present at all times, permitting continuity and spread of infection. Hyphal connections between individual plants of the same and different species have been demonstrated (Magrou, 1936; Stahl, 1949; Hirrell and Gerdemann, 1979; Heap and Newman, 1980a,b; Whittingham and Read, 1982) indicating that cross-infection does occur.

The extreme susceptibility of most species of plant to vesicular-arbuscular mycorrhizal fungi may be very important in this context. There is good evidence that these fungi do not exhibit close species or strain specificity to particular host plants of the kind that is found with biotrophic pathogens of shoot systems (e.g. Rusts, *Phytophthora*, etc.), although there may be some "host-preference". This means that a vesicular-arbuscular mycorrhizal fungus can maintain itself both in space and time by infecting almost any plant that grows sufficiently close to a "propagule" or infected living root for infection to take place. Such absence of specificity and wide host susceptibility may dramatically reduce the time period that a resting propagule may have to survive between the appearance of two generations of a single species of host plant. It also makes possible continued survival in a mixed plant community where individuals of a particular plant species are widely separate in space. These conditions are known to reduce the numbers of individuals of highly specific shoot pathogens, even though they may have very effective modes of dispersal (see Burdon, 1978). Under these conditions vesicular-arbuscular mycorrhizal inoculum is provided by living roots, and infection takes place as hyphae growing in soil make contact with uninfected roots of individuals of the same or another plant species. Infection of one individual by more than one species or strain of vesicular-arbuscular mycorrhizal fungus is not only theoretically possible but has been shown to occur. Molina *et al.* (1978) found that a *Festuca* plant could have two or more species of fungus associated with it and that *F. idahoensis* associated on average with 5 fungal species per site.

Although it is difficult to distinguish between naturally occurring infections caused by "coarse" endophytes, such as *Glomus* sp. and *Acaulospora* sp., it is common to find both "coarse" and "fine" endophytes present in the same root system. Abbott and Robson (1979) have developed a series of quantitative characters by which it should prove possible to identify the individual fungal species of a mixed population within a single root. This technique would allow confirmation of numbers of different fungal species simultaneously able to infect a single root system. It also may provide a means of studying the fluctuations and changes of populations of fungi within the root brought about by different environmental conditions. For example Abbott and Robson (unpublished results) have shown that there may be competition between mycorrhizal fungi in colonizing roots, so that the extent to which a particular fungus infects a root may depend upon the infectivity of other fungi present in the soil as well as upon the inoculum potential of individual fungi. At present, the growth of different fungi can only be compared in separate plants. Thus Bevege and Bowen (1975) showed that development of internal and external mycelium of each of the three species of fungi differed on clover and onions, while Sanders *et al.* (1977) showed that four species of mycorrhizal fungi had different growth characteristics in the same host (onion). It would be most interesting to determine whether competition between fungi in the same root affected such results. A plant community composed of a number of different species and individuals, together with their mycorrhizal fungi, must be envisaged as having a below-ground continuum, i.e. it is a social organism (*sensu* Buller); this may have considerable physiological implications for nutrient distribution and cycling.

## Spore germination

The time taken for spores to germinate in culture is clearly quite variable and affected by environmental conditions. Both Godfrey (1957c) and Mosse (1959a) reported that germination could be slow and erratic, so that it extended over 2–3 weeks when spores were incubated on damp filter paper. In some cases the total observed germination was as low as 5%. Mosse (1959a) found that germination could be increased to at least 80% if opaque spores were discarded and if the remainder were surface sterilized and incubated at 20°C on cellophane discs placed over the surface of soil agar. The influence of diffusible substances from soil in stimulating germination has been confirmed by Daniels and Graham (1976) using *Glomus mosseae*. The method has been successfully modified and used by Sward *et al.* (1978) for several other species. However, soil, or soil extracts, are not invariably required for aseptic spore germination. Thus, Hepper and G. A. Smith (1976) obtained 90% germination of *Endogone* (*Glomus*) *mosseae* on water and agar. Nor did Koske include soil extracts in his studies of spore germination. Presence of roots of clover and onion did not stimulate germination of spores of several species on buried slides (Powell, 1976). In that investigation 30% and 60% germination of yellow vacuolate and laminate spores respectively had occurred in 16 days but honey-coloured and E3 spores did not reach their maximum germination of 53% and 40% until 56 days. Sward *et al.* (1978) also found considerable variation between species. Graham (1982) showed that exudates of roots of different species varied in stimulation of germination of spores of *G. epigaeus*. Probably nutrients and growth factors are both involved. As we have already mentioned, pH, presence of zinc or manganese and possibly also endogenous inhibitors of germination may contribute to the variability of the results obtained. While failure to obtain uniformly rapid germination may be disadvantageous for experimental purposes, variability of germination may be very important under natural conditions ensuring that only a proportion of spores germinate at any one time and that the remainder survive in a dormant condition.

## Conclusion

Vesicular-arbuscular mycorrhizas are formed by members of all phyla of land plants. The fungal symbionts appear to be restricted to a few genera of the phycomycetous family Endogonaceae. In contrast, the host plants are very diverse, not only in taxonomic position but also in life form and geographical distribution. Herbaceous plants, shrubs and trees of temperate and tropical habitats may all form vesicular-arbuscular mycorrhizas and there is little evidence for specificity between particular fungi and host plants. Only a few families and genera of plants do not generally form vesicular-arbuscular mycorrhizas. A fossil record of vesicular-arbuscular mycorrhizas dates back to the earliest land plants, indicating a very long period of co-evolution between plants and these fungal symbionts.

*Chapter 2*

# The Development of Infection and Anatomy of Vesicular-arbuscular Mycorrhizas

### Morphology of vesicular-arbuscular mycorrhizas

Before considering the infection process in detail, a brief and generalized description of a vesicular-arbuscular mycorrhizal infection of a root system will be given and is illustrated in Plates 2–5. Nicolson (1967), in an influential review article, emphasized that this type of mycorrhiza, although designated as endotrophic, was composed of a two-phase mycelial system; an internal mycelium within the cortex of a mycorrhizal root and an external mycelium in the soil which varies considerably in extent but which may be very extensive in some samples, even obscuring the roots. However, well-defined pseudo-parenchymatous sheath tissue around the roots is never formed in vesicular-arbuscular mycorrhizas.

Descriptions and illustrations of the internal mycelium were made as early as 1897 by Janse. Subsequent observations with light and electron microscopy have increased our knowledge of the details and fine structure of the infection, but the basic observations remain the same. After penetration of the epidermis (or more rarely a root hair), hyphae grow inter- and intracellularly through the cortex, sometimes coiling in the cells and extending the infection longitudinally in the root and penetrating to the inner cortex. Arbuscules are normally formed within the cells of the inner cortex. They arise as lateral branches on intercellular hyphae which penetrate the cell walls and invaginate the plasmalemma. Repeated dichotomous branching occurs, giving rise to a three-dimensional bush-like structure, from which the name arbuscule comes (Plates 2 and 4). The hyphal branches are surrounded by the plasmalemma of the root cell and an interfacial matrix between that and the walls of the hyphal branches. The area of the interface between the two organisms is therefore very extensive in these cells. It is probable that transfer of nutrients between the arbuscule and the host cell via the invaginated plasmalemma occurs. The life span of an individual arbuscule is probably between 4 and 15 days (see Table 5); towards the end of this time the fine hyphal branches progressively collapse and their contents disappear, leaving wall material still surrounded by the plasmalemma of the root cell and the interfacial matrix. Details of the structure of the interface and of changes that take place during the life of an arbuscule are described later.

TABLE 5

Longevity of arbuscules. Time in days from inoculation to the initiation of the first arbuscules and first collapse of arbuscules (data of Bevege and Bowen, 1975, and of Cox and Tinker, 1976)

| Fungus | White reticulate | | E. araucariae[b] | | E. mosseae[b] | | |
|---|---|---|---|---|---|---|---|
| Host | Clover | Onion | Clover | Onion | Clover | Onion | Onion |
| Initiation | 2 | 6 | 0 | 2 | 1 | 4 | — |
| Collapse | 4 | 7 | 14–15 | 14–15 | 7–9 | 10–15 | 4[a] |

[a] Estimate of Cox and Tinker, 1976.
  Remainder of data Bevege and Bowen, 1975.
[b] *Glomus* spp.

Most vesicular-arbuscular mycorrhizal fungi also produce vesicles which are terminal swellings on inter- or intracellular hyphae (Plate 2). The vesicles, which develop later in the process of infection than arbuscules, frequently contain large numbers of lipid droplets and may well have a storage function.

## Penetration of the tissues

As we have said, initial infection of an uninfected root system arises from propagules in soil or from hyphae growing from a nearby mycorrhizal root. Details of the infection process have been studied chiefly using spores or infected segments of root as inoculum, either in axenic culture in agar (Mosse and Hepper, 1975; Mosse and Phillips, 1971; Hepper, 1981) or on slides buried in soil (Powell, 1976). Production of entry points has also been studied in pot experiments (e.g. Carling *et al.*, 1979; F. A. Smith and S. E. Smith, 1981; S. E. Smith and Bowen, 1979; S. E. Smith and Walker, 1981) using sequential harvests, which do not, however, permit continual observation of the same infection.

When spores are used as inoculum, germination is followed by considerable growth of one or several germ-tubes, so that a simple mycelium in which total length of hyphae is a few centimetres is produced. Growth is sometimes increased if susceptible roots are present, so that it was at first thought that exudates from

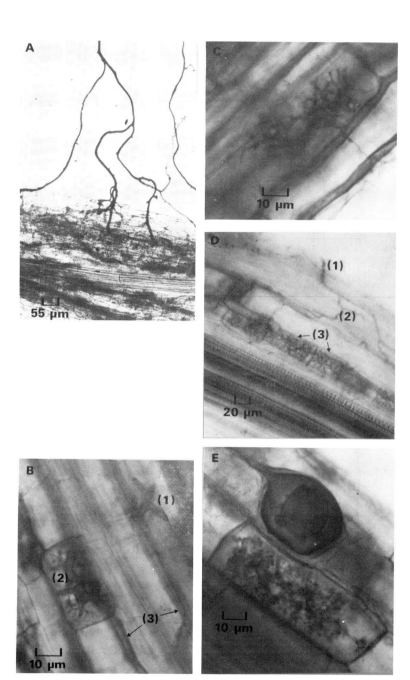

the roots might provide substrates for hyphal growth after the reserves in the spores had been used up. The identity of the important components in the exudates has not been determined, yet continued mycelial growth in culture by hyphae unattached to susceptible roots does not occur and hyphal growth ceases if the spore is excised from the mycelium. The role of root exudates in the development of infection is still receiving considerable attention. In spite of the increased mycelial growth in the presence of roots, hyphae may not appear to make directional growth towards roots until they are very close to them; i.e. within a few millimetres (Mosse and Hepper, 1975; Powell, 1976). Powell showed that the hyphae on slides buried in sterilized soil did not grow towards roots until about 1·6 mm from them, but Koske (1982) observed a reaction by germ tubes 11 mm from roots. The main hypha (diameter 20–30 $\mu$m) then gave rise to characteristic fan-shaped complexes of septate lateral branches (hyphal diameter 2–7 $\mu$m) and infection of the root usually occurred from these narrow lateral hyphae. Similar changes in branching pattern preceding root infection have been observed in axenic excised root culture by Mosse and Hepper. Mycelium developing from a spore in the presence of susceptible roots thus shows striking dimorphism.

The extent of development of these "pre-infection fans" seems to vary with the source of inoculum. Hyphae from honey-coloured spores show directional growth and "fan" production when as much as 3·4 mm from the root. On the other hand, when infection occurs from infected root segments rather than from spores, the pre-infection branching may be limited or absent. It was also absent when infection of *Trifolium pratense* by *Glomus mosseae* was studied on slides buried in non-sterile soil, rather than in axenic root organ culture. Absence of lateral branching did not necessarily result in failure of infection. Nicolson (1959) in one of the earliest studies of the infection process emphasized that grass roots can become infected from thick- or thin-walled hyphae. In either case, hyphae finally make contact with the epidermal cells (or in a few cases, root hairs) and

PLATE 2. Mycorrhizal infection in the roots of *Trifolium subterraneum* and *Medicago truncatula* growing in non-sterile soil.

A. Multiple entry points developing from one main hypha on a root of *Trifolium subterraneum*.

B. Early stages in the development of arbuscules in *M. truncatula*.

1. The arbuscule trunk has branched twice within a cortical cell.
2. Further branching has given rise to a somewhat larger arbuscule.
3. Intercellular hyphae.

C. Older arbuscule within a cortical cell of *M. truncatula*.

D. Entry point (1), inter- and intracellular hyphae crossing the outer cortex (2), and arbuscules in the inner cortex (3) of *M. truncatula*.

E. Arbuscule showing signs of collapse and a vesicle in *M. truncatula*.

a, courtesy C. A. Chambers; b–e, by S. E. Smith and G. D. Bowen.

produce slightly swollen appressoria from which infection pegs arise. Penetration of epidermis or root-hair cells follows.

There is some evidence that regions of a root system differ in their susceptibility to mycorrhizal infection, although the results are not clear-cut. Mosse and Hepper (1975) observed that infection of *Trifolium* spp. in root organ culture by "yellow vacuolate" spores (presumably *Glomus mosseae*) occurred most commonly 0·5 to 1 cm behind the root tip. They also noted that typical arbuscules were formed in young root cultures, but that when the cultures aged the arbuscules were "stumpy" with only a few branches. Holley and Peterson (1979) recorded that infections on *Phaseolus* sp. occurred 1–1·5 cm behind the apex. Subsequent work in axenic culture (Hepper, 1981) also indicated that 21-day-old clover roots did not become infected immediately when inoculated with *Glomus mosseae* or *G. caledonius* but infection did occur when new roots had formed. It is not clear how close behind the root tip infections actually occur, for root growth continues after the formation of an appressorium and subsequent growth of an infection unit. Plate 3 shows entry-points close to the tip of raspberry root and internal hyphae of the infection unit advancing behind the differentiating and elongating zones (Gianinazzi-Pearson *et al.*, 1980). S. E. Smith and Walker (1981) measured both the rate of growth of *Trifolium subterraneum* roots and the distance from the root apex of the first entry point under conditions where a large number of propagules were distributed uniformly and randomly in the soil. They showed that the median distance of the most apical entry point from the root tip, $\hat{l}$, is

$$\hat{l} = [(2 \ ln \ 2)(v/A)]^{1/2}$$

where $v$ is the rate of growth of individual root apices and $A$ is the frequency of infection, and calculated that, in the soil used, $A$ was $360 \ \mathrm{m}^{-1} \ \mathrm{d}^{-1}$ in the apical region. This may be compared with a calculated value of $A$ as an average for the whole root system of $41 \ \mathrm{m}^{-1} \ \mathrm{d}^{-1}$ estimated from counts of total entry points ($N$) and infected ($L^*$) and total ($L$) root length where

$$\frac{dN}{dt} = A(L - L^*).$$

It thus appears that the apical region is about ten times more infectible than the average for the root system as a whole, even in young (up to 35-day) clover plants. More work on this subject is urgently required, for it might lead to a much greater understanding of how the fungi get into the cells and establish typical mycorrhizal infections and the state of the cells and their walls when infected. There is no doubt that arbuscules can be found in cells some way behind the meristem and elongating zone, for infections continue to spread longitudinally both apically and proximally in the root up to distances of 0·5 to 1 cm from the initial point of infection, and given that hyphal growth in the root cortex is of the order of $6 \cdot 1 \times 10^{-4}$ m d $^{-1}$ in each direction (S. E. Smith and Walker, 1981), this means that the youngest regions of such an infection unit must be growing in roots up to about 16 days old. We need to know a great deal more about cell division and elongation in different regions of a root in order to determine at what stages the different cell types can be penetrated and infected. Buwalda and his colleagues (1982) are also of the opinion that some characteristics of the curve of percentage infection against time would be explicable if each root, having passed through an infectible stage, were to become uninfectible.

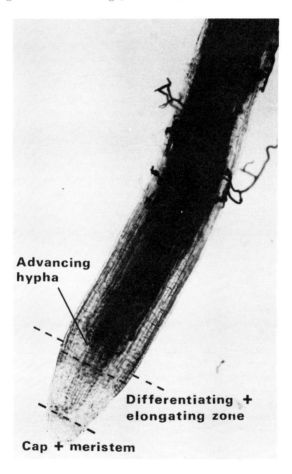

PLATE 3. Development of infection close behind the apex of raspberry root infected by *Glomus tenuis*. Several multiple entry points have given rise to infection units. The leading hyphae growing through the cortex are advancing behind the elongating and differentiating zones of the apex of the root. × 100. (Photograph from Gianinazzi-Pearson *et al.*, 1980.)

Within each infection unit branches pass into the cortical cells. These may form dichotomously branched arbuscules (Plates 2 and 4) or in some plants (e.g. *Ornithogalum*) first form hyphal coils in the outer cortical cells and arbuscules in more deep-seated layers. After a relatively short period of days (Table 5) the intracellular arbuscules or coils degenerate to form a clump or clumps, called by Janse (1897) "sporangioles", whilst the host cell remains alive (Plate 2). A cell may become reinfected by the fungus and come to enclose several digestion clumps.

There is considerable consistency in the relationship observed under the electron microscope in many diverse vesicular-arbuscular mycorrhizas in which the hosts may be angiosperms, gymnosperms, woody or herbaceous. Some of those studied are given in Table 6.

TABLE 6

Examples of vesicular-arbuscular mycorrhizas used in fine structure research

| Host plant | Fungus | Reference |
|---|---|---|
| *Ornithogalum umbellatum* | unknown | Scannerini and Bellando (1967)<br>Scannerini (1972)<br>Scannerini *et al.* (1975)<br>Scannerini and Bonfante-Fasolo (1975, 1976, 1979)<br>Bonfante-Fasolo and Scannerini (1977) |
| *Vitis* | unknown | Bonfante-Fasolo (1978) |
| *Nicotiana tobacum* | unknown | Kaspari (1973, 1975) |
| Grasses | unknown | Old and Nicolson (1975) |
| *Allium cepa* | *Glomus mosseae* | Cox *et al.* (1975)<br>Cox and Sanders (1974)<br>Dexheimer *et al.* (1979)<br>Gianinazzi *et al.* (1979) |
| *Rubus idaeus* | *Glomus tenuis* | Gianinazzi-Pearson *et al.* (1981) |
| *Taxus baccata* | unknown | Strullu (1976a, 1978) |
| *Phaseolus vulgaris* | *Glomus* spp. | Holley and Peterson (1979) |
| *Glycine max* | *Glomus caledonius* | Carling *et al.* (1977) |
| *Liriodendron tulipifera* | *Glomus mosseae* | Kinden and Brown (1975a,b,c; 1976) |

PLATE 4. Stages in the development of arbuscules of *Glomus mosseae* within cells of *Liriodendron tulipifera*.

A. Young arbuscule, showing penetration point and dichotomous branching.

B, C. Later stages in arbuscule development, showing how the hyphal branches come to fill the cytoplasmic volume.

From Kinden and Brown (1975b).

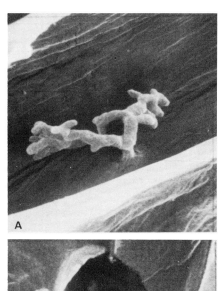

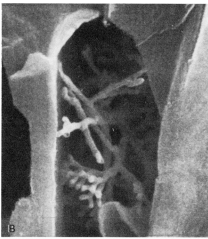

The extracellular hyphae within the tissues have a typical hyphal structure, peripheral cytoplasm, with mitochondria, nuclei and other organelles and vacuoles. Lipid materials occur in the cytoplasm and dense granules (or "globules") probably of polyphosphate occur in the vacuoles. In some cases, bacteria-like organelles, which often appear to be separated from the cytoplasm by a granular zone and fungal plasmalemma are present in the hyphae (Macdonald and Chandler, 1981; Macdonald *et al.*, 1982). This granular zone is probably similar to that surrounding the intracellular hyphae, the interfacial matrix, which will be described later.

The hyphal walls are described as locally somewhat thickened and lamelliform and usually of one layer in *Glomus mosseae*, but in *G. tenuis* they were shown by Gianinazzi-Pearson *et al.* (1981) to be bilayered throughout. It is therefore considered that the "fine endophyte", *G. tenuis*, might really belong to a different genus from *G. mosseae*. This appearance of having a bilayered hyphal wall was different from that observed by Strullu (1978) in mycorrhizas of *Taxus baccata* formed with an unknown endophyte. In that case the walls of the intercellular hyphae consisted of an inner electron-dense layer and an outer granular more electron-lucent layer. In the intracellular fungal structures, however, the walls were of one electron-dense layer only. He concluded that the outer layer of the walls of intercellular hyphae was derived by modification of the middle lamella of the host through which they passed. Holley and Peterson (1979) described the extramatrical hyphae of a fungus (probably a species of *Glomus*) invading *Phaseolus* to have a two-layered wall and fibrillar layers separated by electron-dense bands. The hyphae within the tissues had a single wall-layer and as they spread through the tissue, singly or as groups of two or three hyphae, their walls were fused with or were closely adjacent to the walls of the host cells. A similar description of the relationship of the hyphal walls and those of the cells of the host is given by Scannerini (1975) for infection in *Ornithogalum*. This evidence of modification of the walls of the host is not a consistent feature of the

TABLE 7

Time of appearance of extramatrical mycelium (days after infection) of "white reticulate", *Endogone* (*Glomus*) *mosseae* and *E. araucariae* and rate of spread of internal fungal structures within the cortex of the roots of clover and onion (results of Bevege and Bowen, 1975)

| | White reticulate | | *E. mosseae* | | *E. araucariae* | |
|---|---|---|---|---|---|---|
| | Clover | Onion | Clover | Onion | Clover | Onion |
| External hyphae: time of appearance (days) | 1–2 | 6 | 1–2 | 0 | 1–2 | 6 |
| Rate of spread within root | fast | slow | fast | slow | fast | slow |

interpretation of electron-microscopic preparations. Some describe the walls of host and fungus to lie side by side in close juxtaposition, others describe the middle lamella to be modified and pushed aside.

The intercellular hyphae are described in *Ornithogalum umbellatum* to run parallel and form cords in which some hyphae appear to have a conducting function whilst some are rich in food material (Scannerini and Bellando, 1967). In *Taxus baccata* they appear to be few in number and most of the spread of infection seems to take place from cell to cell (Strullu, 1978). A spreading of the fungus both along the surface of the root and within its tissues occurs in all vesicular-arbuscular mycorrhiza but the extent of the two processes seems to vary according to the host as well as the fungal species (see Table 7).

## Development of infection

Formation of an appressorium on the root epidermis is normally followed rapidly by penetration of the epidermal and cortical cells by hyphae and development of typical mycorrhizal structures within the root. As with other biotrophic symbionts, the means by which hyphae of vesicular-arbuscular mycorrhizal fungi actually penetrate the cell walls is unclear. Electron microscopy has revealed some details of the way in which the hyphae enter cortical cells, but investigations have not extended to events taking place beneath the appressorium, when fungal hyphae first breach the epidermis. Common to both processes is the production of a narrow hypha, or infection peg, which appears to push into the cell wall. The latter bulges round the hypha and, in cortical cells, becomes much thinner (Cox and Sanders, 1974). This bulging implies the exertion of pressure by the growing hypha and a degree of extensibility, existent or induced, in the cell wall. Whether enzyme production is also involved is not known, but it seems unlikely that the hyphae can generate much hydrolytic activity in view of the poor saprophytic ability of the fungi concerned. Nevertheless, changes in the middle lamella, as seen by electron microscopy (Kinden and Brown, 1975b), when intercellular spaces are colonized by hyphae, might be thought to give some credence to the suggestion that fungal enzymes may be important.

Mosse (1962) suggested that fungal entry into the root might be facilitated by the presence of pectinases but it is possible that initial infection can only occur in young parts of the root before wall deposition is complete. Holley and Peterson (1979) describe a peg-like projection of the hyphae to cause the wall of the host cell (in *Phaseolus*), and with it the plasmalemma, to invaginate. No papilla like that described by Kaspari (1973, 1975) in tobacco was seen, but the wall of the host was stretched thin and finally perforated by a narrow hypha, leaving it surrounded by the plasmalemma of the host which grew to keep pace as the hypha developed into an arbuscule. As the wall of the host stretched it kept its staining properties, so that it was suggested that the penetration was non-enzymatic. This whole question is considered later together with cognate problems connected with other types of mycorrhiza and with the problems of specificity and recognition.

Once inside the root, hyphae branch and spread within and between outer cortical cells and hyphal coils may also be formed within them. Frequently 2 or 3 appressoria may occur fairly close together so that an "infection unit", composed of several hyphae penetrating the epidermis and spreading in the outer cortex, in a fan-shaped way, may be formed (Plate 2). Hyphal growth outside the root may be extensive and it is interesting that it occurs *after* primary infection of the cortex has taken place from a propagule (Mosse and Hepper, 1975; I. R. Hall, personal communication). Hepper (1981) confirmed this finding and made the important observation that the mycelial growth outside the root could precede the formation of any arbuscules within the cells. If nutrients from the host are required for this mycelial growth it seems likely that transfer of them must occur across the interface between intercellular hyphae and the cells of the host. Therefore at this stage transfer across the arbuscular interface need not be involved. Secondary mycorrhizal infections, with typical appressoria and cortical hyphae, then develop from the external mycelium. This behaviour is so similar to that of the parasite *Gäumannomyces graminis* on wheat roots that the description of that fungus provided by Garrett (1956) could be applied to vesicular-arbuscular mycorrhizas with very little change.

Fungal spread within and between the cells of the outer cortex is normally restricted, so that fungal growth and differentiation chiefly occur in the middle and inner cortex. It is frequently stated (e.g. Holley and Peterson, 1979) that the apical meristem also remains uninfected—but whether or not it is actually immune to infection remains in doubt. No mycorrhizal colonization of tissues within the endodermis or of root nodules in leguminous plants normally takes place, although nitrogen-fixing nodules in the non-leguminous *Ceanothus* are colonized (Rose and Youngberg, 1981). An anatomical study of the endodermal and stelar anatomy of the two types of nodule, in relation to mycorrhizal infection, might be helpful here.

In the inner cortex the typical mycorrhizal structures, arbuscules and vesicles, are formed. The process of development of arbuscules within the cells in *Allium cepa* infected by *Glomus mosseae* was described by Dexheimer *et al.* (1979). Their description applies with little modification to other mycorrhizas investigated. As the hypha penetrates to form the main trunk of the arbuscule, the host plasmalemma is not breached but grows so that the invading hypha and all its branches remains surrounded by it. At the base of the trunk of the arbuscule a layer of host wall is laid down around it (Plate 5). This seems to consist of the same material as the primary wall of the host, and it is thick at the base where it is continuous with the wall of the host. Higher up the trunk hypha it becomes

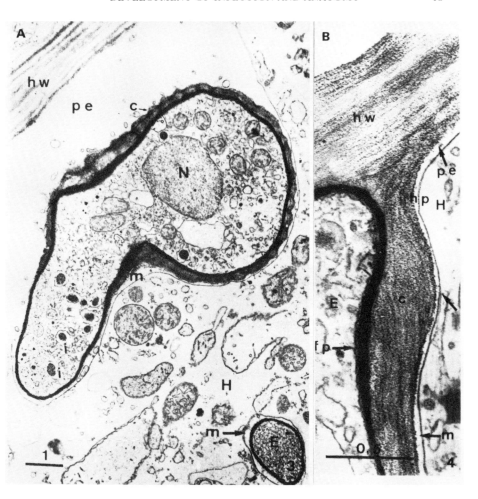

PLATE 5. Ultrastructural details of a vesicular-arbuscular mycorrhiza of *Allium*.

A. Section through an arbuscule trunk which is giving rise to a branch. The interfacial matrix (m) can be seen to be an extension of the host periplasm (pe). The densely stained fungal wall is clearly distinguishable from the surrounding coating of fibrils and also from the host wall (hw). Features of fungal and host cytoplasm can be distinguished. Both contain mitochondria and a nucleus (N) is present in the arbuscule trunk. (Scale 1 μm.)

B. Detail of the penetration point of a hypha (E) into a host cell (H). The host plasmalemma (hp) is invaginated by the arbuscule trunk so that the endophyte (E) is surrounded by a continuation of the host periplasm (pe), this is the interfacial matrix (m). Within the matrix an apposition layer (c) of fibrillar material has been laid down which is continuous with the host wall (hw) but clearly distinguishable from the darkly stained fungal wall. (Scale 0·5 μm.)

From Dexheimer *et al.* (1979).

gradually thinner and it is often believed to be absent from the finest branches of the arbuscule. Similar sheathing zones of wall material are formed around the coils in *Ornithogalum* as well as around its arbuscules. There the penetrating hyphae are enclosed in a "collar" which joins with the cell wall of the host and the plasmalemma of the host surrounds the coiling hypha. A collar is not evident, however, in the photographs of *Taxus* given by Strullu (1976a, 1978) nor is it shown by those of Holley and Peterson (1979) of *Phaseolus*.

Between the plasmalemma of the host and the wall of the hypha is a matrix which is continuous with the periplasm (or *éspace membranaire*) of the host. This matrix is somewhat electron-lucent and contains membranous vesicles apparently derived from the plasmalemma of the host, and scattered polysaccharide fibrils. An interfacial matrix of this kind surrounds both whole intracellular arbuscules including their finest branches and the intracellular coils (Plate 5). The arbuscule and its branches occupy much of the volume of the cell of the host and provide an extremely large area of contact between the fungus and the cytoplasm of the cell, for the tips of the branches are about $0.5–1.0 \mu$m in diameter as compared with $4–5 \mu$m in the intercellular hyphae. The area of contact between the symbionts has been calculated to be increased twofold by the formation of arbuscules. This estimate seems rather low and was made for infected onion cells in which the arbuscule occupied only about $0.64$ of the length of the cell (Cox and Tinker, 1976). In more completely colonized cells the increase would be considerably more. Estimates based on electronmicrographs of the amount of interfacial matrix derived from the root cells which surrounds the hyphal walls of the arbuscule differ. This may be because different combinations of host and fungus have been studied, or because there have been differences in preparation of the material for electronmicroscopy.

The host cell and the arbuscule or coil show every sign of physiological activity at first. They both contain active organelles and nuclei. The arbuscule is highly vacuolate, but in the protoplast, especially of the trunk, organelles, glycogen granules, and lipid droplets occur. In the vacuoles polyphosphate granules are present in the material prepared for electron microscopy (Cox *et al.*, 1975) but a considerable soluble component is probably present in fresh material, as it is believed to be in ectomycorrhizas (Loughman, personal communication).

At this stage the mycorrhiza can be envisaged as an association between metabolically active fungal structures and living root cells, as Dexheimer *et al.* (1979) pointed out. There is indeed good evidence of the physiological activity of the host cells, perhaps increased activity, because the nucleus and nucleoli are increased in size and the volume of the cytoplasm, which contains a full complement of organelles, is also increased. Table 8 shows the increased size calculated by Cox and Tinker (1976) of onion cells containing arbuscules of *Glomus mosseae*.

Great interest is attached to the matrix between the host plasmalemma and the arbuscule because this is the main interface between the symbionts. It has been observed by most workers on fine structure and has been given a number of names: matrix, extra-haustorial matrix, interfacial matrix, electron-lucent zone, isolation zone, sheath, layered zone, etc. The interpretation of this zone by Dexheimer *et al.* (1979) is extremely compelling. They suggest that the host plasmalemma produces membranaceous vesicles which are associated with the formation of the primary wall and that these are the membranous vesicles found in the interfacial matrix. The plasmalemma of the host may be viewed as retaining the ability to produce polysaccharide fibres which are visible within the interfacial matrix but it loses the capacity to organize them into a wall in the presence of the fungus. This view is completely in accordance with those put forward in several contexts in this book, that the explanation of the ability of the fungal hyphae, lacking suitable hydrolytic enzymes, to penetrate cells and tissues depends upon their ability to inhibit the enzymes of the host which build these walls. Such an activity would afford in the present context an explanation of the ability of the intercellular hyphae to pass along the middle lamella and then penetrate the cell walls. This explanation depends on penetration taking place in a zone where the host tissues are not yet fully mature or at least in places where the cell-wall material is being synthesized.

TABLE 8

Increases in size of cellular components in the infected region of a cortical cell of onion when infected by an arbuscule of *Glomus mosseae* (data of Cox and Tinker, 1976)

| Cell component | Increase |
|---|---|
| Host cytoplasm | 23 × uninfected cell |
| Host plasmalemma | 3·1 × cell perimeter[a] |
| Host tonoplast | 2·2 × cell perimeter[a] |

[a] Tonoplast and plasmalemma were assumed to be uninvaginated and therefore have the length of the cell perimeter in an uninfected cell.

## Degeneration of the fungus in the tissues

As time goes on, older regions of an infection show cytological changes, the intercellular hyphae become progressively more vacuolated and their degenerating sections may become separated from younger regions by septa. Similar changes take place in the arbuscules which have a life span of a few days, and cytoplasm progressively "retreats" from the tips of the fine branches and appears to be sealed off by the collapse of the hyphal walls, and possibly also by septa (e.g. Kinden and Brown, 1976). The collapsing arbuscule remains surrounded by the plasmalemma of the root-cell and may also become "encapsulated" by material presumably of host origin deposited in the interfacial zone.

Dexheimer *et al.* showed that the invaginated plasmalemma of the host continues to produce vesicles and fibrils as the arbuscules degenerate. The process of condensation of fibrillar polysaccharide material is much accentuated at that time, and as the collapsed hyphae aggregate they become encased in polysaccharide material. The sequence is also extremely well illustrated in *Taxus* by Strullu (1976a, 1978). It is attractive to suppose that the very process of degeneration causes the inhibitory action of the fungus on the condensation and on the aggregation of polysaccharide fibres to cease, and that the encapsulation of the decaying fungus arises from that. The active period of the arbuscules is undoubtedly short. Holley and Peterson (1979) observed that very few arbuscules in *Phaseolus* showed no signs of degeneration, but actual life periods estimated by Cox and Tinker (1976) and Bevege and Bowen (1975), given in Table 5, range from 2 to 15 days.

The significance of the degeneration of the arbuscules has been, and still is, the subject of some controversy. The process has been termed both "digestion" and "lysis". The former implies host activity resulting in transfer of nutrients from the fungus to root. There is no direct evidence either that host enzymes are responsible for the collapse of the arbuscules or that nutrient transfer occurs as a result. It is conceivable, however, that enzymes produced by the host, or indeed by the degenerating fungus, might release soluble molecules that could be absorbed through the plasmalemma of the host. This kind of active digestion of the arbuscule, i.e. phagocytosis, by the cell of the host has been suggested as a possible mechanism for movement of material from fungus to host. This view, popular in the past, was also held by Scannerini *et al.* in 1975. It seems unlikely because only the contents of the arbuscules, not their wall material and membranes, which are encapsulated, would be absorbed and these, as Cox and Tinker (1976) have calculated, would not go far to explain movement of nutrients from fungus to host.

An alternative view is that arbuscule degeneration is the manifestation of a host defence reaction against progressive fungal invasion. Once again, evidence for host-cell involvement in the process is lacking. There are changes in phosphatase enzymes of the root as a result of fungal infection. A mycorrhiza-specific phosphatase (which is probably formed by the fungus) increases in activity as onion roots become progressively infected by *Glomus mosseae*, but declines as the infection ages, so it is unlikely to be involved in the fungal degeneration (Gianinazzi-Pearson and Gianinazzi, 1978; Dexheimer *et al.*, 1979). Ultrastructural studies have further revealed that acid phosphatases are localized in the terminal branches of young arbuscules and that vacuolar alkaline phosphatases are associated with mature arbuscules and hyphae. The disappearance of these activities as the arbuscules collapse confirms that they are unlikely to be involved in the degeneration of the fungus.

The degeneration of fungal structures in the cells deserves further study in comparison with similar processes in the mycorrhizas of orchids and ericaceous plants. Until unequivocal evidence of its role in the symbiosis is forthcoming, non-commital terms like "degeneration" seem preferable to "digestion". We may be putting too much emphasis on this process, for all fungal hyphae in culture undergo ageing and this frequently results in autolysis, septum formation, etc. (J. E. Smith and Berry, 1977). The fact that degeneration is observed in mycorrhizas but not in parasitic symbioses (but see Pegg and Vessey, 1973) may simply reflect the greater longevity of the host cells in vesicular-arbuscular mycorrhizal symbioses. As will be seen, in ericoid mycorrhizas both host and fungus degenerate together. If it is ageing and autolysis that occur in mycorrhizas we might expect that reabsorption of nutrients back into hyphae, rather than transfer to host, would occur.

An alternative site for transfer of metabolites between the symbionts is across the intact and closely associated membranes in the developing and mature arbuscules. These structures have been likened to the haustoria of parasites and certainly have many similarities to them. Bidirectional transfer, which has been experimentally demonstrated (see below), could certainly take place at these sites and across the faces of intracellular coils or intercellular hyphae, and control could be operated by the organisms. It is the large surface area presented by the arbuscule that suggests that it is most important.

Vesicles are formed on the hyphae in many, perhaps most, vesicular-arbuscular mycorrhizal associations although some fungi seem rarely to produce them (Daft *et al.*, 1975). They consist of intercellular or intracellular swellings which occur terminally on hyphae in the middle or outer cortex. The thickened walls of the vesicles have made electron-microscopic studies difficult but it is clear that they contain very large amounts of lipid (as shown by histochemistry) and are almost certainly storage structures produced in the older regions of infection (Plate 2).

## Quantity of fungus in the tissues and in the soil

It is important to realize that a vesicular-arbuscular mycorrhizal root is a dynamic system and that both the root and fungal components exhibit continuing growth and development. The root grows apically with cell division, elongation and differentiation and it initiates lateral roots. After primary infection has taken place the growth of the extramatrical mycelium gives rise to increased fungal colonization of the soil and also results in secondary infections which increase the number of connexions between the internal fungal structures and the external mycelium. Within the root the internal mycelium also continues to develop so that the root is colonized by a number of "infection units" each connected to the extramatrical mycelium by entry point hyphae. As mentioned before, each infection unit is usually subtended by one main external hypha which may branch before the formation of appressoria on the epidermis, so that a fan-like complex is produced near or upon the surface of the root.

Each infection unit develops longitudinally and to some extent radially in the cortex of the root. Main branches of the longitudinal hyphae give rise to arbuscules in the cells. Hence the oldest arbuscules are closest to the site of penetration, i.e. of the primary infection, and the young and immature ones are progressively further away. The proportion of a root system colonized by a mycorrhizal fungus thus depends upon the rate of root growth, the rate at which the fungus colonizes the root from the soil giving rise to new infection units, and the rate of longitudinal spread of the fungus within the root cortex. In general a graph of the percentage of the root length infected against time has a sigmoid form, and examples from particular experiments are illustrated in Figs 1, 2 and 3.

The amount of fungal tissue within an infected root has been estimated in several ways. Routine determinations usually involve measurement (or estimation) of the percentage of the root length infected by fungal hyphae. Many workers, following Nicolson (1960), have used a "root-slide" method in which the proportion of infected root segments (of predetermined length) is counted.

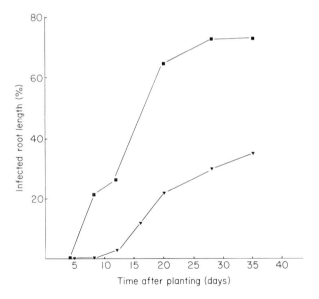

Fig. 1. The effects of differences in propagule density on the progress of infection in *Trifolium subterraneum*. Plants were grown in non-sterile soil with a propagule density (determined by the most probable numbers method) of 4·0 propagules per g (■) or in soil diluted with steamed sand to provide a propagule density of 0·4 propagules per g (▼). Percentage of the root length infected was determined by a line intersect method. (Results of F. A. Smith and S. E. Smith, 1981.)

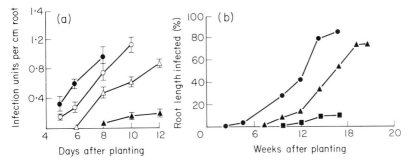

FIG. 2. The effect of temperature on development of mycorrhizal infection. (a) Production of infection units on the roots of *Medicago truncatula* grown in non-sterile soil at four different root temperatures. ▲, 12°C; △, 16°C; ○, 20°C; ●, 25°C. (Redrawn from S. E. Smith and Bowen, 1979.) (b) Progress of infection by *Endogone (Gigaspora) calospora* in the roots of onions under three temperature regimes: ●, 26°C day/21°C night; ▲, 21°C day/26°C night; ■, 16°C day/11°C night. Thirteen-hour photoperiod. (Redrawn from Furlan and Fortin, 1973.)

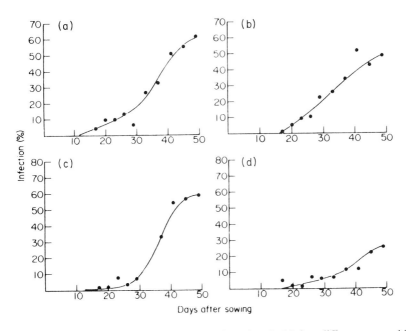

FIG. 3. The progress of infection of onion roots inoculated with four different mycorrhizal fungi. (a) *Glomus mosseae* (YV); (b) *Glomus macrocarpus* var. *geosporus* (LAM); (c) *Gigaspora calospora* (BR); (d) *Glomus microcarpus* (Mic). (Results of Sanders *et al.*, 1977.)

The precision of the method depends upon the length of the segments used. In some investigations relatively long segments (e.g. 1 cm) are not only scored for presence or absence of infection but ranked according to the intensity of infection, thus giving a better estimate over the root system as a whole. In other cases, short lengths of 1 or 2 mm are used. Recently methods for measuring root length by grid intersect (e.g. Newman, 1966; Marsh, 1971; Tennant, 1975) have been adopted for the microscopical examination of both mycorrhizal roots (e.g. Hirrell *et al.*, 1978; S. E. Smith *et al.*, 1979) and lengths of hyphae associated with roots (Tisdall and Oades, 1979). The methods used to estimate the percentage of root length infected have been reviewed and tested by Giovanetti and Mosse (1980). They found that "root-slide" methods gave consistently higher estimates of percentage infection as well as having higher standard deviations. This difference may be important when detailed comparison of estimates by the two types of method are made and various pleas for standardization (e.g. by Bierman and Lindermann, 1981) have been made.

Abbott and Robson (1978, 1979) have presented estimates of the volume or weight of infected roots compared with the weight of the whole root system. These were calculated from estimates of the volume of the root colonized by fungal structures. However, full details of how these estimates were made and how they relate to the percentage of the root length infected as obtained by other methods are not clear. Several attempts to determine the extent of infection by chemical methods have been made. Becker and Gerdemann (1977) made colorimetric determinations of the amount of the yellow pigment found in mycorrhizal roots. The amount of pigment extracted by hot water from onion roots correlated well with the percentage weight of yellow roots in the sample and also with the percentage of the length infected as determined after staining. The weight of glucosamine derived from chitin (by alkaline digestion) per unit weight of yellow roots was also found to be constant. Hepper (1977) made a more extensive investigation of the chitin content of mycorrhizal roots and compared the amounts that could be extracted from roots with the amounts found in weighed samples of extramatrical mycelium. On the basis of these measurements she was able to obtain estimates of the dry weight of the fungus within the roots, which ranged from 4% to 17% of the total (root and fungus) dry weight. Bethenfalvay *et al.* (1982) estimated the chitin content of extramatrical and internal hyphae of *Glomus fasciculatus* in artificial culture with *Glycine max* and calculated their biomass. This increased as the plants matured and was 12·3% of the total dry weight of the root at senescence. They suggested that the ratio of extramatrical to internal hyphal weight could be viewed as the index of the importance ("usefulness") of the fungus to the host. These estimates of the proportionate weight of the fungus may be compared with estimates of 39% for the sheath of ectomycorrhiza of *Fagus sylvatica* (Harley and McCready, 1952b). Determinations are usually underestimates, for recovery of extramatrical mycelium in *Fagus* mycorrhizas is difficult and the Hartig net and the fungal fruit bodies were not estimated.

Length or weight of hyphae associated with infected roots is important, for the extramatrical mycelium extends the root system (see later) and we need to know how extensive it may be and how it is attached and related to the internal mycelium in the root. Bethenfalvay *et al.* in the paper quoted estimated that the extramatrical mycelium per g dry weight of root increased 1·9 times during 10 weeks while the internal mycelium increased over 32 times. This represents a decrease of ratio of extra- to intraradical mycelium from 7·4 to 0·4. It is not clear, however, how much of the mycelium was metabolically active especially when the plant was mature or senescent. Direct measurement of length of hyphae extracted from soil by a grid intersect have given values of 1·36 and 1·29 m cm$^{-1}$ infected root length of ryegrass and clover respectively (Tisdall and Oades, 1979). These values are considerably higher than the 0·80 m cm$^{-1}$ calculated for onion after sieving and weighing the extramatrical mycelium, during which some hyphae would certainly have been lost (Sanders and Tinker, 1973). Estimates of the dry weight of hyphae associated with onion root infected by several different fungi were subject to similar errors but in a comparative experiment appeared to be constant at 3·6 μg per cm length of infected root (Sanders *et al.*, 1977). If the ratio of fresh weight to dry weight and the specific gravity are assumed to be 10 : 1 and 1 respectively, and the hyphae cylindrical of mean diameter 10 μm, this would give a hyphal length of about 0·5 m cm$^{-1}$ infected root length. These measurements of hyphal length can be related to counts of entry points or infection units on the root epidermis.

Statements of the number of entry-points per unit length of root vary considerably. Most recent estimates are of about 0·2 per mm length of root (e.g. Cox and Sanders, 1974; Smith and Bowen, 1979; Jasper *et al.*, 1979). This contrasts with the much larger numbers counted by Mosse (1959b) and some previous observers, in sectioned root material. She observed between 2·6 and 21·1 entry hyphae per mm. It has become apparent that the lower values are usually counts of "infection units" composed of a main external hypha subtending several appressoria (see Plate 2A), and that they may also be average values for the whole root system, including uninfected regions. Mosse herself counted actual numbers of hyphae penetrating the epidermis (normally several per infection unit). If we take the values for infection units assuming a single main hypha for each, then the length of hyphae associated with each unit might be as high as 60 cm (see also Read and Stribley, 1975).

If increased uptake of nutrients by mycorrhizal roots is mediated by the exploitation of the soil by extramatrical mycelium, uptake over the hyphal surface and by translocation to the root, then measurements of the extent of this mycelium may be very important in the interpretation of the differences in behaviour of different combinations of fungus and host, or in the efficiency of nutrient uptake in different external conditions. Comparisons could also be made of the relative importance of the external ramification of the fungus and the extent of its internal colonization in different conditions if the appropriate data on living hyphae could be obtained. This information is important in determining how the symbiosis operates and at what stage of development physiological and anatomical adjustments in the organisms are made.

Environmental conditions and species of host plant certainly affect the colonization of the cortex of the root. Comparisons of the percentage of the length of the root infected and sometimes of the relative numbers and conditions of arbuscules, vesicles and hyphae (which together have been called the "quality of infection") have been made under a wide variety of circumstances. Some attempts have been made to distinguish effects which may operate on fungal growth in the soil from those which may be mediated via the plant. The former would include the germination of propagules, primary infection of the root system and also some aspects of the development of extramatrical hyphae and the formation of secondary infection units. Effects mediated by the plant might include structural and other features affecting tissue and cell penetration and colonization, exudates from the root and the establishment of integrated transfer systems.

## The rate and extent of infection

The rate and extent of initial infection of a root system, and to some degree the duration of the lag phase in the typical sigmoid curve of percentage infection against time, depend partly upon the numbers of propagules in the soil (see Fig. 1). It must also depend upon the rate at which the propagules germinate and initiate primary infections on the root. One of the earliest investigations to consider this relationship was that of Daft and Nicolson (1969a). They inoculated tomato plants with different numbers of spores (3–225) per plant of *Endogone macrocarpa* var. *geospora* (*Glomus macrocarpus*) and made successive measurements of growth and leaf numbers over a period of 12 weeks. All inoculated treatments grew better than uninoculated controls, but the time at which the growth response was first apparent was earlier the higher the number of spores used. Mycorrhizal plants also produced more leaves and retained those that they had for longer times. The percentage of the root length colonized by mycorrhizal endophytes was only determined after 12 weeks, by which time there was no difference between plants grown with different numbers of spores. They concluded that the speed of infection was important in the onset of the growth response and also in the extent to which phosphorus concentration in the tissues was affected by infection. Subsequent work, in which the extent of infection has been measured at intervals throughout the growth period, has confirmed this. Rich and Bird (1974) emphasized that early mycorrhizal infection had an important accelerating effect on the appearance of growth responses in crop plants. Carling *et al.* (1979) inoculated root systems of soybean plants with varying amounts of a mixture of infected roots and soil containing 12 chlamydospores of *Glomus fasciculatus* per gram. The inoculum was mixed throughout the soil in 3 kg pots. The relationship between the amount of inoculum and the number of infection units formed on the roots of 21-day-old plants was linear between 0 and 0·006 g inoculum per gram of soil in the pots (0 and 0·08 spores per g). Inoculation with 0·013 g inoculum produced no further increase in numbers of infection units at 21 days. Similar results have been obtained by S. E. Smith and Walker (1981) in experiments with *Trifolium subterraneum* in which the inoculum consisted of soil containing up to 4·0 propagules per g (estimated by the most probable numbers method) diluted with sterile sand. *T. subterraneum* becomes infected more rapidly than soybeans, so the plants were harvested at 12 days. A linear relationship between the number of infection units formed and the numbers of propagules was obtained between 0 and 2·4 propagules per g soil. Again, saturation of the relationship occurred at higher propagule densities (Fig. 4). The reason for this has not been investigated but we may speculate that there may be a limited infectible length of root at any time and that this may limit the numbers of infections that can occur. Mosse (1978) has previously emphasized the relationship between infectivity of the soil (propagule density) and root density.

Once primary infections from propagules in the soil have become established, there is, as we have already mentioned, extensive growth of extramatrical mycelium. Hyphae grow along the surface of the root epidermis and initiate secondary infection units. Rapid increase in the numbers of infection units per plant and in the percentage of root length infected occurs (Figs 1 and 5). We know little about the relative importance of primary and secondary infections of the root system in determining the final extent of root colonization, although it is possible to predict that the relationship will be very much influenced by the density of propagules in the soil. It appears that low numbers of propagules in field soils (e.g. in eroded sites) may result in low levels of infection (Hall and Armstrong, 1979; Moorman *et al.*, 1979; Reeves *et al.*, 1979; Powell, 1980). There is no doubt that differences in the rate of infection, resulting from differences in the numbers of available propagules, may affect the rate of phosphate uptake from soil and the growth response to infection. Thus S. E. Smith *et al.* (1979) showed that slow infection was associated with a delay in the rate at which increased phosphate inflow into mycorrhizal plants developed with consequent effects on nodulation, $N_2$ fixation and growth.

Differences in rates of infection engendered by other factors have similar effects on plant growth. Sanders *et al.* (1977) made a detailed comparison of infection by four mycorrhizal fungi on plant growth and phosphorus uptake.

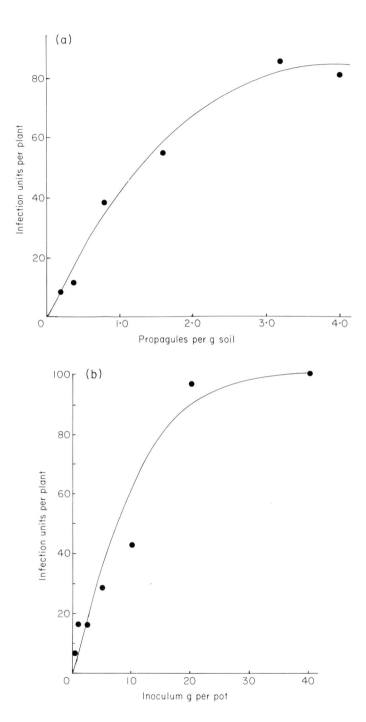

They found that one fungus, probably *Glomus microcarpus*, which was very slow to colonize roots and achieved only about 25% infection of root length by 50 days, had little or no effect on the behaviour of onion plants, whereas infection by three other fungi (*Glomus mosseae*, *Glomus macrocarpus* var. *geosporus* and *Gigaspora calospora*), which colonized more quickly, resulted in the usual increases in phosphorus uptake and growth of the onions (Fig. 3).

It appears that at later stages of infection the rate of colonization of a root system and the rate of root growth come into equilibrium, for the curve of the percentage of the root length colonized against time almost invariably shows a plateau, following the phase of rapid root colonization. The sigmoid shape of this curve may explain why a close relationship between the growth response of the plant to infection and the percentage infection is not always observed if infection is measured only at a final destructive harvest, as for example in the experiments of Daft and Nicolson (1969a) described above and in many other studies. Sanders *et al.* (1977) stressed the importance of the development of the external mycelium and the problem has been examined by Graham *et al.* (1982). They used the dry weight of the soil closely adhering to the roots of *Citrus* as a measure of the development of external mycelium. Using several strains of species of *Glomus* including 4 of *Glomus fasciculatus* and 2 of *G. macrocarpus*, they showed that the growth of the host during 4 months was related to the extent of development of extramatrical mycelium in conditions where the percentage of the root system infected only varied between 95 and 98% at the final harvest. Nevertheless, the percentage infection is a very commonly used measure of the amount of mycorrhizal fungus associated with the roots, so that much information on the factors affecting colonization is presented on this basis. An understanding of the processes underlying the form of the curve relating it to growth is therefore most important to any interpretation of the results.

As we have seen, the extent of root colonized by a mycorrhizal fungus depends upon the rate of formation of infection units from extramatrical hyphae, the rate of spread of the fungus within the root and the growth of the root itself. Sutton (1973) was one of the first to emphasize the contribution of root growth to percentage infection. Any factor which affects the rate of growth of the root will almost certainly affect the measured percentage infection and it is most difficult to separate this effect from any other, more direct, effects upon fungal

FIG. 4. The effect of different amounts of inoculum on the development of infection units in *Trifolium subterraneum* and *Glycine max*. (a) Numbers of infection units in the roots of *Trifolium subterraneum* grown in mixtures of non-sterile soil and steamed sand to provide a range of inoculum densities from 0 to 4 propagules per g (determined by the most probable numbers method) and harvested after 12 days (redrawn from S. E. Smith and Walker, 1981). (b) Numbers of infection units in the roots of *Glycine max* grown for 21 days in autoclaved soil with inoculum of *Glomus fasciculatus* in the form of root/soil/fungus mixture blended throughout the soil (redrawn from Carling *et al.*, 1979).

colonization or growth. In some investigations, microscopical examination of fungal structures, such as the numbers and state of vesicles and arbuscules, has supplemented the measurements of percentage infection and may give indications of the state of development of the symbiosis.

Environmental factors which have been investigated with respect to their effects on development of mycorrhizas include the quantities of nutrients in the soil, light supply, temperature and soil pH. Most of the work has concerned the effects of phosphorus in the soil. Daft and Nicolson (1966) reported an inverse relationship between percentage infection of tomato plants and the relative level of phosphate supplied as bone meal. Later (1969b) they investigated the relationship between infection of maize by *Endogone (Glomus) macrocarpa* var. *geospora* and the source and duration of application of fertilizer phosphate. There was an inverse relationship between amounts of soluble $KH_2PO_4$ supplied and infection of the roots estimated as percentage infection or by the numbers of

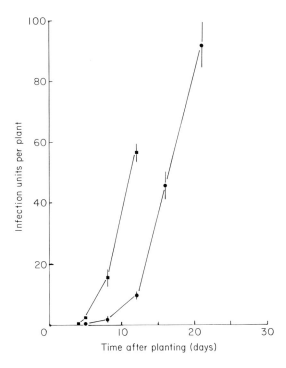

Fig. 5. Formation of infection units in *Trifolium subterraneum* grown at two propagule densities. ■, non-sterile soil containing 4·0 propagules per g soil; ●, non-sterile soil diluted with steamed sand to give 0·4 propagules per g. Incomplete data for the higher propagule density arise because infection units became too dense to count with accuracy. (Redrawn from F. A. Smith and S. E. Smith, 1981.)

chlamydospores formed in the soil around each plant. This relationship held whether $KH_2PO_4$ was applied at one application or over a period of time. Similar results were obtained by Baylis (1967), by Mosse (1973), who also described changes in the detailed anatomy of the infections at high levels of $Ca(H_2PO_4)_2$ and by Sanders and Tinker (1973, see Fig. 6) and many others. The numbers of arbuscules in onions were much lower with high phosphate, an observation which has been made repeatedly since. However, Abbott and Robson (1979) found that while phosphate did decrease the extent of infection of the roots of *Trifolium subterraneum*, *Erodium botrys* and *Lolium rigidum*, it had little effect on arbuscule development. There are also reports (e.g. Mosse, 1973) of the failure of infection units to develop normally in the cortex of onion roots grown with high levels of soluble phosphate. This may explain, as Mosse suggested, the difficulty in obtaining mycorrhizal infections in agar culture when phosphate in the medium was about 1 mM (Mosse and Phillips, 1971).

There is no doubt that the increased growth in length of roots when soil phosphate supplies are good must contribute to the reduction in the percentage of the root length infected. There may also be effects on fungal growth mediated

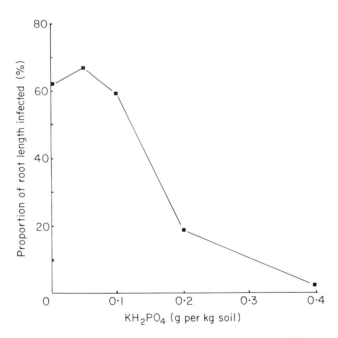

FIG. 6. The effect of $KH_2PO_4$ additions to soil on the percentage of the root length of onions infected by vesicular-arbuscular mycorrhizal fungi after 8 weeks' growth (redrawn from Sanders and Tinker, 1973).

via the root or directly on fungal growth in the soil. Mosse's (1973) observations apparently discount the second possibility, for in many cases hyphae growing on the surface of the root looked normal. Sanders (1975) investigated this problem by injecting phosphate solution into the hollow leaves of onion plants. Phosphate was translocated from shoot to root. At the final harvest the percentage of the root length infected was reduced and this could have been a result of slower growth of the hyphae along the cortex of roots of high phosphate content. The amount of external mycelium produced per cm of infected root was reduced from 3·5 mg to 2 mg, and changes in the anatomy of the infections, like those seen by Mosse (1973), were also recorded. Sanders therefore concluded that the effects of soil phosphate in reducing infection were mediated via the root and need not involve any direct effects upon fungal growth in the soil.

Transplanting experiments and split-root techniques have been used in the same way as foliar application of phosphate. Menge, Steirle *et al.* (1978a) showed a reduction in numbers of chlamydospores produced by *Glomus fasciculatus* on both halves of the split root system of sudan grass, even though only one half received high levels (750 mg per kg soil) of phosphate. Additional experiments indicated that the reduction in numbers of arbuscules and external hyphae, as well as chlamydospores, was more closely related to the phosphate concentration in the roots than to that in the soil. Azcon *et al.* (1978) found that the percentage infection of lettuce was reduced as the phosphate content of the soils into which plants were transferred was increased. Similarly Jasper *et al.* (1979) observed that a lower percentage of the root volume was infected and also that fewer entry points per 100 mm of root were produced when *Trifolium subterraneum* was transplanted from sterilized soil low in phosphate to a soil high in phosphate containing spores of *Glomus monosporus*. Such an effect was not observed when plants from sterilized high phosphate soil were transplanted to infective low phosphate soil. In this experiment high phosphorus concentrations were associated with lower soluble carbohydrate concentrations in the roots and the authors suggested that carbohydrate availability to the endophytes might be important in determining fungal establishment. However, the results of transplanting experiments are usually hard to interpret and it is essential to be wary of making too much of correlations between phosphate and carbohydrate concentrations and growth of infections. There is nevertheless some evidence that other factors which might affect carbohydrate status of the roots can affect mycorrhizal colonization. Mycorrhizal infection has long been believed to be related to the light intensity. Schrader (1958) believed that infection in peas was related to carbohydrate status because it was correlated with light supply. Boullard (1957, 1959, 1960) reached similar conclusions after experimentation with several species. Peuss (1958) experimented with reduction of light intensity and defoliation and showed that both these treatments reduced infection rating in tobacco plants. Hayman (1974) found that infection was higher at higher light intensities and that this was correlated to sugar concentrations in the roots. Ratnayake *et al.* (1978) have formed the hypothesis based on experiments with *Citrus* that exudation of substances from the roots of plants growing in low phosphate conditions is increased owing possibly to the decrease in phospho-lipids and an increase of permeability of the cell membranes. They found that there was a much greater leakage of amino acids and sugars from the roots and suggested that these might stimulate the growth of the fungus and the development of mycorrhizal infection. They have followed this up (Johnson, Menge *et al.*, 1982; Johnson, Graham *et al.*, 1982) by studying other conditions where increase of exudation results in an increase of mycorrhizal infection. The difficulty of this hypothesis as with others depending on exudation of common classes of substances is that the mechanism is not specific to the stimulation of mycorrhizal fungi so that it must be incomplete.

Daft and El Giahmi (1978) found that both defoliation and either shading or short daylength reduced infection by *Glomus macrocarpus* var. *geosporus* or *G. mosseae* in a variety of host plants. Numbers of secondary spores produced on the extramatrical mycelium were also lower. The development of arbuscules and vesicles may also be affected. Gunze and Hennessy (1980) recorded increases in the numbers of arbuscules in roots of shaded or defoliated cowpeas, while the numbers of vesicles fell. However, when the shoot apex alone was removed, numbers of arbuscules declined, but development could be restored by application of indolyl acetic acid (IAA) at 1 mg g$^{-1}$ to the shoot apex. Lanoline alone and lower concentrations of IAA (0·1 and 0·5 mg g$^{-1}$) had no effect in restoring mycorrhizal development. This apparent intervention of the hormone IAA in the development of the fungus inside the tissues is not yet explicable, but must be complicated.

Temperature, not surprisingly, affects the rate and extent of the development of vesicular-arbuscular mycorrhizal symbiosis. In general, there is an increase in percentage infection up to about 30°C. Furlan and Fortin (1973) found that the lag phase in the development of *Endogone (Gigaspora) calospora* on onion was shortest, the rate of infection most rapid and spore production greatest when temperatures were 26°C (day) and 21°C (night) compared with either 21°C/16°C or 16°C/11°C (see Fig. 2). Final spore production per plant was 2600, 1800 and 50 for the three temperature regimes. Shenck and Schroeder (1974) controlled root temperatures between 17°C and 41°C, while exposing the shoots of soybeans to the same (28/34°C) temperature regime. After 60 days the percentage infection and the development of arbuscules were both at a maximum at a root temperature of 30°C. Root growth was also highest at this temperature. Above 30°C both root growth and percentage infection were reduced and no infection occurred at 41°C. Temperatures (particularly moist heat) above 40°C also inhibit germination of *Acaulospora laevis*, *Glomus caledonius* and *G. monosporus* (Tommerup and Kidby, 1980), so that absence of infection in soil above this temperature could be explained by failure of initial germination or germ-tube growth. The optimum temperatures for spore germination, on agar at least, vary with fungal species. For instance, *Gigaspora coralloidea* and *G. heterogama* have optima at 34 °C, considerably higher than that for *Glomus mosseae* at 20 °C (Schenck *et al.*, 1975). *Glomus mosseae* was a major component of the mixed soil inoculum used by S. E. Smith and Bowen (1979) in their study of temperature effects on formation of infection units in *Trifolium subterraneum* and *Medicago truncatula*. The roots were kept as 12 °C, 16 °C, 20 °C and 25 °C and all plants were exposed to shoot temperatures of 20 °C (day) and 15 °C (night). Infection units appeared earlier at higher root temperatures and the biggest increase in their numbers was between 12 °C and 16 °C. When plants at the same stage of development regardless of age (spade leaf fully emerged and root length about 12 cm) were compared, there were no significant differences in the numbers of infection units per plant at 16 °C, 20 °C and 25 °C. Higher temperatures were not applied in this experiment (see Fig. 2).

Interactions between nutrients in the development of infection have also been observed. Many nitrogenous fertilizers have been reported to decrease infection; these included ammonium nitrate applied to *Pisum sativum* (Lanowska, 1966), calcium nitrate applied to winter wheat (Hayman, 1970) and urea, in association with high soil phosphorus (Bevege, 1971). Chambers *et al.* (1980) distinguished between ammonium and nitrate ions in their study of infection in *Trifolium subterraneum*. Formation of infection units and percentage infection in young plants was reduced more in the presence of ammonium sulphate than with sodium nitrate. Sodium sulphate had a small effect on the development of infection. There may be interactions between the effects of nitrogen and phosphorus on plant growth and their effects upon infection. For example, Mosse (1973) observed that the inhibitory effects of high phosphate were reduced when onions were given additional nitrogen. Plant growth was increased and phosphate concentration in the tissues was decreased, giving a further indication that internal nutrients may play a part (direct or indirect) in the development of infection. The results of Mosse's experiment are explicable if nitrogen became the factor limiting plant growth at high levels of phosphate.

## Mathematical models of the infection process

Implicit in much of the preceding discussion is a concern about distinguishing between factors affecting initial infection of roots from the soil and effects mediated via changes in root physiology and possibly transfer of nutrients between the organisms. Several of the investigations have included a study of the formation of infection units, which gives an indication of the completion of the phase of colonization from soil. In other cases development of extramatrical hyphae and spores has been measured, which also gives an idea of the way in which factors may affect hyphal growth outside the root. However, this must also be affected by the transfer of nutrients between the symbionts. Growth within the root has been measured as percentage of the root length infected and by frequency of arbuscules. Recently attempts have been made to calculate parameters from infection data which may be useful in predicting the effects of internal and external factors on the development of infection. Tinker (1975a) used measurements of total root length infected and relative growth rate of plants to calculate a constant $S$, using the equations shown below, which he considered might be a "specificity" or "susceptibility" parameter, useful in determining the way in which the two organisms grew relative to each other. The equations are:

$$L_t = Ae^{Rt}, \tag{1}$$

$$\frac{dL_i}{dt} = SL_i(nAe^{Rt} - L_i), \tag{2}$$

where $L_t$ = total root length, at time $t$; $L_i$ is the length of infected root; $R$ is the relative growth rate of the host, and $n$, $A$ and $S$ are constants. Tinker found that these equations fitted his own curves of percentage infection against time quite well, but he did not extend the approach to results obtained under various conditions likely to affect the development of the symbiosis.

S. E. Smith and Walker (1981) have also used equations to calculate two infection parameters, the frequency of infection of a root system from the soil ($A$ m$^{-1}$ d$^{-1}$) and the rate of linear extension of the fungus within the root cortex in both directions from an infection point ($B$ m d$^{-1}$). One of these equations has already been mentioned, but both are given here, for the sake of completeness:

$$\frac{dN}{dt} = A(L - L^*), \tag{1}$$

$$\frac{dL^*}{dt} = BN\left(L - \frac{L^*}{L}\right), \tag{2}$$

where $N$ is the number of infection units per plant; $L$ is the total root length and $L^*$ is the infected length of root. As in Tinker's equation, the rate of formation of new infections is considered to be proportional to the length of uninfected root available for infection. This may be an over simplification, as we have already indicated, because the apical region may be more "infectible" (p. 36). However, equations such as these give us a means of calculating separately rates of different phases of infection under different conditions. S. E. Smith and Walker (1981) have already shown that $NaNO_3$ and $NH_4NO_3$ act by reducing both the frequency of formation of infection units ($A$) and the rate of growth of the fungus within the cortex ($B$), but that NaCl appears to affect only the frequency of infection. Tinker's work has recently been extended by J. G.

Buwalda and his colleagues (Buwalda, Ross Stribley and Tinker, 1982a,b). While assuming that infection occurs from a single and localized source of inoculum rather than from numerous, randomly dispersed sources, these workers have modelled the progress of formation of infected root length with an equation which is quite similar to that of S. E. Smith and Walker. Propagule density is not taken into account nor are numbers of infection units included among the parameters measured. The equation is

$$\frac{dL}{dt} = S'L_i \left( 1 - \frac{L_i}{nL_t} \right)$$

where $L_i$ and $L_t$ are infected and total root length respectively, $S'$ is a parameter related to spread of infection from a source of inoculum both within and outside the root system and $n$ is the asymptotic value in the curve of per cent infection of the root system against time, which is introduced as a correction factor to account for the fact that per cent infection rarely reaches 100. The equation modelled changes in infected root length of wheat and clover at different rooting densities more closely than Tinker's equation, particularly at early harvests. If infected root length is proportional to numbers of infection units (which it may be under some conditions: see S. E. Smith and Walker 1981) then calculation of $S'$ would provide a means of determining the susceptibility of a root-system to infection under different conditions and in different hosts. It does not allow a distinction to be made between factors affecting infection from the soil and factors operating via the root, nor does it permit consideration of propagule density in the soil. However the model does have considerable practical advantages in that enumeration of infection units (a very time-consuming activity!) is not required and that it can be used in experiments in which plants were inoculated with localized inoculum as in many investigations.

The modelling approach could certainly be extended to a study of the effects of a wide range of factors on mycorrhizal infection, provided the necessary parameters were measured. The most important factors appear to be effects of nutrients, especially phosphate and those arising in different autobiont/fungus combinations. For example, Bevege and Bowen (1975) observed that formation of appressoria on onion roots by *Endogone (Glomus) mosseae* and *G. araucariae* was generally slower than that on clover, but they did not relate this to root growth in the two species. Similarly, spread of infection within roots was slower in onions than in clover. Differences have been observed in other hosts, so that Koske *et al.* (1975) found that while most sand dune plants had between <1 and 92% of their root length infected, it was only 10% at maximum in *Xanthium* (Koske and Halvorson, 1981) and Koske (1981b) also found an indication that extramatrical spores increased in numbers in particular host-fungus combinations. The use of equations like those which have been given above could help to determine the bases for these differences.

## Conclusions

Roots become infected with vesicular-arbuscular mycorrhizal fungi by hyphae growing from propagules present in soil or from nearby roots. Soil-borne propagules may be spores or dormant fungal structures present in dead, previously infected roots.

Infection units in the cortex of the root develop following initial penetration of the epidermis. Characteristic structures are arbuscules which are much-branched, intracellular fungal hyphae, and vesicles which are intracellular or intercellular and probably have a storage function. The interface between the symbionts is usually composed of host plasmalemma, an interfacial matrix, fungal wall and fungal plasmalemma. Modifications of this interface may occur at different stages of mycorrhizal development and in different host/fungus combinations.

Spread of infection in a root system occurs by growth of infection units and by the formation of new infection units from extramatrical hyphae. Environmental conditions, such as nutrients, light, pH and temperature influence the extent and quality of infection by affecting colonization from the soil, fungal growth and development in the cortex of the root. Growth of the root itself also affects the proportion of the root which becomes mycorrhizal.

*Chapter 3*

# Carbon Metabolism and Growth

## Introduction

Some insight into the physiology of vesicular-arbuscular mycorrhizal associations can be gained from nutritional characteristics of the symbionts. The host plants are almost all autotrophic for carbon. The exceptions, such as the prothalli of some pteridophytes, certain Burmanniaceae and other non-chlorophyllous ("saprophytic") hosts should, perhaps, be reinvestigated, particularly since some "saprophytic" plants have been found to be epiparasitic on green plants via a common mycorrhizal fungus. Indeed, since we have no evidence that vesicular-arbuscular mycorrhizal fungi can exist vegetatively, except for a small period of time in the saprophytic mode, epiparasitism might explain the existence of vesicular-arbuscular mycorrhizal "saprophytes".

Green host plants, although normally infected by mycorrhizal fungi in the field, are usually capable of satisfactory growth in the absence of such infections as long as mineral nutrient supplies are adequate. In contrast, the fungi of vesicular-arbuscular mycorrhizas have a very limited capacity for independent growth, and are likely to be ecologically obligate symbionts. They have little or no capacity to produce enzymes capable of degrading complex carbohydrate polymers, such as cellulose or pectin, even when growing satisfactorily in association with the roots of plants. Absence of these enzymes is characteristic of biotrophic fungi and it has been suggested that, even if the fungus had the genetic information to code for their synthesis, catabolite repression might prevent expression in symbiotic systems (Lewis, 1973). Cellulase and pectinase enzymes are certainly subject to such repression (see Keen and Horton, 1966; Goodenough and Maw, 1975; Goodenough and Kempton, 1977, to name a few sources only). However, evidence for the importance of such a mechanism in disease development is conflicting, for, while Horton and Keen (1966b) reported that high tissue sugar content repressed development of the symptoms of pink root of onions by *Pyrenochaeta terrestris*, Goodenough and Kempton found that cultivars of tomato resistant to *Pyrenochaeta lycopersici* had lower sugar concentrations in their tissues than susceptible varieties. There is, however, as yet no evidence of the production of these hydrolytic enzymes by vesicular-arbuscular mycorrhizal fungi. The problems involved in explaining the penetration of cell walls by hyphae apparently unable to degrade complex carbohydrate need separate consideration; here let it be said that the local degradation of cell walls is a far cry from the production of enzymes in quantities capable of providing soluble carbon compounds for hyphal growth. In this group of biotrophic associations there is, as already described, intimate contact between the living cells of the symbionts, which is maintained for considerable time. Damage to the root cells apparently does not occur, although some changes such as increased cytoplasmic volume and plasmalemma area do take place (Table 8). Degeneration of arbuscules could be considered as tissue damage to the fungus, but it is not clear whether this is brought about by fungal or plant enzymes. It has been suggested that if one of the symbionts dies or degenerates during the symbiosis, the surviving symbiont must be a necrotrophic parasite. However, total death of the degenerating partner does not occur, and the view ignores the fact that a symbiotic union between physiologically active partners exists at the same time as degeneration is occurring. An additional feature of vesicular-arbuscular mycorrhizal symbiosis, which is common to many biot-rophic unions, is the increase in nuclear and nucleolar size. As Callow (1976) has emphasized, increase in nuclear size may be the result of either increased RNA or DNA synthesis. The latter is most common when changes in the differentiation of host cells occurs, such as in the production of galls or root nodules. We know little about the changes that actually occur in the nuclei of vesicular-arbuscular roots, but it may well be that increases in nucleic acid synthesis are important where particular proteins or enzymes are involved in the regulation of the symbiosis. There is evidence for this in vesicular-arbuscular mycorrhiza. Strullu (1976a) has pointed out that increase in nuclear size occurs in those kinds of mycorrhizas in which the fungus penetrates the cells of the host, i.e. "endomycorrhizas" in the old sense, but not in ectomycorrhizas, and this difference may well provide a point worthy of investigation.

With regard to the movement of materials between symbionts there are two experimental physiological questions to be answered: (1) does the degeneration of the hyphae result in nutrient transfer? (2) Does nutrient transfer occur across intact hyphal and cell boundaries? These are questions also applicable to all other mycorrhizal symbioses where degeneration of hyphae occurs. The answers to them may depend on whether organic or mineral nutrients are being considered, but they cannot both be negative. Moreover it is impossible to believe that fungal degeneration can result in movement from the host cells into the hyphae, so that the answer to the second question must be positive for movement in that direction. It therefore follows that if movement of carbon compounds formed in photosynthesis occurs from host to fungus, it must occur via the arbuscules or the less specialized intra- or inter-cellular hyphae. That is, it must occur across living plasma membranes.

A feature of vesicular-arbuscular mycorrhiza atypical of most (pathogenic) biotrophic associations is the lack of close specificity between host species and fungal species or strain. This is discussed in greater detail in a later chapter, but it may be that specificity is a necessary consequence of the damaging parasitic habit and results from selection for resistant hosts and consequent mutation of efficient pathogens, particularly in crop monocultures. Mutualistic associations, such as vesicular-arbuscular mycorrhizas, are subject to quite different selection pressures and may have evolved quite differently in this respect. An inescapable corollary to the non-specificity of the relationship is that very diverse hosts can set up prolonged physiological relationships with species of several genera of Endogonaceae.

## Carbon nutrition

The assumption was made in the early days of mycorrhizal research that movement of carbohydrate is predominantly from autotroph to heterotroph in this mycorrhizal system and still seems eminently reasonable. Very much increased growth of extramatrical mycelium takes place once infection is established. The increase of fungal biomass presumably depends upon a supply of carbohydrate from the roots, although availability of hormones or growth factors might have an important effect if carbon in substrate quantities could be obtained from another source. However, available evidence suggests that a source other than the host is unlikely. The stimulus to mycelial growth which occurs once infection has occurred is highly suggestive of a transfer of nutrients, most especially because it takes place in sand or vermiculite culture where little carbon is present in the substrate.

Direct evidence of carbohydrate transfer from root to fungal cells has been rather difficult to acquire. The close and complex association of the two organisms within the root, and relatively scanty external fungal tissue in artificial systems have meant that techniques involving their separation cannot be easily applied. Such separation has given very useful results for lichens, powdery and downy mildews and for ectomycorrhizas. However, methods are now available for the production of large amounts of external mycelium associated with roots in solution culture (e.g. Howeler *et al.*, 1981; St. John *et al.*, 1981; Macdonald, 1981) and this should facilitate future physiological and biochemical studies.

In several investigations the destinations in roots and extramatrical hyphae of photosythetically incorporated $^{14}C$ have been traced. Ho and Trappe (1973) performed the earliest experiments which demonstrated that, over a period of weeks, small amounts of labelled photosynthate appeared in extramatrical hyphae and spores of vesicular-arbuscular mycorrhizal fungi. The more recent work of Bevege *et al.* (1975) and of Cox and his colleagues (1975) has established that there is rapid translocation of photosynthate to root systems of mycorrhizal plants and that some of this passes into intracellular fungal structures (demonstrated by autoradiography) and into hyphae outside the root (shown by direct counting of harvested mycelium). Hayman (1974) and Bevege *et al.* (1975) analysed uninfected roots and mycorrhizas of several species of host plant, but failed to reveal compounds specific to the mycorrhizal fungus such as mannitol and trehalose which are found in the fungi of other mycorrhizas. They appear to act as sinks or markers for carbohydrate transfer where they are formed. The carbohydrate composition of vesicular-arbuscular mycorrhizal and uninfected roots appears very similar, with no evidence of rapid conversion of photosynthate into soluble or insoluble fungal metabolites which, being unavailable to the root cells, would maintain directional carbohydrate transfer towards the fungus. Such carbohydrates, trehalose, mannitol and glycogen, have been identified in sufficient quantity in ectotrophic mycorrhizas by Lewis and Harley (1965a,b,c) to lend credibility to a transfer mechanism in which they constitute a sink in the fungus or a "biochemical valve". D. C. Smith *et al.* (1969) have elaborated this idea with respect to many other biotrophic systems.

There is some evidence that lipid synthesis in the fungal component of vesicular-arbuscular mycorrhizas may play a similar role to polyols, glycogen and trehalose in other systems (Bevege *et al.*, 1975; Cooper and Lösel, 1978). Total lipid levels in mycorrhizas may be considerably higher than in uninfected roots of onion, clover and ryegrass (see Table 9). In these hosts infected by *G. mosseae*, the lipid fractions did not appear to differ qualitatively. Increased deposition of lipid droplets in older mycorrhizal hyphae and arbuscules, as well as increased amounts of membrane lipids (derived from both fungus and root) associated with arbuscule formation and proliferation of plasmalemma and cytoplasm, presumably contributed to increased total levels. Cox *et al.* (1975) demonstrated autoradiographically that lipid droplets in the fungus do become labelled in $^{14}$C-labelling experiments, and Nagy and co-workers (1980) also

TABLE 9

The effect of infection by *Glomus mosseae* on the total lipid content of the roots of onion, clover and ryegrass (results of Cooper and Lösel, 1978)

| | Onion | Clover | Ryegrass |
|---|---|---|---|
| | mg lipid per g fresh weight root | | |
| Infected | 3·4 | 4·1 | 6·6 |
| Uninfected | 2·7 | 3·3 | 3·7 |

The lipid content of mycelium of *Glomus mosseae* was 43·8 mg lipid per g fresh weight.

detected increases in amount of triglycerides and phospholipids in mycorrhizal roots of *Citrus* spp. infected by the same fungus. There was in *Citrus* a definite qualitative difference between the lipids of mycorrhizal and non-mycorrhizal material. Three unidentified fatty acids, which made up 31 to 41% of the total lipids in mycorrhizal roots, were not present in uninfected roots. In view of these results it is surprising that increased incorporation of $^{14}C$ into the lipid fraction of mycorrhizal roots was not detected by others. Bevege *et al.* (1975), besides suggesting lipids as a possible destination for carbohydrates in the fungus, emphasized that a significant sink is related to the growth of the hyphae and is represented by $^{14}C$-labelling of organic and amino acids and of proteins and cell wall material. They showed that in clover activity per unit weight of extramatrical hyphae was four times that of the subtending root and represented 24% of the total $^{14}C$ activity in the localized part analysed. As they point out, the proportion of the total activity translocated to the root-system which is found in the extramatrical hyphae will depend on the intensity and distribution of infection throughout the whole root-system. In this instance, where infection was intentionally localized, it was only 4% of the total but in normally infected clover it would be expected to be higher. Snellgrove *et al.* (1982) showed that in leek plants with 60–70% of the length of their roots infected, 8% more of the carbon fixed in photosynthesis was diverted to the root system than in uninfected plants. Silisbury *et al.* (in press) showed that both uninfected and mycorrhizal clover respired 30% of $CO_2$ fixed in photosynthesis when phosphate supply was adequate. Experiments at low phosphate levels are needed.

Carbon transfer between roots and hyphae must be across the living plasmalemmae of both organisms. Arbuscules are assumed to act as haustoria (see Harley, 1969; Bushnell, 1972; Cox and Tinker, 1976), that is to provide a large area of interface between the symbionts and facilitate nutrient exchange between them. However, there appears to be no theoretical reason why transfer to and from living inter- and intra-cellular hyphae of unspecialized kinds should not also take place. It would be interesting to know whether the fungi have the capacity to promote leakiness of the root cell membranes, such as is observed in freshly isolated algae of some lichens (D. C. Smith, 1978). It has been suggested to occur by the action of mannitol (Wedding and Harley, 1976) in ectomycorrhiza. On the other hand, the natural leakiness, as shown by exudation from intact roots, might be enough. Diversion of cell-wall material of the root to the fungus would provide a supply of carbohydrate without invoking any new mechanisms of secretion by the cells. It would also explain the apparently poor wall development in cells adjacent to arbuscules and the appearance of the interfacial matrix, which have been described.

The proportion of the total photosynthate which is used by the fungus is not known. There is no unequivocal evidence for large-scale diversion of photosynthate to infected root cells, such as is observed in the altered translocation patterns of plants infected by some biotrophic leaf pathogens and ectomycorrhizal fungi. Interpretation of experimental data is complicated by the difference in size between mycorrhizal and non-mycorrhizal plants and reduction in root/shoot ratio which occurs in many experiments. Mycorrhizal plants may also exhibit elevated rates of photosynthesis per unit leaf area (Dehne, 1978; Levy and Krikun, 1980).

Depression in the rate of growth of young plants (as measured by dry weight) as they develop mycorrhizas has been attributed to competition between the symbionts for a limited supply of carbohydrate, although some think that the amounts of fungal tissue present are too small. Considerable synthetic activity must take place during mycorrhizal development, since the cytoplasm of the root cells increases in volume and this, together with the considerable ramification of the fungus both within and without the root, might require considerable amounts of material and respiratory energy. Calculations of the minimum demand for carbohydrate in the synthesis of fungal biomass could be based upon estimates of the amount of fungal tissue within the root of about $10\%$ root weight ($4$–$17\%$: Hepper, 1977; $10\%$: Tinker, 1978), of external mycelium at $0.9\%$ root fresh weight (Bevege *et al.*, 1975) or $3.6$ $\mu$g dry weight per cm infected root length (Sanders *et al.*, 1977) and spore numbers up to several thousands per plant (Daft and Nicholson, 1969a). The total fungal weight as estimated by Bethenfalvay *et al.* (1982) in soybeans increased from $2.3$ to $12.3\%$ of the dry weight of the root during growth. Hence an increased respiratory demand for carbohydrate well in excess of $10\%$ of that used by the root might be expected.

Stribley *et al.* (1980) have made a useful contribution to this problem by pointing out that mycorrhizal plants normally contain higher internal concentrations of phosphate in their shoots than non-mycorrhizal controls and that this difference does not depend on size, phosphate supply or species of plant. Increased demand for carbon in the infected root system was put forward by them as an explanation. They point out that Pang and Paul (1980), using $^{14}$C-labelling, showed a very great increase of $CO_2$ loss ($74\%$) from roots of soybeans as a result of mycorrhizal infection. These views, as they note, agree with the proposal of Harley (1969a) that the growth of vesicular-arbuscular mycorrhizal plants may depend upon the balance between the growth-promoting effect of increased nutrient supply to the plant by the fungus and the growth-decreasing effect of the drain of carbon compounds to the fungus (see below).

## Respiration

Information about the respiratory metabolism of mycorrhizal roots would be helpful in determining the effect of fungal infection on metabolic activity and carbohydrate utilization. Unfortunately, such information is as yet very scanty. Pang and Paul (1980) measured the photosynthetic incorporation of $^{14}CO_2$ into

*Vicia faba*, using a growth chamber which separated the atmosphere surrounding the shoots from that surrounding the roots. They found that at $6\frac{1}{2}$ weeks mycorrhizal and non-mycorrhizal plants incorporated 44 and 36 mg C per g dry matter per day respectively. At this stage the mycorrhizal plants were slightly but insignificantly larger than the non-mycorrhizal plants ($2\cdot70 \pm 0\cdot19$ g versus $2\cdot40 \pm 0\cdot07$ g). Despite the fact that the mycorrhizal plants had lower root:shoot ratios ($0\cdot79$ versus $0\cdot86$) their roots respired 30% of the incorporated $^{14}CO_2$ as against 18% in the non-mycorrhizal plants (Table 10). However, S. E. Smith and M. D. Ramsey (unpublished) showed that although oxygen uptake ($\mu l\ O_2\ g^{-1}$ fresh weight $h^{-1}$) of freshly harvested mycorrhizal clover roots is frequently somewhat higher than of non-mycorrhizal roots of the same age, it is not invariably so (Table 11). Their results certainly do not show that infected roots respire much faster than uninfected ones, as suggested by Pang and Paul. RQs were the same in all the Smith and Ramsey experiments ($1\cdot06 \pm 0\cdot02$ for both treatments), so there was no indication that lipid from a fungal storage pool was being respired by freshly excised mycorrhizal roots, although ageing might increase utilization of endogenous lipid. As mentioned, Snellgrove *et al.* (1982) have also employed $^{14}C$-labelling techniques to investigate the carbon distribution and use in mycorrhizal and non-mycorrhizal leek plants, which were grown under conditions of different phosphate fertilization in order to obtain plants with similar growth rates. In their investigations *Glomus mosseae*, colonizing about 60–70% of the root length, caused a diversion to the roots of about 8% of the total $^{14}C$ fixed from $^{14}CO_2$ in photosynthesis. The additional carbon in the below-ground fractions appeared first in the mycorrhizal root (presumably in the fungal component) and later as an increase in respired $^{14}CO_2$. In the leek plants this increased drain of carbohydrate was probably compensated for by a greater assimilation rate per unit leaf dry matter. Their results thus agreed closely with those of Pang and Paul (1980). Snellgrove *et al.*

TABLE 10

The effect of mycorrhizal infection by *Glomus mosseae* on the incorporation of $^{14}CO_2$ into tissues and respiration from roots of *Vicia faba*

*Plants were supplied with 240 mg C as $^{14}CO_2$.* (Results of Pang and Paul, 1980)

|  | C per plant | | | |
|  | Mycorrhizal | | Non-mycorrhizal | |
|  | mg | % | mg | % |
|---|---|---|---|---|
| Leaves | $65 \pm 14$ | $30 \pm 4$ | $70 \pm 12$ | $36 \pm 3$ |
| Stems | $47 \pm 5$ | $23 \pm 4$ | $56 \pm 8$ | $28 \pm 1$ |
| Roots and nodules | $36 \pm 7$ | $17 \pm 2$ | $33 \pm 4$ | $18 \pm 2$ |
| Respired (below ground) | $61 \pm 4$ | $30 \pm 2$ | $35 \pm 3$ | $18 \pm 2$ |
| TOTAL | 209 | 100 | 194 | 100 |

were able to calculate that if all the additional $CO_2$ produced by mycorrhizal root systems was the result of fungal respiration, then the fungal metabolic rate was $11·0$ mg $CO_2$ g $^{-1}$h $^{-1}$, which is very close to the value for the sheath of *Fagus* mycorrhizas obtained by Harley *et al.* (1956).

Cytochemical techniques have been used to demonstrate that enzymes characteristic of the Embden-Meyerhof pathway, Krebs cycle and hexose

TABLE 11

Respiration of freshly excised roots of *Trifolium subterraneum*. Mycorrhizal inoculum was provided by non-sterile soil mixed with steamed sand in the ratio 1 : 10. Non-mycorrhizal plants were grown in sterilized soil/sand mixture (unpublished results of S. E. Smith and M. D. Ramsey)

| Batch of plants | Plant age (d) | Fresh weight g per pot | | % infection | Gas exchange $\mu$l g $^{-1}$h $^{-1}$ | | RQ | $O_2$ $\mu$l per pot h $^{-1}$ |
|---|---|---|---|---|---|---|---|---|
| | | Root | Shoot | | $O_2$ | $CO_2$ | | |
| I | 36 | 2·25 | 2·0 | 62 | 107·8 | 116·3 | 1·08 | 242·6 |
| | | 2·06 | 1·09 | <1% | 81·9 | 87·5 | 1·07 | 180·3 |
| | 37 | 2·12 | 2·12 | 63 | 134·9 | 144·1 | 1·07 | 284·1 |
| | | 1·80 | 1·23 | 0 | 140·4 | 142·9 | 1·02 | 252·0 |
| | 40 | 1·44 | 2·22 | 73 | 211·3 | 233·2 | 1·10 | 304·3 |
| | | 1·73 | 1·10 | 0 | 100·8 | 104·9 | 1·04 | 174·4 |
| | 48 | 3·25 | 3·35 | 63 | 89·3 | — | — | 290·2 |
| | | 3·35 | 1·15 | 0 | 41·0 | — | — | 137·4 |
| II | 26 | 1·23 ±0.12 | 0·83 ±0·09 | 46 ±8·4 | 120·6 | — | — | 148·3 |
| | | 1·37 ±0·09 | 0·70 ±0·06 | 0 | 111·3 | — | — | 152·5 |
| | 28 | 1·47 | 0·97 | 37·2 | 127·1 ±3·1 | — | — | 186·7 |
| | | 1·9 | 0·9 | 0 | 75·7 ±16·9 | — | — | 143·8 |
| | 34 | 1·63 ±0·08 | 1·47 ±0·03 | 57·3 ±6·1 | 115·9 ±3·4 | — | — | 188·9 |
| | | 1·63 ±0·12 | 1·03 ±0·03 | 0 | 110·7 ±4·7 | — | — | 180·4 |
| | 47 | 3·03 ±0·07 | 3·23 ±0·19 | 48 | 140·0 ±5·8 | — | — | 424·2 |
| | | 2·47 ±0·2 | 1·33 ±0·08 | 0 | 114·5 ±11·6 | — | — | 282·8 |

monophosphate shunt are active both in germinated spores of *G. mosseae* and in mycorrhizal roots (MacDonald and Lewis, 1978). A cyanide-insensitive respiratory pathway has been demonstrated in mycorrhizal roots of *Salix nigra* (Antibus *et al.*, 1980). Uninfected roots were not investigated, and its significance is not understood. It may be noted that a similar cyanide-insensitive pathway alternative to the cytochrome pathway is present in ectomycorrhizas of *Fagus* (Coleman and Harley, 1976).

## Growth of host plants

Association between the development of vesicular-arbuscular mycorrhiza and improved growth of the host was made by Asai (1944) in his studies of mycorrhizal infection and nodulation in a large number of legumes. He concluded that infection was important both in plant growth and in the development of nodules. Subsequently a large number of experiments by many investigators have been carried out which in general demonstrate that infection is followed by considerable stimulation of growth. Much of this early work was reviewed by Harley (1969a) who considered both problems of soil sterilization and soil phosphate content as they influenced the results obtained by different investigators. Daft and Nicolson (1966, 1969b, 1972) were among the first to demonstrate that development of mycorrhizal roots and their effect on plant growth is greater in soils of low or imbalanced nutrient status, particularly if phosphate is in short supply. They grew tomato plants in sterilized sand culture with the addition of three different levels of bone meal as a phosphate fertilizer. Their results, which are presented in Table 12, illustrate three features of mycorrhizal effects on plant growth which have subsequently become commonplace. First, that the increase in dry weight following infection was greatest at the lowest level of the bone meal, and secondly that the percentage of root length infected was reduced as bone meal was increased; thirdly that the phosphate concentration in the tissues was increased. Figure 7 illustrates the effect of phosphate fertilization on shoot growth of mycorrhizal and non-mycorrhizal *Trifolium subterraneum*. Improved growth of the sort illustrated has been demonstrated for a very wide variety of host plants including many crop plants and trees, and is manifest as increased root as well as shoot growth, in increased vascular development, flower production and yield in some hosts (e.g. Daft and Okusanya, 1973a; Daft and El-Ghiami, 1974; Schoenbeck and Dehne, 1979; Berthau *et al.*, 1980; etc.). Nodulation and nitrogen fixation in mycorrhizal legumes and in at least one example of dually infected actinorrhizal plants (*Ceanothus velutinus*) are also increased (Asai, 1944; Crush, 1974; Mosse *et al.*, 1976; Daft and El-Ghiami, 1975; S. E. Smith and Daft, 1977; Rose and Youngberg, 1981; and others), and have been shown to result in increased tissue nitrogen concentrations in some experiments. Reduction in root : shoot dry

TABLE 12

The effect of phosphate fertilizer (bone meal) and mycorrhizal infection by *Endogone* *(Glomus)* sp. on the growth, phosphate concentration in the tissues and mycorrhizal infection of tomato (results of Daft and Nicolson, 1966)

| Experiment | Relative phosphate level in sand | | Mean dry weight g | Phosphate concentration $\mu$mol 100 mg$^{-1}$ dry tissue Roots | Leaves | % infection |
|---|---|---|---|---|---|---|
| 1 | 0·25 | + M[a] | 0·393 | 6·2 | 4·7 | 90·5 |
|  |  | − M[b] | 0·088 | 3·3 | 3·2 | — |
|  | 0·5 | + M | 0·383 | 6·8 | 5·6 | 82·4 |
|  |  | − M | 0·127 | 2·5 | 2·9 | — |
|  | 1 | + M | 0·344 | 9·3 | 5·4 | 77·2 |
|  |  | − M | 0·115 | 3·6 | 3·3 | — |
|  | 2 | + M | 0·483 | 14·0 | 3·7 | 63·9 |
|  |  | − M | 0·355 | 2·6 | 1·8 | — |
| 2 | 2 | + M | 0·516 | 13·6 | 9·2 | 73·4 |
|  |  | − M | 0·301 | 3·8 | 5·0 | — |
|  | 4 | + M | 0·730 | 6·1 | 3·3 | 72·4 |
|  |  | − M | 0·707 | 1·8 | 3·0 | — |
|  | 8 | + M | 0·737 | 6·2 | 3·9 | 61·4 |
|  |  | − M | 0·702 | 2·7 | 2·6 | — |

[a] + M, inoculated with 50 spores of *Endogone* sp. per plant.
[b] − M, not inoculated.

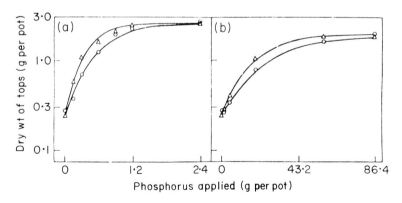

FIG. 7. The effect of phosphate fertilization and inoculation with *Glomus mosseae* on dry weight of tops of *Trifolium subterraneum* after 7 weeks' growth. (a) Superphosphate fertilizer. (b) C-grade rock phosphate. ○, uninoculated controls; △, inoculated with *G. mosseae*. (Results of Pairunan *et al.*, 1980.)

weight ratio has also been observed in a few plant species (Sanders, 1975; Hayman and Mosse, 1971; S. E. Smith and Daft, 1978; S. E. Smith, 1982) and in one experiment, illustrated in Table 13, it was correlated with the percentage of the root length that was colonized by mycorrhizal fungi (Becker and Gerdemann, 1977).

TABLE 13

Effect of different amounts of inoculum (spores of *Glomus fasciculatus*) on per cent infection of the roots of onion (measured by pigment extraction), on plant growth and root:shoot ratio (results of Becker and Gerdemann, 1977)

| Spore number in inoculum | % infection | Fresh weight g per plant | Root:shoot ratio |
|---|---|---|---|
| 0 | 5·7[a] | 6·72 | 0·91 |
| 450 | 58·7 | 12·02 | 0·57 |
| 900 | 72·1 | 12·21 | 0·49 |
| 1350 | 78·5 | 12·65 | 0·43 |
| 1800 | 79·6 | 11·81 | 0·49 |

[a] Positive value probably caused by yellow colour from adhering soil particles and phenolics in dead roots.

All these effects on growth can be attributed, directly or indirectly, to improved mineral nutrition and in many cases similar changes have been shown to take place in response to fertilizer application in the absence of mycorrhizal infection. The involvement of different nutrients is discussed below. However, it must be stressed that it is essential to distinguish direct mycorrhizal effects from those which would follow inevitably on changes in size, form or nutrient content.

As has been stated above, reduction in plant dry weight following mycorrhizal infection may also occur (e.g. Winter and Peuss-Schönbeck, 1963; Baylis, 1967). These growth depressions are probably due to carbohydrate utilization by the fungi, as suggested by Harley (1969a), Furlan and Fortin (1973), S. E. Smith (1980) and Stribley *et al.* (1980). Although this view has been questioned on the grounds that there is insufficient fungal tissue associated with the roots to provide a vigorous sink for photosynthate (e.g. Cooper, 1975), the fungus, as has been pointed out, may constitute about 10% of the dry weight of the root and have a considerable demand for carbon compounds. Moreover low light intensities and short day lengths, which would be expected to reduce photosynthesis, sometimes depress or reverse positive growth responses to infection as shown in Table 14 (see Furlan and Fortin, 1973; Daft and El-Ghiami, 1978) as well as lowering the intensity of infection in the roots. Increased competition between the symbionts for photosynthate would certainly be one reasonable explanation for such results and would provide a mechanism to explain differences in tissue concentrations of phosphate between mycorrhizal and non-mycorrhizal plants of the same dry weight (Stribley *et al.*, 1980). Other suggestions include competition between the symbionts for meagre phosphate supply (Crush, 1976). This would apply only at extremely low phosphate concentrations, if ever. Where soil and plant phosphate supply is very high, phosphate toxicity might also cause a decrease of growth rate (Mosse, 1973; Cooper, 1973). Further experimental work is required to establish whether these latter suggestions could constitute possible mechanisms for growth depressions. Here it is necessary to emphasize that measurement of growth by dry weight essentially measures carbon in the plants. A depression of growth means either (1) less photosynthesis goes on; (2) more carbon is respired per unit time, or (3) more carbon is transported to extramatrical hyphae or fruit-bodies not included in the estimate of plant dry weight. If photosynthesis increases in mycorrhizal plants, as it has been reputed to do, then effects (2) or (3) or both must operate in mycorrhizal plants to offset it if a depression of growth in weight is observed.

TABLE 14

The effect of daylength and irradiance on growth and mycorrhizal infection in maize. + M plants inoculated with *Glomus mosseae* (results of Daft and El Giahmi, 1978)

| Day length h | Irradiance W m$^{-2}$ | Dry weight g | | % increase in dry weight | % infection |
|---|---|---|---|---|---|
| | | − M | + M | | |
| 18 | | $0.60 \pm 0.1$ | $1.0 \pm 0.1$ | 67 | 81 |
| 12 | 26.1 | $0.60 \pm 0.1$ | $0.8 \pm 0.8$ | 33 | 65 |
| 6 | | $0.40 \pm 0.1$ | $0.5 \pm 0.1$ | 25 | 27 |
| | 26.1 | $0.55 \pm 0.06$ | $0.84 \pm 0.12$ | 53 | 62 |
| 12 | 19.7 | $0.33 \pm 0.04$ | $0.62 \pm 0.12$ | 88 | 40 |
| | 6.7 | $0.15 \pm 0.02$ | $0.49 \pm 0.08$ | 227 | 21 |

## Conclusion

The carbon metabolism of vesicular-arbuscular mycorrhizal plants is understood in outline only. There seems to be no doubt that the fungus is supplied by the host. The knowledge of the quantities of carbon involved in the growth and activity of the symbiotic fungus and of the compounds and mechanisms involved in the transfer is as yet very sketchy. Perhaps the efforts to compare uninfected and mycorrhizal plants which demand carefully prepared equivalent samples have daunted experimenters. However, the carbon metabolism of mycorrhizal plants in all its aspects can be studied without such quantitative comparison and experiments on it alone will yield, as it has done in other cases, a framework of understanding which will help in later quantitative comparisons. The demonstration that vesicular-arbuscular mycorrhizal infection may increase the growth of the host plant, especially when nutrient supply in the soil is low, can be accepted almost universally. It does not require further demonstration in a simple form. What is required is further examination of the effect of infection on the relative growth of various organs of the host, on its reproductive efficiency and on its growth in competition. Fitter's experiments (1977) which show that although *Lolium perenne* and *Holcus lanatus* are both increased in growth by infection when grown alone, but that *Holcus* dominates *Lolium* in mixtures, is an example of the kind of information needed to evaluate the ecological effects of vesicular-arbuscular infection.

*Chapter 4*

# Mineral Nutrition

Mineral nutrition of vesicular-arbuscular mycorrhizal plants, together with measurements of dry weight, has been the subject of more research than any other aspect of their physiology. The increased growth, just discussed, is accompanied by higher total amounts and frequently higher concentrations of some mineral nutrients in the tissues. In 1957 Mosse published the results of an experiment with apple seedlings which clearly demonstrated increased amounts of potassium, iron and copper per unit weight of tissue in mycorrhizal plants, compared with uninoculated controls. Later, Gerdemann (1964), Daft and Nicolson (1966) and Baylis (1967) established that tissue concentrations of phosphate (which were not measured by Mosse) were higher in mycorrhizal plants. Subsequent work has concentrated most effort on the absorption of phosphate because it is commonly deficient in soil. It is now well established that mycorrhizal roots take up phosphate and some other nutrients from soil more efficiently than uninfected root systems do. Before techniques and results are discussed, it is worth noting that comparison of mycorrhizal and non-mycorrhizal plants is beset by problems of interpretation which include (1) the provision of adequate uninfected control plants; (2) the difficulty of comparing nutrient uptake by whole plants which differ not only in dry weight but also in root : shoot ratio; and (3) the effect of "dilution" of nutrient concentrations in the tissues as a result of the increased growth of the plants following relief of stress.

## Absorption of phosphate

Phosphate has received the most attention and our knowledge of mycorrhizal influence on its uptake highlights our ignorance of other nutrients, which have not been the subjects of detailed experiments. Not only is phosphate required in relatively large amounts by plants, but it is also frequently in very low (micromolar and below, rather than millimolar) concentrations in the soil solution. One reason for this is that inorganic phosphate ions rapidly become bound to soil colloids, or fixed as iron and aluminium phosphates, rendering them relatively immobile (see Tinker, 1975b). In addition, a large proportion of total inorganic phosphorus is normally in insoluble form not readily available to plants. In humic soils including forest soils, much of the phosphate (up to about 80% or more in the extreme) present within the rooting depth of the plants may be in the form of phytates (inositol phosphates) which are also insoluble, but may be brought into solution by phosphatases positioned on the surfaces of roots or fungal hyphae. Slow diffusion of phosphate ions in the soil solution contrasted with rapid absorption of phosphate by roots and other absorbing organs results in the development of depletion zones around them (Nye and Tinker, 1977). Uptake may be limited by the rate of diffusive movement of ions into these depletion zones, rather than by the rate of the transport across living membranes into the root or mycorrhiza. The longer a segment of root remains actively absorbing from soil at a rate greater than that of movement to it, the wider will be the phosphate depletion zone. Conversely, roots or hyphae growing into uncolonized soil can absorb phosphate at a rate more closely related to their uptake capacity, until a depletion zone becomes established. It is clear that any absorbing system which can economically colonize undepleted soil will have advantages over other systems whose exploitation of new soil volumes requires greater expenditure. In addition, uptake mechanisms with very high affinity for phosphate (i.e. a low $K_m$) will be able to maintain uptake from low concentrations in the soil solution. The exploitation of soil by hyphae involves a smaller expenditure per unit absorbing area than by roots and their affinity for phosphate is at least as high or higher per unit area.

Indirect evidence that mycorrhizal roots are more efficient comes from the fact that mycorrhizal plants are frequently not only larger but also contain higher concentrations of phosphorus in their tissues than uninfected control plants. There are problems in this approach. Despite reductions in root:shoot ratio, increased total root length in mycorrhizal plants would certainly contribute to increased total uptake, but this would not necessarily lead to increased tissue concentrations. If growth kept pace with phosphate uptake as it would if phosphate supply were the limiting factor, tissue concentrations would be constant, for they are clearly dependent upon the relative rates of uptake and growth. If tissue concentrations rise, some other factor than phosphate must be limiting growth except in extreme starvation conditions. Typical results for mycorrhizal and non-mycorrhizal leek and subterranean clover plants are illustrated in Fig. 8 (Stribley et al., 1980; Pairunan et al., 1980) in which the percentage of phosphate in the shoot tissue is plotted as a function of shoot dry weight. Mycorrhizal plants have considerably higher concentrations of phosphate than non-mycorrhizal plants of the same dry weight. Stribley et al., (1980)

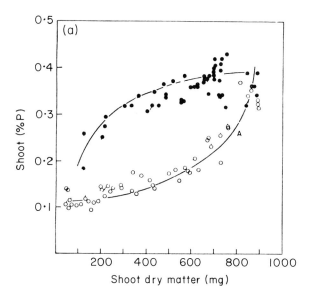

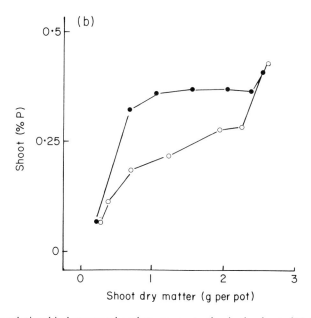

Fig. 8. The relationship between phosphate concentration in the shoots (% dry weights) and dry weight of shoots in non-mycorrhizal plants, ○, and plants infected with *Glomus mosseae*, ●. (a) Leeks grown on ten different γ-irradiated soils differing in initial P content and receiving five different levels of added phosphate to give the 50 soil treatments used in the experiment (redrawn from Stribley *et al.*, 1980). (b) *Trifolium subterraneum* grown in soil with the addition of superphosphate (redrawn from Pairunan *et al.*, 1980).

discuss this anomaly and suggest that increased carbohydrate utilization in mycorrhizal plants, together with increased phosphate uptake, means that mycorrhizal plants are carbon-limited. With respect to the earlier discussion on carbohydrate utilization, they estimated that the "hypothetical dry-weight loss" due to mycorrhizal fungi might be between 40 and 60%. Such speculation apart, the elevated tissue concentrations of phosphate in mycorrhizal plants certainly alerted investigators to the possible role of mycorrhizas in phosphate nutrition.

More direct evidence about the efficiency of phosphate absorption can be obtained if uptake is expressed on the basis of the amount of absorbing tissue. It seems likely that surface area of the root system, either the whole or that part involved in uptake, would be the most suitable estimate of absorbing tissue to be related to uptake rates when nutrients are not diffusion limited. In practice this area is difficult to determine and most results have been expressed on the basis of root length. Brewster and Tinker (1972) defined *inflow* as uptake in mol cm$^{-1}$ s$^{-1}$. Other results have been based upon root weight, to give specific absorption rate in mol g$^{-1}$ s$^{-1}$ (Hunt, 1975). Inflow, based on root length, gives a realistic basis for comparison of very immobile ions, the uptake of which is diffusion limited, for the contribution of root radius to the estimate of absorbing tissue is relatively unimportant. Here linear extension of the root system into undepleted soil may be more important in determining uptake than the surface area presented to a depletion zone (Brewster and Tinker, 1972).

Increased inflow of phosphate into both onion and clover roots infected by several different mycorrhizal fungi has been demonstrated (Sanders and Tinker, 1971, 1973; Sanders, 1975; Sanders *et al.*, 1977; S. E. Smith *et al.*, 1979; S. E. Smith, 1982). Table 15 shows clearly that the inflow into mycorrhizal roots (of which up to about 50% of the root length was infected) was, on average, about 3–4 times greater than into uninfected onion roots. The explanation for this

TABLE 15

Calculation of the inflow of phosphate into onion roots via the hyphae of *Glomus* sp. and the phosphate flux in the entry point hyphae

*Cross-sectional area of hypha entering 1 cm root length taken as $4.7 \times 10^{-6}$ cm$^2$ (results of Sanders and Tinker, 1973)*

| Experiment | Infected root length % | Inflow mol cm$^{-1}$ s$^{-1}$ | | | Flux mol cm$^{-2}$ s$^{-1}$ |
|---|---|---|---|---|---|
| | | Total | | Hyphal | |
| | | Mycorrhizal | Uninfected | | |
| 1 | 50 | $13 \times 10^{-14}$ | $4.2 \times 10^{-14}$ | $17.6 \times 10^{-14}$ | $3.8 \times 10^{-8}$ |
| 2 | 45 | $11.5 \times 10^{-14}$ | $3.2 \times 10^{-14}$ | $18.5 \times 10^{-14}$ | |

increased efficiency of uptake has been sought, and one important contributing factor is the existence and continued growth of extramatrical mycelium into soil. This hyphal system can be envisaged as extending beyond the phosphate depletion zone around the root (and root hairs) and exploiting a greater, and less depleted, volume of soil than the roots alone would be able to do. There is no doubt that depletion zones would also form around the hyphae, but uptake would be maintained as long as the hyphae continued to colonize undepleted soil. It is possible to calculate the apparent contribution of the fungus to the total uptake process and the flux which might be expected to occur through mycorrhizal entry points. Figure 9 shows the calculated inflow due to hyphae as a function of the proportion of the root length of onion infected. The relationship appears to be linear above about 10% infection, with those strains of fungi which improved phosphate uptake. These produced a fairly constant amount of external mycelium per cm of infected root ($3.6\ \mu g\ cm^{-1}$) so that it is possible to conclude that, once the fungus has become established, the rate of phosphate uptake by hyphae is related to the length of the external mycelium (Sanders *et al.*, 1977; Graham *et al.*, 1982).

Translocation of phosphate in the extramatrical mycelium and a flux of the order of $3.8 \times 10^{-8}$ mol P $cm^{-2}$ $s^{-1}$ through entry point hyphae would be required for this system to provide sufficient phosphate to account for the

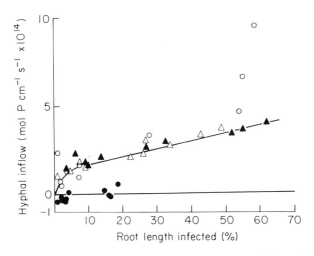

Fig. 9. Inflow of phosphate to onion roots attributable to inflow via hyphae of mycorrhizal fungi. ▲, *Glomus mosseae*; △, *Glomus macrocarpus* var. *geosporus*; ○, *Gigaspora calospora*; ●, *Glomus microcarpus*. Values calculated by subtracting inflow into uninfected roots from inflow into mycorrhizal roots. Note that the ineffective fungus *Glomus microcarpus* which colonized the roots very slowly did not increase inflow of phosphate over the rate in controls. (From Sanders *et al.*, 1977.)

observed increase in efficiency of uptake by mycorrhizal roots over uninfected ones. Hyphae of *Glomus mosseae* and *G. fasciculatus* have been shown to translocate [32]P-labelled orthophosphate up to 7 cm through soil to roots of onion. When hyphae between the source of [32]P and the root were cut, no translocation to the root was observed (Hattingh *et al.*, 1973; Rhodes and Gerdemann, 1975; Hattingh, 1975). This distance of 7 cm is quite credible when compared with measured lengths of hyphae associated with roots. The actual flux of phosphate in hyphae of *G. mosseae* to *Trifolium repens* and *Allium cepa* plants has been measured in split agar plates by Pearson and Tinker (1975) and Cooper and Tinker (1978, 1981), but the values obtained range from $1.02 \times 10^{-10}$ to $2 \times 10^{-9}$ mol cm$^{-2}$ s$^{-1}$ which are one or two orders of magnitude lower than the theoretical flux calculated for entry-point hyphae mentioned above. The tracer measurements of flux involved counting and measuring the hyphae traversing a diffusion barrier in a split agar plate and the discrepancy between measured and calculated fluxes has been attributed to three factors (Cooper and Tinker, 1978): (1) the possiblity of higher fluxes in entry-point hyphae than in those of the general extramatrical mycelium; (2) lower growth rates and phosphorus uptake of plants growing in agar culture, compared with plants growing in soil; and (3) the fact that some of the hyphae used to calculate cross-sectional area for flux determinations may have been dead. The second point may be of particular importance, since it has now been shown that a mass flow component of the translocation process is increased by rapid transpiration rates in associated host plants (Cooper and Tinker, 1981). Plants grown in agar might well have lower transpiration rates than plants in pots. There is a fourth possibility: the calculation of hyphal fluxes involves the assumption that inflow from the soil into the non-mycorrhizal roots of wholly non-mycorrhizal plants (the controls) is the same as that into uninfected regions of the roots of mycorrhizal plants, in spite of observations that uptake into the latter appears to be faster, at least from solution (Gray and Gerdemann, 1969; Bowen, Bevege and Mosse, 1975). Uptake from soil is clearly complicated by the establishment and maintenance of phosphate depletion zones and the rate of diffusive movement of phosphate through them. Nevertheless, any increased uptake by uninfected regions of mycorrhizal roots over wholly uninfected roots would result in overestimate of hyphal inflow, and bring calculated and measured values of phosphate fluxes in hyphae closer together.

    All in all it is reasonable to accept that much of the increased rate of phosphate uptake by mycorrhizal roots is due to improved exploitation of a given volume of soil, coupled with more rapid translocation of phosphate through hyphae to roots than diffusion of phosphate through soil to the root surface. However, apart from this physical extension of the root system there are at least two other mechanisms which would result in increased efficiency of uptake by infected roots. These are: (1) the possibility that mycorrhizal hyphae are able to absorb phosphate effectively from a lower concentration in the soil solution than the associated roots; and (2) that infected plants can exploit sources of soil phosphate not available to uninfected ones (e.g. rock phosphate fertilizer or fixed inorganic or organic soil phosphate).

Mosse *et al.* (1973) suggested that mycorrhizal infection might alter the threshold concentration from which plants were able to absorb phosphate, and this has been investigated by Cress *et al.* (1979) using tomato and by Jintakanon *et al.* (1979) and Howeler *et al.* (1979) using cassava (*Manihot* sp.). Phosphate uptake from solution by tomato roots was studied at concentrations between 1 and 100 $\mu$M $KH_2PO_4$. The lower part of this range corresponds realistically with the concentration in soil solutions, an important point if we are to use the results (Table 16) to help interpret mycorrhizal effects on plant growth in soil. Both non-sterile and axenically grown non-mycorrhizal roots were used for comparison with mycorrhizal roots in consideration of the possible effects of rhizosphere microorganisms on the uptake processes. Moreover the initial internal phosphate concentrations of the roots used were fairly similar. This was achieved by growing mycorrhizal plants on $Ca_3(PO_4)_2$ and non-mycorrhizal plants on $NaH_2PO_4$. Non-mycorrhizal plants grown on $Ca_3(PO_4)_2$ had tissue concentrations about half those shown in Table 16 ($1\cdot25 \pm 0\cdot2$ $\mu$mol g$^{-1}$ fresh weight). Over the concentration range 1–20 $\mu$M, $V_{max}$ was not very different in mycorrhizal and non-mycorrhizal plants, but the $K_m$ indicates that the affinity of the uptake sites for phosphate was much higher in the mycorrhizal roots. This applied whether or not the non-mycorrhizal roots were grown axenically.

TABLE 16

Tissue phosphate concentration and apparent kinetic constants of phosphate uptake into mycorrhizal and non-mycorrhizal tomato roots. Phosphate uptake was measured as initial uptake rates from solutions with concentrations ranging from 1 to 20 $\mu$M $KH_2PO_4$ at pH 4·6 (results of Cress *et al.*, 1979)

| Infected root length % | Culture conditions | Internal phosphate concentration $\mu$mol P g$^{-1}$ fresh weight | $V_{max}$ $\mu$mol P g$^{-1}$ fresh weight h$^{-1}$ | $K_m$ $\mu$M |
|---|---|---|---|---|
| 79–91 | Sand culture *Glomus fasciculatus* $Ca_3(PO_4)_2$ | $2\cdot37 \pm 0\cdot27$ | $0\cdot10 \pm 0\cdot04$ | $1\cdot6 \pm 0\cdot8$[a] |
| 0 | Sand culture soil filtrate $NaH_2PO_4$ | $3\cdot17 \pm 0\cdot21$ | $0\cdot10 \pm 0\cdot06$ | $3\cdot9 \pm 1\cdot4$ |
| 0 | Axenic root organ culture $NaH_2PO_4$ | $2\cdot87 \pm 0\cdot34$ | $0\cdot13 \pm 0\cdot06$ | $4\cdot2 \pm 1\cdot60$ |

[a] Significantly different from non-mycorrhizal treatments at $P < 0\cdot05$.

The work with cassava illustrates two points. This species appears to have a very high phosphate requirement, coupled with a very inefficient phosphate uptake system in the absence of mycorrhizal infection. Despite this, it is well known for its growth on low fertility soils and clearly its efficiency of uptake is markedly increased when roots are infected by mycorrhizal fungi. Thus the importance of mycorrhizal infection to a particular species or variety of host plant may depend on the concentration of phosphate in the soil, the relative affinities of the root and fungal systems for phosphate and also upon the phosphate requirement of the host. This latter point was appreciated in the early investigations into phosphate nutrition of mycorrhizal plants carried out by Daft and Nicolson (1969b). It is possible to argue that if phosphate uptake is limited by the rate of diffusion of ions through a depleted zone around a root (or indeed a hypha), then uptake characteristics of the root, mycorrhiza or hypha are irrelevant to considerations of phosphate uptake from the soil. Two things deserve mention in this context:

1. The rate of development of and concentration gradient across a zone of soil depleted of phosphate (or indeed of any other ion) are affected by the actual rate of uptake at the absorbing surface. When soil phosphate is low, the rate of uptake will depend upon the $K_m$ of the uptake sites.

2. Mycorrhizal hyphae, with a low $K_m$ for phosphate uptake, may well be in a better position than roots to compete with other soil organisms for transitory and dilute supplies of phosphate.

## Insoluble sources of phosphate

The suggestion that mycorrhizal roots can exploit sources of phosphate in soil not normally available to plants is based on results such as those illustrated in Table 17 (data of Murdoch et al., 1967). It can be seen that growth of mycorrhizal maize responded to the application of rock phosphate or tricalcium phosphate, whereas these fertilizers had no effect on the growth of non-mycorrhizal plants at the rate of application used in the experiment. In contrast, both mycorrhizal and non-mycorrhizal plants responded to monocalcium or superphosphate with no significant difference between them. Similar results have been obtained for insoluble phosphate fertilization of a variety of host plants, usually in soils of low pH.

In all the investigations, comparisons were made at one or two rates of fertilizer application and in many cases the results indicate that phosphate from the fertilizer was available to mycorrhizal but not to non-mycorrhizal plants. Pairunan et al. (1980) have suggested that this conclusion is invalid, and that it is essential to compare growth over a wide range of fertilizer levels, so as to encompass complete phosphate response curves. If this is done as illustrated in

Fig. 7, the curves for the application of super-phosphate and rock phosphate are of similar form. However, there are notable quantitative differences between the treatments. First, the amount of phosphate as rock phosphate which has to be added to soil to achieve maximum growth of *Trifolium subterraneum* was about 40 times greater than the amount of phosphate from superphosphate that was required both for mycorrhizal and non-mycorrhizal plants. Secondly, the maximum growth with rock phosphate was less than that with superphosphate, irrespective of mycorrhizal infection. These differences apart, and assuming that the rock phosphate contained no soluble phosphate, nor indeed toxic substances, the results do suggest that there is no absolute difference in the availability of rock phosphate to mycorrhizal and non-mycorrhizal plants. Nevertheless, at moderate and realistic levels of application, and at any level of phosphate equivalent to the superphosphate range (0 to 0·8 g P per kg soil, see Fig. 7), mycorrhizal plants were clearly more effective at extracting phosphate from the fertilizer. The mechanisms underlying the increased uptake might depend upon hyphal exploitation of the soil volume or a lower $K_m$ of mycorrhizal roots. In addition, both synergistic action between mycorrhizas and phosphate-solubilizing bacteria (Azcon *et al.*, 1976), and the possible excretion of $H^+$ or hydroxyacids by hyphae which would increase the availability of rock phosphate (Johnston, 1956; Johnston and Miller, 1959; S. E. Smith, 1980) have been suggested.

Although the stimulus for recent experiments designed to obtain direct evidence of mycorrhizal involvement in uptake of less readily available forms of

TABLE 17

The effect of phosphate sources of different availability on the growth and internal phosphate concentration of non-mycorrhizal ( − M) and mycorrhizal ( + M) maize (results of Murdoch *et al.*, 1967)

| Experiment | Phosphate source | | Dry weight g | Phosphate concentration % dry weight |
|:---:|---|:---:|:---:|:---:|
| 1 | Rock phosphate | − M | 8·73 | 0·068 |
| | | + M | 20·85 | 0·094 |
| | Superphosphate | − M | 40·17 | 0·070 |
| | | + M | 40·80 | 0·079 |
| 2 | No phosphate | − M | 7·54 | 0·061 |
| | | + M | 12·41 | 0·065 |
| | Tricalcium phosphate | − M | 8·24 | 0·067 |
| | | + M | 23·10 | 0·088 |
| | Monocalcium phosphate | − M | 40·43 | 0·075 |
| | | + M | 42·03 | 0·080 |

phosphate was provided by the work of Daft and Nicolson (1966) and Murdoch *et al.* (1967) with tricalcium and rock phosphate, the experiments themselves have not usually investigated uptake from these fertilizers, but rather from less well defined "fixed phosphate" sources. Phosphate uptake into plants has been followed after either the labile phosphate fraction (Sanders and Tinker, 1971; Hayman and Mosse, 1972; Mosse *et al.*, 1973; Powell, 1975; Pichot and Binh, 1976; Gianinazzi-Pearson *et al.*, 1981), or the non-labile fraction (Swaminathan, 1979) was labelled with $^{32}P$. The results are contradictory. Although mycorrhizal onions, rye grass and soybeans took up more total phosphate from soil with the labile fraction labelled than non-mycorrhizal plants, there was no difference in specific activity of $^{32}P$ in the two groups of plants. If we assume that there was no extensive exchange of $^{32}P$ with the fixed phosphate fraction in the soil, the results indicate that mycorrhizal plants had no direct access to fixed phosphate sources, as this would have diluted the $^{32}P$ and lowered the specific activity. In contrast, mycorrhizal potatoes apparently did take up fixed phosphate from the latosols on which they were grown, the evidence being that the specific activity of $^{32}P$ in mycorrhizal plants was higher than in uninfected plants when they were absorbing $^{32}P$ from the fixed phosphate fraction. However, different combinations of hosts and fungi seem to respond differently to supply of insoluble phosphate. For instance, in the experiments of Gianinazzi-Pearson *et al.*, soybeans did not respond at all to $Ca_3(PO_4)_2$ whether mycorrhizal or not. In unpublished work, S. E. Smith also found no increase in phosphate inflow into mycorrhizal and non-mycorrhizal *Trifolium subterraneum* following fertilization with $Ca_3(PO_4)_2$ in short-term experiments, although there was a slight growth response in long-term experiments.

Another attempt to provide direct evidence of uptake of "fixed phosphate" (not insoluble phosphate fertilizer) involved heating soil to provide a range of concentrations of fixed phosphate and following growth and phosphate uptake by plants in these soils. The results gave no indication of significant differences which could be attributed to mycorrhizal uptake from fixed phosphate sources (Barrow *et al.*, 1977). All in all there is still considerable confusion about the uptake from fixed phosphate and insoluble phosphate fertilizers. Most results which do indicate higher uptake by mycorrhizal plants could probably be interpreted in terms of improved hyphal exploitation of the soil volume, and competitive ability of the hyphae to absorb localized and dilute sources of phosphate. There has been a little work on the ability of root and fungal enzymes to hydrolyze organic phosphates. Mosse and Phillips (1971) found that phytates were satisfactory sources of phosphate for plant and fungal growth in agar cultures, and that calcium phytate stimulated fungal growth. Gianinazzi-Pearson *et al.* (1981) have also shown that both mycorrhizal and non-mycorrhizal soybeans could hydrolyze calcium phytate in soil, but there was no differential growth response.

## The fate of phosphate in the fungus

Rapid absorption of phosphate by mycorrhizal fungi is followed by the synthesis of inorganic polyphosphate in the fungal vacuoles. These have been detected both by metachromasy of granules following staining with toluidine blue, and by chemical methods (Cox and Tinker, 1976; Callow *et al.*, 1978). It seems likely that synthesis of polyphosphates prevents excessive accumulation of inorganic phosphate in fungal cells when external supply is plentiful and uptake is rapid. Such a storage role is common; it is found in ectomycorrhiza, rusts and other obligate fungal parasites, in lichens, in many unicellular algae and in some free-living fungi and in bacteria (see Harold, 1966; Beever and Burns, 1980). In vesicular-arbuscular mycorrhizas formation of polyphosphates might account for the increased phosphate concentrations in the root (Sanders, 1975; S. E. Smith and Daft, 1978; S. E. Smith, 1982), which are associated with mycorrhizal infection. An additional role for polyphosphate in translocation has also been suggested in these mycorrhizas. Calculations based on numbers of granules, amount of phosphorus per granule, and rates of cytoplasmic streaming (Cox and Tinker, 1976; Cox *et al.*, 1980) indicate that movement of the granules within the hyphae might account for measured fluxes into the root. Cytoplasmic streaming certainly seems to be important in translocation. However, whether phosphorus translocation actually occurs by movement of polyphosphate granules remains to be proved. In any event, where granules are present they are likely to be accompanied by soluble polyphosphate molecules of shorter chain length. The problems of translocation in fungal hyphae and in determining the mechanisms by which it occurs are discussed in a later chapter. However, it is quite clear that vesicular-arbuscular mycorrhizal fungi have considerable ability to translocate phosphate. Polyphosphates might also act as a store of energy as they have been shown to do in some bacteria. However, Beever and Burns (1980) have calculated that in *Saccharomyces cereviseae* breakdown of all the stored polyphosphate via polyphosphate kinase would generate only 300–400 $\mu$mol ATP per g dry weight and would therefore supply sufficient energy for only a few minutes' growth in the cells which normally synthesize ATP at rates between 62 and 67 $\mu$mol per g dry weight per min during active growth.

## Transfer of phosphate to the host

[32]P-labelling experiments (Bowen *et al.*, 1975) have shown that transfer of phosphate from fungus to root cell and thence to the shoot occurs rapidly. The site and mechanism of transfer remain the subject of discussion and speculation, but not of very much experimentation. There is no doubt that the interface between the living cells of the symbionts in the arbuscules would provide, as Cox and Tinker (1976) re-emphasized, a relatively large surface area across which

transfer of phosphate and other nutrients also could occur. Calculated phosphate fluxes across the interface are of reasonable magnitude ($1\cdot3 \times 10^{-14}$ mol mm$^{-2}$ s$^{-1}$) and of the same order as for phosphate fluxes in algal cells (F. A. Smith, 1966; Raven, 1974). "Mycorrhiza specific" alkaline phosphatases are localized in the young developing arbuscules while acid phosphatases are characteristic of mature arbuscules and intercellular hyphae, and are located in the fungal vacuoles (Gianinazzi *et al.*, 1979). These findings indicate that the fungal hyphae and arbuscules may be involved in the phosphate metabolism of a mycorrhizal root, but the actual roles of the enzymes have not been determined. There is evidence that distribution of membrane-bound ATPases on the plasmalemma of the root cells is altered as arbuscules develop. Their possible role in phosphate metabolism is discussed in Chapter 17 (see Plate 15). There remains the possibility that hyphal and arbuscular degeneration may release nutrients which can be absorbed through the host plant plasmalemma, as discussed earlier. Direct evidence to distinguish between these possible transfer mechanisms is lacking.

## Other nutrients

Clear evidence for mycorrhizal involvement in uptake of nutrients other than phosphate is scanty. Analysis of the elemental composition of mycorrhizal and non-mycorrhizal plants at the end of relatively long experiments has failed, not surprisingly, to reveal any clear trends because total amounts and concentrations depend so much on what factors are actually limiting growth. In addition, many of the analyses are of shoot tissue alone and therefore cannot provide a reliable picture of uptake by the whole plant. Two micronutrients have, however, been shown to be consistently at higher concentration in mycorrhizal plants. These are zinc and copper. Increased uptake of $^{45}$Zn by mycorrhizal *Araucaria* roots (Bowen *et al.*, 1974) and translocation of $^{45}$Zn[1] along *Glomus mosseae* hyphae into *Trifolium repens* (Cooper and Tinker, 1978) have also been observed. The rate of $^{45}$Zn translocation was $2\cdot1 \times 10^{-12}$ mol cm$^{-2}$ s$^{-1}$, considerably lower than the rate of phosphorus translocation, but it appears to be adequate for a micronutrient such as this, because zinc deficiency symptoms in peach disappear as mycorrhizas develop (Gilmore, 1971).

Interactions between phosphate fertilization and deficiencies of trace elements, particularly copper and zinc, are well known in several species of potentially mycorrhizal plants. In general, when the availability of phosphate is increased, phosphate uptake and plant growth also increase. Concentrations of copper and zinc in the tissues fall sometimes to levels at which deficiency symptoms become

---

[1] It should be noted that these, and indeed all, estimates of rates of translocation through fungal hyphae include uptake of the compound as well as its translocation (see Chapter 16).

apparent. Mycorrhizal infection has been shown to affect these interactions (Lambert *et al.*, 1979; Timmer and Leyden, 1980), so that at moderate levels of phosphate fertilization deficiency symptons are alleviated as tissue concentrations of the trace elements rise. At very high levels of phosphate, mycorrhizal infection itself may be reduced with consequent reappearance of the deficiency symptoms. Both copper and zinc are required for satisfactory nodulation in legumes and their uptake as well as that of phosphate may be involved in the improved nodulation and nitrogen fixation of mycorrhizal legumes, though direct investigations have not been reported.

Translocation of sulphur and calcium as well as phosphate has been investigated by Rhodes and Gerdemann (1975, 1978a,b), using onions grown in soil chambers composed of two compartments (X, Y, Fig. 10). The onion roots were restrained within the outer compartment, while the hyphae of *Glomus mosseae* were able to colonize both chambers. Tracers of $^{35}S$ and $^{32}P$ injected into the inner compartment at 4·5 cm from the roots appeared in the mycorrhizal but not in the non-mycorrhizal plants, indicating that hyphae might be involved in

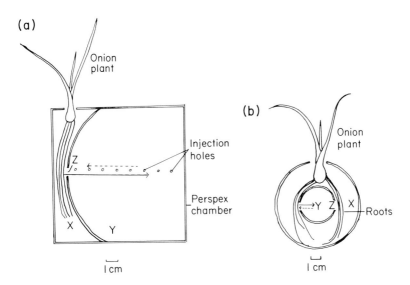

FIG. 10. Two types of "soil chamber" used in studies of translocation in mycorrhizas. Both were made of perspex and had appropriate drainage holes which are not shown in the diagrams. Hyphal growth in the direction of the solid arrows; translocation in the direction of the broken arrows. (a) Tracer was injected at different distances from the onion plant and translocation monitored by sampling onion roots and shoots (redrawn from Rhodes and Gerdemann, 1975). (b) Tracer was injected in compartment Y and translocation monitored by sampling onion plants. Hyphae were severed in some experiments at the junction between the two compartments, Z. (Redrawn from Hattingh, 1975.)

translocation of sulphur. However, in a second experiment three types of plant were used, mycorrhizal and non-mycorrhizal as before, with the addition of mycorrhizal plants with all hyphal connections with the soil severed. In this experiment the tracers were applied at different distances from the roots. Table 18 shows that mycorrhizal plants became more heavily labelled than non-mycorrhizal plants, regardless of whether the hyphae had been severed. If injections of $^{35}$S were made close to the root then no differences were observed.

TABLE 18

The effects of mycorrhizal infection and of severing extramatrical hyphae on the uptake of $^{35}$S (applied as $Na_2{}^{35}SO_4$) by onion plants grown in compartmented soil chambers (see Fig. 10) (results of Rhodes and Gerdemann, 1978b)

| Condition of plants | $^{35}$S absorbed (counts per min per mg dry weight) | |
| --- | --- | --- |
| | Shoots | Roots |
| Mycorrhizal, hyphae intact | 34·5 | 179·6 |
| Mycorrhizal, hyphae severed | 22·7 | 154·0 |
| Non-mycorrhizal | 4·4 | 14·8 |

Sulphate may be present at relatively high concentrations in soil and is also fairly mobile compared with phosphate, so that hyphal translocation may not be very important in sulphate uptake. However, infection clearly has an effect on the absorption of this ion by roots. In a subsequent paper, Rhodes and Gerdemann (1978c) showed that phosphate fertilization as well as mycorrhizal infection resulted in increased $^{35}$S uptake from soil and by excised roots. They suggested that improved phosphate nutrition mediated by mycorrhizal infection was the most likely explanation of their results. Translocation of $^{45}$Ca was also compared with $^{32}$P translocation. As before, no $^{45}$Ca or $^{32}$P appeared in non-mycorrhizal plants grown in the soil chambers. Of fourteen mycorrhizal plants used, thirteen became labelled with $^{32}$P via the hyphae, while $^{45}$Ca was detectable in nine root systems and five shoots. Translocation of $^{45}$Ca therefore appears to be less efficient than translocation of $^{32}$P. Strullu *et al.* (1981) and White and Brown (1979) have identified calcium in the granules of polyphosphate in mycorrhizas of *Taxus* and hyphae of *Glomus mosseae*. Together, these results argue against the translocation of phosphate in granule form; however, we need to know more about whether polyphosphate is largely in granules *in vivo* and about the relative amounts of phosphate and calcium in such granules.

Analyses of potassium concentrations in plant tissues have failed to demonstrate unequivocally that mycorrhizas are directly involved in $K^+$ uptake, despite the relative immobility of this ion in soil. Possingham and Groot-Obbink (1971) observed that mycorrhizal vine plants grown on Hoagland's solution had higher $K^+$ concentrations in their shoots than non-mycorrhizal plants of similar dry weight. Comparable results have been obtained with strawberry (Holevas, 1966), *Lilium* (Vanderploeg, 1974) and apples (Mosse, 1957). However, in the majority of investigations, including one which was a direct investigation of $K^+$ nutrition (Powell, 1975b), $K^+$ was found to be at lower concentrations in the tissues of mycorrhizal than in those of non-mycorrhizal plants. This does not necessarily mean that mycorrhizas are not involved directly or indirectly in $K^+$ nutrition. In Powell's investigation $K^+$ deficient *Griselinia littoralis* plants showed a growth response to mycorrhizal infection, which could not be interpreted in terms of improved phosphate nutrition. However, phosphate nutrition may interact with $K^+$ uptake. S. E. Smith *et al.*, (1981) observed elevated concentrations of $K^+$ in shoots (but not roots) of mycorrhizal *Trifolium subterraneum* when plants were grown on phosphate deficient soils. If sufficient phosphate was supplied to soil to remove any mycorrhizal growth response, then $K^+$ concentrations in both groups of plants were very similar. This suggests an indirect effect of mycorrhizas on $K^+$ uptake in phosphate-deficient plants similar to the effect of sulphate discussed above. It is worth mentioning that $K^+$ uptake by any plant is strongly influenced by the form of nitrogen available (whether $NO_3^-$ or $NH_4^+$), as well as by other cations, particularly $Na^+$. It might also be expected to be influenced by the metabolism and polymerization of phosphate. More experiments designed to investigate $K^+$ nutrition are required, as analysis of concentrations of this nutrient in tissues of plants from experiments designed for other purposes has, not surprisingly, failed to give results which can be readily interpreted.

## Nitrogen uptake

Increased nitrogen concentrations have been reported in vesicular-arbuscular mycorrhizal plants. Of course, where they are also symbiotic with nitrogen-fixing bacteria or actinomycetes this can be attributed to increased rates of nitrogen fixation induced secondarily, e.g. by increased phosphate uptake, rather than to direct uptake of nitrogen compounds from the soil. In spite of earlier suggestions to the contrary, there is no evidence that mycorrhizal fungi, or any fungi for that matter, can fix atmospheric nitrogen, so that in most mycorrhizal plants when increased concentrations of nitrogen have been recorded they must result from increased uptake from the soil. In agricultural soils the source of nitrogen is often more likely to be nitrate than ammonium because of the rapid nitrification of ammonium in them. By contrast, in those natural soils which are on the acid side of neutrality the ammonium ion, which is, like phosphate, relatively immobile in soil, is likely to be the most important source of nitrogen for plant growth. Ammonia itself is not an important source because it is only present at impossibly high pH values. Since the plant's demand for nitrogen exceeds by about a factor of ten the demand for phosphate, depletion zones of ammonium, owing to its relative immobility, will develop as readily or more readily than those of phosphate. Indeed, even in rapidly stirred solutions similar in ammonium concentration to soil solutions, Walker *et al.* (1979) showed that unstirred layers affected uptake by cells of giant algae. We do not know the affinity of the uptake mechanisms of mycorrhizal fungi for ammonium, but in *Neurospora crassa* the apparent $K_m$ for this ion is $1–2~\mu$M (N. A. Walker, personal communication; Slayman, 1977). This is of the same order as the $K_m$ for uptake of phosphate from dilute solutions (see Beever and Burns, 1980). Hence if mycorrhizal fungi had similarly low $K_m$ values and rates of uptake to those fungi, their hyphae could be envisaged as an extremely active means of scavenging ammonium in the soil.

Nitrate is much more mobile in the soil than ammonium but owing to the great demand for nitrogen by plants it is also likely to be deficient in the immediate root environment but less so perhaps than ammonium or phosphate. Its absorption and assimilation involve reduction mediated by nitrate and nitrite reductases. Mycorrhizal interactions with the synthesis of nitrate reductase, when these occur, may be related to the improved availability of phosphate to vesicular-arbuscular mycorrhizal plants, and it is possible that the fungi themselves are concerned in reduction of nitrate. Ho and Trappe (1975) found that nitrate was reduced to nitrite by isolated spores of *Glomus mosseae* and *G. macrocarpus* so there are grounds for such an assumption. However, it is not unusual for fungi to vary even within a species in the ability to reduce nitrate.

Since most fungi that reduce nitrate have been shown to have an NADP-dependent enzyme, Oliver *et al.* (in press) assayed NAD- and NADP-dependent activity in mycorrhizal roots of *Trifolium subterraneum* (80% infected). They observed that although there was no NADP-dependent activity, NAD-dependent nitrate reductase activity was present. This enzyme, more characteristic of higher plants, varied in activity with the phosphate nutrition and mycorrhizal status of the *Trifolium* plants. Figure 11 shows the effects of infection or phosphate supply in increasing the activity of the enzyme, not only in the roots but also in the shoots, where a direct fungal contribution would be precluded. These results confirm those of Carling *et al.* (1978) who, however, only assayed nitrate reductase in non-mycorrhizal tissues of soybeans, i.e. their nodules and shoots. There remains the possibility that mycorrhizal fungi might have an NAD-dependent enzyme which might contribute to nitrate reductase activity. This will have to be ascertained by techniques which produce a large amount of harvestable extramatrical mycelium or by cytochemical techniques. It is of course not absolutely impossible that the main site of nitrate reductase is always in the host and that nitrate may be absorbed by the fungal hyphae and translocated to its cells.

Direct investigations of $NH_4^+$ uptake by vesicular-arbuscular mycorrhizas have been few. Haines and Best (1976) reported that loss of $NH_4^+$, $NO_3'$ and $NO_2'$ from soil by leaching with water was retarded when plants of *Liquidambar styraciflua* growing in it were mycorrhizal with *Glomus mosseae*. However, the root systems of the mycorrhizal plants were considerably larger than the uninfected systems, so that these results do not necessarily indicate increased uptake of these ions by the mycorrhizal fungi. The size of the root system is also affected by the form of nitrogen. F. A. Smith and S. E. Smith (unpublished results) found that roots of mycorrhizal and non-mycorrhizal *Trifolium subterraneum* grown with $NH_4^+$ as a nitrogen source were considerably shorter than $NO_3'$-fed plants. Mycorrhizal infection certainly stimulated growth of the ammonium-fed plants more than that of the nitrate-fed plants. This could have been either because infection increased ammonium uptake directly, or because the fungal hyphae compensated for shorter root length and maintained the uptake of other nutrients, particularly of phosphate.

The assimilation of gaseous $N_2$ in rhizobial root nodules is certainly increased when plants, growing in low phosphate soils, are also infected with mycorrhizal fungi. This effect was probably first observed by Asai (1943, 1944) who made detailed observations of growth, nodulation and mycorrhizal status of a large number of legumes. More recently, nodulation and $N_2$ fixation by mycorrhizal and non-mycorrhizal legumes have been the subject of many experiments. In

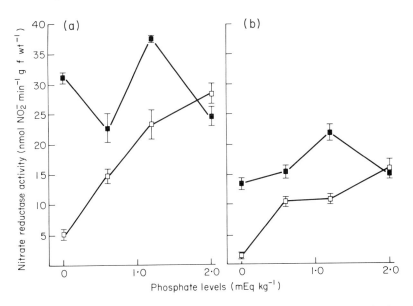

FIG. 11. Nitrate reductase activity in shoots (a) and roots (b) of mycorrhizal (■) and non-mycorrhizal (□) *Trifolium subterraneum* grown at four different levels of phosphate fertilization (data of Oliver *et al.*, in press).

most cases, improved nodulation and nitrogen fixation in mycorrhizal plants appears to be the result of relief from phosphate stress, which results in both a general improvement in growth, and indirect effects upon the $N_2$ fixing system. The differences between mycorrhizal and non-mycorrhizal plants disappear if the latter are supplied with a readily available phosphate source.

## Water relations

Mosse and Hayman (1971) observed that mycorrhizal onions did not wilt when transplanted, but that non-mycorrhizal plants did. Subsequently several similar observations have been made, consistent with the suggestion that mycorrhizal fungi may increase water uptake, or in some other way improve the water relations of associated plants. The problem was first investigated systematically by Safir et al. (1971, 1972). They estimated the resistance of the whole plant and of detached shoots to water transport by measuring the rate of recovery of the water potential of a water-deficient leaf on an intact plant. Mycorrhizal plants had lower resistances to water transport than uninfected plants, and it appeared that most of the difference was attributable to changes in root resistance, for shoot resistances were small and did not differ in the two groups of plants. Safir et al. (1972) concluded that the effect was probably due to improved nutrition for three reasons. First, the effect was apparent in plants with root systems of similar weight and volume although shoots of mycorrhizal plants were larger; secondly, resistance to water movement in whole non-mycorrhizal plants was reduced if the plants were grown at high nutrient levels; and thirdly, application of the fungicide PCNB had no effect on the resistance of mycorrhizal plants. There have been several subsequent attempts to elucidate the role of mycorrhizas in plant water relations. Levy and Krikun (1980) used *Citrus jambhiri* and a fungus similar to *Glomus fasciculatus*. Their growth and fertilizer conditions permitted the comparison of mycorrhizal and non-mycorrhizal plants of similar size and growth rate. Measurements of water stress (xylem pressure potential and leaf proline concentration), transpirational flux, stomatal conductance and photosynthesis were made during a period of increasing soil moisture stress followed by recovery. In this experiment the major effect of mycorrhizal infection was an increase in transpirational flux and stomatal conductance, both during stress and recovery, and an increased rate of photosynthesis per unit leaf area during recovery. There were apparently no differences between mycorrhizal and non-mycorrhizal plants in terms of resistance of the root to water movement. Somewhat similar results were obtained by Allen and co-workers (1981) and Allen (1982) using the grass *Bouteloua gracilis* and *Glomus fasciculatus*. Again, transpiration rates were increased in mycorrhizal plants while leaf resistance to water vapour diffusion was greatly reduced, as were the stomatal and mesophyll resistances to $CO_2$ uptake, with consequent increase in rate of photosynthesis.

The results could not be explained in terms of differences in gross anatomy or morphology. Furthermore, there was no change in size of mesophyll cells, bundle sheath cells or stomata, nor was the stomatal density altered following mycorrhizal infection. Resistance to water transport was not separated into contributions of root and shoot by Allen *et al.* but they did comment that increased branching of the roots in mycorrhizal plants could lead to substantial increases in root surface area without changes in root biomass, and that this might reduce the root resistance. Changes in root morphology were considered in greater detail by Hardie and Leyton (1981). They found that the hydraulic conductivities of the root systems of *Trifolium pratense* (red clover) were much higher when infected by *Glomus mosseae*, but that this could only be partly attributed to increased surface area or length of the mycorrhizal roots. The conductivities were 2–3 times higher per unit length of mycorrhizal root, and they therefore concluded that hyphal growth in the soil was important in reducing root resistance to water flow. Again the leaves of mycorrhizal plants had slightly lower diffusion resistances with respect to water vapour. They also had very much larger surface areas which, together with the effect on root conductivity, resulted in higher transpiration rates. Transpirational flux (per unit leaf area) was also higher in the mycorrhizal plants when there was adequate soil water. Both Allen *et al.* (1981) and Hardie and Leyton (1981) found that at low soil water potential mycorrhizal plants generated lower leaf water potentials. In clover, but not in *Bouteloua*, this was associated with lower transpirational flux and an ability to extract water from soils of lower water potential. Thus mycorrhizal clover plants wilted at lower soil water potentials that non-mycorrhizal plants; they also recovered faster (within 30–60 min compared with 120 min) on re-wetting. Several conclusions can be drawn from this work.

Stomatal physiology is affected by mycorrhizal infection, as shown by decreased stomatal resistances to water and $CO_2$ movement and by increased transpirational fluxes and rates of photosynthesis. The effect could be mediated by increased stomatal opening, consequent perhaps on increased $K^+$ concentrations in mycorrhizal plants or in plants given additional phosphate. Allen *et al.* and Levy and Krikun also canvas the idea that hormonal changes in mycorrhizal plants could have an effect on stomatal opening, particularly as Allen *et al.* (1980) have observed changes in cytokinin levels in *B. gracilis* following mycorrhizal infection; this idea remains to be substantiated. Lowered osmotic potentials of leaf cells could be brought about, as Hardie and Leyton emphasize, by accumulation of ions or organic molecules in the vacuoles of leaf cells. Again, $K^+$ may be important here, as also may the synthesis of organic anions. All such changes in the leaf would contribute to the observed effects of mycorrhizas on water relations. However, they cannot explain the lowered resistance to water uptake in the root system. It seems quite likely that water uptake by fungal hyphae or reduced resistance in the root cortex could arise from mycorrhizal infection, but more experiments are required.

## Effects of different combinations of host and fungus

We have already noted that all plants are, with a few exceptions, generally susceptible to mycorrhizal infection. With vesicular-arbuscular mycorrhizas there is no evidence for any well-defined specificity in the ability of particular fungal taxa to infect particular host species or varieties. Nor indeed is there any evidence for a broader specificity of the type observed in *Rhizobium*-legume cross inoculation groups. Nevertheless, some differences in performance of particular combinations of host and fungus have been observed. This may be compared with "symbiotic effectiveness", a term used to describe differences in nodulation and nitrogen fixation in *Rhizobium*-legume symbiosis. In some of the mycorrhizal examples the differences can be correlated with differences in root morphology and root hair production by the host. The idea that mycorrhizas might replace root hairs was first mooted in the nineteenth century and more recently Baylis has discussed the evolution of root systems and their associated mycorrhizas (Baylis, 1972, 1975). His general thesis has been that plants with thick unbranched roots and few root hairs (e.g. *Allium, Coprosma, Citrus*) are apparently more dependent on mycorrhizal infection when growing in low phosphate soils than plants with finely branched roots and long or numerous root hairs, although all may be susceptible to infection. He described them as being more "mycotrophic", using that word therefore in a special sense of "dependent on infection". Baylis would envisage that the long root hairs and much-branched root systems of the Cruciferae and Juncaceae compensate for lack of mycorrhizas in these groups and may have evolved because these families were not readily susceptible to mycorrhizal infection. The hypothesis certainly fits the general way in which vesicular-arbuscular mycorrhizal fungi are thought to extend the root system. However, it might be expected that other types of mycorrhiza with similar general physiology (e.g. ericoid and ectomycorrhizas) would fit the same scheme. This is not the case. Ericaceous plants such as *Calluna* and *Vaccinium* not only have very finely branched ("hair") roots but also have very large numbers of hyphal connections with the soil. They do, however, grow in impoverished soils where diffuse ramification both of the hair roots and the hyphae together may be of selective advantage. In any event, the simple theory of Baylis is undoubtedly complicated by variation of soil supplying power and plant demand for soil-derived nutrients.

Apart from general considerations of mycorrhizal interactions with plants of differing root geometry, there have been a few more detailed investigations of mycorrhizal response to infection in different cultivars of the same species or group of closely related species. In studies of this kind it is most important to put measurement of responses on a sound comparative basis. We have already seen that the magnitude of the difference in (for example) dry-weight between mycorrhizal and non-mycorrhizal plants often depends upon the availability of nutrients in the soil (see Table 12, for example). The rate of development of infection, the strain or species of fungus and the total amount of infection may also be important. Hence any statement about response to infection must be qualified according to the conditions. For this purpose "mycorrhizal dependency" was defined by Gerdemann (1975) as "the degree to which a plant is dependent on the mycorrhizal condition to produce its maximum growth at a given level of soil fertility". Menge *et al.* (1978) calculate mycorrhizal dependency as the dry weight of the mycorrhizal plant expressed as a percentage of the dry weight of the non-mycorrhizal plant grown under the same experimental conditions. This gives a similar measurement of difference to "mycorrhizal efficiency ratio", used by Powell and Daniel (1978).

The number of investigations into varietal responses to infection is few, so that generalizations would be dangerous at this stage. It seems best to summarize some of the findings. Hall (1978a) found that a variety of *Zea mays* (cv PX 610), which had a rapid growth rate, extensive root system and low tissue phosphate concentrations did not respond to mycorrhizal infection (with a mixed inoculum of several fungi), whereas *Z. mays* cv 415 and *Z. mays × robusta* cv Golden Cross Bantam, both with slower growth rates and root development, and higher tissue phosphate concentrations, were responsive to infection (see Table 19). Differences could not be attributed to differences in the percentage infection of the root system measured at the final harvest, although different rates of fungal

TABLE 19

Response of three varieties of maize to inoculation with vesicular-arbuscular mycorrhizal fungi (results of Hall, 1978a)

| Maize variety | % infection of root | Shoot dry weight g per pot | Ratio of root:total plant weight | Concentration of phosphate in leaves % dry weight |
|---|---|---|---|---|
| *Zea mays* | | | | |
| PX 610 | 0 | 12·02 | 0·288 | 0·069 |
| | 28 | 11·73 | 0·178 | 0·069 |
| W 415 | 0 | 5·08 | 0·186 | 0·089 |
| | 30 | 10·19 | 0·174 | 0·085 |
| *Zea mays × robusta* | 0 | 5·90 | 0·110 | 0·090 |
| Golden Cross Bantam | 33 | 8·63 | 0·131 | 0·083 |

colonization of the root systems cannot, as we have already noted, be ruled out as a factor contributing to the differences. Similar correlations between root growth and morphology in a range of *Citrus* species and their dependency on infection by *Glomus fasciculatus* for dry-weight production have also been found (Menge *et al.*, 1980). Here again details of the colonization of the roots by the fungus were not given. This is unfortunate because differences in behaviour of a single fungus within the roots of different species do occur (Bevege and Bowen, 1975) (see Table 7), and might be important in the growth responses of different cultivars as well as different species, especially when the responses are determined at a single harvest. Recently Daniels *et al.* (1981) have used the most probable numbers method to calculate the inoculum potential of spores of six vesicular-arbuscular mycorrhizal fungi. The results are illustrated in Table 20. Differences in inoculum potential were not correlated with differences in spore germination nor with spore size, except that *Glomus mosseae* had both the largest spores and the highest inoculum potential. This fungus also colonized the roots of sudan grass more rapidly than the other fungi.

TABLE 20

Spore germination in non-sterile soil, relative inoculum potential and root infection of sudan grass by six vesicular-arbuscular mycorrhizal fungi (results of Daniels *et al.*, 1981)

| Fungus | Spore germination % at 3 weeks | Relative inoculum[a] potential (by 6 weeks) | % infection[b] sudan grass |
|---|---|---|---|
| *Glomus mosseae* | 86·1 | 100·0 | 70 |
| *G. constrictus* | 66·7 | 19·5 | 25 |
| *G. fasciculatus* | | | |
|    isolate 92 | 93·4 | 12·1 | 20 |
|    isolate 0–1 | 73·4 | 8·1 | 0 |
|    isolate 185 | 84·2 | 2·5 | 0 |
| *G. epigaeus* | 94·0 | 2·4 | 0 |

[a] Compared with *Glomus mosseae*.
[b] After inoculation with 600 spores.

Varieties of two species of wheat (*Triticum aestivum* and *T. durum*) also responded differently to inoculation with *G. mosseae* (Berthau *et al.*, 1980). Three types of response were recorded, which could not be correlated with infection levels in the roots of the host plants. Some varieties showed increased dry-weight of tops or yield of grain or both; in others mycorrhizal infection had no effect on growth and in a third group reductions in dry-weight and yield of mycorrhizal plants were observed. Similar results were obtained by Lambert *et al.* (1980) using clones of *Medicago sativa*. The underlying reasons for the differences between mycorrhizal and non-mycorrhizal plants were not determined. However, in wheat and in the other hosts mentioned above, there appear to be grounds for further investigation of varietal responses. The genotype of the host may be an important determinant in mycorrhizal response, as it is with the *Rhizobium*-legume symbiosis. As already indicated, the morphology of the root and its growth rate may well be involved, together with phosphate requirement, phosphate uptake mechanisms and unknown factors which affect the growth of the mycorrhizal fungus. In two investigations with *Trifolium repens* (Crush and Caradus, 1980) and *Medicago sativa* (O'Bannon *et al.*, 1980), no varietal differences in response to mycorrhizal infection were apparent.

Differences between fungal strains or species in their effect on the growth of the host plant also occur. Here again it is important to consider the growth of the fungus as well as that of the host. For instance, the form and amount of inoculum used in experiments can influence the rate and extent of infection, so that differences in them might either mask real differences between the fungi or suggest differences between the fungi which would not occur if the inoculum were standardized.

While Mosse (1972a,b) did not record quantities of inoculum used, nor give data on development of infection, her work on *Paspalum notatum* did give one of the earliest indications that there might be important differences in the response of a single species of plant to different strains of fungi. Sanders *et al.* (1977) compared the effect of four fungi on growth and phosphate uptake in onions. Their results are illustrated in Fig. 12. Three of the fungi which colonized roots relatively rapidly (Fig. 3), and produced similar amounts of external mycelium, increased phosphate uptake and growth. A fourth isolate was slow to colonize the roots, produced little or no external mycelium and had no effect on phosphate uptake or growth. In this example, fungal growth appears to be the basis of the differences between fungal isolates. Table 21 shows differences in growth of *Medicago sativa* plants which were also correlated with differences in subjective ratings of spores and mycelium associated with roots at the final harvest, after inoculation with several different fungi (O'Bannon *et al.*, 1980). In this experiment inoculum was standardized at 55 spores per pot of three plants, so that our criticisms about inoculation do not apply to it, as long as we assume that the viability of the spores was similar in all species and strains. In descending order of "effectiveness" were *Glomus epigaeus*, *G. fasciculatus* (2 strains), *G. mosseae* (2 strains) and *Acaulospora trappei*. *G. monosporus* was less effective than the other fungi, but the results were not strictly comparable as inoculum was only 25 spores per pot, and although infection levels were similar to the other fungi at the final harvest (10 weeks), the rate of development of infection may have been quite different, with consequent effects on phosphate uptake and growth.

Environmental factors can certainly affect the relationship between host and

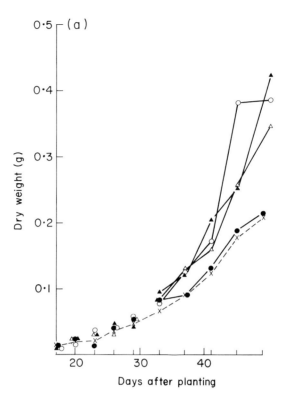

0.5 ⎡ (a)

Dry weight (g)

Days after planting

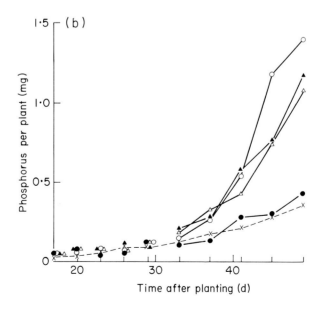

1.5 ⎡ (b)

Phosphorus per plant (mg)

Time after planting (d)

fungus, so that in Mosse's (1972a,b) work the addition of lime to soil altered the apparent effectiveness of the fungi and the results differed according to the soil type. There is insufficient data about differential effects of other changes in soil (e.g. phosphate level, pH, nitrogen fertilizers, temperature, placement of inoculum, etc.) on the effectiveness of different host/endophyte combinations and more experiments are urgently required. Mathematical modelling of infection might well provide useful information in this sphere as the models may give a means of calculating rates of infection or fungal growth under different conditions. It may be possible to extend the approach to a consideration of more physiological aspects of the symbiosis. Sanders (in press), while making assumptions about rates of infection, has successfully modelled aspects of phosphate uptake and plant growth which are certainly important criteria in "symbiotic effectiveness". If selection of "effective" fungal strains for inoculation in agricultural and horticultural ecosystems is to be a realistic proposition, we will need to pay greater heed to the critieria used to determine effectiveness, and be very careful that inocula used for assessment are standardized. Luedders *et al.* (1979) have suggested that "saturating" amounts of inoculum should be used in comparative experiments. This would certainly overcome problems of different inoculum potential of different fungi. We need to

TABLE 21

The effect of inoculation of *Medicago sativa* with 55 spores per pot of seven different mycorrhizal fungi on dry weight of shoots, spore production in soil and root colonization (spore/mycelium rating) after 10 weeks' growth (results of O'Bannon *et al.*, 1980)

| Fungus | Source | Shoot dry weight g per pot | Spore numbers in 100 cm$^3$ soil | Spore/mycelium[b] rating |
|---|---|---|---|---|
| *Glomus epigaeus* | Oregon | 0·79 | 340 | 2·2 |
| *G. fasciculatus* | Oregon | 0·64 | 200 | 2·1 |
| *G. fasciculatus* | Florida | 0·59 | 143 | 2·3 |
| *G. mosseae* | Oregon | 0·48 | 347 | 2·0 |
| *G. mosseae* | Florida | 0·51 | 415 | 1·7 |
| *G. trappei* | Oregon | 0·52 | 110 | 1·8 |
| *G. monosporus*[a] | Oregon | 0·28 | 205 | 2·2 |
| Not inoculated | — | 0·34 | — | — |

[a] Only 25 spores per pot used.
[b] Ratings on a scale of 0–3.

FIG. 12. Dry weight (a) and phosphorus content (b) of onion plants inoculated with one of four mycorrhizal fungi or uninoculated. ×, control; ▲, *Glomus fasciculatus*; △, *Glomus macrocarpus* var. *geosporus*; ○, *Gigaspora calospora*; ●, *Glomus microcarpus*. (From Sanders *et al.*, 1977.)

know much more about why the number of infection units saturates in young plants with high inoculum density in the soil (Carling *et al.*, 1979; Smith and Walker, 1981; see Fig. 4), and why the percentage of the root infected also reaches a plateau value at longer times in older plants. This saturation level of the fraction of the root infected differs with different hosts, fungi and external conditions. In their recent papers, Buwalda *et al.* (1982a,b) introduce a parameter, $n$, into their equations to limit the value of the fraction of root infected in their model. This parameter then becomes the maximum fraction of the root system that can become mycorrhizal in a given set of conditions. Walker and Smith (in preparation) have also considered this problem and show, by recasting their equations (see Smith and Walker, 1981), that such an arbitrary introduction of a parameter may not be necessary and that the maximum fraction of the root system that becomes infected would normally be expected to approach values of less than 1. However, it is certainly necessary to consider why this fraction differs under different conditions. The explanation will almost certainly be found to be complex, being related to the rate of root growth and the age structure and possibly also branching of the root system, that is to the mean relative quantities of susceptible and non-susceptible root tissue present in any given conditions. It will also be related to the rate of infection of the root by the fungus and thus to the inoculum density in the soil and to the rate of fungal spread within and on the surface of the root system. It seems clear that the transplantation of pre-infected seedlings to unsterilized soil should not be used as a means of comparing introduced and resident (indigenous) strains of fungi. The pre-inoculated fungi would almost certainly be at an advantage unless very high levels of rapidly germinating propagules were present in the soil.

In the studies of the symbiotic effectiveness, as well as in many of those on other aspects of the physiology of vesicular-arbuscular mycorrhiza, dry weight has been used as the measure of the performance of the host plant. If the importance of mycorrhizal infection in ecological situations and in crop production is to be more fully comprehended, other more informative measures of the effects of infection will also have to be used.

## Conclusions to Section 1

The last two decades have seen an immense amount of progress in our knowledge of vesicular-arbuscular mycorrhizal infections. The identity of the fungal symbionts has been established and the taxonomy of the Endogonaceae is now on an excellent footing, due largely to the interest in the group which has been generated because of the mycorrhiza-forming ability of many of the genera within it. "Unculturability" remains one of the attributes of these fungi, but some progress towards axenic culture has been made, and the indications that the fungi may have some slight ability to survive saprophytically in dead roots should give impetus to continuing efforts in this direction. Failure to grow the fungi in pure culture means that both the production of pure inoculum and also biochemical studies on the fungi have been somewhat hampered. The problems of producing inoculum for experiments have been overcome to some extent by the use of spore or root material from pot-cultures, but these techniques may raise further difficulties in the provision of satisfactory controls and also because experimental conditions are very different from those met within the field.

Techniques are now available for producing large amounts of sterile external hyphae in association with roots and this should provide material in sufficient quantity for increased investigations of some aspects of the physiology and biochemistry of the fungi. Excised roots have been used in some experiments and there is scope for more work using such material. The coordinated study of the activity of phosphatase enzymes by gel electrophoresis and their location in cells by ultrastructural and ultracytochemical techniques by V. Gianinazzi-Pearson and her associates is an example of the way in which modern methods can be exploited and much information gained using small amounts of material. Investigations such as these, together with knowledge of the ultrastructure of the mycorrhizal associations, should yield information about the physiological and biochemical bases of the growth responses to mycorrhizal infection which have now been so frequently demonstrated. We need to know much more about the identity of carbohydrates transferred from autotroph to heterotroph and the quantities that are involved. The determination of an "energy budget" or a "carbon budget" for vesicular-arbuscular associations would be a worthwhile goal, as it, together with other information, might show why different combinations of host and fungus "perform" differently. Some ecological and epidemiological investigations on this subject have been carried out and have shown that differences in rates of infection, spore production, overall differences in plant dry weight and so on, do occur, but we do not know why. Further studies are required under both laboratory and field conditions before a satisfactory understanding of the symbiosis can be achieved and before assessment of the economic potential of vesicular-arbuscular mycorrhizal fungi can be made.

# Chapter 5
# The Symbionts

## Introduction. Relationship with other kinds of mycorrhiza

Ectomycorrhizal roots are characterized by (1) a fungal sheath or mantle which encloses the root in a fungal tissue, and (2) a Hartig net which is a plexus of fungal hyphae between epidermal and cortical cells (Plate 6A). The hyphae do not normally penetrate the cells. The fungal sheath is usually between 20 and 100 $\mu$m thick, most often in the range 30–40 $\mu$m and comprises 25–40% of the dry weight of the whole organ, according to the few available estimates (Harley and McCready, 1952b; Melin and Nilsson, 1958). The presence of such a large fungal component of the absorbing organs suggests that the sheath must have some selective advantage perhaps in nutrient absorption or storage. This kind of consideration led D. H. Lewis (1973) to suggest that the name of ectomycorrhiza be altered to "sheathing mycorrhiza". As he agreed (Lewis, 1975) there seems no advantage in this change of name, for apart from the conversion of a useful adjective "sheathing" into a technical term, the complex of similar kinds of mycorrhiza in which the sheath varies in development, from near absence to differentiated pseudoparenchyma, together with ectendomycorrhizas in which the hyphae penetrate the host cells, makes a classification based simply on the existence of a sheath inexpedient. Nevertheless, well-developed sheaths have important physiological functions not possessed by mycorrhizas lacking them.

Hyphal connexions run from the sheath between the cells of the epidermis and cortex of the autobiont, forming what is called the "Hartig net". There is usually little hyphal penetration into the cells of the autobiont of young mycorrhizas, but in the senescent parts of a mycorrhizal axis the cortex becomes colonized by hyphae within the cells (Atkinson, 1975).

The presence of the Hartig net has led some authors, especially in Russia (e.g. Lobanov, 1960) to use the term "ectendotrophic" for the organs usually called

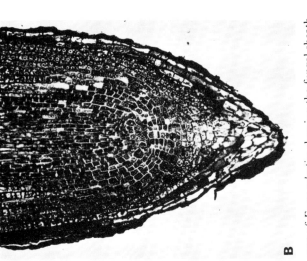

B

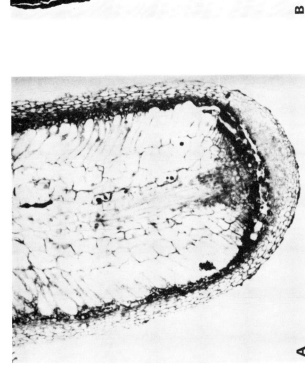

A

PLATE 6. A. Longitudinal section through the apex of a mycorrhizal root of *Fagus sylvatica* showing the fungal sheath enclosing the rootlet. The inner sheath appears dark due to the presence of senescent host cells of the epidermal-cap complex containing phenolic compounds. The epidermal cells of the host are obliquely elongated in a radial direction, and proximally the Hartig net may be seen between their walls. The tissues of the host are mature close to the initials which are small in extent. The root cap cells are few and many senescent cells lie immediately within the fungal sheath. The cortical cells of the host next within the epidermis have tannin-impregnated walls. × 195 approx.

B. LS of the apex of a large uninfected lateral root of *Fagus sylvatica* for comparison with A. The root cap is well developed and the maturation of the tissue occupies a lengthy zone behind the initials. The epidermal cells are not transversely elongated. × 175 approx.

(Photo: F. A. L. Clowes.)

"ectotrophic". The term "ectendotrophic" (now ectendomycorrhiza) was coined to designate mycorrhizal organs with sheaths of variable development (sometimes absent), Hartig net, and extensive intracellular penetration, and is best restricted to that purpose. The mycorrhizas formed by *Endogone* spp. (*sensu stricto*) with *Pinus*, *Pseudotsuga*, *Eucalyptus* and other forest trees have a well-developed Hartig net but often a fairly rudimentary sheath of surface hyphae. These are also extreme variants of ectomycorrhiza. The arbutoid mycorrhizas of some Ericaceae possess a well-formed fungal sheath, a Hartig net and regular penetration of the living cells of the root by fungal hyphae. Similarly members of the Monotropaceae and Pyrolaceae, which include completely or partially achlorophyllous members, have mycorrhizas with well-developed fungal sheaths, Hartig nets and intracellular penetration resembling the arbutoid type, or of a more specialized kind. The mycorrhizal roots of other green Ericaceae such as *Calluna*, *Erica* and *Rhododendron* do not form a fungal sheath, but their roots are usually copiously enmeshed in an external weft of loose fungal hyphae which also penetrates into the cortical cells. It is possible, therefore, to perceive a gradation of structure from the typical ectomycorrhiza of *Fagus* or *Pinus* to the ericoid endomycorrhizas of Ericaceae and Epacridaceae with increasing intracellular colonization and decreasing sheath development. In addition, considerable research on the mycorrhiza of juvenile pines and other conifers, especially in nursery conditions (Mikola, 1965; Laiho, 1965; Wilcox, 1968a,b; Wilcox *et al.*, 1974) has led to the recognition of ectendomycorrhizas, which may merge into the so-called pseudomycorrhizas recognized by Melin, which were believed to be pathological structures. It is noteworthy that almost all the autobionts in all these mycorrhizas are woody perennials.

Among the typical ectomycorrhizas there is also considerable variation in the development of the sheath, the development of the Hartig net, the extent of intracellular penetration and in the connexions between the sheath and soil. In the extreme there may be no Hartig net, as in the mycorrhiza of *Pisonia grandis* recently described by Ashford and Allaway (1982) or in superficial ectomycorrhizal roots described by Clowes (1951) on long roots of *Fagus* and found on other ectomycorrhizal hosts, or no sheath formation but only a Hartig net as in the long roots of *Pinus* described by Robertson (1954). Between these variants, which are often only stages of development or infections of different kinds of root, intermediates occur. This variation has led Warcup (1980) to explain his use of the term: "Ectomycorrhizal associations are considered to include any case of a fungus forming a sheath or mantle with or without a Hartig net on the roots of a plant. The sheath might enclose only part of a root or rootlet. This is a wider definition than is often used for an ectomycorrhiza, but there seems no logical reason why the term should only be used in the restricted sense, i.e. that an ectomycorrhiza should have a Hartig net as well as a well-developed fungal sheath that encloses the apex of the root."

This is an unhelpful view, unless descriptions of detailed morphology and histology are given, for they are necessary to the investigation of physiological function and ecological significance. The investigation of the "typical" ectomycorrhizas has led to some knowledge of their functioning, but the conclusions derived from their study cannot be used to formulate hypotheses of function for other kinds of mycorrhiza lumped as "ectomycorrhizas" whose structure is different. In the past ectomycorrhizas have been contrasted with endomycorrhizas because in the latter the fungus penetrates the cells of the host and in the former this is a relatively rare event in the mature mycorrhizas (see later). It will be realized from the above discussion that many kinds of intermediate mycorrhizas or stages of development occur. Hence, although the term "endomycorrhiza" is not as useful as it once was, it should be realized that fungal penetration into the cell lumen of the host involves activities, processes and properties not met in ectomycorrhizas.

### The hosts: taxonomy and geographic occurrence

According to Meyer (1973) only about 3% of phanerogams are ectomycorrhizal and he gives a table of the genera concerned. Amongst the cryptogams, ectomycorrhizas have been described in some fern sporophytes. Cooper (1976) in an investigation of New Zealand ferns showed that although they mostly developed vesicular-arbuscular mycorrhiza, if they were mycorrhizal at all, ectomycorrhizas might occur on their root systems. These might become more plentiful than other types of mycorrhizas in habitats dominated by ectomycorrhizal trees with no other undergrowth but ferns. Iqbal *et al.* (1981) also described pure ectomycorrhizas on *Adiantum venustum* in stands of *Picea excelsa* and *Abies pindrow*, but mixed ectomycorrhizas and vesicular-arbuscular mycorrhizas in mixed conifer and broadleaf forests. An extended list of phanerogams forming ectomycorrhizas which includes information not at Meyer's disposal is given in Table 22 but it is not of course exhaustive. It includes genera of autobionts in which at least one species has been described as forming ectomycorrhiza, but that does not mean that all or even most of the species have ectomycorrhizal roots, nor does it mean that the ectomycorrhizal organs of the genera exactly conform with that of "typical" mycorrhiza, in structure and function. The list includes only one monocotyledon, *Kobresia*. Most of the plants listed are forest dominants of the north and south temperate and sub-arctic regions. In addition to the well-known genera given by Meyer, the large genus *Eucalyptus* whose species are important components of forest and bush in temperate and sub-tropical Australasia, as well as some other genera of Myrtaceae, are ectomycorrhizal. Evidence is also accumulating that the family Dipterocarpaceae, dominant trees of the monsoon and moist tropical regions of Indo-Malaysia, has a great many ectomycorrhizal members and may be totally ectomycorrhizal. Sal

# TABLE 22

Genera reported to contain at least one species on which Ectomycorrhiza has been described (compiled from various sources)

*This list cannot pretend to be exhaustive but illustrates the wide range of families and genera of Angiospermae and Gymnospermae in which ectomycorrhiza has been observed. A record of the presence of ectomycorrhizal individuals in a genus does not mean that all species are or may be ectomycorrhizal, nor does it mean ectomycorrhizal infection is necessarily consistently present in any species of that genus. Those marked * are herbaceous*

| Family | Genus | Family | Genus |
|---|---|---|---|
| Aceraceae | *Acer* | Euphorbiaceae | *Poranthera* |
| Betulaceae | *Alnus* | | *Uapaca* |
| | *Betula* | Fagaceae | *Castanea* |
| | *Carpinus* | | *Castanopsis* |
| | *Corylus* | | *Fagus* |
| | *Ostrya* | | *Lithocarpus* |
| | *Ostryopsis* | | *Nothofagus* |
| Bignoniaceae | *Jacaranda* | | *Pasania* |
| Caprifoliaceae | *Sambucus* | | *Quercus* |
| Casuarinaceae | *Casuarina* | | *Trigonobalus* |
| Cistaceae | *Helianthemum* | Goodenaceae | *Brunonia** |
| | *Cistus* | | *Goodenia** |
| Compositae | *Lactuca* (*Mycelis*) | Hammamelidaceae | *Parrotia* |
| Cyperaceae | *Kobresia** | Juglandaceae | *Carya* |
| Dipterocarpaceae | *Anisoptera* | | *Juglans* |
| | *Balanocarpus* | | *Pterocarya* |
| | *Cotylelobium* | Lauraceae | *Sassafras* |
| | *Dipterocarpus* | Caesalpinoideae | *Afzelia* |
| | *Dryobalanops* | | *Aldina* |
| | *Hopea* | | *Anthonota* |
| | *Monotes* | | *Bauhinia* |
| | *Shorea* | | *Brachystegia* |
| | *Valica* | | *Cassia* |
| Elaeagnaceae | *Shepherdia* | | *Eperua* |
| Ericaceae | *Arbutus* | | *Gilbertiodendron* |
| | *Arctostaphylos* | | *Julbernardia* |
| | *Chimaphila* | | *Monopetalanthus* |
| | *Gaultheria* | | *Paramacrolobium* |
| | *Kalmia* | | *Swartzia* |
| | *Ledum* | Mimosoideae | *Acacia* |
| | *Leucothoë* | Papilionatae | *Bartonia* |
| | *Rhododendron* | | *Brachysema* |
| | *Vaccinium* | | *Chorizema* |
| | | | *Daviesia* |

| Family | Genus | Family | Genus |
|--------|-------|--------|-------|
| | *Dillwynia* | Rubiaceae | *Galium** |
| | *Eutaxia* | | *Opercularia* |
| | *Gompholobium* | | *Rubia* |
| | *Hardenbergia* | Salicaceae | *Populus* |
| | *Jacksonia* | | *Salix* |
| | *Kennedya* | | |
| | *Mirbelia* | Sapindaceae | *Allophylus* |
| | *Oxylobium* | | *Nephelium* |
| | *Platylobium* | Sapotaceae | *Glycoxylon* |
| | *Pultenaea* | | |
| | *Robinia* | Saxifragaceae | *Ribes* |
| | *Vicia* | Sterculiaceae | *Lasiopetalum* |
| | *Viminaria* | | *Thomasia* |
| Myricaceae | *Comptonia* | Stylidiaceae | *Stylidium* |
| | *Myrica* | Thymeliaceae | *Pimelia* |
| Myrtaceae | *Angophora* | Tiliaceae | *Tilia* |
| | *Callistemon* | Ulmaceae | *Ulmus* |
| | *Campomanesia* | | *Celtis* |
| | *Eucalyptus* | | |
| | *Leptospermum* | Vitaceae | *Vitis* |
| | *Melaleuca* | Cupressaceae | *Cupressus* |
| | *Tristania* | | *Juniperus* |
| Nyctaginaceae | *Neea* | Pinaceae | *Abies* |
| | *Torrubia* | | *Cathaya* |
| Oleaceae | *Fraxinus* | | *Cedrus* |
| Platanaceae | *Platanus* | | *Keteleeria* |
| Polygonaceae | *Coccoloba* | | *Larix* |
| | *Polygonum** | | *Picea* |
| Pyrolaceae | *Pyrola** | | *Pinus* |
| | | | *Pseudolarix* |
| Rhamnaceae | *Cryptandra* | | *Pseudotsuga* |
| | *Pomaderris* | | *Tsuga* |
| | *Rhamnus* | | |
| | *Spyridium* | Gnetaceae | *Gnetum* |
| | *Trymalium* | | |
| Rosaceae | *Chaembatia* | | |
| | *Circocarpus* | | |
| | *Crataegus* | | |
| | *Dryas* | | |
| | *Malus* | | |
| | *Prunus* | | |
| | *Pyrus* | | |
| | *Rosa* | | |
| | *Sorbus* | | |

(*Shorea robusta*), the important tree of northern India, is an example (Bakshi, 1974), as is *Monotes elegans*, an outlying species from Africa (Högberg, 1980) and many genera and species from Ceylon, Malaysia and Thailand (Singh, 1966; Alwes and Abeynayake, 1980).

Tropical and subtropical trees may be ectomycorrhizal especially in areas where vegetative activity is restricted during some period of the year for edaphic, climatic, or other environmental reasons. Many of the arborescent Leguminosae, especially Caesalpinoideae, have been found to be ectomycorrhizal, although others, e.g. *Bauhinia, Cassia, Scorodophloeus* and *Tamarindus* are believed usually to form vesicular-arbuscular mycorrhiza, as do most species of Mimosoideae and Papilionatae. Singer and Morello (1960), discussing the occurrence of communities of ectomycorrhizal forest trees in Central America, emphasize that they are particularly to be found outside the region of 25°N and 35·5°S of the equator at all altitudinal levels, but within that zone they only occur at higher altitudes, and on the equator itself ectotrophs occur only above 3000 m (see Fig. 13). Singer has explained to us that in his work he has examined microscopically all the species that he recorded as ectomycorrhizal, e.g. *Gnetum* (Gnetaceae), *Neea* (Nyctaginaceae), *Glycoxylon* (Sapotaceae), *Allophyllus* (Sapindaceae), etc.

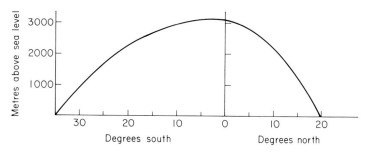

FIG. 13. Diagram of the lower altitudinal limits of communities of ectomycorrhizal trees in tropical regions (redrawn after Singer and Morello, 1960).

He also mentions that Fiard had found a species of *Torrubia*, another Nyctaginaceous plant, to be ectomycorrhizal. The reports of Pegler and Fiard (1979) about species from the Lesser Antilles must, however, be viewed with reservation, for although they state, for instance, that a species of *Coccoloba* is "unquestionably" mycorrhizal, they do not mean that, for they did not (personal communication) examine the root system. Their conclusions are mainly based on ecological and taxonomic probabilities, hence their suggestions are not included in Table 22. Although their ecological contention that ectomycorrhizas are most prevalent in seasonally variable areas agrees with that of other observers, it cannot carry much additional weight.

It must not be supposed that the presence of ectomycorrhizas upon any species indicates that its members are always infected in that way; indeed, many species may also produce other kinds, especially of the vesicular-arbuscular sort. This occurs, for instance, in some species of *Ulmus*, *Tilia*, *Cupressus*, *Juniperus*, *Leptospermum*, *Acacia* and *Casuarina* to name a few. Table 61 shows the kinds of mycorrhiza described on various members of the Ericales. There is a considerable variation and typical ectomycorrhizas have been described on a few as well as the more usual types of mycorrhizas found in Ericales. In addition to trees and shrubs, a very few herbaceous species are said to form ectomycorrhizas (Table 22). Hesselman (1900), as mentioned by Rayner (1927) and Kelly (1950), described the rhizomatous perennial *Polygonum viviparum* as typically ectomycorrhizal. This has been confirmed by Fontana (1977) who showed that several types of ectomycorrhizas might be present on the plants, including some formed in association with *Cenococcum* and *Russula emetica* var. *alpestris*. The latter had a typical sheath 20 $\mu$m thick and a well-developed Hartig net. The herbaceous species mentioned by Warcup (1980) of the genera *Goodenia* and *Brunonia*, had "patches of sheath, sometimes with a Hartig net on some of the fine roots". In other cases a part of the root might show ectomycorrhizal structure as a sleeve of infection.

It is of interest to test whether ectomycorrhizal hosts belong to advanced or primitive families of dicotyledons. Figure 14 gives a comparison of the Sporne (1980) advancement indices of ectomycorrhizal families with the distribution of advancement indices in dicotyledonous families as a whole. The two sets of histograms are similar and it cannot therefore be concluded on that basis that ectomycorrhiza is either a primitive or advanced character nor correlated other than with the woody habit. A comparison with other theories might be rewarding.

## The fungi of ectomycorrhiza

The fungi of ectomycorrhiza include members of many families of Basidiomycetes, some Ascomycetes with hypogeous or bulky fruit bodies (e.g. in the Eurotiales, Tuberales and Helotiales) and sterile Fungi Imperfecti such as *Cenococcum* and many sterile unnamed basidiomycetous and ascomycetous mycelia isolated into culture by various workers. A valuable list was given by Trappe in 1962, but since then many additions have been made, notably the Phycomycetes, *Endogone flammicorona*, *E. eucalypti* and *E. lactiflua* which will form ectomycorrhiza with various Pinaceae and *Eucalyptus* (Fassi, 1965; Fassi and Palenzona, 1969; Fassi *et al.*, 1969; Gerdemann and Trappe, 1974; Warcup, 1975) and many Ascomycetes (see Trappe, 1971).

Many of these fungal associates of ectomycorrhizas (see Table 23) have been identified by field observations, that is by the consistent association of fruit bodies with a particular species of tree, often combined with the confirmation that the fungal hyphae were attached both to the fruiting body and a mycorrhizal root. In many cases such observations have been the basis for the isolation of the fungus from the fruit body into pure culture and its inoculation onto a potential host in axenic conditions to form a recognizable mycorrhizal organ. Isolation of mycelia from surface-sterilized mycorrhiza or fragments of fungal sheath and back-inoculation on an autobiont has also been successful, but the determination of the identity of the fungus is very difficult because very few indeed can be made to form fruit bodies in culture. Their identity remains problematical, except in so far as hyphal fusion with known species, or immunological, serological or chromatographic methods have allowed identification (see Chilvers, 1972; Seviour *et al.*, 1974). The visual similarity of mycelia or hyphae to those of known fungi may be helpful, as also is the structure of the mature mycorrhiza.

In addition to those ectomycorrhizal species of fungi that have been identified, many mycelia of uncertain taxonomic affinity have been shown, after isolation and re-inoculation, to be mycorrhiza-formers, and it is very likely that lists, such as those of Trappe (1962), only include a minority of the ectomycorrhizal species and their potential hosts. Moreover, the failure of an isolate to form mycorrhiza

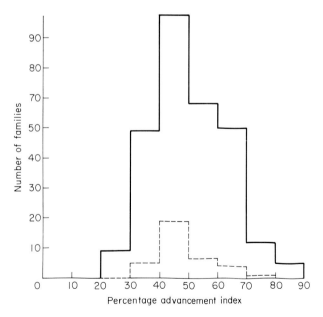

Fig. 14. Comparison of the Sporne advancement indices (Sporne, 1980) of the families in which ectomycorrhiza is found (– – –) with those of all dicotyledonous families (———). Number of families having a given advancement index plotted as a histogram.

## TABLE 23

Some of the important fungal genera in which one or more species have been reported to be ectomycorrhizal

*This list is not exhaustive for it is increasing every year. An entry does not imply that all species of that genus are mycorrhizal. Generic names as in published papers*

| Family | Genus | Family | Genus |
|---|---|---|---|
| **Basidiomycetes** | | **Gasteromycetes** | |
| Agaricaceae | *Lepiota* | Hymenogastraceae | *Alpova* |
| Amanitaceae | *Amanita* | | *Rhizopogon* |
| | *Amanitopsis* | Geastraceae | *Geastrum* |
| Boletaceae | *Boletinus* | | *Astraeus* |
| | *Boletus* | Lycoperdaceae | *Calvatia* |
| | *Fistulinella* | | *Lycoperdon* |
| | *Gyrodon* | Phallaceae | *Clathrus* |
| | *Gyroporus* | | *Phallus* |
| | *Krombholzia* | | |
| | *Leccinum* | Pisolithaceae | *Pisolithus* |
| | *Suillus* | Sclerodermataceae | *Scleroderma* |
| | *Tilopilus* | | |
| | *Xerocomus* | **Ascomycetes** | |
| Cortinariaceae | *Alnicola* | Balsamiaceae | *Balsamia* |
| | *Cortinarius* | Elaphomycetaceae | *Elaphomyces* |
| | *Hebeloma* | | |
| | *Inocybe* | Geneaceae | *Genea* |
| | *Rozites* | Geoglossaceae | *Cudonia* |
| Gomphideaceae | *Gomphidius* | | *Spathularia* |
| Hygrophoraceae | *Hygrophorus* | Helvellaceae | *Helvella* |
| Paxillaceae | *Paxillus* | Hydnotryaceae | *Bassia* |
| Rhodophyllaceae | *Clitopilus* | | *Choiromyces* |
| | *Rodophyllus* | | *Hydnotrya* |
| Russulaceae | *Lactarius* | Otidiaceae | *Otidia* |
| | *Russula* | Pyronemaceae | *Geopora* |
| Strobilomycetaceae | *Boletellus* | | *Lachnea* |
| | *Strobilomyces* | | *Sepultaria* |
| Tricholomataceae | *Laccaria* | Rhiziniaceae | *Gyromitra* |
| | *Leucopaxillus* | Terfeziaceae | *Mukomyces* |
| | *Lyophyllum* | | *Picoa* |
| | *Tricholoma* | | *Terfezia* |
| Cantharellaceae | *Cantharellus* | | *Tirmania* |
| Hydnaceae | *Hydnum* | **Phycomycetes** | |
| Thelephoraceae | *Corticium* | Endogonaceae | *Endogone* |
| | *Thelephora* | **Imperfecti** | |
| | | | *Cenococcum* |

with a given autobiont in culture does not prove it to be unable to do so under natural conditions. Again, Lamb and Richards (1970) have shown that, amongst many fungi tested by them for mycorrhiza formation and their effect on the growth of *Pinus radiata* and *P. elliottii*, some unidentified mycelia isolated from the roots of adult trees proved more efficient than known species of mycorrhizal fungi in terms of their effects on the growth of the host.

Although there is much variation in detail, ectomycorrhizal fungi have been found to be somewhat selective in nutrition, especially in their inability or relative inability to use cellulosic or lignified materials as carbon sources for growth. They rely in consequence on simpler carbohydrates as primary sources of carbon compounds and this sets them apart from other soil fungi with permanent mycelia which make use of resistant carbon compounds. The conclusion that as a group they are dependent on their hosts for carbohydrates explains the observation that many of them are unable to form fruit bodies if detached from their autobiont partners (see Romell, 1938; Laiho, 1970). Those proved mycorrhiza-forming species that have been reported to do so, e.g. *Paxillus involutus*, *Boletus subtomentosus*, *Laccaria laccata*, *Leccinum scabrum*, etc. have been described sometimes as facultative mycorrhiza-formers. However, the variability of many species in culture suggests that this description should be treated with caution.

It is much more important to consider genetic and phenotypic variation in the background of ecological behaviour. Bowen and Theodorou (1973) and Theodorou and Bowen (1971b) have shown that a mycorrhizal fungus, *Rhizopogon luteolus*, could colonize the rhizospheres of *Lolium perenne*, *Phalaris tuberosa* and *Trifolium subterraneum* as well as the ectomycorrhizal trees *Eucalyptus leucoxylon* and *E. camaldunensis* (Table 24). This indicates the possibility that ectomycorrhizal fungi may be able to exist at least temporarily in some ecological situations where mycorrhizal symbiosis does not occur. This would be a different kind of facultative mycorrhizal behaviour allowing persistence in the vegetative state in the absence of suitable hosts in special habitats. The existence of cellulolytic or ligninoclastic strains in a dicaryotic mycorrhizal fungus could also allow for persistence in the absence of suitable autobionts, even though these strains themselves might not be capable of forming mycorrhiza.

The lists of potential hosts which have been found to form ectomycorrhizas with a single species of fungus (see for instance Trappe, 1962) indicate that most species of fungus are not host-specific, but they do not show whether strains of that species are specific to a single or group of species of host. This question will be considered in detail below, but the present evidence suggests a very wide range of autobionts may be associated with a single strain of a given species of fungus and vice versa.

## Mycorrhiza in the Cistaceae

The mycorrhiza of *Helianthemum chamaecistus* (*nummularium*) was described by Boursnell (1950) as involving a systemic infection of the whole plant and a general resemblance in behaviour to that of many Ericaceae as described by Rayner (see 1927). In 1969 Harley reconsidered the evidence available about *Helianthemum* and suggested that a further examination should be made for there appeared to be doubt whether systemic fungal infection really occurred in this species. At that time Thérèse Stelz, working in Professor Boullard's laboratory, had (1968) made an examination of several species of *Helianthemum* (*H. polifolium*, *H. canum* and *H. vulgare*) in the course of her work on mycorrhizas of plants of calcareous soil. She observed that *H. polifolium* had an "endotrophic" infection by septate hyphae, "appartement au groupe des Eumycètes ou champignons supérieurs", which seemed to produce dichotomously branched arbuscules in the cells, but no vesicles. *H. canum* and *H. vulgare* had in addition ectomycorrhizas with well-developed mantles and Hartig nets between the outer two series of host cells. These ectomycorrhizas could be referred to Dominik's sub-type F (1955). Read *et al.* (1977) reassessed the mycorrhizal status of *H. chamaecistus* (*nummularium*). They pointed out that of seedlings obtained in the field, some 22% of those observed up to the stage of producing four foliage leaves were infected with vesicular-arbuscular mycorrhiza, and about 13% by an unknown brown fungus resembling *Cenococcum* and forming ectomycorrhizas. Ectomycorrhizas became more evident as the seedlings aged and *Cenococcum* became the dominant fungus in the mature plant, although in the early stages

TABLE 24

Colonization of the surface of roots by *Rhizopogon luteolus*

*The length of the root (mm) colonized and the intensity of colonization (arbitrary rating) in 4 weeks (or 6 weeks\*) at 24 C day, 15 000 lm m$^{-2}$, 16 C night (data of Theodorou and Bowen, 1971b)*

| Species of root | Mean root length of plant | Mean length of root colonized | Number of plants in each intensity rating | | | | | |
| --- | --- | --- | --- | --- | --- | --- | --- | --- |
| | | | 4 | 3 | 2 | 1 | 0 | Mean ± SE |
| *Pinus radiata* | 81 | 24 | 8 | 4 | 4 | 1 | 3 | 2·65 ± 0·33 |
| *Lolium perenne* | 52 | 29 | 7 | 3 | 5 | 3 | 1 | 2·63 ± 0·30 |
| *Phalaris tuberosa* | 35 | 24 | 1 | 4 | 3 | 5 | 1 | 2·00 ± 0·29 |
| *Trifolium subterraneum* | 63 | 12 | 1 | 2 | 5 | 6 | 3 | 1·53 ± 0·27 |
| *Eucalyptus leucoxylon* | 72 | 32* | 1 | 5 | 1 | 1 | 0 | 2·75 ± 0·31 |
| *E. camaldulensis* | 45 | 37* | 2 | 1 | 4 | 0 | 0 | 2·71 ± 0·34 |
| Glass fibres | (20) | 5·3 | 0 | 0 | 4 | 5 | 11 | 0·65 ± 0·18 |

Intensity rating: 0 absent, 1 light, 2 moderate, 3 high, 4 very high.

both vesicular-arbuscular mycorrhizas and ectomycorrhizas might occur in the same plant. No sign of systemic infection was observed and no consistent fungal infection of any part of the plant except the roots was present in their preparations. They put forward the suggestion that the importance of *Cenococcum* as a mycorrhizal fungus of *Helianthemum* might arise from its being able to withstand drought and grow in soils of low water potential, as had been suggested by many observers and experimentally verified by Mexal and Reid (1973). Kianmehr (1978) examined the effect of inoculation with *Cenococcum* on the growth of *H. chamaecistus* (*nummularium*) on two types of soil. The growth of infected plants on renzina was 7·2 times as great as that of the uninoculated controls, whilst on agricultural soil the increase was only 1·4 times as great. *H. chamaecistus* must therefore be included amongst the plants having ectomycorrhiza and as one capable of developing vesicular-arbuscular mycorrhiza also.

Fontana and Giovanetti (1978–1979) reviewed the knowledge of the association of Cistaceae with fungi. They pointed out that for a very long time it had been known that their roots were associated with the hypogaeous truffles. The species of the genera *Helianthemum*, *Tuberaria* and *Cistus* are found associated with the species of *Terfezia*, *Tuber*, *Hydnocystis*, *Delastria* and *Tirmania*. They demonstrated by inoculation that *Tuber melanosporum*, a well-known mycorrhizal fungus, forms typical ectomycorrhiza with *Cistus incanus* ssp. *incanus* and showed photographically that there was a considerable stimulation of growth in height after inoculation. They also pointed out that Bruchet (1973) synthesized ectomycorrhizas on three species of *Cistus* and on *Helianthemum nummularium* with *Hebeloma cistophilum* and that Chevalier *et al.* (1975) synthesized mycorrhizas between *Tuber melanosporum*, *T. hiemale* and *T. aestivum* and *Cistus salvifolius* and *Furmana procumbens*, in semi-axenic conditions.

We can conclude therefore that the Cistaceae as a family is ectomycorrhizal as adults but that *Helianthemum nummularium* (*chamecistus*) and *H. polifolium* at least may produce vesicular-arbuscular mycorrhizas in their seedling states.

# Chapter 6
# Structure of Ectomycorrhiza

## Introduction

Frank (1885) named the actual absorbing organs composed of fungal and host tissues "Mykorrhizen" and since that time the word has been used for the absorbing rootlet, the feeding root and its fungal partner. In vesicular-arbuscular mycorrhizal plants individual mycorrhizal organs are not so recognizably different from uninfected roots as are ectomycorrhizas, so that vesicular-arbuscular mycorrhizal root systems as a whole have been investigated. Ectomycorrhizas have received rather separate consideration from the whole root-systems which bear them, but more emphasis should be put on the root-systems of ectomycorrhizal plants and relatively less upon the ectomycorrhizas themselves. Such a reorientation of emphasis, which agrees with the view of Wilcox (1968a,b), encourages study of the whole underground absorbing and anchoring system, and of the production, development, maturation and senescence of the root axes including the permanent as well as the short-lived branches. This outlook is also implicit in recent papers by Chilvers and Gust (1982a,b) on the initiation and development of ectomycorrhizal root systems of *Eucalyptus*, and by Sohn (1981) on *Pinus resinosa*. It increases rather than decreases the understanding of the functions of the fungal symbiosis in the plant and in the ecosystem of which it is a part. In the following account therefore the root-systems of the ectomycorrhizal trees will be shortly described before the details of the mycorrhizal organs themselves are discussed.

Besides the ectomycorrhizas of forest trees which form the central or typical kinds of ectomycorrhiza, there are variants which depart conspicuously in structure from them. The ectendomycorrhizas described by Mikola and Laiho and by Wilcox and his colleagues are considered later. These usually have a reduced sheath of surface hyphae, a well-developed Hartig net and intracellular penetration of living hyphae into living cells. The mycorrhizas formed by species of *Endogone* with forest trees first described by Fassi and his colleagues may also have a rather rudimentary sheath but a well-developed Hartig net. Some of these almost sheathless types of structure bear a considerable resemblance to some of the kinds of mycorrhiza found in the Ericales.

## The root systems of mycorrhizal trees

In the seedling stage most trees and shrubs possess a tap-root which usually gives way in time to a series of main axes which form the permanent part of the root system. The apex of each of these main axes grows to extend the root system, and as it ages each axis may undergo secondary thickening and become woody. In the mature parts of the permanent axes, carbohydrates and other compounds may be stored so that the growth and development becomes less dependent on current photosynthesis, although in the early stages and at times of rapid growth it is supplied almost directly by translocation from leaves.

The branches of the system are of two kinds, roots of potentially unlimited extension in length, and roots of restricted growth and life-span. Where the majority of ultimate laterals are of limited growth the root system is called 'heterorhizic''. In most ectomycorrhizal species, both angiosperm and gymnosperm, the long and short roots differ only in degree. All apices grow for a period and all may become infected in some way by a mycorrhizal fungus so that the whole root system is mycorrhizal. Those apices which become fully invested by a sheath of fungus are usually short and continue to grow slowly and give rise to racemose systems of branches (see Plate 7). Those apices which are not permanently invested by a sheath or which remain uninfected either maintain very active growth and are the leading apices of the system, or especially if of small diameter, may abort or become dormant.

The genus *Pinus* departs from this general form because its ultimate laterals— its short roots—are sharply differentiated from the axes which bear them. They are simpler in stelar construction, have a restricted apical meristem and root cap, and soon abort if not infected. If infected they continue to grow, branch dichotomously and form mycorrhizal systems. Dichotomy, a characteristic of *Pinus*, is rare or absent in other genera. Nevertheless, as Wilcox (1968b) has shown, long roots of *Pinus resinosa* may have the structure of a typical ectomycorrhiza and they may produce dichotomous mycorrhizas as do the more usual mother roots. This shows that there may be, even in *Pinus*, the kind of hierarchy of root found elsewhere. Although the mycorrhizal organs described by Frank and most later observers were the ultimate short laterals of the root system enclosed in a fungal sheath with fungal hyphae penetrating between the epidermal and cortical cells, it is important to stress again that all the roots of the system, long or short, may be infected by the mycorrhizal fungi in some way. Infection of the root-system is permanent although the mycorrhizas themselves may, as will be seen below, have short lives of months or years.

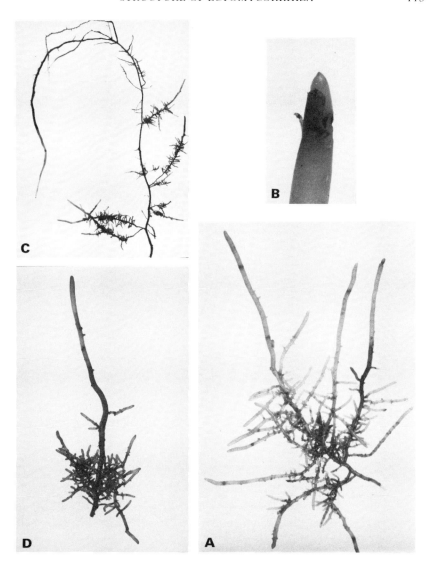

PLATE 7.  Mycorrhizal roots of *Fagus sylvatica*.

A.  Mycorrhizal system showing the hierarchy of branching. Some of the main apices have broken through the sheath. The darker colour shows the older regions of the system in which senescence of the tissues may be occurring. × 0·75.

B.  Apex of a leading root which has broken through the sheath. × 3.

C.  Growing long roots in spring which bear some long laterals and some mycorrhizal systems. × 0·25.

D.  Mycorrhizal systems such as are seen proximally on C, totally enclosed in a sheath. × 0·75.

Although we have some knowledge of the root systems of some of the temperate ectomycorrhizal trees, very little is known of those in the tropics, subtropics or semi-desert regions. In temperate forests the active roots of the ectomycorrhizal trees are intensely developed in the surface and subsurface layers of the soil. In the humic layers especially, ectomycorrhizas which form the main component of the feeding system, often lie below the F layer in great quantities (Werlich and Lyr, 1957; Mikola and Laiho, 1962; Lyr, 1963; Meyer and Göttsche, 1971; Meyer, 1973). The roots tend to grow in a lateral or upward direction colonizing the newly accumulated humus and litter (Harley, 1940). This behaviour has led to the suggestion that the absence of geotropic response of mycorrhizas is due to the relatively small development of the root-cap of mycorrhizal roots and the lesser accumulation of starch in them. This does not seem to be valid, because the long roots of *Fagus* and other genera, whose apices are free from fungal infection and equipped with well-developed root-caps (Plate 6) are also diageotropic in the litter and humus layers.

All the apices of the root system grow by means of their apical meristems, the rapidity of division of which depends upon their size and position in the hierarchy of branches, as well as upon external conditions. In the extreme, short roots such as the smallest short roots of *Pinus* may from a very early time have only a very small meristematic region, grow little and differentiate mature stelar and cortical tissues close to the apex. Between these and the main long roots of the root system of most ectomycorrhizal trees there is a range of root behaviour in respect of the rates of meristematic division and of differentiation. Indeed, most of the root axes eventually become differentiated close to the apex and in this state they may become dormant and the apices metacutinized, or more often they may become infected by fungi to form a typical mycorrhizal structure (Warren Wilson and Harley, in preparation). Behind the distal region of primary structure, secondary thickening of the stele and the development of phellogen in the cortex may occur both in mycorrhizal and uninfected axes. In consequence there is a progressive development behind the apex of the symbiotic structures, their maturation and senescence, in each phase of which the actual histological, as well as physiological relationships between fungus and host, may change.

## Ectomycorrhizal organs

In spite of variation in detail, the common types of ectomycorrhiza of most trees are similar in general structure. Plate 8A,B shows parts of mycorrhizal root systems of beech and of pine, collected from the upper humic layers of forest soil. All the branches shown are mycorrhizal and covered with a fungal mantle or sheath, so that the main axes are the only parts uncovered. As seen in Plate 7, uninfected growing apices occur on such root systems especially at certain times of year when root growth is active.

Plate 6A,B shows longitudinal sections of mycorrhizal and uninfected apices contrasted. Besides the presence of the fungus on the surface of the mycorrhiza and between the cortical cells, the lesser development of the root-cap and the closer differentiation of the mature tissues behind the meristem are the most conspicuous differences. The features of the mycorrhiza are therefore those of a slow-growing organ, in contrast to the uninfected root (Clowes, 1950, 1951, 1954, 1981). Recently Chilvers and Gust (1982b) have compared the rates of growth of mycorrhizas and uninfected roots of *Eucalyptus* seedlings. By photographic recording and by measurement of seedling roots in culture they confirmed that mycorrhizal roots grow slower than uninfected roots in the same category of the hierarchy. Uninfected roots grew approximately five times as fast as mycorrhizas (Table 25). Although mycorrhizas grow slowly they continue to grow steadily, whereas uninfected roots grow very fast at times and slower at other times. The differentiation of the root systems is indeed into organs of growth, the main axes especially when uninfected, and organs of absorption, the mycorrhizas. Even when the main axes may become ensheathed, as in *Fagus*, *Eucalyptus* and other genera, their apices break through the sheath and grow rapidly at some periods of the year and become ensheathed again when their growth rate falls. This differentiation is not quite sharp because some lateral mycorrhizal apices, usually of the largest diameter with the largest meristems, may also, at times conducive to rapid root growth, escape the sheath and renew free growth for a time. The mycorrhizal apices of the ultimate orders of the hierarchy do not, however, break through the sheath, but both host and fungal tissues grow slowly in unison.

The obliquely transverse elongation of the epidermal cells of the mycorrhiza seen in Plate 6A is usually present, especially in mycorrhizas of large diameter. It may be absent, but its existence may be missed in transverse sections if the elongation is very oblique. Its oblique orientation was explained by Clowes (1951) in terms of the stresses involved in the greater elongation of the stelar cells

## TABLE 25

Growth rates of various orders of uninfected and mycorrhizal roots of *Eucalyptus st. johnii*

*Seedlings grown in culture. Rates in mm per week (data of Chilvers and Gust, 1982b)*

| Order of root in hierarchy | Uninfected Roots | | Mycorrhizas | |
|---|---|---|---|---|
| | mean | maximum | mean | maximum |
| 1st | $49 \cdot 41 \pm 1 \cdot 98$ | 84 | — | — |
| 2nd | $2 \cdot 78 \pm 0 \cdot 61$ | 17 | $0 \cdot 55 \pm 0 \cdot 15$ | $1 \cdot 8$ |
| 3rd | $0 \cdot 79 \pm 0 \cdot 09$ | $4 \cdot 3$ | $0 \cdot 16 \pm 0 \cdot 02$ | $0 \cdot 6$ |
| 4th | — | — | $0 \cdot 10 \pm 0 \cdot 02$ | $0 \cdot 4$ |
| 5th | — | — | $0 \cdot 07 \pm 0 \cdot 01$ | $0 \cdot 1$ |

PLATE 8. A. Mycorrhizal roots of *Fagus* in the litter layer of the soil exposed by the removal of the loose leaves above them. Note the large apices completely ensheathed and the paucity of mycelium. Natural size.

B. Mycorrhizal roots of pine in the litter layers of the soil. Note the dichotomous branching and the abundant mycelium. ×2. (Photo: G. D. Bowen.)

Inset C. Close up of *Pinus* mycorrhizas. ×20. (Photo: G. D. Bowen.)

Inset D. SEM of two adjacent mycorrhizal roots of *Fagus*. ×40. (Photo: David Kerr.)

Inset E. SEM of pine. ×40. (Photo: J. Duddridge.)

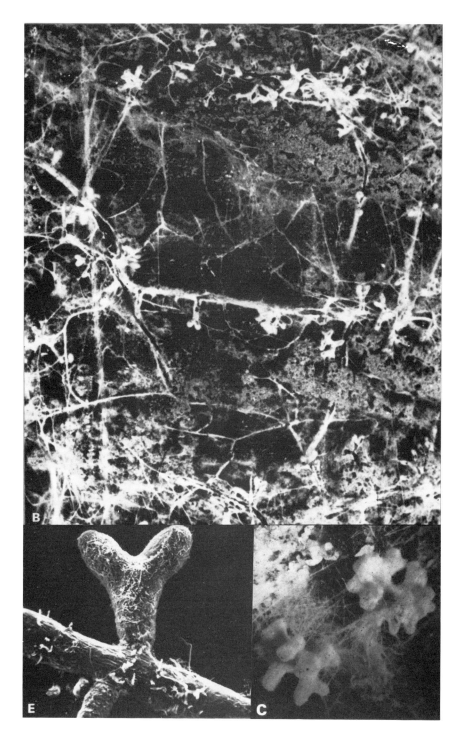

than of those of the cortex and epidermis, and the consequent strain put on the fungal sheath.

In longitudinal sections and in transverse sections (Plate 9), the penetration of the fungus in the cell walls of the epidermis and cortex to form the Hartig net gives the impression of cellular hyphae between the adjacent walls. This appearance results from the fact that fan-like or complicated branch systems of

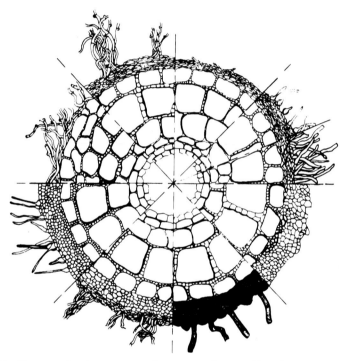

Fig. 15. Diagram showing a range of structure of mycorrhizas in transverse section in *Pinus maritima* (after Mousain, 1971). Note the differences in connexions with the soil, in the sheath and in the Hartig net.

Plate 9. A. Transverse section of mycorrhiza of *Fagus sylvatica* showing the sheath, dark phenolic inner zone of the sheath containing senescent cells, the transversely elongated epidermal cells with Hartig net between them and the layer of the cortex with tannin impregnation of the walls. × 350. (Photo: F. A. L. Clowes.)

B. Detail of the inner sheath and Hartig net of *Fagus*. tl, tannin layer; ec, host cortical cells; fh, fungal hyphae of Hartig net. Incomplete septa arrowed. × 3160. (Photo: Atkinson, 1975.)

C. Detail of the Hartig net of *Pinus contorta* to show the involving layer, IL. HC, host cytoplasm; FC, fungal cytoplasm of the Hartig net (HN); P, host plastid; CC, cortical cell; V, vacuole; M, mitochondrion. × 8420. (Photo: J. Duddridge.)

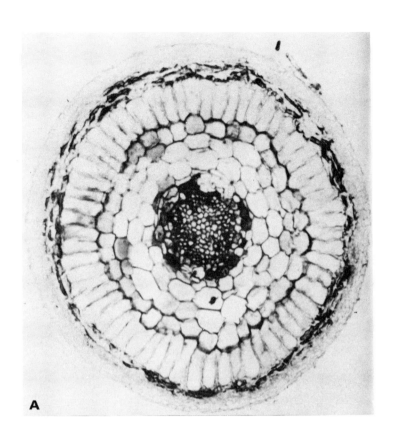

A

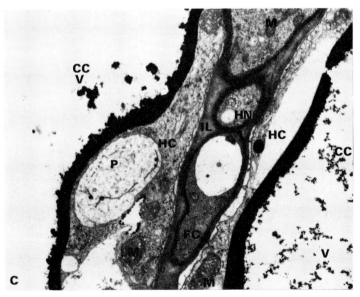

C

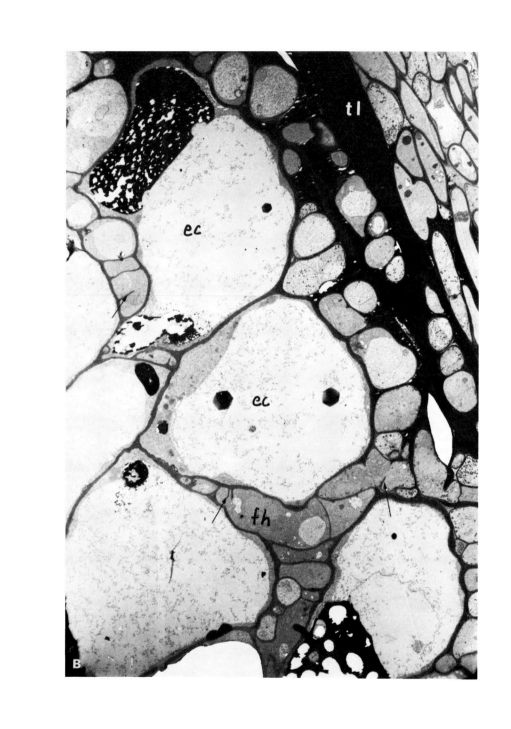

hyphae penetrate the walls and form a large area of contact between fungal and root tissues. The formation of the Hartig net is discussed below. It should however be noted that the Hartig net is not present in the region of division of the meristem but is formed behind this region.

Connexions between the fungal sheath and the substrate by hyphae, hyphal strands or rhizomorphs are variable in number and extent. The apical regions of some mycorrhizas may be smooth; others are amply provided with hyphae connecting them with the soil (Plate 8). Clearly the hypotheses of function of ectomycorrhizas depend in some measure upon the extent of such connexions with the soil, and indeed emanating hyphae of some sort would be expected to be formed by all those fungi which produce fruit bodies external to the root system. Nevertheless, the actual surface of the mycorrhiza in the young state is very variable (Mousain, 1971, see Fig. 15).

## *The fungal sheath*

The fungal sheath or mantle of ectomycorrhizas varies in both constitution and thickness. It may, in face view or in section, appear to be of one layer and constructed of coherent hyphae. On the other hand it may be pseudoparenchymatous throughout and often of two layers, the outer usually more dense, compact and pseudoparenchymatous than the inner which is frequently hyphal and loosely aggregated. The variations in structure of the sheath, the ornamentation of its surface by hyphae, rhizomorphs, setae or cystidium-like structures and its colour have been used, together with chemical, immunological and other tests and the structure of the Hartig net to classify ectomycorrhizas (Melin, 1927; Dominik, 1955; Chilvers, 1968a; Zak, 1971, 1973). There is a possibility that eventually a system of classification can be perfected so that, on the analogy of the lichens, taxa can be recognized and particular combinations of autobiont and fungus grouped together.

The inner layers of the sheath, as Clowes (1951) showed, have more deeply staining protoplasts. Under the electron microscope they are less vacuolate, and possess well-developed organelles, glycogen rosettes and polyphosphate granules. In the outer sheath the cytoplasm becomes restricted to the periphery of the cells which contain large vacuoles. Foster and Marks (1967) observed that the outer cells of the sheath in *Pinus radiata* were senescent or dead and became colonized by bacteria, but this is not always the case, as scanning electronmicroscopic investigations by Strullu (1973, 1976a,b, 1977, 1979) and Seviour *et al.* (1978) have shown. Strullu made the point that although the outermost hyphae were free from one another, those below were cemented together with an interhyphal matrix. The outermost often had soil particles enmeshed in them. The various kinds of hyphal structures from complex strands and simple hyphae growing into the soil to cystidia and other hyphal ornaments which are observed on the sheath surface are important in respect of absorption of nutrients as well as in classification. There is, however, not always a development of extra-matrical mycelium such as is found in vesicular-arbuscular mycorrhiza (see Fig. 15).

The main tissues of the sheath, as suggested by Marks and Foster (1973) may be reminiscent of those of the fruit bodies of the fungi; and indeed Scannarini (1968) had pointed out that sheaths formed by species of *Tuber* had both active and storage hyphae in them, and Luppi and Gautero (1967) had shown that those of *Lactarius* contained lactifers. It was Strullu's belief that the interhyphal matrix, probably of carbohydrate polymers, that binds the hyphae of the sheath together differed in ascomycetous and basidiomycetous sheaths, being electron-dense in the former and electron-lucent in the latter. However, Duddridge (1980) and others do not agree that there is a consistent difference of this sort.

The sheath may age and show either a zonation of age from apex to proximal region (Chilvers, 1968a), or change with time throughout, when, for instance, the outer layers change colour and the outer hyphae or cells form thickened walls or lose their contents (Strullu, 1976a). The contact between the sheath and host tissue changes from the apical to the proximal zones of most mycorrhizas. At the apex the sheath is over the root cap. Further back it lies over the senescent cells of the root cap, and further back still over the epidermal cells, the senescent cells having been incorporated into the body of the sheath.

The sheath varies in thickness but is typically 20–40 $\mu$m thick and comprises 20–30% of the volume of the rootlet. Harley and McCready (1952b) estimated that the sheath contributed $39.1 \pm 0.5$ mg per 100 mg of dry weight of mycorrhiza and Brierley (1953) estimated that it was $37.0 \pm 1.9$ mg in *Fagus*. Edmonds (unpublished) found that the sheath of Nothofagus was 34% of the weight of the mycorrhiza. The values calculated from the data of Melin and Nilsson (1958) are very variable, but for seedlings of *Pinus sylvestris* are generally in the range of 25–30 mg per 100 mg. In any event, the fungal layer is a very significant part of the mycorrhizal organ in volume and dry weight and insulates the root tissue from the soil.

## The Hartig net

Although in section the Hartig net appears under the light microscope as if it were made up of frequently septate hyphae, its structure is in reality much more complicated, as was realized by Mangin (1910) who described digitate structures, "palmetti", formed by the fungus between the cells. In spite of this and of figures showing the fungus in the walls of the cortex in face view illustrating the works of many authors, the Hartig net has usually been considered to be essentially hyphal (e.g. Rayner, 1927). Strullu (1976a), however, describes the Hartig net as consisting of "lames fungique", and examinations of its structure by Atkinson (1975), Duddridge (1980) and Nylund (1981) using both light and electron microscopy, have demonstrated that it consists of complicated fan-like or labyrinthine branch systems which provide a very large surface of contact between cells of the two symbionts (Plate 9). Atkinson pointed out that there appeared often to be incomplete septa in the hyphae of the net and indeed these had also received comment from others (e.g. Marks and Foster, 1973). Atkinson likened them to the wall ingrowths in "transfer cells", but Duddridge pointed out that they were in fact double and represented the adjacent walls of hyphae under compression. This agrees with the appearance in face view and also with Nylund's (1981) observations of the very complicated labyrinthine structure in artificially synthesized ectomycor-rhizas between *Piloderma croceum* (*Corticium bicolor*) and *Picea abies*. Nylund (1980) has also made the very important observation that in these and in natural spruce mycorrhizas, the formation of the Hartig net did not disrupt the plasmodesmata between the cortical cells of the host, so that symplastic continuity was retained in the cortex.

The method of penetration of the fungus between the cell-walls of the host has been much discussed because most of the fungi in culture seem unable to use cellulose or pectin for growth. That does not itself prove that such enzymes are absent, but only that insufficient quantities or activities are produced to allow sufficient hydrolysis to take place to provide an adequate carbon source for growth. Hence it has been suggested that the cultural conditions used for growing the fungi may have been inimical to enzyme production (e.g. Giltrap and Lewis, 1982, on phosphate effects); that enzymes may be attached to the fungal walls especially locally at the apex (e.g. Lindeberg and Lindeberg, 1977, on pectinase); that the host provides some factor that allows or promotes enzyme production and so on. It now has become clear as a result of electron microscopy that penetration occurs primarily by mechanical means. The leading edge of the fungal elements in section is often wedge-shaped (Foster and Marks, 1966, 1967; Hofsten, 1969; see Marks and Foster, 1973; Strullu, 1974; Atkinson, 1975; Duddridge, 1980). Foster and Marks (1967) suggested that carbohydrate hydrolysis within the penetrating hyphae might form an osmoticum increasing hydrostatic pressure, but there is no agreement about this. In any event if penetration is by pressure the cortical cells must either be diminished in size, that is, water must be lost by them, or they must change shape—that is, their walls must be plastic. It is, however, widely believed that some degree of pectic hydrolysis of the middle lamella also occurs (Marks and Foster, 1973; Hofsten, 1969).

In some types of ectomycorrhiza the interface between the fungus and host in the region of the Hartig net is described as between unmodified hyphal wall and host cell-wall (e.g. Atkinson, 1975, for *Fagus*; Foster and Marks, 1966, for *Pinus radiata*). A second type of interface is described as consisting of a modified external layer of the wall of the host being apposed to the hyphal wall or a modified hyphal wall. This type of interface was observed by Atkinson in synthesized mycorrhizas of *Betula* with *Amanita muscaria*, and by Strullu (1974) and Strullu and Gerault (1977) in *Pseudotsuga menziesii* and *Betula* in association with ascomycetous fungi. The latter authors described it as a "contact zone". A similar layer was described by Scannerini (1968) in mycorrhizas of *Pinus strobus* with *Tuber albidum* as an "involving layer" probably of polyuronides. All these reports indicate a degree of modification of the walls of the symbionts which was at one time believed to be characteristic of ascomycetous rather than basidiomycetous fungal symbionts; but the work of Atkinson and of Duddridge suggests that both types of fungi may have similar effects on the walls. Duddridge (1980) observed modified (see Plate 9) as well as unmodified interfaces in mycorrhizas between Basidiomycetes and *Pinus sylvestris*. She suggested that these were stages of development, modification of the cell walls occurring as the association aged in both ascomycete and basidiomycete associations. The modification of host walls to give the "contact zone" or "involving layer" is assumed to be brought about by activity of enzymes produced by the fungi. This again calls into question why enzymes acting on cellulose or carbohydrate polymers have not been identified in the mycelium or cultures of mycorrhizal fungi, apart from very minor exceptions.

Although it is widely supposed that the Hartig net provides a surface of great size and allows for the interchange of material between host and fungus, Scannerini (1968) suggested that most of the host cells surrounded by it were dead. This is not the view of the other workers quoted above, but Strullu (1976a) pointed out that the death of cells in the Hartig net region is centripetal as the mycorrhiza ages. The demonstration by Nylund (1980) that the symplastic connections via plasmodesmata are intact between cortical cells of the Hartig net region indicates that the cells are alive and that normal movement of materials between them is possible. The Hartig net is variously described as uniseriate or multiseriate. The development according to Duddridge often depends upon the conditions of culture in artificially produced mycorrhizas. In carbohydrate-rich systems both the Hartig net and the sheath are more heavily developed than on less rich media. This may explain the differences sometimes seen between artificially produced and natural mycorrhizas (e.g. Atkinson's descriptions of *Betula* mycorrhizas). The fungal cells of the Hartig net are usually described as appearing metabolically active with full complements of organelles, glycogen rosettes and polyphosphate granules. The cells around them contain either no starch or less starch than in other host parenchyma. This together with the lesser secretion of the intercellular mucilage in the root-cap is often viewed as evidence of carbohydrate movement from host to fungus (Foster and Marks, 1966; Scannerini, 1968; Marks and Foster, 1973; Strullu, 1976a; Hofsten, 1969; Ling Lee, Ashford *et al.*, 1977b).

The Hartig net of mature mycorrhizas is therefore undoubtedly composed of active hyphae in contact with active cortical or epidermal cells (see below). It appears that many of the disagreements mentioned, and others to be described, arise from the lack of full appreciation that like other roots ectomycorrhizas age, and as they do so the relationship of the symbionts changes.

## The host tissue

As shown in Plate 6 the meristematic region of the ectomycorrhizal apex is small in extent compared with that of uninfected roots of similar diameters. Clowes (1949, 1950, 1951, 1954, 1981) by studying the apices of *Fagus* showed that the fundamental architecture of mycorrhizas and uninfected roots is the same and their divisions similar in pattern, but those of the mycorrhizas were reduced in rate. This results in the differentiation of mature tissues, stele, cortex and epidermis, closer to the apex, which might be brought about by factors reducing the rate of cell division. In spite of the widespread belief that hyperplasia, i.e. increase of the number of cells, occurs in the cortex of mycorrhizas, it was not observed by Hatch and Doak (1933) in *Pinus*, nor by Clowes in *Fagus*. Nor, again, did Clowes find any consistent increase of cell volume in the cortex or epidermis, i.e. hypertrophy of the cells. There was, however, a change of shape of the cells, greater radial and smaller axial diameter, so that the cell volume was often found to decrease because the decreased axial elongation was not compensated by the increase in tangential or radial dimensions of the cells in the mycorrhizal roots. Clowes observed, however, an increase of diameter of the host tissue in mycorrhizas, which was indicated by the ratio of stelar to total diameter. It was 0·3 in mycorrhizas as opposed to 0·4 in uninfected roots. Therefore, on the assumption that the stele is unaffected by the infection, there is an increase of total host volume per unit length of more than one half, and of surface area a third. Others (e.g. Chilvers and Pryor, 1965, with *Eucalyptus*) have found slight increases in the volume of mycorrhizal epidermal cells but they depend on the extent of the increase of radial elongation of the epidermal or outer cortical cells (see Table 71). These cells are obliquely orientated because they retain the direction that they had when first formed by the dividing cells on the flanks of the quiescent centre. This is due, according to Clowes, to the fact that the cells of the stele, which elongate more rapidly in an axial direction, effectively push on the sheath, which carries the outer walls of the epidermis forward. The cells of the inner cortex show, as might be expected, the reverse orientation.

In a recent paper, Clowes (1981) used continuous labelling of DNA with tritiated thymidine to study the organization and rate of cell proliferation in the meristems of *Fagus*. In particular he compared apices completely enclosed in the fungal sheath to those which had burst through and escaped from it. There were higher rates of mitosis of the initials of each of the tissue systems in escaped apices; nevertheless in the fully enclosed mycorrhizal apices as many as 50 products of division of the root-cap initials were produced per day. Since there were on average only 160 cells in the root-cap of these mycorrhizas, a rapid consumption or elimination of cells might seem to occur. Clowes' method, however, estimated nuclear divisions, so it does not follow that mature cells are formed and consumed at the rates estimated, for the fungus may directly or indirectly prevent synthesis of cell walls and other structures and make the precursors available for absorption by its own hyphae. The extent of the cap varies, being large in leading apices which break through the sheath at times, and small in the smallest short roots, but the extent to which the fungus might gain nutrients in this way would depend upon the rate of division of the initials rather than upon the number of cells in the root-cap. In both enclosed and free apices there may occur zones or regions of cells of small dimensions which indicate a periodic slowing or cessation of growth followed by recovery, such as one would expect from a seasonality of the growth of the roots.

Behind the dividing meristem in the region where the cells increase in size no Hartig net is present. It develops, however, in the region where the cells have attained or almost attained their full dimensions in the cortex and epidermis. This corresponds, according to Nylund (1981), to a change in wall chemistry. Penetration by the fungal hyphae actuated by internal hydrostatic pressure might be assisted by the development of intercellular spaces during tissue maturation, by enzymic digestion or by inhibition of the formation of cell-wall polymers. The relative importance and probability of operation of these aids to penetration will be discussed elsewhere.

The cells of the cortex in the region of the Hartig net are usually fully active and contain mitochondria, golgi, endoplasmic reticulum and plastids in the peripheral cytoplasm which surrounds the large vacuole. The nucleus is not increased in size as it is in mycorrhizas where penetration of the cells occurs (Strullu, 1976a). The cells seem to be little affected by the fungus except for the lesser accumulation of starch. The epidermal cells as well as some cortical cells may have phenolic compounds in their vacuoles which Ling Lee, Chilvers *et al.* (1977a) thought were produced by fungal influence in *Eucalyptus fastigata*. There is no real agreement about the effect of the fungus on the production of phenols nor on their function. Piché *et al.* (1981) for instance found no difference between the intracellular contents of phenols in mycorrhizal and non-mycorrhizal short roots of *Pinus strobus* produced in culture with *Pisolithus tinctorius*.

Atkinson (1975) divided the apical region of *Fagus* mycorrhiza into regions as follows:

1. *The root-cap region.* The sheath overlies the cap cells.
2. *The pre-Hartig net region.* The sheath overlies the epidermis and no intercellular penetration occurs.
3. *The Hartig net region.* The sheath overlies the epidermis and the Hartig net penetrates between the cells of the epidermis and cortex of the host.

4. *The late Hartig net region* where intracellular penetration of the cortical cells by the fungus occurs.

Each axis in its growth goes through this sequence except that the complete sequence, including the late Hartig net region, only develops fully in those axes of relatively long life. This latter region where intracellular penetration occurs was first described by Harley (1936) as arising following the initiation of secondary thickening of the stele, and phellogen and secondary cortex formation in *Fagus*. Atkinson's work confirmed and extended this. The linear dimensions of the regions estimated by Atkinson were: root-cap region 0·4 mm, pre-Hartig net region 0·25 mm, Hartig net region 25 mm, late Hartig net region > 100 mm. The values are of course approximate and apply to a main long-lived infected axis. In short-lived ultimate mycorrhizal branches the zones are much foreshortened and the late Hartig net region may not exist. Atkinson did not find a late Hartig net region in *Betula* mycorrhizas—merely a zone of fungal invasion of the cortical cells in the oldest part of the Hartig net region. A similar observation was made by Nylund (1981) in *Picea* mycorrhizas. This sequence of growth and of the development of infection will be further discussed after the structure and behaviour of the long roots have been discussed.

## Long roots

As was mentioned above, the distinction between long roots and short roots which is so clear in the genus *Pinus* is much less so in other genera. Wilcox (1964, 1967, 1968) has explained the separation of the categories in *Pinus* and has described their developmental differences: "Roots which come to be recognized as long roots are those which have maintained cell division long enough to elongate appreciably and to augment the size of their meristems. A meristem that is below a critical size . . . develops as a short root which is unable to maintain cell division . . . and consequently produces few cells for elongation." (Wilcox, 1967.) It is the latter which, when infected by fungi, resume cell division and persist as mycorrhizal organs. Nevertheless the long roots may also be infected. Sohn (1981) in his study of *P. resinosa* with *Pisolithus tinctorius* suggested that there was a threshold growth rate of the root axes below which mycorrhizas could form. A Hartig net often occurs in the cortices of long roots growing normally, as reported for instance by Laing (1923), Möller (1947) and Robertson (1954) for *Pinus*, Masui (1926) for *Abies* and Linneman (1969) for *Pseudotsuga*. Wilcox (1967) says that in *Pinus* all roots show some signs of infection; the largest have a non-continuous intercellular infection of the cortex, but "progressively narrower roots displayed fungal features which increasingly resembled those in typical mycorrhizae".

The same can be probably said of other species but in most cases inadequate investigation has been made of them. In *Fagus sylvatica* Clowes described long roots as enclosed in a fungal sheath forming, as he called it, "a superficial ectotrophic mycorrhiza". Under the sheath, root hairs could be observed. "Superficial ectomycorrhizas", perhaps better called "perirhizic mycorrhizas" (Nylund, Unistam, *in litt.*), have also been described by Morrison (1956) in *Nothofagus*, by Ashton (1976) in *Eucalyptus regnans* and as a phase in *Eucalyptus st. johnii* by Chilvers and Gust (1982b). The apices of such roots, according to Clowes, escape the sheath during periods of rapid growth and may form uninfected axes of several centimetres' length. Some long roots may remain uninfected till their growth rate decreases, when they may be converted to typical or to perirhizic mycorrhizas at the apex or become dormant or perish. Between these kinds of long root and those short roots, which are permanently enclosed in the sheath from their initiation as laterals beneath a sheath, is a series of increasing permanence of infection of the apex and decreasing longevity. The distinction between long and short roots is hence one of degree. Clearly those axes which are completely ensheathed and equipped with Hartig net yet grow for a moderately long period, e.g. several seasons, will show in their structure a well-developed late Hartig net region, whereas short-lived fully mycorrhizal axes may not. This may explain both the variability of the extent of the late Hartig net region and its apparent absence in some genera, species or growth stages. The onset of senescence of the host cells and the colonization of them by fungal hyphae is a general phenomenon taking place sequentially behind the apex and also, according to Strullu, centripetally in the adult region. The occurrence of senescence may go far to explain some of the instances where observers have described "ectendomycorrhiza" or "parasitism" in a host known to be ectomycorrhizal. The so-called "saprophytism" of the mycorrhizal fungus (Nylund, 1981) in senescent host cells must be seen in the same perspective. These are examples of nutrient circulation within the symbiotic system. The contents of the senescent cells of the host are not lost but, by the activity of the penetrating fungus, are retained, thus obviating competition for them with free saprophytes.

## The spread of mycorrhizal infection in adult trees

Robertson (1954) studied the distribution of mycorrhizal systems in the long roots of *Pinus sylvestris*. He observed that the development of mycorrhizal infection in the short roots depended on the distribution of the Hartig net in the parent long root. The lateral root initials became infected if they passed through a region of the cortex in which a Hartig net was present, for then they rapidly became ensheathed and fully infected as they emerged. Wilcox (1968a,b) agreed with this, but he also showed that fully ensheathed long roots might be present on *P. resinosa*, the branch initials of which became infected in the cortex and never escaped the sheath. Wilcox's description of the spread of the fungus in the long roots after a period of root dormancy is especially relevant here. The fungus grows acropetally along the new growth both on the surface and between the cortical cells. That on the surface may be in advance of that within the cortex. It becomes the source of infection of new laterals which, as in *P. sylvestris*, may avoid infection if the cortex or surface is devoid of fungus in their neighbourhood. In *Fagus* many long roots are fully ensheathed except for their fast-growing escaped apices. Their branch initials do not escape the sheath but become directly ensheathed by it.

Hence the ectomycorrhizal infection of a host is a permanent feature and the fungus in or on the long roots, i.e. associated with the permanent root axes, often acts as a reservoir of infectivity ensuring that most of the new laterals become infected. It is an important feature that sporadic laterals escape infection for they provide new actively growing apices for the spread of the root system. Such apices may at a later time become infected from without, but this process is better considered along with the discussion of the primary infection of seedlings in a later section. The fungus in the cortex of long roots also, like that in the senescent mycorrhizal organs, penetrates cells when they age. Hence the total recycling of material by this process may possibly be large and deserves investigation.

## Longevity of mycorrhizal rootlets

The question of the length of life of mycorrhizal roots was discussed by Harley (1969a). Clearly a period of activity can only be ascribed if a unit mycorrhizal rootlet can be recognized. This is more possible in *Pinus* spp. which are extremely heterorhizic than in other species of ectomycorrhizal trees. Estimates for *Pinus* have usually been of the order of several months to a year but some estimates exceed these values. The life period may well depend upon climatic conditions, or location in the soil horizons. In climates with extreme winters, root growth may cease in the surface layers during the coldest period and there may be considerable mortality of roots. By contrast, in less extreme areas or under snow cover root growth may continue throughout most winters, and active young mycorrhiza may be found during the whole year.

It is much more difficult to define a unit mycorrhiza in species with racemose branching which are not extremely heterorhizic. *Fagus sylvatica* in Britain is such a species, and moverover its root system continues active growth to some extent throughout most seasons. The extreme branches of its root systems are recognizable units whose life period can be measured, but since all axes are usually infected to some extent general estimates of longevity give problems. Clowes and Harley (see Harley, 1969a) estimated photographically the life period of the ultimate mycorrhizal rootlets as approximately 9 months. The estimates which Orlov (1957, 1960) made on *Picea* depended on following the growth of lateral mycorrhizal systems contained in small glass-topped frames.

The systems persisted for four years but as they aged they became dark and the cortex of the older parts senesced. Clearly in such a case the axes as a whole are long-lived, although the apical regions only are actively metabolic. This behaviour is in line with the description of the stages of development and ageing of mycorrhizal axes given by Atkinson (1975).

An alternative measure of longevity of mycorrhizal rootlets may be obtained from measurements of biomass and net primary production of mycorrhizas in an ecosystem. As an example, Vogt *et al.* (1982) made observations on *Abies amabilis* ecosystems which allow estimates to be made (see Table 26). The life period of mycorrhizal roots in this case would be 10–14 months. The significance of the considerable turnover of root material and of fungus shown by these estimates will be discussed later.

TABLE 26

Estimates of longevity (months) of mycorrhizas of *Abies amabilis* based on measurements of living biomass and net primary production of mycorrhizas (data of Vogt *et al.*, 1982)

| Stand age | 23 years | 180 years |
|---|---|---|
| Living biomass kg ha$^{-1}$ | 1130 | 1850 |
| Net primary production kg ha$^{-1}$ year$^{-1}$ | 1320 | 1570 |
| Calculated life period (months) | 10 | 14 |

## Primary infection in seedlings

Primary infection in seedlings differs from the infection of roots and root initials in adult plants because the source of infection is always from outside the plant in the early stages. In nursery beds or in culture it depends on the presence of spores or propagules in organic particles such as root material, or on inocula in artificial media. The extent to which hyphae emanating from adjacent roots of adults are important in primary infection in natural conditions has not been investigated in detail but is always assumed to be great. Another difference is that the root system of the seedling is in the process of development and differs in the proportion of its parts from that of the adult. Nevertheless much valuable information is obtainable from the study of primary infection about the factors that promote mycorrhiza formation and about the details of the process of penetration of the tissues and the formation of the sheath by the fungus. The introduction by Fortin and his colleagues of the use of growth pouches in this study (Fortin *et al.* 1980) seems likely to be a landmark in the progress of work on the sequence of infection and on the means by which the symbionts recognize and react to one another. The technique has potentialities for precise inoculation by known propagules, direct light microscope observation of effects and the possibility of removing, preparing and observing exactly selected material for transmission or scanning electron microscopy.

The early development of the root system of *Fagus sylvatica* has been described by Warren Wilson (1951) and Warren Wilson and Harley (in preparation); that of *Pinus resinosa* by Wilcox (1968a,b). The results of these two investigations are remarkably similar. Roots of all categories grow at a maximal rate after a brief accelerating phase and then grow slower until they cease growing at a limiting length (see Fig. 16). The maximal elongation rate is correlated with the diameter of the root, that is presumably with the size of its meristem. Higher maximal rates tend to be correlated with longer periods of active growth and hence with the differentiation of long and short roots. Since in *Fagus* the elongation rate is correlated with the distance of the youngest root hairs, and in *Pinus* with the distance of the first lateral root, from the apex, it is a reasonable presumption that the rate of differentiation of the tissues is relatively more constant than, and independent of, the elongation rate. Indeed when growth ceases mature tissues differentiate close to the apex and the meristem becomes greatly restricted in extent. This process is illustrated in Plate 10. The root tissues then bear a near resemblance to those of the host tissues in mycorrhizas.

In *Fagus* this process occurs even in the absence of mycorrhizal fungi so that the changes in the apices of the roots are not dependent on any effect of the

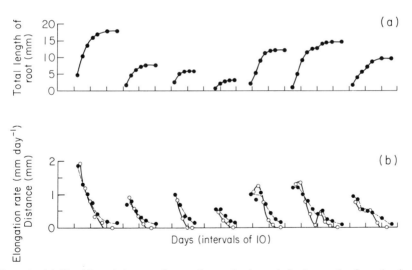

Fig. 16. (a) Total length in mm of seven first-order laterals (at intervals of ten days) of first year seedlings of *Fagus sylvatica*. (b) Elongation mm per day (○) and distance of the youngest root hairs from the apex (●). Higher-order laterals behave similarly but grow at a lower rate. (Data of Warren Wilson, 1951.)

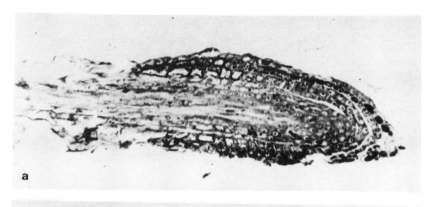

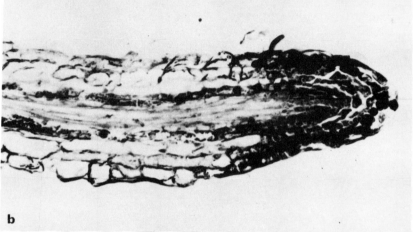

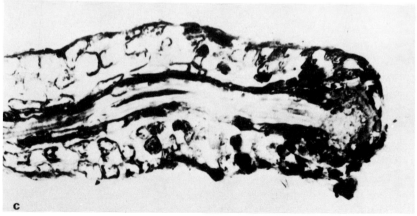

PLATE 10. Longitudinal sections of ultimate laterals of the root systems of *Fagus sylvatica* to illustrate the change to the apical region described in Table 27. a, Stage 1; b, Stage 2; c, Stage 3. ×300. (From J. Warren Wilson, 1951.)

mycorrhizal fungi, but are an internally generated part of root development. A similar conclusion was reached by Faye *et al.* (1981) who have made a study of the short roots of *Pinus pinaster* in culture using seedlings, roots produced by axillary buds and apical meristems of roots as sources of root material. Short roots similar morphologically and histologically to the host tissues of mycorrhiza were formed in culture in the absence of fungi. Clones differed consistently from one another in the potential of their short roots to dichotomize. It was concluded that the formation of "mycorrhizogenic" short roots was genetically determined and did not depend on fungal intervention. The sequence of development of the roots of seedlings shown in Plate 10 was classified by Warren Wilson into stages 1, 2 and 3, the characteristics of which are shown in Table 27. These results of experiments with *Fagus* and *Pinus* have a very important bearing on the hypotheses which concern the effect of mycorrhizal fungi on the structure and histology of the host, full discussion of which has been relegated to Chapter 15 on the causal anatomy of ectomycorrhiza.

TABLE 27

Characteristics of uninfected lateral root tips of *Fagus* at three stages of development illustrated in Plate 10.

*Distances in μm from apex. Limits given in brackets (results of Warren Wilson, 1951)*

| Characteristic | Stage of development | | |
| | Stage 1 | Stage 2 | Stage 3 |
| --- | --- | --- | --- |
| Shape of apex | Pointed | Rounded | Blunt |
| Root cap | Thick, extending behind apex | Thinner, of lesser extent | Much reduced or indistinguishable |
| Meristematic region | Extensive | Small, region near apex | Almost indistinguishable |
| Distance from apex of: | | | |
| root hairs | 223(130–345) | 89(45–135) | 45(0–75) |
| cortical differentiation | 101(93–112) | 62(50–82) | 5(0–47) |
| endodermal differentiation | 106(100–117) | 67(57–85) | 42(27–63) |
| xylem differentiation | 1000(750–1340) | 405(25–425) | 142(87–250) |

The spread of the fungus within the rooting volume of the seedling is greatly increased after primary infection. This is described in the experiments of Robertson (1954) where, after being subjected to a rain of spores, the seedlings of *Pinus sylvestris* became equipped with extensive hyphae and hyphal strands connecting centres of mycorrhizal infection in the root ball. Similar descriptions are given with figures in the work of Bowen and Theodorou (see 1973; and Bowen, 1973). Direct observation of the spread of fungus in a similar fashion is recorded in the experiments of Piché *et al.* (1981 and personal communication) and Piché and Fortin (1982) carried out in growth pouches with *Pinus strobus* and *Pisolithus tinctorius*. Not only are the short roots in the neighbourhood of the inoculation infected but there is also a rapid spread of the fungus along and around the surfaces of all roots. These observers describe that in *P. strobus* with *Pisolithus* the initiation of the sheath on short roots occurred 4–5 days after inoculation, thereafter penetration of the tissues took place through discontinuities in the cellular layer, and after nine days a full sheath and Hartig net had been formed. The Hartig net was absent from the meristematic region although the apex was enclosed by the sheath. Dichotomy rapidly followed and was obvious on the 12th day. Piché and Fortin observed that the timetable of the process of infection was different with the different species of fungus that they used and that nitrogen and phosphorus supply altered the intensity of infection and markedly affected the extent of external hyphae and the formation of sclerotia. In this work of Piché *et al.* two features of the host usually ascribed to the effects of the fungus upon it, i.e. the presence of phenolic substances in and the absence of starch grains from the short root, were present before it came under fungal influence. Duddridge (1980), using *P. sylvestris* with a number of fungal isolates, agrees that the sheath often forms or begins to form before the Hartig net, but the Hartig net may form long before a fully developed sheath exists. Duddridge makes the point that the details of the infection process may be unreliable if the fungus is provided with excessive carbohydrate in the medium. Robertson (1954), using *Pinus sylvestris* in soils, showed that the Hartig net might form simultaneously with or before the sheath, and Melin (1923), Laiho and Mikola (1964) and Nylund (1981) essentially agree with this sequence. In all their descriptions the formation of the Hartig net follows an initial colonization of the surface of the root. It might seem that much depends on whether the colonization is thought to be thick enough to be called a sheath. Nylund puts forward the view that the Hartig net, as it forms, generates a factor which stimulates the surface hyphae to differentiate into a real sheath. In some of the experiments of Duddridge, the sheath hyphae are described as growing initially between and within the outer senescent or broken cells of the root. This is an important observation because the root-cap initials are producing root-cap cells at a considerable rate. They may well be an important initial source of nutrient to the fungus. Infection of the short roots encourages their growth and branching, as described by Hatch and Doak (1933), Hatch (1937) and others. This has been re-emphasized by the direct observations of Piché *et al.* who observed that the dichotomy of the short roots of *P. strobus* in growth pouches began between the 8th and 12th days after inoculation, about seven days after colonization of their surfaces had occurred.

Chilvers and Gust (1982a,b) investigated the ways in which infection was initiated and spread in the whole root system of the seedlings of *Eucalyptus*. They showed that there were five constituent processes: (1) a pre-infection colonization of the root surface by fungal hyphae; (2) primary infection from primary inocula of spores or propagules in the soil; (3) perpetuation of infection by fungal growth following up the growth of the root; (4) proliferation of subsidiary branches of infected axes which do not escape the sheath; (5) secondary infections by hyphal strands or rhizomorphs to bring about infections of neighbouring roots.

These processes together bring about the colonization of the root system and the formation and spread of mycorrhizas within it. Simple mycorrhizal apices were formed by primary infections. Clusters of conjoined mycorrhizas form from proliferations of primary infections, and colonies of mycorrhizal clusters are formed by hyphal spread from the clusters. Chilvers and Gust found that the superficial spread of the fungus was a gradual process and that the spread of hyphae and rhizomorphs to form colonies of mycorrhizas depended in efficacy on the particular soil type. It was most effective, in their experiments, in the moist organic conditions of peat.

## Factors affecting the intensity of infection

Ectomycorrhizal trees are abundantly infected in their natural habitats and it is doubtful whether even in the richest garden soil any is virtually uninfected. Elias Melin in his earliest researches showed that the roots of conifers were entirely mycorrhizal in raw humus soils, but less so in mull soils. Hatch and Doak (1933) agreed with this, as did Björkman (1942), and Harley (1940) also showed it to be true for the broad-leaf *Fagus*. Of course there is a tendency to exaggerate the difference because in raw humus soils the fine roots are accumulated in the subsurface regions where the uninfected or unsheathed apices may be subject to damage by drought and cold to which they are more susceptible than fully ensheathed mycorrhizas. By contrast the more uniform distribution of roots in mull soils and the more equable water (see Worley and Hacskaylo, 1959) and temperature regimes in the depths of the mineral soil encourage persistence of the apices of long roots. Nevertheless the results of experiments with seedlings have agreed in showing that in soils, or on sites, moderately deficient in mineral nutrients infection is more complete and intense than on those rich in nutrients. Hatch (1937), using *Pinus strobus*, concluded that the internal nutrient status of seedlings was the prime factor in determining the intensity of infection; a moderate deficiency of nitrogen, potassium or phosphorus promoted mycorrhiza formation, but a great deficiency gave stunted growth and was detrimental to infection. Björkman also found that "A severe lack of available nitrogen or phosphorus hampers the formation of mycorrhizas as well as growth, but

moderate scarcity of one or other of the nutrients is a condition for infection." It
should be noted that this result relates to pine and spruce, and that a soil giving
them "a moderate deficiency" might give a "severe deficiency" to many
nutrient-demanding broad-leaved trees (refer to Baumeister, 1958, and Rennie,
1955). This may be the explanation for some of Meyer's observations (1962,
1973) that mycorrhizal development in *Fagus* was greater in mull than in mor
soils both in the field and in experimental conditions. However, the matter is
complicated, as shown by Table 28. The actual number of mycorrhizal tips per
unit volume of mor soil greatly exceeds that in mull soil although the intensity of
infection was less. Indeed, in mor soil the tree puts a greater proportion of
its growth effort into its root system and mycorrhizal equipment.

TABLE 28

Abundance of root tips and of mycorrhizas of *Fagus sylvatica* in the upper layers of mull and
mor soils (figures of Meyer, 1973)

|  | Number of root tips per 100 ml soil | Percentage of mycorrhizal tips | Total mycorrhizas per 100 ml soil |
|---|---|---|---|
| Mull soil | 500 | 88 | 400 |
| Mor soil | 45 600 | 51 | 23 256 |

Inverse correlations have been observed between the availability of nitrogenous substances in the soil and the frequency of mycorrhizas in the root system
(Richards and Wilson, 1963; Richards, 1961, 1965, etc.). Sometimes the
experiments have been done with nitrate as a source of nitrogen and sometimes
with ammonium. In the latter case some conversion of ammonium to nitrate
would be expected except on very acid soils. The results have been attributed to
the influence of increased nitrogen supply on infection and correlations of
increased internal nitrogen with decreased infection, as measured by number of
seedlings in a treatment infected or percentage of roots converted to mycorrhizas, have been published. However, Richards (1965) demonstrated that in
both his and in Björkman's experiments a better correlation was obtained
between infection and the ratio of soluble reducing carbohydrates to total
nitrogen.

This result is a modification of the conclusion of Björkman (1942) that the
intensity of infection was correlated with the presence in the root tissue of readily
soluble reducing sugars. This hypothesis had been criticized by Handley and
Sanders (1962), by Meyer (1962) and by Schweers and Meyer (1970) on the
grounds that analysis of mycorrhizal roots could not be taken as evidence that
their higher concentration of readily soluble reducing sugars was an important

factor in, or a cause of, increased mycorrhizal infection, because part of the carbohydrates came from the fungus. Harley (1969a) and Harley and Lewis (1969) also pointed out that the methods of carbohydrate analysis used in this work were subject to considerable errors due to reducing impurities, but that the important soluble carbohydrates of the fungal tissues, trehalose and mannitol, were unlikely to have been estimated at all by the methods used by Björkman.

The dependence of infection on internal carbohydrate is further suggested by the fact that primary infection of seedlings follows the maturation of the first leaves (Huberman, 1940; Harley, 1948; Warren Wilson, 1951; Robertson, 1954; Boullard, 1960, 1961; Laiho and Mikola, 1964; Warren Wilson and Harley, in preparation) and that increased mycorrhizal infection occurs in high light intensities (Harley and Waid, 1955). This correlation is also in line with the fact that the fungus depends on its host for carbohydrate compounds. Björkman (1970) repeated his experiments, using improved methods of analysis, on *Pinus sylvestris* mycorrhizal with two species of *Boletus* and obtained similar results to his previous ones. In 1977 Marx, Hatch and Mendicino published an experimental reassessment of the problem, using *Pinus taeda* and *Pisolithus tinctorius*. The pines were grown for two weeks uninfected in a series of nutrient regimes. Samples of them were then analysed and the remainder inoculated with the fungus and left for 19–21 days, when differences of intensity of infection could be recognized. The internal nutrient status was estimated by analysis of the needles of the seedlings, and the initial carbohydrate of the short roots before inoculation was determined after purification of the extracts on Dowex columns. Variation in the sucrose content of the short roots accounted for 85% of the variation in the intensity of mycorrhiza formation, although there was also a negative correlation between levels of nitrogen and phosphorus in the rootlets and their susceptibility to *Pisolithus*. Björkman's hypothesis was therefore "supported in principle" and it may be concluded, as suggested by Harley (1969a), that Björkman's analytical methods may well have led to an hydrolysis of sucrose during extraction or preparation, so giving him a correlation with reducing substances. These results are important not only in themselves as being relevant to an understanding of mycorrhizal physiology, but also in their importance in practical forestry. They provide guidance in the matter of producing nursery stock suitable for outplanting.

One further point of importance to both these considerations is a recent discovery by Bigg (1981) that the nitrate ion itself, apart from acting as a source of nitrogen, may inhibit mycorrhiza formation. Bigg used a strain of *Paxillus involutus* which are capable of nitrate reduction and one of *Lactarius rufus* unable to use nitrate. In the presence of nitrate no mycorrhiza was formed by spruce in culture with the *Lactarius*, and with the *Paxillus* only very poor mycorrhiza which had defective sheaths. The addition of nitrate to ammonium cultures also had a disruptive effect on the formation of mycorrhizas. The effect was not due to salt addition for other ions were ineffective, but it was a nitrate ion effect. Clearly correlations of diminished intensity of mycorrhizas with nitrification in soil require re-examination as also do the use of nitrate in the production of plants for forestry and the testing of fungi for mycorrhiza formation in media containing nitrate. In a similar way, Giltrap and Lewis (1981) have shown that concentrations of phosphate of the order of 50 mM in culture affected the growth of *Cenococcum geophilum* and *Rhizopogon roseolus*. Although they tentatively suggest that this effect may be connected with the inhibition of mycorrhiza formation by high phosphate concentrations in soil, more information is required before the suggestion can be accepted. Even soils which are rich in phosphate have concentrations which are at least two orders of magnitude lower than these. Nevertheless the work of Bigg and of Giltrap and Lewis emphasizes that nutrients may affect infection in ways other than in nutrition.

## Chapter 7

# Growth Physiology of Ectomycorrhizal Fungi

## Introduction

Unlike the Endogonaceae of vesicular-arbuscular mycorrhizas, many of the fungi of ectomycorrhiza have been isolated into culture and the physiology of their growth has been studied. Their taxonomic diversity poses in acute form the question of what common property enables them to produce composite organs of similar morphology, histology and function. The ease with which many of them have been grown in culture and the fact that some of them can spread in the rhizosphere of plants that are not potentially mycorrhizal partners show that they are not obligately mycorrhizal in a fundamental sense, although they may be ecologically so. They have been studied a great deal in culture but it must be admitted that the results have been disappointing. Much has been learned about factors which affected the growth of their mycelia but little about their intermediary metabolism, biochemistry or genetical variation. No positive characteristic that might explain their mycorrhizal habit has yet been discovered.

The methods used to investigate their growth have been the usual cultural techniques employing solid and liquid culture media, in which the initial concentrations of nutrients are usually excessively high, and there are changes in concentration, osmolarity, pH, and gaseous condition of atmosphere and solution during growth. The changes on solid media are not uniform owing to the restriction of convection, but vary in degree from the centre of the culture outward (see Harley, 1934). Usually growth, as measured by dry weight increment or lateral spread or rarely by nitrogen content of the mycelium (Norkrans, 1950), has been used as an index of the suitability of a medium or substrate, or of the need for trace elements or vitamins. Few experiments have considered intermediary metabolism, respiration, secretion of metabolites and similar activities except in the sphere of the production of auxin-like substances. Within the limitations of the cultural methods used, information on growth requirements relevant to the ecological activities of mycorrhizal fungi in their associated and free states has been obtained, but little else.

## Carbon sources

Frank assumed from the first that the source of organic nutrient for mycorrhizal fungi was from the autobiont, and much later Melin (1925) began to examine experimentally their requirements for carbon. He found that most strains had little ability to grow on complex carbon polymers such as might be found in litter and humus, and that they could not use lignin or cellulose. They therefore seemed dependent for growth on simple sugars such as might be produced by or released from the roots with which they were associated. Rommell (1938, 1939a) showed that a number of ectomycorrhizal fungal species depended upon being associated with living roots to produce fruit bodies—a reasonable expectation if they depended on the roots for carbon supplies. Recent work has confirmed the view that the majority of the ectomycorrhizal fungi require simple carbohydrates as carbon sources, although there is much variation. Hacskaylo (1973) reviewed this work (e.g. Mikola, 1948; Jayko *et al.*, 1962; Rawald, 1962; Lyr, 1963; Palmer and Hacskaylo, 1970), in which many of the experimental conclusions are open to criticism for two important reasons. First, the inability of a fungus to grow on a particular compound as sole carbon source does not show that it is not absorbed (or cannot be absorbed in suitable conditions) and used in some aspect of metabolism. It only shows that the compound cannot act as sole source of carbon. Secondly, since enzymes for the breakdown, absorption and use of a compound may be adaptive, the conditions must be such as to allow enzyme synthesis. In normal circumstances of culture when this is so, a second source of carbon is usually necessary in addition to available nitrogen and other nutrients during the period of adaptation. For this purpose Norkrans (1950) and since then others have provided a small amount of "start" glucose in the medium when testing complex carbon compounds as substrates for growth. Table 29 gives an example from her work on *Tricholoma* growing on starch and inulin.

The Table indicates that some species need to adapt, although they vary in this regard. For instance, *T. brevipes* adapted to neither starch nor inulin breakdown, *T. vaccinum* adapted to both and *T. flavobrunneum*, *T. imbricatum* and *T. pessundatum* adapted only to starch breakdown in different degrees. Norkrans further showed that her strain of *T. fumosum* utilized lignin and cellulose to a significant extent. Cellulolytic activity has also been recorded by Ritter (1964) and by Lyr (1963) to occur in some mycorrhizal fungi but usually at a very much lower level than in litter and wood-decaying fungi. The strain of *Boletus subtomentosus* that Lindeberg (1948) found could decompose cellulose and lignin

## TABLE 29

Relative growth of *Tricholoma* species upon starch and inulin (compared with growth on glucose) with and without a trace (tr) of glucose ("start glucose"). *T. brevipes* and *T. nudum* non-mycorrhizal, the rest mycorrhizal (data of Norkrans, 1950)

| | *T. brevipes* | *T. nudum* | *T. fumosum* | *T. flavobrunneum* | *T. imbricatum* | *T. pessundatum* | *T. vaccinium* |
|---|---|---|---|---|---|---|---|
| Starch | 18 | 136 | 93 | 11 | 9 | 72 | 11 |
| Starch and glucose (tr) | 54 | 133 | 83 | 93 | 45 | 117 | 50 |
| Inulin | 11 | 92 | 105 | 3 | 2 | 1 | 2 |
| Inulin and glucose (tr) | 44 | 101 | 107 | 49 | 25 | 13 | 27 |
| "Start" glucose control | 49 | 47 | 28 | 64 | 22 | 12 | 12 |

and grow on sterilized litter was shown by Björkman (Lindeberg, personal communication) not to be a mycorrhiza-forming strain. Strain variation within *B. subtomentosus* was well illustrated by Lundeberg (1970) who showed that his strain C30 (non-mycorrhizal), which grew rapidly, produced cellulase and pectinase, whereas strain C118 (mycorrhizal), which grew much more slowly, produced none (see Table 30). This may be, however, an aggregate species (Watling, 1970).

TABLE 30

Growth, enzyme activity and mycorrhiza formation on *Pinus* spp. of strains of *Boletus subtomentosus* on 2·5% malt agar (from Lundeberg, 1970)

| Isolate | Growth rate mm day$^{-1}$ | Mycorrhiza formation | | | Enzyme activity | |
|---------|---------------------------|----------|--------------|---------------|---------|-----------|
| | | *P. mugo* | *P. sylvestris* | *P. virginiana* | Laccase | Cellulase |
| C29 | 5 | — | — | — | + + | — |
| C30 | 16 | — | — | — | ( + ) | + + |
| C31 | 1·0 | — | — | — | — | — |
| C36 | 0·5 | — | — | — | — | ? |
| C118 | 1·5 | + | + | + | — | |

Ferry and Das (1968) studied the carbon nutrition of four species of *Boletus* (*Suillus*), (*B. elegans*, *B. bovinus*, *B. luteus* and *B. variegatus*), all known mycorrhiza-forming species. In each case they recorded the growth of 3–5 isolates of each species upon fifteen soluble carbon sources giving mean values with standard errors of the performance. Some of the results are given in Table 31. Threefold differences in growth rate between strains of the same species were commonly observed, e.g. *Boletus bovinus* on sucrose, *B. variegatus* on maltose, or *B. elegans* on l-arabinose. They stated: "slight responses (over control with no added carbon compound) to pectic compounds and to starch were observed in some species", whereas Hacskaylo (1973), quoting Palmer and Hacskaylo (1970), stated (after studying strains of *Amanita rubescens, Cenococcum graniforme, Rhizopogon roseolus, Russula emetica, Suillus cothurnatus* and *Suillus punctipes*): "of the soluble and insoluble polysaccharide and carbohydrate derivatives, pectin alone increased the growth of all fungi." Similarly Lamb (1974), using a group of 21 fungi, some named (including *Suillus grevillii* and *S. luteus*) and some unidentified isolates from mycorrhizas of conifers, found that pectin was a suitable carbon source for all of them. By contrast, Lindeberg and Lindeberg (1977) studied pectinolytic abilities of strains of the mycorrhizal fungi *Boletus edulis, B. (Suillus) grevillii, B. (Suillus) luteus, B. (Suillus) variegatus* and *B. subtomentosus* and compared them with those of four saprophytes (species of *Marasmius* and a strain of *Boletus subtomentosus* which is not mycorrhizal). None of the mycorrhizal species (including, be it noted, their strain of *S. grevillii*) grew better on 0·5% pectin + 0·1% glucose than on 0·1% glucose alone, and they gave the same economic coefficient referred to glucose (30–40%) and occasioned no change in viscosity of the solution. The saprophytes, including the non-mycorrhizal strain of *Boletus subtomentosus*, all grew better on pectin with increased economic coefficient and decreased the viscosity of the medium. Galacturonic acid which was utilized by the saprophytes was not used by mycorrhizal fungi. As can be seen, the results of Lamb appear to be in direct contrast with those of Lindeberg and Lindeberg but different strains of *Suillus grevillii* and *Suillus luteus* were used.

There are clearly two possible sources of the variability in behaviour: first a genetic variation for which there is much important and convincing evidence but as yet no real investigation; secondly a variability arising from different cultural treatment, including impurities in the constituents and differences of constitution. Recently Giltrap (1979) and Giltrap and Lewis (1981) have shown that phosphate at concentrations present in buffers (e.g. 40–50 mM) might significantly affect the growth of some mycorrhizal fungi on various carbon sources. Using *Cenococcum geophilum*, *Rhizopogon roseolus* and *Suillus bovinus* they showed that growth on all carbon sources and adaptive growth on disaccharides

TABLE 31

Strain variation in the ability of ectomycorrhizal fungal species of *Boletus* (*Suillus*) to grow on various soluble carbohydrates

*mg per culture in 8 days (15 days for* B. elegans*) (results of Ferry and Das, 1968)*

|  | Sucrose | Xylose | Arabinose | Maltose | Cellobiose |
|---|---|---|---|---|---|
| *Boletus elegans* | | | | | |
| 102 | 11·4 ± 1·4 | 4·0 ± 0·4 | 17·7 ± 0·4 | 15·2 ± 2·4 | 15·8 ± 0·9 |
| 109 | 10·8 ± 0·5 | 7·2 ± 0·5 | 8·7 ± 0·5 | 5·9 ± 0·7 | 9·9 ± 0·3 |
| 128 | 10·0 ± 1·0 | 12·3 ± 0·2 | 3·4 ± 0·5 | 7·0 ± 0·5 | 13·3 ± 1·5 |
| *Boletus bovinus* | | | | | |
| 110 | 6·1 ± 0·2 | 5·4 ± 0·2 | 2·6 ± 0·2 | 3·1 ± 0·2 | 4·5 ± 0·5 |
| 111 | 7·9 ± 0·7 | 3·4 ± 0·3 | 3·1 ± 0·1 | 6·4 ± 0·8 | 6·7 ± 0·5 |
| 121 | 13·3 ± 0·9 | 3·3 ± 0·5 | 3·3 ± 0·3 | 5·2 ± 0·5 | 8·4 ± 1·0 |
| 123 | 6·2 ± 0·8 | 3·1 ± 0·3 | 2·6 ± 0·2 | 3·1 ± 0·2 | 6·4 ± 1·0 |
| 130 | 4·3 ± 0·6 | 4·6 ± 0·1 | 2·1 ± 0·4 | 7·2 ± 0·9 | 10·2 ± 0·9 |
| *Boletus luteus* | | | | | |
| 116 | 7·9 ± 0·8 | 2·4 ± 0·4 | 2·1 ± 0·1 | 11·8 ± 0·1 | 9·3 ± 1·5 |
| 122 | 10·4 ± 0·6 | 3·4 ± 0·2 | 5·6 ± 0·1 | 10·9 ± 1·5 | 14·4 ± 2·3 |
| 131 | 15·2 ± 0·5 | 2·9 ± 0·3 | 2·7 ± 0·5 | 14·2 ± 1·1 | 14·9 ± 0·9 |
| *Boletus variegatus* | | | | | |
| 100 | 5·0 ± 0·2 | 3·1 ± 0·3 | 3·6 ± 0·5 | 4·1 ± 0·4 | 10·6 ± 0·8 |
| 112 | 7·4 ± 0·8 | 2·9 ± 0·2 | 2·2 ± 0·1 | 5·2 ± 0·3 | 9·7 ± 0·9 |
| 113 | 7·7 ± 0·3 | 2·9 ± 0·6 | 2·8 ± 0·3 | 4·4 ± 0·4 | 5·6 ± 0·1 |
| 127 | 7·4 ± 1·1 | 1·8 ± 0·2 | 2·6 ± 0·2 | 15·3 ± 1·2 | 7·7 ± 0·1 |

might be diminished by the presence of 50 mM phosphate buffer as compared with MES buffer (2-[*N*-morpholino]ethane sulphonic acid) and that the three species differed in their susceptibility. Others have noted somewhat similar effects of high phosphate concentration; Marx and Zak (1965) with *Laccaria laccata, Suillus luteus, Cenococcum geophilum* and Jayko *et al.* (1962) with several species of *Lactarius*. Many of those working on the utilization of carbon compounds by mycorrhizal fungi have not used strong phosphate buffers in their media (e.g. Mikola, 1948; Ferry and Das, 1968; Palmer and Hacskaylo, 1970; Laiho, 1970) and the phosphate concentration in them has been of the order of 5 mM or less, as is usual in fungal media. It is possible that a source of variability in their experiments may have been pH change, especially when $NH_4Cl$ was the nitrogen source and there may have been a rapid decrease of pH below that of maximum buffer capacity of the potassium phosphate added.

Giltrap and Lewis follow up their observations of the inhibitory effects of high concentrations of phosphate on growth with a discussion of the relevance of these effects to those of high soil phosphate upon the susceptibility of the autobiont to mycorrhizal formation or the virulence of the fungus in mycorrhiza infection. Their view is that high phosphate supply may limit growth by the preemption of ATP in polyphosphate synthesis. This may be extremely important in tests of carbohydrate utilization where buffers are used, but it does not seem to explain the variability of all results, for neither Mikola (1948) nor Norkrans (1950) used buffer except for experiments on pH in which carbon uptake was not studied. Lamb (1974) certainly did, and surprisingly the growth that he recorded in phosphate buffer (strength unstated) did not seem to have been inhibited. On the other hand it is not very likely that the effects of phosphate concentrations even of the order of 1 mM are relevant to mycorrhiza formation in the field where soil phosphate in solution is of the order of 0·001–0·1 mM.

We may conclude that most of the ectomycorrhizal fungi have very limited ability to use lignin and cellulose. They vary within and between species in the ease with which starch, glycogen and inulin, the simpler oligosaccharides and the disaccharides sucrose and trehalose are used. The monosaccharides glucose, mannose and fructose are usually good sources of carbon. Pectic substances can be used for growth by some but not by others.

Interest has centred on the ability of the fungi to hydrolyse pectic substances and hemicelluloses because of the need to explain the penetration by their hyphae into the middle lamellae of the epidermis and cortex of the autobiont to form the Hartig net. There is, as Lindeberg and Lindeberg (1977) imply, a world of difference between the production of the appropriate enzymes for a restricted and local hydrolysis or softening of the cell-wall or middle lamella and the enzymic dissolution of cellulose, pectin and other wall constituents sufficiently rapidly and in sufficient quantity to support growth using the products as a sole source of carbon, and an investigation of the location of enzymes on the hyphae by histo-chemical methods is desirable. The strain variation within species in respect of the carbon compounds that can be used as sole sources for growth is wide, but in all cases studied proved mycorrhiza-formers have negligible or very limited cellulolytic activity and there is no consistency except that simple mono- and oligo-saccharides can usually be used.

## Nitrogen sources

The ectomycorrhizal fungi are similar to other fungi in the kinds of nitrogen compounds which they can use for growth. Nothing has been discovered which differentiates them as a group. Most of them grow fastest on ammonium compounds amongst the inorganic sources, and some can use nitrate but others not. The experimental work on nitrogen absorption is beset with the common difficulties concerning pH change. The usual media are somewhat buffered by their phosphate content; but since their initial pH is usually a little below neutrality, they are less well buffered against changes to greater acidity than towards alkalinity. Absorption of ammonium results in a change to acidity which is especially marked if absorption is rapid and it may be followed by a sharp cessation of growth as the pH falls to low values. On the other hand, the absorption of nitrate, as well as the release of ammonium from amides or other readily hydrolysed compounds, causes a change to a higher pH which is usually slower and may have a less marked effect on growth. Changes in pH have also an effect on the solubility of constituents of the medium, on the solubility of $CO_2$, and on the activity of surface-attached or exocellular enzymes. Some workers have used buffered media but these introduce their own problems.

In spite of these defects, common to much work on fungi in culture, there is not much doubt of the broad results, although there is much variation between different species and strains of mycorrhizal fungi in nitrogen requirements. Norkrans (1950) found that none of the species of *Tricholoma* that she used — except the non-mycorrhizal *T. nudum* — could grow on nitrate as a sole nitrogen source. Rawald (1963) by contrast found that six of the nine *Tricholoma* spp. that he tested grew well on nitrate. Lundeberg (1970) found that of the 32 species and strains of Basidiomycetes of many ecological kinds that he grew in culture, about two-thirds could not grow freely on nitrate, but five, *Boletus (Suillus) elegans (grevillii)*, *B. (S.) granulatus*, *B. (S.) luteus* and *Tricholoma pessundatum*, which are mycorrhizal, as well as a non-mycorrhizal strain of *B. subtomentosus*, C30, could use nitrate as efficiently as ammonium for growth. Most of the species, mycorrhizal as well as others, used ammonium more readily than any other nitrogen source that was tried, except *Boletus (Suillus) bovinus*, two strains of *B. edulis* (C18 and C9s) and *B. (S.) luteus* among the mycorrhizal fungi and also the non-mycorrhizal *B. subtomentosus* (strain C30), which grew better on asparagine and very well on glycine. Certain mycorrhizal fungi, *B. subtomentosus* strains

C118, and *B. edulis*, strains C18 and C93, grew only slowly on inorganic sources and more rapidly on asparagine and glycine, and some, e.g. the mycorrhizal *B. (S.) grevillii*, *Lactarius deliciosus* and *Tricholoma pessundatum* and the non-mycorrhizal *B. subtomentosus* (C30) and *Agaricus campestris* used the amide, acetamide, as well as or better than ammonium or asparagine. The strain of *B. (S.) luteus* used by Luppi and Fontana (1967) in contrast to that of Lundeberg, only used nitrate significantly after a period of adaptation to it, but ammonium sulphate was always a better nitrogen source.

Laiho (1970) published a study of one species, *Paxillus involutus*, in which he examined its cultural characteristics and ecology using many strains. Tables 32 and 33 give examples of the considerable variations between strains that were observed. There is however no case of an organic nitrogen source being very readily used by one strain and being totally useless to another. An interesting point arises in the fact that aspartic acid appears in this experiment to be a very poor nitrogen source for all the strains, whereas glutamic acid tested in another experiment was a source of nitrogen almost equal to or better than ammonium sulphate for most of them. Lundeberg (1970) records that his strain of *P. involutus* did not grow particularly well on asparagine; hence, surprisingly, aspartic acid and its amide (which might indeed have been expected to be hydrolysed externally by an amidase) do not readily seem to be used by this species.

TABLE 32

Strain variation in the use of nitrogen compounds by *Paxillus involutus* grown in medium with phosphate buffer, nitrogen concentration equivalent to $0.3$ g $l^{-1}$

*Growth mg dry weight per culture flask (after Laiho, 1970)*

| Compound | Strain | | | | | | | | | Mean |
|---|---|---|---|---|---|---|---|---|---|---|
| | 1 | 3 | 4 | 5 | 7 | 8 | 12 | 18 | 24 | |
| $KNO_3$ | 13 | 25 | 33 | 6 | 23 | 25 | 27 | 14 | 42 | 23.0 |
| $NH_4 NO_3$ | 44 | 26 | 84 | 13 | 26 | 15 | 18 | 8 | 17 | 27.7 |
| $(NH_4)_2SO_4$ | 46 | 35 | 108 | 12 | 10 | 20 | 29 | 62 | 61 | 43.5 |
| $NH_4$ tartrate | 54 | 47 | 145 | 25 | 18 | 23 | 77 | 46 | 51 | 54.0 |
| Aspartic acid | T | T | 8 | 3 | 2 | 6 | 3 | 1 | 4 | 2.9 |
| Glycine | 1 | 4 | 10 | 1 | 1 | 3 | 1 | 1 | 2 | 2.7 |
| Nucleic acid | 7 | 10 | 9 | 14 | 5 | 7 | 10 | 6 | 12 | 8.9 |
| Urea | 23 | 15 | 5 | 10 | 29 | 11 | 5 | 10 | 1 | 12.1 |
| Casein hydrolysate | 73 | 22 | 108 | 76 | 28 | 80 | 78 | 52 | 56 | 63.7 |
| Peptone | 47 | 45 | 95 | 48 | 73 | 69 | 51 | 45 | 78 | 61.3 |
| CONTROL | — | 1 | — | — | — | 1 | 1 | 1 | — | 0.6 |

T, traces of growth only.

—, no growth.

Lundeberg pointed out that asparagine seemed to be quite unusable, even perhaps inhibitory, to *Amanita citrina*, *A. muscaria*, strain C4 (but not strain C256), *Boletus (Suillus) aeruginescens* and two strains of *Boletus badius* (C8 and C251) and stated that Melin had experienced a similar effect of glutamic acid on *Boletus versipellis (Leccinum veripelle)* and *Lactarius rufus*. At the moment such effects have no explanation.

To test the ability of mycorrhizal fungi to use organic nitrogen compounds, Lundeberg prepared humus agar in which the inorganic nitrogen compounds were reduced in quantity and the organic nitrogen of the humus was labelled with $^{15}N$. He allowed fungi to grow from an inoculum which straddled the interface between nitrogen-free glucose agar and the humus agar. The subsequent analyses showed that none of the mycorrhizal fungi had absorbed organic nitrogen from the humus in significant quantities, although five other fungi which produced some or all of the hydrolytic enzymes, cellulase, pectinase, proteinase and laccase, were able to do so. If this be a true assessment of the ability of mycorrhizal fungi (and it agrees with the results of Mosca and Fontana, 1975, on the use of protein nitrogen by *Boletus luteus*) it must be concluded that they must compete with other organisms for such nitrogen as is released by mineralization or present as simple amino acids in the soil.

A nitrate reduction system has been demonstrated in a number of mycorrhizal basidiomycetes by Ho *et al.* (1977) and Ho and Trappe (1980) but, as mentioned

TABLE 33

Strain variation in the use of amino acids by *Paxillus involutus* grown in medium with nitrogen concentration equivalent to $0.15\,g\,l^{-1}$

*Growth mg dry weight per culture flask (after Laiho, 1970)*

| Compound | Strain | | | | | | | | | Mean |
|---|---|---|---|---|---|---|---|---|---|---|
| | 1 | 3 | 4 | 5 | 7 | 8 | 12 | 18 | 24 | |
| Leucine | 2 | 4 | 6 | 3 | 3 | 4 | 2 | 2 | 4 | 3·4 |
| Tyrosine | T | 0 | 3 | 2 | 2 | 2 | 1 | 1 | 2 | 1·5 |
| Glutamic acid | 59 | 72 | 61 | 31 | 61 | 27 | 53 | 40 | 112 | 57·4 |
| Proline | 1 | 2 | 1 | 2 | 1 | T | 1 | 2 | 2 | 1·2 |
| Valine | 6 | 12 | 5 | 3 | 3 | 2 | 2 | 2 | 6 | 4·4 |
| Arginine | 44 | 68 | 79 | 16 | 63 | 9 | 16 | 31 | 79 | 44·8 |
| Histidine | 2 | 2 | 4 | 2 | 2 | 3 | 2 | 2 | 2 | 2·5 |
| Lysine | 3 | 4 | 2 | 2 | 1 | T | 2 | 2 | 1 | 1·9 |
| All amino acids | 41 | 11 | 97 | 47 | 58 | 33 | 35 | 25 | 87 | 48·2 |
| $(NH_4)_2SO_4$ | 50 | 20 | 72 | 24 | 34 | 25 | 23 | 48 | 57 | 39·0 |
| CONTROL | T | 1 | 2 | 2 | 2 | T | 1 | 1 | 1 | 1·4 |

T, traces of growth only.

above, it is not likely to be present in all of them. Bigg (1981), working in Alexander's laboratory, made a special study of the change in economic coefficient of a strain of *Paxillus involutus* when it grew in culture on either nitrate or ammonium as sources of nitrogen. He pointed out that the requirement for reducing power to produce ammonium from nitrate should result in an increased consumption of carbohydrate per mol of nitrogen assimilated. His analyses, although based upon the growth of the fungus in culture till the carbohydrate supplied was exhausted, gave reduction in economic coefficient in general accordance with theoretical expectation.

None of the work on nitrogen requirement and uptake by ectomycorrhizal fungi indicates any particular way in which they may differ as a group from other fungi. The experimental results neither help consideration of the problems of the absorption and metabolism of nitrogen compounds by ectotrophs nor show the mycorrhizal fungi to be an unique group, but it must be stressed once again that investigators have only examined the effects of nitrogen compounds upon vegetative growth. The mode of absorption and the metabolism of nitrogen compounds have not been studied. It cannot be assumed that growth rate in culture is a good indicator of the physiology of absorption and metabolism of the fungus in the mycorrhiza where the fungal sheath and Hartig net are not rapidly growing organs. It would seem that researches upon uptake of nitrogenous substances by and their metabolism within resting adult mycelium and their excretion in different conditions were essential to progress.

## Phosphate supply

A supply of phosphate is essential to all mycorrhizal fungi, as to other organisms. It is normally present in artificial media as inorganic potassium phosphate but since in the normal habitat of the fungi, the humus layers of the soil, phosphate may be present also as organic phosphate, commonly to the extent of 40–50% of the total but sometimes up to 80%, experimental observations have been made on their ability to make use of organic compounds such as inositol phosphate. Theodorou (1968, 1971) demonstrated that mycorrhizal fungi such as strains of *Rhizopogon luteolus*, *Boletus (Suillus) luteus* and *Cenococcum graniforme (geophilum)* were able to hydrolyse phytates (inositol hexaphosphate and lower phosphate esters of myo-inositol) and to use the soluble phosphate but not the carbon residue for growth. Similarly Ho and Zak (1979) showed that strains of six mycorrhizal fungi, *Hebeloma crustuliniforme*, *Laccaria laccata*, *Amania muscaria*, *Thelephora terrestris*, *Piloderma bicolor (Corticium croceum)*, *Rhizopogon vinicolor*, hydrolysed *p*-nitrophenyl phosphate by surface phytases, although the last two were less active than the others under the conditions used. As will be seen, these findings may be related to the demonstration of phosphatases on the surfaces of ectomycorrhizas (Woolhouse, 1969; Bartlett and Lewis, 1973; Williamson and

Alexander, 1975; Alexander and Hardy, 1981) and also in vesicular-arbuscular mycorrhiza.

The effects of concentrations of phosphate on phosphatase activity have been studied by Calleja *et al.* (1980) using strains of the three species, *Hebeloma edurum*, *Suillus (Boletus) granulatus* and *Pisolithus tinctorius*, which had been confirmed as mycorrhiza forming. In a medium containing both ammonium and nitrate as sources of nitrogen, and phosphate as inorganic orthophosphate, at zero, 2mM, 5 mM or 7·5 mM, at pH 5·7, inoculated with 8 mm discs of agar cultures containing 10 $\mu$M $KH_2PO_4$ as a mineral supply, they estimated, using *p*-nitrophenyl phosphate hydrolysis, total externally accessible phosphatase activity of entire mycelia, and excreted phosphatase. Homogenization of the mycelium gave a further estimate of total phosphatase, which could be divided by differential centrifugation into that on the hyphal walls, that in the membranes and that soluble. The results were expressed on a mycelial or protein basis.

TABLE 34

Acid phosphatase activity of different fractions of the homogenate of *Pisolithus tinctorius*

*nmol of* p-*nitrophenyl phosphate hydrolysed per culture per minute, inocula from agar cultures containing 10* $\mu$M $KH_2PO_4$ *(Calleja* et al., *1980)*

| mM phosphate in medium | 0 | 2 | 7·5 |
|---|---|---|---|
| Total homogenate | 3583 | 1592 | 502 |
| Walls | 2100 | 480 | 115 |
| Eluted from walls | 86 | 14 | 35 |
| Cytomembranes | 101 | 107 | 34 |
| Soluble | 590 | 421 | 134 |

Tables 34 and 35 give examples of the results with *P. tinctorius* which indicate that the total accessible phosphatase activity of the mycelium is greatly increased in conditions of phosphate deficiency, and that this is mainly the resultant of the increase in the activity of the wall-attached phosphatase.

TABLE 35

Acid phosphatase activity in *Pisolithus tinctorius* cultures containing different concentrations of $KH_2PO_4$

*nmol* p-*nitrophenyl phosphate hydrolysed per mg protein per minute. Inocula from agar cultures containing 10* $\mu$M $KH_2PO_4$ *(Calleha* et al., *1980)*

| mM phosphate in medium | 0 | 2 | 7·5 |
|---|---|---|---|
| Phosphatase in medium | 209·0 | 14·0 | 22·8 |
| Accessible phosphatase | 386·8 | 9·6 | 9·9 |
| Total phosphatase | 1678·7 | 114·8 | 63·8 |

In view of these results there is an urgent requirement that the ability of ectomycorrhizal fungi to use insoluble organic and inorganic sources of phosphate for growth should be investigated together with the mechanism by which they are absorbed and metabolized. Indeed the whole question of the phosphate nutrition of these fungi is made more interesting by the effects of inorganic phosphate both upon phosphatase activity and also upon growth rate at the levels used by Giltrap and Lewis.

## Other inorganic nutrients

Nothing has been observed about the requirement of ectomycorrhizal fungi for potassium that differentiates them from other fungi in this regard. It is clearly an essential element and the counter-ion in the absorption and release of anions. Since it appears to be readily absorbed and released from some ectomycorrhizas (Harley and Wilson, 1959; Edmonds et al., 1976) further investigation of potassium in the physiology of the fungi would be of value.

Calcium is required in traces by almost all fungi and it appears that Ascomycetes and Basidiomycetes may require greater quantities than others (Treschow, 1944; Lindeberg, 1944) but far below those required by green plants. The recent analyses of the polyphosphate granules in mycorrhizal hyphae by electron probe analysis show them to contain calcium. This has been demonstrated in the granules of vesicular-arbuscular mycorrhiza and ectomycorrhiza (e.g. Strullu et al., 1981; Strullu et al., 1982) and an estimate of the importance of calcium in fungi forming these granules is therefore desirable.

The ecological experiments of Hora (1959) and Garbaye et al. (1979) recorded the effects on fruit body production of application of mineral nutrients to woodland soil. Amongst the fungi observed in both these researches were a number of ectomycorrhizal fungi. The kinds of application made were of lime, phosphate, ammonium, nitrate and complete fertilizer to a variety of soil types. The fruiting of some mycorrhizal fungi was stimulated by some of the applications and diminished or unaffected by others. Clearly the results of this kind of experiment have complex explanations because the changes brought about by fertilizer treatments depend on direct effects on the fungi, on secondary effects via associated microorganisms, and on direct and indirect effects on the growth and nutrient status of the root-system of the host. They may however be relevant to problems concerning the restriction of fruiting of some mycorrhizal fungi and to apparent sporocarp specificity to certain hosts or habitats.

## Physical factors and the growth of mycorrhizal fungi

### Temperature

Ectomycorrhizal fungi are affected by temperature in much the same way as other fungi but they show much variation between the strains of a single species (see Table 36). For most of them the optimum temperatures for growth lie

TABLE 36

Strain variation in the growth of mycorrhizal fungi at different temperatures

*Growth in diameter (mm) of colonies in 26 days (18 days Rhizopogon) on Melin-Norkrans agar (data from Theodorou and Bowen, 1971)*

| Fungus | Temperature | | | LSD |
| | 16°C | 20°C | 25°C | |
|---|---|---|---|---|
| *Rhizopogon luteolus* | | | | |
| strain 5 | 36·8 | 87·6 | 88·6 | 11·3 (p. 05) |
| 3 | 30·2 | 70·7 | 73·4 | 15·3 (p. 01) |
| 2 | 31·9 | 69·9 | 82·8 | 20·3 (p. 001) |
| *Suillus granulatus* | | | | |
| strain 8 | 7·0 | 41·0 | 44·8 | 8·2 (p. 05) |
| 5 | 63·2 | 66·7 | 57·5 | 11·2 (p. 01) |
| 2 | 17·1 | 22·6 | 46·0 | 15·0 (p. 001) |
| *Suillus luteus* | | | | |
| strain 2 | 23·8 | 49·4 | 48·5 | 11·7 (p. 01) |
| 3 | 37·7 | 60·7 | 61·3 | 15·6 (p. 001) |
| 1 | 8·9 | 70·9 | 63·8 | |

between 8 and 27°C, but the form of the curve of growth rate against temperature may show either a sharp or an indistinct optimum. A very interesting example is that of *Paxillus involutus* of which Laiho (1970) investigated a number of strains. All of his nine strains grew within the range 5 to 32°C with optima of very variable sharpness between 15 and 25°C and very different rates of growth at the optimum (see Table 37). Two strains of *P. involutus* were also included by Moser (1958) in his survey of the ability of higher fungi (including mycorrhiza-formers) to grow and survive at low temperatures. One strain from low ground in Austria had a minimum growth temperature of 2–8°C and failed to survive even five days at −11 to −12°C, whereas another strain from a high mountain area had a minimum growth temperature of −2 to +4°C and survived two months at −11 to −12°C. Ability to withstand low temperatures may, as Moser showed, be of importance both in the ecological activities of the fungi and of the mycorrhizas which they form and also in the practical problems of preservation and distribution of strains for inoculation of seedlings for forestry. France and Reid (1979) examined the ability of 18 species comprising 73 isolates of mycorrhizal fungi to withstand the low temperature $-10°C$ for 48 hours. Of these 71 survived although the time required for recovery after the period of low temperature as indicated by the beginning of active growth in culture varied very greatly. Of the species of which a number of different isolates were tested, all strains of *Cenococcum geophilum (graniforme)* and *Suillus (Boletus) granulatus* survived and resumed their growth between 3 and 39 days after being returned to laboratory temperature. Most of the strains of *Pisolithus tinctorius* recovered, but there was an even greater range of time before growth resumed. Only one of the two strains of *Suillus (Boletus) punctipes* slowly recovered growth after a prolonged period.

TABLE 37

Growth of 9 strains of *Paxillus involutus* in liquid culture at different temperatures. Relative weight produced in 28 days (results of Laiho, 1970)

| Strain | Temperature Limits (°C) | Optimum | | Relative rate of growth at optimum (approximate) |
|---|---|---|---|---|
| 1 | 5–30 | 22·5 | sharp | 75 |
| 3 | 5–32 | 25 | sharp | 97 |
| 4 | 5–32 | 25 | sharp | 120 |
| 5 | 5–30 | 15–22·5 | indefinite | 12 |
| 7 | 5–30 | 22·5 | sharp | 70 |
| 8 | 5–30 | 20 | definite but not sharp | 15 |
| 12 | 5–27·5 | 15 | definite but not sharp | 10 |
| 18 | 5–30 | 15–20 | plateau | 12 |
| 24 | 5–32 | 22·5 | sharp | 42 |

Moser has been using "low temperature" strains of *Suillus (Boletus) plorans* and other fungi for the inoculation of conifers, *Pinus cembra* in particular, in high level plantings almost at the tree limit in the Alps, and there is much interest in the possibility that "high temperature" fungi may be important as mycorrhizal fungi in tropical regions. This problem has been examined by Ivory (personal communication; Table 38) who is of the opinion that the temperature reactions of the fungi for tropical pines, e.g. *Pinus caribaea*, are not greatly different from those of temperate regions, although they do have higher thermal death points. In temperate regions of Europe and in much of North America, soil temperatures within the forest do not reach the levels of the optima for the growth of many mycorrhizal strains, except perhaps in the surface horizons. The fungi in temperate forest soils are usually therefore well below their optimum temperature, so that they might be expected to be efficient also in tropical and subtropical regions except at extreme temperatures.

It will be essential in the future to learn more of the effects of temperature upon relevant processes like absorption, excretion of products and translocation of nutrients which are vital to the functioning of the fungus in mycorrhizal association. Hacskaylo *et al.* (1965) showed that in only one case out of six, that of *Suillus (Boletus) punctipes*, did the optimum temperature for growth correspond with the optimum temperature for respiration. Sometimes the optimum temperatures for nutrient uptake by mycorrhizal organs may be below the optima of fungal growth in culture. For instance, Harley and Wilson (1959) and Edmonds *et al.* (1976) observed actual losses of potassium from *Fagus* mycorrhizas into KCl solutions at temperatures as low as 20 to 25 °C. Such scanty examples as are available emphasize the need to determine the effects of temperature on other processes than total growth rate.

TABLE 38

The effect of temperature on the radial growth of tropical ectomycorrhizal fungi on buffered yeast agar (results of M. H. Ivory, 1981, unpublished)

| Fungus | Origin | Cardinal temperatures °C | Growth rate at optimum temperature mm d$^{-1}$ | Mycelial thermal death point[a] °C |
|---|---|---|---|---|
| *Pisolithus tinctorius* | Nicaragua | 5 (33) 41 | 1·7 | 41 |
| | Brazil | 6 (32) 41 | 1·4 | 41 |
| *Fistulinella conica*[c] | Belize | 10 (28) 35[b] | 3·0 | ? |
| *Rhizopogon nigrescens* | Belize | 7 (28) 38 | 1·4 | 41 |
| | Bahamas | 5 (27) 39 | 2·1 | 41 |
| | P. Rico | 5 (27) 38 | 2·6 | 41 |
| *Scleroderma texense* | Belize | 10 (27) 32 | 0·3 | 34 |
| *Amanita muscaria* | Nicaragua | 8 (24) 31 | 0·4 | ? |
| *Suillus granulatus* (introduced from Europe?) | Zambia | 10 (26) 31 | 0·8 | 34 |
| *Paxillus involutus* | UK | 4 (20) 28 | 0·4 | 30 |

[a] No regrowth after 14 days at this temperature.
[b] Shows a marked tendency to sector into fast and slow-growing forms at 25–35 C.
[c] *Boletus* pp = *Fistulinella* P. Henn.
Figures in brackets under cardinal temperatures are optima.

## pH of the substrate

pH affects the growth of mycorrhizal fungi in culture and there is nothing very memorable about its effect on them. The conclusion is reached, whether they are grown on normal media (Melin, 1925; Theodorou and Bowen, 1969) or strongly buffered media (e.g. Mikola, 1948; Modess, 1941; Norkrans, 1950; Laiho, 1970) that they usually have pH optima for growth between 3·0 and 7·0 (i.e. on the acid side of neutrality), but may exhibit a range of tolerance which can be wide or narrow. In these respects they possess no characteristic properties as a group, and indeed there may be great strain variation within the various species. Laiho found that strains of *Paxillus involutus* might have optima between pH 3·1 and pH 6·4, with very great differences in the rate of growth at the optima and also great differences in the shape of the optimum curve. He pointed out that the range of variation in pH optimum of the strains of this one species was as wide as that for all the mycorrhizal species examined by all the workers before him.

Rarely are the pH changes in culture different from those predictable from the effects of the absorption of nitrogen compounds, i.e. increase in pH as nitrate is absorbed, decrease in pH as ammonium is absorbed. This suggests that it is unlikely that large quantities of organic acids are released into the substrate by most of the fungi under the cultural conditions used. The possible importance of a secretion of acids has been suggested (Johnston, 1959; Johnston and Miller, 1959). Coleman and Harley (1976) and Harley (1978) mentioned that linkage with acid secretion was one possible function of the cyanide-insensitive non-phosphorylating pathway in *Fagus* mycorrhizas which is alternative to the phosphorylating cytochrome pathway and has now been reported in other ectomycorrhizas (Antibus *et al.*, 1980) and may possibly be a common property of them, but few reports of acid secretion have been made. Preliminary experiments with *Fagus* mycorrhiza have not indicated rapid excretion of common acids related to the TCA cycle. Cromack and his colleagues (1979) have, however, demonstrated the deposition of large quantities of calcium oxalate as crystals in the hyphal nets of *Hysterangium crassum* (a putative mycorrhizal fungus of *Pseudotsuga menziesii*), in the surface layers of the forest soils. The possibility of organic acid production by mycorrhizal fungi needs examination because of their property of complexing with or precipitating metals such as calcium, iron or aluminium with a consequent increase of phosphate dissolution in soils (Johnston, 1959; but see S. E. Smith, 1980). Such an action could also provide some explanation of the tolerance of some tree species to alkaline soil when mycorrhizal, but intolerance in the uninfected condition (Clement *et al.*, 1977) which is discussed below.

## Water supply

Uhlig (1972) and Malabari (1979) have examined the effects of decreasing osmotic potential of the nutrient medium on the growth of mycorrhizal fungi. Using *Tricholoma saponaceum*, *T. pessundatum*, *Hebeloma crustuliniforme*, *Scleroderma aurantium*, *Amanita rubescens* and *A. citrina*, Uhlig showed that all except the last two grew fastest in a solution containing sucrose at 34–35 atmospheres[1] osmotic pressure and could make some growth at 93–116[1] atmospheres. In soil, the fungi could survive a level of 2–7% of water capacity and grow to some extent at 10–15% water capacity. Malabari, working with *Cenococcum geophilum (graniforme)* showed that both of the strains that he used increased in growth rate down to −10 bars (with polyethylene glycol). Below that level, growth of an isolate from *Betula* was diminished somewhat at −15 bars on some media, but not that of an isolate from *Helianthemum*. The ability to withstand low water potentials may be ecologically important, indeed Uhlig pointed out that the fungi he tested, which were mycorrhizal with spruce, survived water deficits below the wilting point of the spruce. *Cenococcum* is a fungus with a reputation for forming mycorrhizas with roots under dry conditions. These results agree with older observations (see Bowen, 1973) that mycorrhizal fungi vary much in their ability to withstand low water potential. He related this to the observations that ectomycorrhizal rootlets and seedlings have been observed to be more drought-resistant than their uninfected counterparts. Sands and Theodorou (1978) compared mycorrhizal and non-mycorrhizal plants with respect to transpiration rate, leaf water potential and resistance to water through-put in the plant–soil system. Similar transpiration rates were observed but the leaf water potential of the mycorrhizal plants was lower so that the calculated resistance to water flow was greater in them. They considered that this was due to a greater soil resistance due to a different root geometry in the mycorrhizal plants. Clearly much more experimentation of this kind is needed for the results seem to run contrary to what is widely believed.

It has been tacitly assumed in the past that the hyphae, hyphal strands and rhizomorphs of mycorrhizal fungi which emanate from mycorrhizas in soil absorb and translocate both water and nutrients to the host. Recently Duddridge *et al.* (1980) have examined the functioning of the rhizomorphs of *Suillus bovinus* in the transport of water to *Pinus sylvestris*. They showed that the rhizomorphs contained differentiated "vessel" hyphae between 20 and 60 $\mu$m in diameter in which the transverse walls had broken down. Experiments with tritiated water showed convincingly that it moved readily in the rhizomorphs into the mycorrhizal roots and so to the young needles. Since not all ectomycorrhizal fungi produce rhizomorphs the rates of movement estimated are not generally applicable. Nevertheless, as they point out, perhaps fungi which produce rhizomorphs might be best adapted as symbionts for hosts growing in dry habitats. It remains to be determined in what circumstances the intensive exploitation of the soil by separate hyphae may be more efficient in terms of water uptake and carbon expenditure on the upkeep of the fungus than extensive exploitation by rhizomorphs.

[1] 1 atmosphere $\approx 10^5$ Pa; 1 bar $= 10^5$ Pa

### Requirements for growth factors

There is general agreement that the requirement for vitamins and other growth factors by fungal species is not necessarily characteristic of any taxonomic or ecological group. Such requirements are found in all the fungal taxa, there is even strain variation within single species, and similar requirements may be exhibited by saprophytes, parasites and mutualistic symbionts. All ectomycorrhizal fungi are dependent on a supply of thiamine or one or both of its moieties—pyrimidine and thiazole, and this indicates that in their habitats a supply is likely to be available. The requirement is not solely theirs and there is some interspecific and intraspecific variation amongst them in the degree of dependence (Melin and Nyman, 1940, 1941; Melin, 1963). Studies of this variation have been made, for instance by Laiho using strains of *Paxillus involutus*. All strains required thiamine for growth in culture and some could attain maximum growth when it was the sole vitamin presented, but others grew faster if supplied in addition, with biotin, choline or niacin. Since these fungi flourish in the root region, where exudates or breakdown products of cells provide a variety of compounds, there is unlikely to be a selection pressure against strains incapable of, or slow in synthesis of, these essential enzyme or co-enzyme precursors. The same applies to amino acids and ketoacids, small quantities of which have been found to increase the growth of many ectomycorrhizal fungi in culture. Norkrans (1950 and 1953) showed that none of the species of *Tricholoma* that she tested had an absolute requirement for, but grew faster with, glutamic acid, aspartic acid, proline, hydroproline, histidine or tryptophane, in the presence of ammonium as a nitrogen source. Results such as these may well explain the effects of small quantities of fruit-body extracts on growth of mycorrhizal fungi which were observed by Modess (1941) using *Amanita muscaria* and by Melin and Norkrans (1948) using species of *Lactarius*. The active principle was found to be heat stable.

Other materials including leaf litter, humus, seeds and roots have been found to contain substances which may increase the growth of mycorrhizal fungi in culture containing supplies of the known vitamins and amino acids (Lindeberg, 1944; Melin, 1946; Mikola, 1948). Ash constituents of these materials, especially calcium and perhaps manganese, may exert a stimulating effect, but over and above this one or more of the organic constituents is active. Not a great deal of importance can be attached to this kind of effect for the extracts are of complex composition and may include amino acids, vitamins and other growth factors, micronutrient elements or even nutrients. However, high concentrations of such extracts may exert an inhibitory effect on mycorrhizal fungi and this fact tends to distinguish them from humus and litter fungi. Mikola (1948) showed that extracts of fresh leaves or litter of birch, aspen and pine had such an effect on strains of *Cenococcum graniforme* and on *Lactarius torminosus*, stimulating growth in low and inhibiting it in high concentrations. Melin (1946) used litters from broadleaved species and from *Pinus sylvestris* and applied them to both mycorrhizal and non-mycorrhizal fungi. He found that the inhibitory substances were soluble in 88% ethanol and heat stable.

Olsen *et al.* (1971) identified the inhibitory substances in aspen (*Populus tremula*) leaves. Both green and yellowing leaves, the extracts of which both contained inhibitors, were dried, ground and extracted in distilled water. After dialysis, the extract was fractionated with ethyl acetate in which the inhibitors were soluble and the stimulating substances were retained in the aqueous phase. Using thin-layer chromatography and gas chromatography, two compounds were separated which were identified as benzoic acid and catechol. These in pure state exerted the expected inhibition which could be offset in the case of wood and litter fungi by the addition of the equivalent amount of the aqueous extract. Mycorrhizal fungi were inhibited by both of them in the presence and absence of the aqueous extract. Olsen *et al.* concluded that mycorrhizal fungi differed from the others in their reactions because they do not usually produce extracellular phenoloxidases (Lindeberg, 1948). Such results help to explain the inability of mycorrhizal fungi to colonize some kinds of litter and humus and also the influence of lichens on the growth of mycorrhizal fungi and on mycorrhizal development which was studied by Brown and Mikola (1974). They showed that extracts of lichens—*Cladonia alpestris*, *Cetraria islandica* and *Sterocaulon paschale*—or of the humus below their thalli contained inhibitors of mycorrhizal fungi which are as yet unidentified. Perhaps a similar explanation might also apply to the susceptibility of mycorrhizal fungi to extracts of *Calluna* peat described by Handley (1963). These aspects of allelopathy in the widest sense are receiving increased interest. Some species and strains of mycorrhizal fungi (e.g. species of *Lactarius*, Giltrap, 1982) do produce extracellular phenolases and these clearly require further attention in this context.

## Root exudates and the M-factor

Melin and Das (1954) and Melin (1963) demonstrated that excised roots when introduced into complete culture media greatly increase the growth rate of several mycorrhizal fungi. The kinds of roots used were not only those of potential ectotrophs, such as *Pinus* spp., but included *Medicago sativa*, *Cannabis sativa*, *Triticum aestivum* and *Lepidium sativum*. These observations were a great advance on those of Harley (1939) who had simply shown that aqueous extracts of *Fagus* roots stimulated the growth of an imperfect fungus (*Mycelium radicis fagi*) from *Fagus* roots. However, it must be realized that the active stimulating substance, called the "M-factor" by Melin, is not solely produced by potentially ectomycorrhizal roots but also by many others, including some which do not form any kind of mycorrhiza, e.g. *Lepidium*, and that it may stimulate other fungi besides ectomycorrhiza-formers. The work of Melin (1959, 1963) showed that although all the mycorrhizal fungi tested reacted in the same way to root exudates, by increased rate of growth, some, like *Russula xerampelina, R. aeruginosa* and *R. sardonia* were virtually unable to grow on the "complete" medium unless root exudates were added to them. Melin believed that there were two active stimulating principles in root tissue, a diffusible one appearing in the exudate, and an indiffusible one only extracted by homogenization of the tissues. The latter was associated with an inhibitor of fungal growth.

By using agar media "seeded" with homogenized mycelium of mycorrhizal fungi, usually *Boletus (Suillus) variegatus*, Melin was able to determine the location of the release of the diffusible factor and the inhibitor from various parts of the root system of *Pinus* seedlings which had been raised in aseptic conditions. The growth of the fungus was reduced or inhibited in the neighbourhood of secondarily thickened roots and greatly stimulated around primary rootlets. The combination of inhibition and stimulation has been considered to go some way towards explaining why full colonization of the roots by the fungi and the subsequent formation of mycorrhiza seem to be restricted to the primary rootlets. It might also explain why different fungi react differently to equivalent doses of exudate in culture because of a greater or lesser sensitivity to concentrations of the stimulative or inhibitory compounds present. In assessing these points it must, however, be remembered that almost all root apices of a root system are infected in some way, although fully developed mycorrhizas may be found only on certain rootlets.

Nothing very notable has yet been learned about the M-factor, although influences of the kind described seem very likely to be needed to generate the onset of infection. Melin (1963) reported that H. Nilsson had concluded that the M-factor "could be replaced in its effect on hyphal growth" by nicotinamide adenine dinucleotide (NAD) which is, of course, a hydrogen acceptor essential to many dehydrogenase reactions. The results of Benedict *et al.* (1965) although based on studies with a single concentration, do not fully agree with this. Only *Leucopaxillus (Clitocybe) amarus f. roseibrunneus sf. majusculus*, out of eight fungi tested, was stimulated by NAD. If the M-factor were NAD. perhaps the non-specificity of its production by roots and the fact that it stimulates the growth of other fungi besides ectomycorrhizal symbionts would be expected. The same expectation of non-specificity would apply if the M-factor were a cytokinin. Gogala (1970) identified cytokinin activity in the exudate of the roots and germinating seeds of *Pinus sylvestris*. They were extracted by the method of Carr and Burrows (1966) and estimated by the *Avena* test of Kende (1964). Gogala compared their activity on the growth of *Boletus edulis* var. *pinicolus* with pure kinetin, tryptophane, gibberellic acid (GA$_3$) and indoleacetic acid. Kinetin and the extracted cytokinins stimulated growth at low concentrations whereas the other compounds uniformly inhibited growth between concentrations of $10^{-6}$

and $10^{-3}$ g $1^{-1}$. Both extracted kinins and kinetin stimulated growth between $10^{-6}$ and $10^{-1}$ g $1^{-1}$ but at higher concentrations sharply inhibited it. Gogala was of the opinion that the M-factor of Melin might well be a cytokinin and that both its stimulative and its inhibiting action at high concentrations might therefore be explained. However, the demonstration by Fortin (1967, 1970) that some mycorrhizal fungi are inhibited by low concentrations of IAA and that inhibition can be reduced by inclusion of living roots in cultures, puts the M-factor in a new light. Part of the stimulation may be due to reduction of auxins rather than the presence of a factor.

## Spore germination

Numerous experiments on the growth of young coniferous and broadleaved trees have shown that they become mycorrhizal by apparently airborne sources of infection. The effectiveness of spores in this regard has been experimentally demonstrated by Robertson (1954), Tharpar *et al.* (1967), Marx and Ross (1970), Theodorou and Bowen (1973) and Stack *et al.* (1975). Robertson showed, using a spore rain of *Boletus (Suillus) granulatus*, that seedlings of *Pinus sylvestris* could become infected by spores falling onto the soil surface. Table 39 gives an example of the results of one of his experiments. Centres of infection composed of mycorrhizas and spreading hyphae were observed on the surface of the soil-core when it was removed from the pot and these were counted. Infection had arisen from the spores which had been carried down through the soil and germinated close to the roots. Mycorrhizal infection was similarly obtained by Marx and Ross (1970) using spores of *Thelephora terrestris* applied to *Pinus taeda* seedlings in a vermiculite-moss peat substrate. These spores could not be made to germinate in media containing glucose, malt extract with vitamins, or containing cold water extract of roots in agar, or in the neighbourhood of detached roots placed on the culture media. Palenzona (1969), using the Ascomycetes, *Tuber aestivum*, *T. brumale* and *T. melanosporum*, was able to observe the germination of spores in contact with the rootlets of *Corylus avellana*.

TABLE 39

Airborne infection of *Pinus sylvestris* by spores of *Boletus (Suillus) granulatus*

*Seeds of* P. sylvestris *sown July 1950, subjected to spore rain from fructifications September 1950, pots examined April 1951 (data of Robertson, 1954)*

|  | Control | Exposed |
|---|---|---|
| Number of pots | 43 | 43 |
| Total centres of infection | 15 | 292 |
| Average centres per pot | 0·37 | 6·8 |

In spite of such experimental demonstrations of the efficiency of spores as infective agents in conditions where a proportion of them must have germinated, it has been found that the spores of mycorrhizal Basidiomycetes have an extremely low rate of germination in artificial media. Indeed, although those of the other ecological groups of Basidiomycetes such as coprophilous, wood-destroying, or litter-inhabiting forms usually germinate fairly readily, only 1% or less of the spores of mycorrhizal species usually germinate in equivalent conditions. Fries (1941, 1943, 1976, 1978, and see 1966) showed that in special conditions the percentage germination may be significantly raised. He first showed that germination activators or stimulators may be produced by actively growing yeasts such as *Torulopsis sanguinea* or mycelia of *Cladosporium*, *Mycelium radicis atrovirens* or *Cenococcum graniforme*, so that if spores were set to germinate in their environs in culture an improvement in the percentage germination resulted. *Torulopsis* affected many kinds of spore, the other organisms were more restricted in their activity. In a similar way Melin (1959, 1962) showed that his M-factor might stimulate spore germination of some mycorrhizal Basidiomycetes, for pine and tomato roots if placed in the neighbourhood of spores encouraged germination. The fungi whose spores germinated included species (e.g. of *Russula*) of which germination had not previously been obtained. However, Benedict *et al.* (1965) did not find that tomato roots (nor NAD) had a like effect on the spores of a few species of *Amanita*, *Tricholoma* and *Xerocomus (Boletus) chrysenteron*; nor did Fries (1981a) find the roots of herbs and grasses effective. Lamb and Richards (1974) found that the percentage germination of oidia and chlamydospores of mycorrhizal basidiomycetes might be higher than that of basidiospores. For instance, the oidia of *Xerocomus (Boletus) subtomentosus* (the strain was not actually proved to be mycorrhizal) germinated 98%. Because of the variation of percentage germination observed, they expressed the results as "germination index"—the germination percentage observed as a proportion of the maximum value observed for the species—when studying the effects of factors on the germination and survival of spores. Using strains of seven proved mycorrhizal fungi including three imperfect strains isolated from mycorrhizas, they concluded that chlamydospores showed a minimum of tolerance to temperature and humidity of storage, whereas basidiospores and oidia were more resistant. Many of their results are most relevant to the storage of propagules for use in practical forestry and add little to the solution of the problems of the physiology of spore germination, but they do underline once again the low rates of germination of mycorrhizal propagules. This was re-emphasized by Fries (1978) who pointed out that the spores of most ectomycorrhizal Basidiomycetes of genera like *Amanita*, *Cortinarius*, *Gomphidius*, *Paxillus*, *Lactarius*, *Russula*, had rarely, if ever, been brought to germinate in artificial conditions, although their mycelia could grow in the very same cultural conditions. He used for his later experiments media low in ammonium ($1 \text{ g l}^{-1}$

ammonium tartrate, for in 1965 he found higher concentrations to be inhibitory), and containing vitamins, malt extract and nutrient salts. The specific treatments applied in attempts to encourage spore germination were: charcoal (to absorb possible inhibitors), and living *Rhodotorula glutinis* growing near the spores. Alone these treatments had little effect upon the spores tested, but when applied together they raised the germination of those of *Laccaria laccata* from zero to as high as 20%. *Paxillus involutus* germinated to about 1% in similar conditions but higher percentages were obtained in the presence of its own hyphae, derived from spore germination or inoculated close to the spores. At best only about 0·001% of *Lactarius helvus* and 0·01% of *Amanita muscaria* spores germinated. Fries also tested the effect of isovaleric acid on germination because he suspected that the effects of the associated *Rhodotorula* or *Paxillus* hyphae were due to a volatile agent. He did not believe it to be $CO_2$ as a variety of casual contaminants did not have a like effect. Lösel (1964) and Rast and Stäuble (1970) had found isovaleric acid to be effective on spores of *Agaricus bisporus*, but it was found by Fries to be without effect on any of the spores of mycorrhizal fungi tested. The effect of volatile products of its own hyphae upon the germination of *Paxillus involutus* was paralleled by the effects of a non-volatile product produced by the hyphae of *Leccinum scabrum* upon the germination of its own spores. In a further study of this (Fries, 1979a) a number of isolates and collections of spores of *L. scabrum* and *L. aurantiacum* were obtained from fruit bodies. A high percentage germination of the spores (often 50–100%) was obtained in the vicinity of taxonomically related mycelium only. The mycelium of one species never induced germination of spores of the other.

In a later analysis of the way in which spores and vegetative hyphae recognize one another in these reactions, Fries (1981b) examined species of *Leccinum* in greater detail. He observed that the proximity of a suitable hypha induced the spore to germinate to form a vesicle. The apex of the hypha then grew towards the vesicle to fuse with it. The fusion was either lethal or non-lethal to the hyphal tip and vesicle. He therefore concluded that three factors were operating: an inductor formed by the hyphal tip, a homing factor formed by the vesicle, and a lethality factor in the vesicle cytoplasm. Table 40 illustrates the process and the specificity of the factors as found in the experiments. Fries was of the view that these reactions could have selective advantage if they aided suitable dikaryotization or the variation of dikaryotic or nuclear combinations in the inducing hypha. A further paper by Birraux and Fries (1981) has carried the matter of specific stimuli of germination further. Basidiospores of *Thelephora terrestris* were found to be stimulated by the growing roots of very young seedlings of *Pinus sylvestris*, *Picea abies*, *Betula verrucosa* and *Alnus glutinosa* as they grew on agar near the spores. Seedlings of ten species not forming mycorrhiza were ineffective, and one very slightly effective in this regard. The stimulating principle was able to diffuse through a dialysis membrane and was not very volatile but stable over a period of days. This is an important observation for future research because, as the authors remark, *T. terrestris* is a widely distributed mycorrhizal fungus which can pass its whole life cycle from spore to fruit body in its host in laboratory conditions for it may fruit in conjunction with young hosts.

TABLE 40

Recognition reactions between basidiospores and hyphae in the genus *Leccinum* (after Fries, 1981b)

| Origin factor | Inductor | Homing | Lethality |
|---|---|---|---|
| Inducer | Hyphae | Spore vesicle | Vesicle cytoplasm |
| Reactor | Basidiospore | Growing hyphae | Hyphal tip cytoplasm |
| Effect | Germ vesicle is telomorphotic | Chemotropic | Metabolic |
| Specificity | High within subgenus | Low within genus | Very high within species |

Clearly the physiology of spore germination of these fungi is greatly in need of more investigation. It should, however, be noted that the results of Melin (see 1959), of Fries and Birraux (1980) and of Birraux and Fries (1981) showing a positive effect of root exudates are extremely suggestive of the probable biology of spore germination of ectomycorrhizal fungi. As pointed out by Garrett (1970), there is a selective advantage in the possession of a dormancy of spores which prevents simultaneous germination of a given crop in possibly unsuitable conditions. The dormancy may be an exogenous nutrient-imposed dormancy which prevents "suicidal spontaneous germination in the absence of a substrate stimulus". He contrasts such dormancy with that often possessed by thick-walled spores which may have a complex maturation process and period, and a constitutional dormancy which may break naturally or in response to dormancy-breaking treatments such as temperature-shock or specific chemicals. He emphasizes, following a discussion of Blackwell's work with *Phytophthora cactorum* (1943a,b), that if the spores of a population vary widely in their period of dormancy, i.e. they pass through maturation stages at different rates, the longevity of the infective period of the population is increased and in no "single set of environmental conditions will more than a minor proportion of the population germinate at any one time".

It might be thought that in the forest or in the immediate environs of adult ectomycorrhizal plants, the initiation of infection in seedlings would be by way of hyphae from them. If that were so the importance of spores would be in primary infection of seedlings at some distance and it would be of selective advantage to both symbionts if the population of spores in or on the soil could lie in wait for the arrival of seeds of the host plant. On the other hand, the reactions observed by

Fries between hyphae and spores might provide a potential mechanism of variation. For new dikaryotic combinations might arise when basidiospores germinate in the vicinity of existing dikaryons.

In view of both the ecological and the economic importance of spores as a source of mycorrhizal infection of seedlings, more knowledge of the physiology of those of mycorrhizal fungi is needed.

## Ectomycorrhizal *Endogone* in culture

As mentioned, *Endogone lactiflua* and *E. flammicorona* form ectomycorrhizas with *Pinus* and *Pseudotsuga* and *E. eucalypti* with species of *Eucalyptus*. *E. flammicorona* and *E. eucalypti* have been isolated into culture by Tandy (1975a,b) and Warcup (1975a). Some of the nutritional demands of *E. eucalypti* were investigated in a preliminary way. Warcup was able to grow it upon medium containing sucrose and nutrient salts (Czapek-Dox) with 15 g yeast extract per litre. On such a medium it did not stand repeated subculturing and clearly was nutritionally selective. In the course of investigation Warcup found that impurities in distilled water might inhibit its growth, which was in any event variable perhaps due to impurities in the chemicals used in the media. The fungus thrived best in dilute nutrients and was able to use glucose, sucrose, starch and ethanol as carbon sources. Glycerol was less good. It was unable to use nitrate or ammonium salts as sources of nitrogen, but the nitrogenous constituents of yeast extract, hydrolysed soya protein, and some kinds of peptone were used. Trace elements $Fe^{3+}$, $Zn^{2+}$ and $Mn^{2+}$ improved growth. However, lima bean $(0 \cdot 1–0 \cdot 5\%)$ agar prepared in double distilled water with starch and yeast extract gave good growth. Further work will no doubt provide fuller information on the physiology of these ectomycorrhizal phycomycetes and perhaps lead, as Warcup suggested, to methods of isolating endomycorrhizal Endogonaceae.

## Conclusion

The implications of this work on the growth physiology of ectomycorrhizal fungi will be discussed at the end of the next chapter. It will be realized, however, from the work that has been described so far that the use of the rate of growth of mycorrhizal fungi as a measure of rates of physiological processes has served its turn for most purposes. By its use little has been discovered which differentiates ectomycorrhizal fungi from other fungi in their physiology. Indeed this chapter has served essentially to record results rather than indicate profitable leads. As yet, however, little is known of the nutrient requirements of other ectomycorrhizal fungi than the Basidiomycetes. Increasing numbers of ectomycorrhizal Ascomycetes are being discovered and it would seem imperative that more be learned of them.

## Chapter 8
# Production of Metabolites by Ectomycorrhizal Fungi

### Introduction

The activity of ectomycorrhizal fungi in setting up and maintaining a close association with the root tissue of trees must involve a release of metabolites active both in the approach and infection stages, as well as in the later maturer stages of mycorrhiza formation when compounds are known to pass between the two symbionts. Considerable interest therefore attaches to experimental work on the release of metabolic end-products into culture media as an approach to the interpretation of their activities in natural conditions of symbiosis. The kinds of compounds that immediately come to mind as potentially important are substances having a growth-regulating or hormonal effect on the development of host tissue, compounds having an effect on the release of substances from the host cells, substances which might assist the competition of the mycorrhizal fungus with other microorganisms of the rhizosphere and root region, substances which might allow the host cells to recognize the fungus as a suitable mycorrhiza-former, in addition to nutrients or growth factors required by the host. Far too little experimental work has yet been done on the kinds and quantities of substances secreted by ectomycorrhizal fungi and on factors affecting their release, so that few hypotheses concerning their production and functioning can be put forward. Moreover many of the kinds of products which have been found to be released by mycorrhizal fungi are not solely produced by them but may be formed and released by many other kinds of microorganisms. This is especially relevant when hypotheses of function are being erected. Again, the release of substances into culture media containing a great excess of nutrient compared with the amounts present in natural conditions does not necessarily imply that similar substances are likely to be produced in significant quantities in the wild (see Foster, 1949).

## Hormonal substances

*Indole compounds*

Indole acetic acid (heteroauxin, auxin, IAA) has been known to be produced by fungi for a very long time. It was called rhizopine by Nielson in 1930 and in the following years many observers showed it to be produced by numerous fungi and released into media. Thimann (1935) explained its formation by the oxidative deamination of tryptophane present in the medium. It is not therefore very surprising that many mycorrhizal fungi also release indole compounds into media containing tryptophane.

Interest in this subject was particularly aroused by Slankis (1948) when he showed that the mycelia of some mycorrhizal species of *Boletus* especially *B. (Suillus) variegatus* if grown with excised pine roots in the same culture, encouraged the development of larger numbers of short laterals, many of them dichotomous like pine mycorrhizas. Since the filtrate of the media in which *B. (S.) variegatus* had grown had a like effect, it was assumed that a growth-regulating substance had been released by the mycelium and that this affected the *Pinus* roots. Slankis (1949, 1950, 1951 and 1958) showed that pure auxins, indole acetic, indole propionic, indole butyric and naphthalene acetic acids at concentrations of 0·05 to 50 $\mu$g l$^{-1}$, produced similar reactions in roots in culture. At the higher end of his concentration scale, $\beta$-indole acetic acid caused increased numbers of short roots, often forked. There was therefore a strong presumption that auxins were produced by the fungi in culture. Moser (1959), using a medium with 2·04 g l$^{-1}$ tryptophane as a nitrogen source, grew the mycelia of both mycorrhiza-formers, including several strains of some species, and a few species of fungi of other ecological groups. Most of the mycorrhiza-formers produced IAA in his cultural conditions but three species of *Phlegmaceum (Cortinarius)* did not. These did, however, release indolepropionic acid. Some of the other species of *Phlegmacium* also produced it together with IAA and two species of *Suillus (Boletus)* produced indolebutyric acid and IAA. In all these cases tryptophane was the sole source of nitrogen in the medium. If another source was present also, the production of indole compounds was reduced. Alanine, asparagine, aspartic acid, glutamine and glycine were all inhibitory to IAA production although they increased mycelial growth. This might be interpreted as if the other sources of nitrogen were used in preference to tryptophane, the breakdown of which depended on the requirement of the fungus for nitrogen. Species and strains of mycorrhizal fungi differed greatly in rate and quantity of auxin formation, but Moser formed the opinion that their activity in this regard usually exceeded that of fungi of other ecological groups.

The subject was further examined by Ulrich (1960a). She showed that *Boletus (Suillus) variegatus, B. (S.) granulatus* and *B. (S.) luteus* could form IAA on Hagem's medium (ammonium chloride as nitrogen source) without added tryptophane. The rate of formation of IAA by these species and by others was greater if tryptophane was present. Species of *Amanita* differed in producing other indole compounds from tryptophane, and *Boletus badius* and the non-mycorrhizal *Coprinus comatus* produced IAA at first and then oxidized it.

Horak (1964) found that species of *Phlegmaceum (Cortinarius)* derived from *Picea abies* were able to synthesize indole from *p*-aminobenzoic acid and then with alanine to produce tryptophane from which IAA might be derived. Two species produced 5-hydroxy indole acetic acid. By contrast, Shemakhanova (1967) tested the effects of whole fungal cultures and fungal metabolites on the development of excised pine rootlets. Three species of *Boletus*, *B. (Suillus) luteus*, *B. (S.) bovinus* and *B. subtomentosus* and the non-mycorrhizal *Lycoperdon gemmatum*, *Phallus impudicus*, together with some unnamed species of the genera *Aspergillus*, *Penicillium* and *Stysanus* and two sterile mycelia isolated from pine mycorrhizas, were used. All these fungi produced mycelium which colonized the surfaces of the roots of pine in Slankis' medium rather than developing at large in the medium. Growth of the fungi was more profuse in the presence than in the absence of the roots, but the growth of the roots was somewhat inhibited. *B. (S.) bovinus* and *B. (S.) luteus* encouraged the rootlets to dichotomize but the other fungi did not. Filter-sterilized media from the same two fungi also encouraged root-branching. In further experiments it was found that sterilized filtrates from all the fungi increased the degree of branching of the roots but in no case was IAA or other indole compound detected in them. This failure to detect auxins in used culture media does not seem to be attributable to the methods used nor to strain variation in the fungi, for the strain variation observed by Moser (1959) in IAA production did not involve total abolition. However, according to Ulrich (1960a,b) quoting Foster (1949), IAA may be destroyed by acid conditions in culture, or again it may be subject to enzymic destruction or oxidation in some conditions of culture (see later). In any event Shemakhanova's results do at least confirm that metabolic products released into the media by various fungi, including mycorrhiza-formers, had morphogenic effects upon pine rootlets and also upon root formation by bean cuttings and therefore presumably included substances with auxin-like activity. Pachlewski and Pachlewska (1974) examined the ability of a large number of fungi to produce IAA, including tested and proved mycorrhiza-formers with *Pinus sylvestris*, potential mycorrhiza-formers which did not infect *Pinus sylvestris* in their conditions, and non-mycorrhizal Basidiomycetes and Ascomycetes. On a simple agar medium with ammonium tartrate as nitrogen source, about half of the 55 fungi which they tested produced IAA which was identified within the mycelium or in the medium or in both. A group of eight fungi, including *Collybia butyracea*, were grown in liquid media with tryptophane. All formed IAA including *Scleroderma aurantium*, *Collybia butyracea*, *Tricholoma flavovirens* and *Amanita citrina* which failed to form IAA in the absence of tryptophane; the first three did not form mycorrhizas in these experiments.

Tomaszewski and Wojceichowska (1973) cultivated a number of mycorrhizal basidiomycetes in a medium containing maltose and glucose as carbon sources, nutrient salts, biotin and thiamine, with ammonium nitrate $5 \mathrm{g}^{-1}$ as nitrogen source. They observed that a number of Boleteaceae produced dark pigments, and these species on the whole produced most auxin. Of 26 other species, including members of *Amanita*, *Russula*, *Cortinarius*, *Hygrophorus*, none produced the pigments or auxin. Two species of *Scleroderma* produced yellow pigment but no auxin. In a subsequent test using a medium containing tryptophane, they showed that the addition of polyphenols increased the production, presumably by the prevention of the enzymic destruction of indole compounds by both *B. (Suillus) bovinus* and *B. variegatus* and two species *Scleroderma aurantium* and *Amanita muscaria*, which could not form auxins in the absence of tryptophane. They identified IAA, tryptophol, *N*-acetyltryptophane and indol-3-carboxylic acid as the products and indicated that the first three were common in basidiomycete cultures on tryptophane. The pigment produced by seven *B. (Suillus)* species were found to be variegatic acid excreted into the medium, and the quinones, $\beta$-boviquinone-4, diboviquinone-4,4, methylene diboviquinone-4,4 and atromentin were identified in the ether extract of the mycelium of one species, *B. (S.) bovinus*.

The quantity of variegatic acid and other polyphenols that were released into the medium increased with nitrogen supply up to about $350 \mathrm{mg\,l}^{-1}\mathrm{NH_4NO_3}$ and parallel with it the release of indole compounds increased. At higher levels of $\mathrm{NH_4NO_3}$ there was a continual increase in growth rate but a fall in the release of polyphenols and indole compounds, so that the latter were no longer released in media containing $1000 \mathrm{mg\,l}^{-1}\mathrm{NH_4NO_3}$.

We must conclude from all this work that most of the known ectomycorrhizal fungi can produce auxins in culture. The quantities must vary both according to the rate of oxidative deamination brought about by the species concerned and also according to the presence of enzymes oxidizing or destroying the indole compounds and the production and presence of phenols and other substances inhibiting these enzymes.

The production of indole compounds in the absence of tryptophane has not been found to be so widespread, but whether this is due to the absence of the added precursor alone or to the enzymic or acidic destruction of the smaller quantities produced is not clear.

*Cytokinins*

Some, but by no means all ectomycorrhizal fungi so far tested have been shown to produce substances capable of promoting the growth of cytokinin-dependent soybean callus. Miller reported in 1971 that his interest in this possibility was aroused by the similarity of the enlarged cells of the root cortex of ectomycorrhizas to those of other tissues of plants which had been treated with kinetin. He tested the ability of mycorrhizal fungi to produce cytokinin in culture by growing colonies upon agar in the neighbourhood of soy callus deprived of cytokinin. This method works for many fungi but some may produce growth inhibitors active on the callus and others even be inhibited in some degree by the callus itself. In such cases the fungus may be grown in liquid culture and the products purified on Dowex columns, or on paper or by solvent fractionation, and the resultant solutions finally concentrated.

In these ways Miller (1971) and Crafts and Miller (1974) were able to show that *Rhizopogon roseolus*, *Suillus (Boletus) cothurnatus*, *Suillus (B.) punctipes*, *Amanita rubescens* and an ectendomycorrhizal fungus coming originally from Mikola in Sweden produced cytokinins and there were signs that some others might form traces in these conditions. On the other hand with some fungi, which they had confirmed to be mycorrhizal with pine, the callus did not grow significantly more than the control. Examples were *Suillus (Boletus) grevillii*, *Cenococcum graniforme (geophilum)* and *Xerocomus (Boletus) chrysenteron*. The callus did not survive at all on media treated with mycorrhizal strains of *Amanita pantherina*, *Cortinarius armillatus*, *C. mucosa*, *C. vibratilis*, *Hebeloma crustuliniforme*, *H. mesophaeum*, *H. radicosum*, *Laccaria laccata* and *Tricholoma imbricatum*. Ng *et al.* (1982) have confirmed that *R. luteolus* produces cytokinin in culture as did their strains of *Suillus luteus* and *Boletus elegans (Suillus grevillii)*. The latter produced a callus inhibitor which might explain the failure of Miller (1971) to find cytokinin activity in its culture filtrates.

Although Miller (1971) reported that a negative assay for cytokinins had been obtained with several Basidiomycetes and Ascomycetes that were non-mycorrhizal, this must not be held to imply that cytokinin production is an especial property of ectomycorrhizal fungi, for this is very far from the truth. The valuable work of Miura and Hall (1973) and Laloue and Hall (1973) on the actual chemical nature of the active compounds produced by *Rhizopogon roseolus* must therefore remain of limited value to mycorrhizasts pending a more complete survey of the relevance of cytokinins to mycorrhizal symbiosis.

In a similar way the reports quoted by Slankis (1973) that Gogola observed "gibberellin related" compounds in *Boletus edulis* culture media and fruiting bodies is of some interest but requires further investigation before gibberellin-dependent activities can be attributed to all the fungi of ectomycorrhizas.

### Ethylene and other volatile substances

Ethylene may be present in soils of low aeration in sufficient concentration to affect the growth of some roots (Lynch and Harper, 1974). It is probable that a number of fungi are able to produce the gas and it has been identified as a product of *Mucor hiemalis* and of *Penicillium digitatum*, *Blastomyces dermatidis*, *Fusarium oxysporum* and certain yeasts (Lynch, 1972; Lynch and Harper, 1974; and see Cochrane, 1958). Lynch, and Lynch and Harper, found that ethylene formation only occurred in soils containing a carbohydrate source and methionine.

Graham and Linderman (1979) reported the production of ethylene by mycorrhizal fungi in liquid culture containing 10 mM methionine. They made a particular study of *Hebeloma crustuliniforme*, *Laccaria laccata* and *Cenococcum graniforme (geophilum)* which readily produced ethylene, and of *Pisolithus tinctorius* which produced it if the medium was frequently renewed. In a further examination of 5 isolates of the pathogen *Fusarium oxysporum f. pini* and 19 ectomycorrhizal species, they observed considerable variation of ethylene production. However, some of the ectomycorrhizal species equalled the very active *Fusarium* strain in its production. These are very important observations because the hormonal activities of ethylene resemble those of IAA in many respects, yet as a gas it is extremely mobile, and does not have as localized an action.

Krupa and Fries (1971) examined the production of volatile substances by a strain of *Boletus (Suillus) variegatus* from a medium which included glucose, potassium phosphate and ammonium tartrate. After growth of the fungi, the medium was extracted with ether and the components identified by gas chromatography. The main volatile substances present were ethanol, isobutanol, isoamyl alcohol, acetoin and isobutyric acid (they do not mention ethylene). The maximum rates of production of these compounds occurred in the rapidly growing phase. In a subsequent paper Krupa and Nylund (1971) reported that at concentrations of 1% (v/v) in agar media, isobutanol, isoamyl alcohol and isobutyric acid inhibited the radial growth of the mycelia of both *Phytophthora cinnamomi* and *Fomes annosus*. Ethanol by contrast was stimulatory at such concentrations although inhibitory at 5% (v/v). As an aside it is interesting that low concentrations of aliphatic alcohols encourage rhizomorph production in some Basidiomycetes. Whether these volatile substances are formed by *B. (S.) variegatus* and other mycorrhizal fungi in natural conditions is unknown, nor is it known whether, if so, they would reach significant concentrations.

## Antibiotic production

The production of bacteriostatic and fungistatic substances by mycorrhizal fungi has been known for a long time. Their discovery was incidental, in the main, to the search for antibiotics for medical purposes after the success of penicillin (e.g. Wilkins and Harris, 1944; Wilkins, 1946; Robbins *et al.*, 1945; and others). These workers mostly examined extracts of fruit bodies of common species of basidiomycetes for active substances. About one quarter of the many hundreds of basidiomycetes examined by Wilkins and Harris produced antibiotics against both fungi and bacteria, but by no means all the mycorrhizal or putatively mycorrhizal species did so, e.g. none of the 43 species of *Russula* did so, but all of the 7 species of *Lactarius* did. Amongst the sampled species of the genera *Cortinarius* and *Tricholoma*, well over half were antibiotic producers. Variation between species in antibiotic activity was considerable, and as in other physiological attributes there was also variation in this regard between strains of a single species. In 1954 Morimoto and his colleagues specifically examined mycorrhizal fungi for antibiotic-producing strains which could be used in inoculation, and Zak (1964) also reviewed the subject, putting forward the hypothesis that one of the important activities of mycorrhizal fungi in relation to their hosts was the control of root disease. This idea had been rather vaguely canvassed before that date by Levisohn (1954) and Björkman (1956), and others had in a similar way ascribed the susceptibility of spruce to *Fomes annosus* on certain soils to be due to the lack of mycorrhiza (see Hyppel, 1968a,b). It may be noted that the protection of the host plant against *Mycelium radicis atrovirens* by mycorrhizal fungi, adumbrated by Levisohn, has been confirmed experimentally by Richard *et al.* (1971).

It was the possibility of protecting spruce from *Fomes annosus* that led Hyppel (1968a,b) to estimate the ability of mycorrhizal fungi to inhibit the growth of a strain of *Fomes* in culture. He found that over one-third of the 85 species and strains of mycorrhizal fungi that he tested inhibited *Fomes* to a significant extent. There was, however, in this regard as with other characteristics, genetic variation within the species. Two isolates of *Boletus (Suillus) bovinus* inhibited (albeit with very different intensities) and three did not. Three strains of *B. edulis* were variously effective and two were not. Using highly effective strains of mycorrhizal fungi he tested their ability to antagonize a range of strains of *Fomes annosus*. A single strain of *Boletus (Suillus) bovinus* or *B. (S.) variegatus* might give an inhibition zone of *Fomes* on an agar culture of 2 to 17 or 5 to 20 mm respectively, indicating a very great genetic variation in *Fomes* to resist the antagonistic principle from both the species of *Boletus*.

Marx and his colleagues have made a considerable examination of the potential of mycorrhizal fungi to inhibit pathogens. His reviews of the subject (1972, 1973, 1975) should be consulted. The strains of some of the fungi, e.g. *Leucopaxillus (Clitocybe) cerealis* var. *piceina* produced substances antagonistic to *Phytophthora cinnamomi* on all media. Other fungi might produce little on some media and much on others. Yet other mycorrhizal fungi, e.g. *Amanita muscaria* and *Pisolithus tinctorius* produced none on any medium. Again *Leucopaxillus* was active against a whole range of fungi—against all but two (*Thanatephorus (Corticium) cucumeris* and *Fusarium oxysporum f. pini*) that were tested—whereas other fungi might inhibit the growth of a few of the species. Again *Pisolithus* inhibited the growth of none. The same range in intensity of inhibition was observed in bacterial antagonism between the very active *Leucopaxillus* and the inactive *Pisolithus*. Differences between species of mycorrhizal fungi in their activity against bacteria were also observed by Krywolap (1971). Of 26 species of mycorrhizal fungi, 6 were active against both Gram-positive and Gram-negative bacteria, 11 were active against Gram-negative only. The others varied, but no inhibitory activity was found in the culture media of 4 species.

The active compound produced by *Leucopaxillus (Clitocybe) cerealis* var. *piceina* was identified by Marx (1967) as diatretyne nitrile which, together with diatretyne-3, was found in the culture media. The nitrile was produced during the active growth phase and was progressively broken down during autolysis. The amide and diatretyne-3 have only a bacteriostatic activity whereas the nitrile is very potent against both bacteria and fungi.

It is clear that such is the variation between species and between strains of species that chemical protection of the host against pathogens is unlikely to be a universal trait of ectomycorrhizas, but it may indeed be extremely important in appropriate ecological conditions.

## Intermediary metabolism

There has been virtually no study of the intermediary metabolism of ectomycorrhizal fungi. We know virtually nothing of the respiration and respiratory pathways and their sensitivity to inhibitors. Except in the cases mentioned elsewhere little is known of pathways of synthesis of nitrogen compounds or of other metabolites. Growth is inhibited by derivatives of humus and leaf litter and it has been shown by Olsen *et al.* (1971) that benzoic acid and catechol were responsible in the case of *Amanita rubescens* and five species of *Boletus* and *Suillus*. There have been a number of reports that terpenes, fatty acids, phenolic and other possible products of host metabolism may cause inhibition of respiration and leakage of substances from ectomycorrhizal fungi, but the knowledge is as yet meagre. The comparative studies of the effect of such compounds on the growth of mycorrhizal and non-mycorrhizal fungi have shown that the latter are far less sensitive, indeed they are often stimulated by extracts of litter and humus (Lindeberg *et al.*, 1980). However, the ecological significance of this is not yet clear nor is their mode of action. The absence of the production of exocellular phenolases by most but not all mycorrhizal fungi (Lindeberg, 1944, 1948; Giltrap, 1982) is in line with their inability to use lignin as a source of carbon, but there have been virtually no other comparative studies of other surface-attached or secreted enzymes of mycorrhizal fungi as opposed to those of the mycorrhizal organs that they form with their hosts.

Catalfumo and Trappe (1970) pointed out that phytochemical studies that might assist the establishment of taxonomic relationships or biochemical pathways of ectomycorrhizal fungi were not available. Since then Melhuish *et al.* and Melhuish and Hacskaylo (1980a,b) have studied fatty acid composition and Krupa *et al.* (1973), Krupa and Bränström (1974) and Booker (1980) have reported on amino acid constitution but it is too early to expect much of metabolic or comparative significance to appear from the studies.

Clearly there is a very regrettable gap in work on ectomycorrhizal symbiosis since we know so little about the intermediary physiology of the fungi. In order to consider more confidently the uptake and loss of materials by the fungal sheath in mature mycorrhizas, and the stimuli, their origin and the reaction to them involved in the building of the symbiosis, this kind of information is absolutely essential.

## Genetics and mating systems of ectomycorrhizal fungi

The variation of physiological behaviour observed in the species of ectomycor-rhizal fungi must be explicable in genetic terms. It is usually assumed that the Basidiomycetes of ectomycorrhiza conform to the usual pattern of nuclear behaviour in being dikaryotic in their vegetative mycelium, but there is no corpus of information about them in regard to cytology and nuclear behaviour. As regards the Ascomycetes, Bonfante-Fasolo and Brunel (1972) examined the nuclear cytology of *Tuber melanosporum*, using (i) monospore mycelium, (ii) cultured mycelium (derived from mycorrhizas of *Corylus*), (iii) mycelium removed from inoculated *Tilia* mycorrhizas, (iv) mycelium isolated from a fruit body into pure culture and (v) the mycelium taken from fruit bodies. Ascospores germinated to give monokaryotic mycelium but the mycelium from fruit bodies or mycorrhizas was bi- or multi-nucleate, presumably derived from fusion of monokaryotic hyphae. A later examination of *T. albidum*, *T. maculatum* and *T. melanosporum* by Bonfante-Fasolo (1973) showed them to be usually binucleate in each cell. The nuclei underwent mitotic division showing probably 5 chromosomes. We may reasonably assume therefore that the mycelium of these ectomycorrhizas is dikaryotic or at least heterokaryotic. The nuclear state of the *Endogone* spp. which may be associated with ectomycorrhizas is however unknown. The probability of the ectomycorrhizal fungi being dikaryotic or heterokaryotic has very considerable genetical implications. It implies that they can store recessive mutant genes in their genome and possess the important properties in regard to variation, selection and evolution that diploids have. However, the subject needs attention for we have as yet no real knowledge of their nuclear behaviour, their mating systems, their ability to form uninucleate oidia, chlamydospores, etc. Nor do we know anything about their ability to dediploidize or dedikaryotize or to produce uninucleate hyphal systems, nor anything about nuclear fusion and division or parasexuality.

The existence of very great variation in their growth rates and other physiological characteristics that have already been observed within taxonomic species indicates that intentional breeding and selection of more suitable fungal symbionts for different hosts in different external conditions is a real possibility, as soon as the mating systems of the fungi are investigated. The absence of fruiting bodies in culture or in the field is not necessarily an insuperable defect in work on breeding systems because haploid oidia, haploid hyphal branches, and solitary basidia on the vegetative hyphae are not unknown in the Basidiomycetes; and in any event research on factors which stimulate fruiting would be cognate with essential research on factors promoting sheath formation and the production of other complex bodies like rhizomorphs sclerotia, etc.

Speculation about the evolution of mycorrhizal symbioses, about the variation of the fungi and its meaning in terms of ecology will remain very unsatisfactory until more about the mating systems of the fungi is known. It should be borne in mind too, that in groups like the Basidiomycetes and Ascomycetes, where sexuality and parasexuality may occur by processes not associated with the usual sex organs or fruiting bodies, gene flow and recombination may still be of prime importance in providing the new material for variation and selection.

## Implications of the work on fungal physiology to mycorrhiza formation

There is very little in the studies recounted in the last two chapters which helps to explain why some species or strains of species of fungi form ectomycorrhizas and others, quite closely related, cannot. The aggregate of the properties of the mycorrhiza-formers shows them to have a community of characteristics which are often negative, they do not possess properties (e.g. enzyme production, vitamin synthesis, etc.) not needed in their peculiar habitat. There are very few indeed peculiar positive characters which they possess. Moreover their taxonomic diversity indicates that the required characters, negative and it must be supposed positive, may be easily derived by selection from the diverse gene pools of different fungal taxa. It is almost possible to conclude that any fungus could be a biotroph unless it possesses some positive characters that make it unsuitable.

There is no reason to change the view that mycorrhiza-formers constitute a "third group among basidiomycete species inhabiting the litter layers of the forest; the white rots causing the breakdown of cellulose and lignin, the brown rots attacking cellulose and hardly affecting lignin and the mycorrhiza fungi dependent on simple carbohydrates and having little or no effect on cellulose and lignin" (Harley, 1969a), but we have as yet too little information about ectomycorrhizal Ascomycetes and Endogonaceae to include them. Some strains of some species of mycorrhizal fungi do produce exocellular enzymes such as pectinases, cellulases and hemicellulases, and phenol oxidases in sufficient quantities in culture to allow them perhaps to derive enough carbon from complex compounds for some purposes (see Melin, 1963; Harley, 1969a; Palmer and Hacskaylo, 1970; Lundeberg, 1970; Giltrap, 1982). Nevertheless the fact that some, even most, produce none or no measurable quantities of such enzymes (e.g. Lindeberg and Lindeberg, 1977, on pectic enzymes) indicates that their production, except in very limited or extremely local quantities perhaps, is not necessary to mycorrhiza formation. It is noteworthy, however, that genera and even species are variable in the extent to which such enzymes are produced (Norkrans, 1950; Lundeberg, 1970), and in that respect are similar to species which normally hydrolyse complex carbon polymers, as shown in *Collybia velutipes* by Norkrans and Aschan (1953).

It has been suggested that some ectomycorrhizal fungi are capable of producing exoenzymes which attack complex carbon compounds and may be facultative mycorrhiza-formers. That is, a given strain of a species may exist either free-living or fully mycorrhizal in association with a host. The only direct evidence of facultative behaviour is that provided by Theodorou and Bowen (1971b) that some proved mycorrhiza-formers could spread in the rhizosphere of non-host plants, but nothing is known of their ability to persist there in natural conditions. There are reports that certain mycorrhizal species, e.g. *Boletus subtomentosus*, *Laccaria laccata* or *Paxillus involutus* could fruit in the absence of host plants. All these species are extremely variable in culture and in the field. The last mentioned has been investigated in detail by Laiho (1970) and he concluded that none of the strains of the species were facultative mycorrhiza-formers because in every case examined there was evidence of a connexion between the fruit body of *Paxillus* and a host. *Boletus subtomentosus* (agg.) has been shown to produce non-mycorrhizal strains which are capable of lignin and cellulose breakdown. The evidence of Theodorou and Bowen is, of course, a description of a fungus spreading in a zone where simple carbohydrates, vitamins, etc. are released—of great ecological significance—but not a great departure from mycorrhizal behaviour since such a spread in the host root-region is an essential early stage of mycorrhizal infection.

Lewis (1973) has considered the question of facultative mycorrhizal fungi in his essay "Concepts in Fungal Nutrition and the Origin of Biotrophy" where he examined the hypothesis that biotrophic fungi, mycorrhiza-formers in particular, originated from necrotrophs. He envisages four stages in the process of evolutionary change:

1. *Catabolic repression* (by the presence of compounds which are the end-products of the enzyme activity within the tissues) of degradative enzymes limits their activity. This presumably occurs in active necrotrophs.

2. At the same time, *localized production of hormones by the fungi or the affected tissues or both* and the transmission of an hormonal message to other parts of the host results in an increased import of photosynthetic products into the infected region.

3. These *incoming simple metabolites maintain catabolic repression* and also nourish the fungus.

4. *A genetic loss of ability to produce enzymes* (now of no selective advantage) follows.

This sequence of events is superficially attractive and indeed some evidence for some degree of catabolite repression of pectinase in a mycorrhizal fungus, *Suillus (Boletus) luteus*, was obtained by Giltrap and Lewis (1982); but it does not really grapple with the problem. First, it makes the assumption that the ability to produce enzymes capable of degrading complex carbon and nitrogen polymers is a general primitive character as compared with the inability to do so, and that is by no means certain. Secondly, it admits that (as a result of Stage 4) mycorrhizal fungi do not require, and indeed may not produce, degradative enzymes. It does not, therefore, help in explaining the ability of fungi which do not produce these enzymes to penetrate the cell walls and produce a Hartig net in ectomycorrhiza formation. It does, however, offer an explanation of the apparent existence of "facultative" ectomycorrhizal fungi on the assumption that one and the same strain is both a mycorrhiza-former and capable of degradation of complex polymers. It will be apparent from what has been written above that that assumption is not at all secure, because of the considerable strain variation in the fungi and the demonstration that polymer-degrading strains may not be mycorrhizal. It is clear that this point must be elucidated fully before other questions relevant to the hypothesis are tested. Hormone formation by mycorrhizal fungi has of course been demonstrated under suitable cultural conditions in many of them. Most are capable of forming indole compounds from tryptophane, but of course this property is not solely theirs. It is possessed by a very large number, perhaps a majority of fungi and bacteria, but whether there is enough tryptophane and sufficient deficit of other nitrogen sources in the tissues of the host for the process to be important needs to be demonstrated and is on the face of it unlikely. This stage is the key to the whole sequence, for the production of hormones must occur in sufficient quantities long before a mature mycorrhiza has been produced. Carbohydrate of a simple kind must be delivered or available to the fungus for its total nourishment, as well as in a sufficient quantity to repress its degradative enzymes. However, those enzymes must be in action sufficiently long for the fungus to develop and release hormones in a quantity adequate to start the translocation of simple carbon compounds to the site of infection.

If, however, we accept the importance of catabolite repression, as suggested by Norkrans (1950) and Lewis (1973), in controlling the activity of species or strains of mycorrhizal fungi which form polymer degrading enzymes, we must also assume that the concentration of carbohydrates initially available in the rhizosphere is high enough to support their growth and yet suppress the enzymes.

It is necessary for reasonable evidence to show that enzyme repression can be brought about by concentrations of glucose or other soluble carbohydrate such as are found in the rhizosphere. There is severe competition for carbohydrate in that habitat and it might be expected, therefore, that, as a stage in mycorrhiza formation by some fungi, degradation of the external cells of the host might be observed before repression occurred. Alternatively, digestion of the cell walls of the sloughed-off cap and epidermal cells might be observable and might result locally in sufficient carbohydrate for repression. In any event the observation under the microscope of the initial stages of colonization of roots by known polymer degrading strains of mycorrhiza-formers should help to put these speculations on a better basis by a simple means. This subject will be further considered in later chapters.

## Chapter 9

# Growth and Carbon Metabolism of Ectomycorrhizal Plants

### Growth of the host

Frank, besides describing the structure and occurrence of both ectomycorrhizas and endomycorrhizas, carried out experiments published in 1894 on the effects of ectomycorrhizal infection on the growth of seedlings of *Pinus*. Those which were grown in unsterilized soil developed mycorrhizas and grew faster than those in sterilized soil. Seedlings in sterilized soil which became contaminated during the experiment developed mycorrhizas and grew at an intermediate rate. Although this experimental design is faulty, because heat sterilization may release both nutrients and toxic substances, the results have in principle been repeatedly confirmed. Refinements aimed at obviating the effects of heat sterilization have been used and the results can be accepted with confidence. An example using the Ascomycete *Tuber* is given in Table 41. In addition, world-wide observations on the establishment of exotic species of ectomycorrhizal trees in many parts of the world have shown that artificial inoculation is usually essential to success although seedlings may be grown uninfected if provided with fertilizers on an adequate scale.

The difficulties of carrying out experiments with trees under controlled conditions have precluded extensive researches except during a small part of their early life. In consequence, comparisons between mycorrhizal plants and controls have been made over one or few seasons of seedling life at most. Much work has also been done in open beds of soil, sometimes inoculated with soil, humus, or chopped mycorrhizas, with controls treated with sterilized inoculum (examples are given by Harley, 1969a). In some cases the soil has been sterilized by drenching or fumigation before planting. The published conclusions are simple: mycorrhizal seedlings are usually taller and have larger root systems, both shoots and roots are of greater dry weight and the ratios of root weight to shoot weight are very frequently smaller. Sometimes additional measurements such as diameter of stems, numbers of branches, numbers of leaves, etc. have been made but there has as yet been inadequate investigation of changes in form and relative growth rates of their organs. It would appear to be very dangerous to imply as some have done that increase of the diameter or even the change of structure of the stem of infected plants in their seedling stages bears any relation to the wood-producing potential of the adult.

TABLE 41

Effect of mycorrhizal infection of *Pinus strobus* by *Tuber maculatum*. Two-year plants (data of Fassi and Fontana, 1967)

|  | Unsterilized Uninoculated Soil + peat + vermiculite | Sterilized Uninoculated Soil + peat + vermiculite | Inoculated Soil + peat + vermiculite |
|---|---|---|---|
| Length of root (cm) | 54 | 171 | 181 |
| Number of short root apices | 250 | 577 | 808 |
| Number of forked short roots | 83 | 58 | 356 |
| Number of single apices | 167 | 519 | 452 |
| Height (cm) | 6·4 | 9·1 | 12·0 |
| Diameter (mm) | 1·1 | 1·6 | 2·0 |

The reduction of the root:shoot ratio also needs further investigation for it diminishes in any event, irrespective of mycorrhiza formation, during the early growth of the young seedling, and is lower in soils of high nutrient availability, especially of nitrogen, and it tends to be reduced in conditions where photosynthesis is reduced. Although a reduction of the ratio might be expected in mycorrhizal plants, both because of a greater ability to absorb nutrients, and an increased concentration of nutrients in their tissues, it would be of great value to know more about the extent to which greater size and age themselves contribute to the form of the seedlings. In a recent investigation Alexander (1981) grew *Picea sichensis* seedlings in axenic culture with *Lactarius rufus* for 14 weeks. The results are shown in Table 42. The usual increase of height and dry weight of both shoots and roots were observed. The root:shoot ratio was however greater (although below $P = 0·05$ significance) in mycorrhizal plants. In discussing this Alexander observed that if the fungal sheath were to comprise 20% of the weight of the roots, which is not impossible, the slight rise in root:shoot ratio might be explained by the fungal infection directly. Other results of Alexander seem to confirm the much older results (e.g. of McComb, 1938) that infection increases the total number of short roots per seedling. In this case the form of the root system changes for there are more short roots both per cm of root length and per mg of root weight on the mycorrhizal seedlings. There is a great need to disentangle the normal changes in form due to growth, those dependent on changes in nutrient supply and absorption and those arising from mycorrhizal infection and its physiological consequences. Alexander points out that in some of his unpublished results and in those of W. L. Bigg, there may be no increased growth rate or even a diminution of growth rate of the host as a result of mycorrhizal infection. A specific example is given by Sands and Theodorou (1978) quite incidentally to their paper on water absorption. The seedlings of *Pinus radiata* which they inoculated with *Rhizopogon roseolus* had 72% of their short roots mycorrhizal after 4½ months but were smaller in weight and height than their uninfected counterparts. In a study of the effects of mycorrhizal inoculation of *P. caribaea*, Vozzo and Hacskaylo (1971) measured lengths of the shoots of seedlings at various times after inoculation. Some of the mycorrhizal plants were shorter than the controls for many weeks although at the end of the experiment all of them exceeded the uninoculated controls. In a similar way, seedlings of *P. strobus* inoculated with *Endogone lactiflua* by Fassi and Palenzona (1969) were smaller in diameter and not significantly greater in length at the end of their first year, although they grew fast and greatly exceeded the control later. By contrast, seedlings of *Pseudotsuga douglasii* inoculated with *E. lactiflua* exceeded the controls even after one year.

TABLE 42

Growth of *Picea sichensis* in axenic culture with *Lactarius rufus* during 14 weeks

*Substrate: vermiculate-peat with nutrient medium. The roots kept below 25° C in greenhouse. Midday light 25 W m$^{-2}$ approximately. (Data of Alexander, 1981)*

|  | Control | Inoculated | Significance of difference $P <$ |
|---|---|---|---|
| Mycorrhizal infection (%) | 0 | $58 \cdot 2 \pm 9 \cdot 9$ | |
| Shoot height (cm) | 5·8 | 9·2 | 0·05 |
| Shoot dry weight (mg) | 47·1 | 100·2 | 0·01 |
| Root dry weight (mg) | 25·5 | 74·0 | 0·001 |
| Total dry weight (mg) | 72·6 | 174·2 | 0·001 |
| Root:shoot ratio | 0·59 | 0·76 | NS |
| Number of lateral buds | 4·4 | 5·9 | NS |
| Length of lateral roots (cm) | 98·2 | 219·0 | 0·01 |
| Total number of root tips | 60·8 | 236·7 | 0·05 |
| Short root cm$^{-1}$ lateral roots | 0·64 | 1·17 | NS |
| Short roots mg$^{-1}$ root weight | 2·43 | 3·31 | NS |

NS, not significant.

Results of this kind which show no effect of mycorrhizal infection or even a decrease of growth rate may be found scattered through the literature but are not often stressed. They are important for two reasons. First, they show that ectomycorrhizal infection does not invariably increase growth of the host in size and weight, and that there is frequently an initial phase, sometimes prolonged, which requires its own explanation, when the growth of infected hosts is similar or slower in rate than that of uninfected ones. Secondly, they often show that different combinations of host and fungus are differently effective in growth. Although in the experiments quoted different species of fungus were compared on a single species of host, there is good evidence that different strains of a single fungal species differ in their effects on a host. Table 43 gives one example from the work of Marx (1979a) of the effect of different strains of *Pisolithus tinctorius* on *Quercus rubra* seedlings. A variability of this kind is exactly what would be expected from the evident physiological and biochemical variability of ectomyc-orrhizal fungi which has already been discussed. In a similar way different biotypes of a host may react differently with a single strain of a fungal species (Marx and Bryan, 1971; Marx, 1979b) although research on this aspect has not yet gone far. Clearly further information on both these topics is essential to the full elucidation of the physiological relationships between the symbionts and to the determination of the most efficient combinations of fungus and host for the production of forest crops. Further experimentation should aim at describing in detail the structure of mycorrhizas which appear to be associated with greater or lesser growth rate. Information concerning Hartig net development, sheath thickness, and particularly the quantity and extent of extramatrical mycelium requires to be collected. These are fairly readily ascertained. Less easily discovered but equally important are physiological properties of the fungal

TABLE 43

Growth of *Quercus rubra* seedings with different strains of *Pisolithus tinctorius*

*Day temperature 28°C, night 22°C. Day length 14·5 h. Growth period 4 months. (Data of Marx, 1979a)*

| Fungus strain | Height cm | Fresh weight g | Ectomycorrhizal percentage of short roots |
|---|---|---|---|
| 138 | 13·3 | 8·5 a | 72 a |
| 136 | 13·5 | 6·5 c | 15 b |
| 145 | 11·5 | 7·0 b | 1 c |
| Control | 12·5 | 6·7 c | 0 |

Height differences insignificant at $P=0·05$. Within other columns common letters denote insignificance.
Percentage of ectomycorrhizas estimated visually.

strains, in particular economic coefficients such as mass of carbohydrate used per mass of mycelium formed. By such analyses the recognition of effective symbionts could be put on a firm basis. It should be further noted that in all this work the growth of the host has been the main preoccupation of the experimenters because they have been concerned, however distantly, with the ecology of the host or its use for human benefit. From the point of view of understanding the physiology of symbiosis, much more knowledge is needed concerning the physiological activities of the fungus in symbiotic union and in culture, and factors that affect them.

The decreases in dry weight of the host which sometimes follow infection appear to be exactly similar to those observed with vesicular-arbuscular mycorrhizal plants. A decrease might be expected in a system where the one symbiont depends on the other for carbon compounds, and the other symbiont depends on the first for nutrients which are essential for growth and photosynthesis. Decreases in growth rate would be expected in light conditions that limited photosynthesis in rate but not the intensity of mycorrhizal infection, or where the supply of soil-borne nutrients is adequate for growth but is not high enough to decrease the intensity of infection. Similar examples where mycorrhizal infection results in no increase or sometimes a decrease in growth rate of the host are also found in ericoid mycorrhiza. Stribley *et al.* (1975) have observed growth diminution in *Vaccinium macrocarpum* in some conditions as did Brook (1952) and Morrison (1957b) with *Pernettea macrostigma* and Bannister and Norton (1974) with *Calluna vulgaris*.

## Source of carbon compounds for the fungus

The view that the host plant is the source of carbon for both host and fungus in ectomycorrhizal plants is of very long standing. It was assumed by Frank, and the later observations of Melin and others, which have been discussed, that ectomycorrhizal fungi required simple carbohydrates which would be in very short supply in soils, provided cogent evidence for the idea. The experiments of Björkman (1942, 1956) on the positive relationship between light intensity and carbohydrate concentration in the root system with the intensity of mycorrhizal infection in spruce and pine further appeared to be consistent with this view. Björkman's conclusions that high internal carbohydrate supply in the roots favoured infection, although subject to recent criticism, seem to have been upheld by Marx *et al.* (1977), as already described.

However, the fact that the root system is more susceptible to mycorrhizal infection when it contains high concentrations of soluble carbohydrates such as sucrose is not in itself a proof that the fungal symbiont derives its supply of carbohydrate in whole or in part from its host. The direct proof of this was given by Melin and Nilsson (1957) who fed $^{14}CO_2$ to the leaves of axenically grown

*Pinus sylvestris* seedlings in combination with either *Boletus (Suillus) variegatus* or *Rhizopogon roseolus*. Photosynthetic products were translocated through the seedlings and found in the roots and in their fungal sheaths. Although this experiment demonstrated very clearly that the products of photosynthesis were translocated directly and rapidly to the fungal symbiont in ectomycorrhiza it did not show anything about the quantities involved. Nor did it prove that all carbon compounds in the fungus were derived from the host. Moreover the control plants in these experiments were decapitated mycorrhizal plants which showed the presence of fixed $^{14}$C in their tissues at the end of the experiment. The quantities amounted to 6%, 11% and 15% of that in the photosynthesizing plants in stem, uninfected roots, and mycorrhizas respectively. Lewis (1963) pointed out that this was explicable in terms of dark fixation of $CO_2$ which commonly occurs in plant material. It was demonstrated to occur in mycorrhizas of *Fagus* by Harley (1964). Its significance will be discussed later.

The change in pattern of translocation in mycorrhizal seedlings has been observed by Shiroya *et al.* (1962) and Nelson (1964) in their experiments with *Pinus resinosa* and *P. strobus*. The size of the root system of the mycorrhizal plants was greater than that of uninfected plants and this constitutes a difficulty in the interpretation of the results. Table 44 gives an example of their results and in the final column their figures are calculated to allow for the larger size of the mycorrhizal root system. On this calculation something between 3 and 4 times as much carbon relative to the quantity fixed was translocated to the mycorrhizal systems. Bevege *et al.* (1975) examined translocation of photosynthates to the root system of *P. radiata* making a comparison of the amounts passed to the mycorrhizas and uninfected roots of the same plants. The plants were given a 2-h pulse of $^{14}CO_2$ in the light and left for 24 h. Eight times as much labelled carbon was translocated to the mycorrhizas as to the uninfected roots on a weight basis. This kind of changed pattern of translocation is similar to that in plants whose leaves are infected with obligate parasites (Holligan *et al.*, 1974). It has not always been observed in ectomycorrhizal seedlings. Ahrens and Reid (1973), for

TABLE 44

Translocation in mycorrhizal and non-mycorrhizal *Pinus strobus*

*12 h translocation in dark after $^{14}CO_2$ in light. (Data of Nelson, 1964)*

| Seedling | % distribution of $^{14}$C | | |
| | In shoots | In roots | In roots relative to weight |
| --- | --- | --- | --- |
| Mycorrhizal | 46 | 54 | 17·4 |
| Non-mycorrhizal | 95 | 5 | 4·6 |

instance, failed to observe any such effect of mycorrhizal infection and this requires some explanation.

The carbon metabolism of ectomycorrhizas of *Fagus* has been examined by Harley and Jennings (1958), Lewis (1963) and Lewis and Harley (1965a,b,c) in a number of aspects which are relevant to the supply of carbohydrate to the fungal sheath by its host. The monosaccharides glucose and fructose are readily absorbed by mycorrhizas from aerated solutions but glucose is selected preferentially from mixtures. Sucrose is hydrolysed by a surface-attached enzyme system and glucose preferentially absorbed from the products. The rate of absorption of hexoses is temperature and oxygen dependent and inhibited by metabolic inhibitors of the cytochrome oxidase pathway and of oxidative phosphorylation. During carbohydrate absorption the respiration of the mycorrhizas increases especially if they have been aged in distilled water. These features show that absorption is a metabolically dependent process in which it is very likely that a phosphorylation step occurs. The analysis of the tissues after uptake from solutions of carbohydrate shows that the change in glucose and other monosaccharides is small and absorbed hexoses are rapidly converted to other compounds (Table 45). Amongst the storage carbohydrates the mycorrhizas contain those characteristic of the host and of the fungus. Glucose and fructose are common to both. Trehalose, mannitol and glycogen are fungal and sucrose and starch are from the host. It may be assumed that sucrose is totally or

TABLE 45

Carbohydrate content of excised mycorrhizas and changes after storage for 20 h at 20 C in water or in 0·5% (w/v) solutions of glucose, trehalose, sucrose, fructose or mannitol

*Values as mg per g fresh weight ( ≡ 150 mg dry weight approximately). (Data of Lewis and Harley, 1965a)*

| | Initial sample | Changes after storage in | | | | | |
| | | Water | Mannitol | Fructose | Sucrose | Trehalose | Glucose |
|---|---|---|---|---|---|---|---|
| Total soluble | 14·69 | −5·36 | −5·66 | −1·53 | −0·10 | +2·02 | +2·91 |
| Total reducing sugars | 4·86 | −1·02 | −0·22 | +0·34 | −0·43 | −0·69 | +0·06 |
| Sucrose | 5·07 | −2·34 | −3·20 | −0·99 | −0·80 | −0·06 | +0·28 |
| Trehalose | 4·74 | −2·00 | −2·24 | −0·88 | +1·13 | +2·77 | +2·57 |
| Mannitol | Nil | Nil | +7·69 | +6·36 | +3·86 | +2·27 | +3·86 |
| Insoluble (glucose units) | 25·94 | −2·84 | +3·87 | +6·83 | +5·20 | +8·10 | +11·14 |
| Total carbohydrate present | 40·63 | 32·43 | 38·84 | 45·93 | 45·73 | 50·75 | 54·68 |

dominantly a host sugar. Both mycorrhizal and uninfected roots absorb sugars from solution but mycorrhizal roots are more efficient in this regard. The typically fungal carbohydrates, trehalose and mannitol, are only slowly absorbed by uninfected roots at rates of one-tenth and one-twentieth of that of mycorrhizas (Table 46). Evidence was obtained that the disaccharides, sucrose and probably trehalose were hydrolysed before absorption and that glucose and fructose had different destinations in the tissues. Although, since they competed in uptake, they would seem to have a common stage or chemical mechanism in their processes of entry, glucose was accumulated mainly as trehalose and fructose as mannitol amongst the soluble carbohydrates of the mycorrhizal sheath. This difference in destination of glucose and fructose led to the question whether it was possible that the enzyme phosphohexoisomerase might be deficient or slow in action in the mycorrhizal roots. It was, however, shown to be present at fairly high activity by Harley and Loughman (1966). One must assume therefore that the systems forming trehalose and mannitol are particularly active.

TABLE 46

Relative rate of uptake of carbohydrate from solution by mycorrhizal and uninfected roots of *Fagus* (data of Lewis and Harley, 1965b,c)

| Sugar supplied | Ratio of uptake rate mycorrhizal:uninfected |
|---|---|
| Glucose | 3·1 |
| Fructose | 2·7 |
| Sucrose | 2·3 |
| Trehalose | 11·8, 10·8 |
| Mannitol | 19·6, 23·6 |

To study the movement of carbohydrate from host to fungus, Lewis and Harley (1965c) carried out experiments in which they fed the cut stumps of *Fagus* mycorrhizas with $^{14}C$-sucrose by placing agar blocks on them. The analysis of the apical region of the mycorrhizas after separation of the sheath and core tissue by dissection showed a movement of carbohydrate into the sheath. In a series of experiments, between 55 and 76% of the $^{14}C$ that had been translocated through the host tissues to the tip region and not released as $CO_2$ was found in the fungal sheath where it was present mainly as trehalose, mannitol and glycogen.

Since trehalose and mannitol were only very slowly absorbed by uninfected roots (Table 46) it was suggested that they, with glycogen, constituted a sink in the mycorrhizal sheath into which carbohydrates were accumulated in a form not readily available to the adjacent host tissues.

This idea was elaborated in a review by D. C. Smith *et al.* (1969) in which they consider allied features of carbohydrate movement between the partners in lichens, pathogenic associations and other symbioses. The hypothesis of a "biochemical valve", ensuring movement in the direction of the fungus has proved to be exceedingly attractive and has been widely accepted, sometimes rather uncritically. Some workers, e.g. Bevege *et al.* (1975) have obtained results quite in keeping with it. Mycorrhizal roots of *Pinus radiata* fed $^{14}CO_2$ in the light not only had higher radioactivity per unit weight than uninfected roots on the same plant, but also had 45–50% of the labelled carbon in trehalose and 15–22% in mannitol. In contrast to this Ahrens and Reid (1973) found no differences that could be attributed to a directed movement of photosynthate to mycorrhizal roots in *P. taeda* and no evidence of the fungal carbohydrates in their analyses. The latter is of course surprising if the fungal sheaths were adequately supplied with carbohydrate. Lewis and Harley (1965b) observed that when excised *Fagus* mycorrhizas were fed with a low concentration of labelled glucose a very large part was incorporated into ionized compounds rather than neutral carbohydrates (Table 47). The labelled carbohydrates absorbed had been metabolized in the carbon turnover of the tissue and much smaller proportions and quantities were incorporated into storage carbohydrates. It therefore follows that if the quantities of carbohydrate synthesized by the leaves in a $^{14}CO_2$ labelling experiment is low, there is much less chance of being able to identify labelled fungal sugars in the root system, even if (as Ahrens and Reed did not do) the mycorrhizal root-tips and uninfected root-tips themselves are separately compared. A similar conclusion was reached if low concentrations of sucrose were fed through agar blocks to excised mycorrhizas of *Fagus* (Table 48). The proportion of the translocated carbohydrate passed to the fungal sheath of the apical region is smaller.

TABLE 47

Distribution of radioactivity in mycorrhizal roots of *Fagus* when $^{14}C$-glucose is absorbed from solutions of high (0·5%) and low ($0·5 \times 10^{-4}\%$) concentrations, in 18 h at 20°C (data of Lewis and Harley, 1965b)

| Sugar concentration | Soluble fraction | | Insoluble fraction | |
|---|---|---|---|---|
| | % neutral | % ionized | % neutral | % ionized |
| High | 83·1 | 16·9 | 96·5 | 3·5 |
| Low | 26·6 | 83·4 | 82·1 | 17·9 |

These experiments may help to explain the lack of success some observers have had in identifying the fungal carbohydrates when analysing whole root systems. However, the hypothesis of a biochemical valve dependent upon the synthesis of carbohydrates peculiar to the fungus has achieved a general, perhaps a too general acceptance. Some have concluded that movement of carbon compounds must be totally in one direction from host to fungus. It is therefore much in the interest of clear thinking if the hypothesis is examined more closely. First of all, carbohydrates like glucose and fructose are present in the fungal cells and they might possibly be absorbed readily by host tissue. Although the absorption of mannitol and trehalose by uninfected roots is small, they are absorbed, and trehalose is metabolized to a small extent, as shown by $^{14}CO_2$ production from it, although mannitol is insignificantly metabolized (Lewis and Harley, 1965c). On these grounds we would expect the valve to be able to leak and the available experiments indicate that it may. In the experiments of Lewis and Harley there is some indication of this. It may be assumed that sucrose is only found in the host tissues of mycorrhizas, so in any sugar-feeding experiment where carbohydrate is applied to the sheath and sucrose is found in the tissues we can assume that carbon compounds have passed into the host. In Table 45 it can be seen that the amount of sucrose in the tissues diminishes during storage in water and in any of the carbohydrates except glucose. Sucrose slightly increases in concentration during storage in glucose. This might be interpreted as evidence of a movement of glucose into the host. Lewis and Harley (1965b) also made the point that some aspects of the $^{14}C$-labelling pattern in their experiments are most easily explained if there is some movement of hexoses from fungus to host. Movement of this kind, although not an extensive movement, has been shown by Reid and Woods (1969) to occur from mycelial strands of *Thelephora terrestris* to the mycorrhizas and host tissues of seedlings of *Pinus taeda*. Movements both from the photosynthetically fixed $^{14}CO_2$ to the fungus, and the reverse flow from $^{14}C$-glucose applied to mycelial strands into the host were observed using the same symbiosis, *T. terrestris* and *P. taeda*. On the other hand, Lewis (1963) records an experiment where *Fagus* mycorrhizas were fed with $^{14}C$-glucose and only mannitol, and trehalose became labelled in the sheath, the host tissues were negligibly labelled.

This discussion has mainly concentrated on carbohydrates, which are the

TABLE 48

Distribution of radioactivity within the apical region of *Fagus* mycorrhizas following translocation from agar blocks containing high (10%) and low (0·001%) $^{14}C$-sucrose in 22 h at room temperature (data of Lewis and Harley, 1965c)

| Tissue | Percentage present | |
| | High sucrose | Low sucrose |
| --- | --- | --- |
| Sheath | 76, 67 | 52, 49 |
| Core | 24, 33 | 48, 51 |

main carbon and energy currency, although mention has been made of ionized carbon compounds. These would include, amongst others, organic acids and amino acids. Harley (1964) showed that the rate of dark fixation of $CO_2$ by the mycorrhizas of *Fagus* was greatly increased if ammonium was being absorbed. Using $^{14}C$-bicarbonate he was able to show that the main destination of the increased fixation was into glutamine. Subsequently Carrodus (1967) confirmed this. Since this provided a simple method for labelling glutamine, Reid and Lewis (see Lewis, 1976) used it to observe the movement of glutamine from the sheath to the host. This demonstrates a way in which carbon compounds may return to the host.

The fixation of $CO_2$ in the dark is a process which in the present context maintains the tricarboxylic acids as they are used in synthesis of amino acids. Clearly if the tricarboxylic acids are used in quantity they must be reformed in order that the TCA cycle may be maintained. The synthesis of glutamine by whatever route places a drain on $\alpha$-ketoglutaric acid. The detailed pathway of fixation has not been worked out in mycorrhizas but it commonly involves phosphoenol pyruvate which, with $CO_2$, forms oxalacetic acid by a reaction mediated by phosphoenol pyruvate carboxylase. This reaction is powered by the high-energy phosphate bond of the phosphoenol pyruvate. Hence one quarter of the carbon of the oxalacetate is derived from $CO_2$ and the rest from carbohydrate via glycolysis. This reaction does not therefore produce carbon compounds which can be an additional source of energy but is an energy-requiring process which synthesizes essential compounds using energy and carbon radicals produced in glycolysis.

## Mechanism of the release of carbon compounds by the host

There are several possible mechanisms by which carbon might pass from the host to the fungus. The contact between the organisms varies with position and age and the mechanisms may therefore vary. First, there is the possible mechanism which received comment earlier, that the root cap of ectomycorrhizas is small but according to Clowes (1981) cells are added to it continually by cell division such that it would be replaced at least every three days. As has been mentioned, these cap cells may not all be fully formed but we cannot doubt that the disappearance of the cells, even before they form a complete wall, must represent a release of carbon compounds which might be absorbed by the fungus. Secondly, the outer cells of the cap–epidermal complex also disappear as complete cells, although they may continue to exist as wall fragments in the inner sheath (Marks and Foster, 1973). Their contents may be available to the fungus. Thirdly, around many roots there exists a layer of mucigel which together with other exudates from the root (Foster, 1981b; and see Scott Russell, 1977) is an important source of carbon for the root-surface and rhizosphere populations of microorganisms. Fourthly, as the mycorrhizal fungus penetrates to form the Hartig net it may alter the behaviour of its host with respect to the production of cell wall polymers, and use their precursors as a source of carbon. Fifthly, the fungus may alter the condition of the cells which the Hartig net surrounds, such that they release to it increased amounts of carbon compounds. Lastly, when the fungal hyphae penetrate senescent cortical cells, especially in the late Hartig net region, the soluble contents of the cells are presumably absorbed.

Certain of these possible sources of carbon compounds for the fungus are very likely to be important. The fungi can absorb a variety of soluble carbon compounds other than carbohydrate, such as some amino acids and organic acids, and it would seem certain therefore that degenerating root-cap and epidermis and senescent cortical cells must supply some carbon, but it is difficult to quantify. The soluble carbon compounds secreted into the rhizosphere certainly include amino acids, organic acids and simple sugars. The mucigel itself might be used by those fungi which are capable of hydrolysing the carbohydrate polymers of which it is composed. The quantity of carbon lost in these forms into the soil by most roots seems to be very great. Quantities in the region of 8% of all $^{14}$C fixed in 21 days have been found in the soil around the roots of oat seedlings (see Scott Russell, 1977) and much higher values have been mentioned in the literature.

The fourth possibility—the change of behaviour of the host cells in the Hartig net region so that wall formation is altered or prevented—depends upon the work on fine structure in which the walls of hyphae of the Hartig net are described as forming, with the modified host cell walls, an "involving layer" or a "contact zone". It is possible that here, as in the "interfacial matrix" of vesicular-arbuscular mycorrhiza the fungus modifies the activities of the plasmalemma of the host. The matter requires much further observation both ultrastructurally and chemically.

The fifth suggestion, that the fungus produces a substance which causes the cells of the host to leak, is a common assumption in a number of symbioses. In lichens (see D. C. Smith, 1974) the algae when first isolated from the lichen thallus, leak carbohydrates readily into the medium, but during growth in culture they become less leaky and eventually the whole of the carbon which once leaked is converted into insoluble compounds within the algal cell. This kind of effect of association has been thought to be brought about by some hormone produced by the fungus. No such compound has been found in the algae of lichens. Wedding and Harley (1976) investigated the possibility in the mycorrhiza of *Fagus*. They argued that it was equally possible that a characteristic fungal product produced in quantity might be the operative factor as a hormonal substance. They therefore looked at the effect of mannitol and other linear polyols on the enzymes responsible for carbohydrate metabolism in the host. Figure 17(a) and (b) is a summary of their findings. They showed that

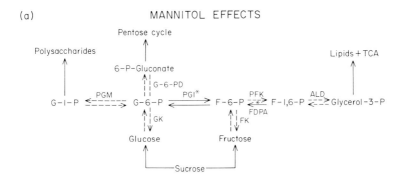

(a)                    MANNITOL EFFECTS

PGM        Phosphoglucomutase          GK     Glucokinase
PGI        Phosphoglucoisomerase       FK     Fructokinase
G-6-PD     Glucose-6-phosphodehydrogenase   FDPA   Fructose diphosphatase
PFK        Phosphofructokinase         ALD    Aldolase

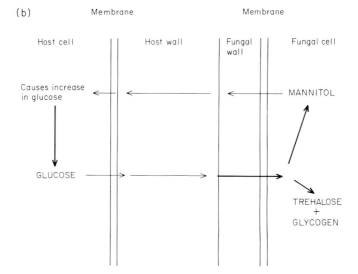

FIG. 17. Diagrammatic representation of the effects of mannitol on the extracted enzymes involved in some aspects of carbohydrate metabolism in beech roots. Sucrose is presumed to be the form in which photosynthate reaches the host tissue of beech mycorrhizas. (a) Enzymes reduced in activity by mannitol, – – ➔; enzymes unaffected by mannitol, ——*➔; enzymes uninvestigated, ———➔. (b) Schematic diagram of the possible effects of mannitol on the movement of carbohydrates from host to fungus. Passive movement, ———➔; active absorption or reactions, ━━━➔

   After Wedding and Harley (1976).

mannitol, by affecting enzyme action, might result in a rise of glucose concentration in the host and hence encourage a leak to the fungus. As they point out, mannitol may be slowly absorbed by the host but it is not apparently metabolized by it. This activity of mannitol should be regarded as a demonstration of the possibility that common, characteristically fungal, products may be important and should be examined before search is made for special hormones.

## Respiration

Comparison of the respiration rates of mycorrhizal and uninfected roots has led to different views. The mycorrhizal roots are frequently said to respire faster than uninfected ones, whereas extensive investigation of the respiration rate of the roots and mycorrhizas of *Fagus* has shown that this is not always true of them. Tables 49 and 50 give examples calculated on a weight basis and the differences must be explained. It is noteworthy that the contention of a more rapid rate of respiration in mycorrhizas is based on measurements using seedling roots of *Pinus*. In these there are two kinds of uninfected root apex. The first, the apices of short roots, are much more abundant than the second, the apices of the long roots of the system. As Hatch and Doak (1933) first showed, the short roots of pine if they remain non-mycorrhizal frequently cease growth and abort, whereas if they become infected they grow and dichotomize. A comparison of mycorrhizal and non-mycorrhizal short roots will show that the infected ones respire faster. In contrast the root systems of other species of tree do not show such a clear difference between long and short roots, and uninfected and mycorrhizal roots even of the same level in the hierarchy of roots differ in that the former may grow rapidly and the latter grow only slowly. The difference was noted by Lewis and Harley (1965b) for uninfected roots when fed with carbohydrate synthesized structural polysaccharides and soluble storage carbohydrate, whereas mycorrhizas accumulated carbohydrates of a soluble and insoluble storage nature. In this case the rate of respiration of uninfected roots varies very much according to

TABLE 49

Respiration rates of mycorrhizal and uninfected roots of *Pinus sylvestris* seedlings. $\mu lO_2 \, g^{-1}$ dry weight $h^{-1}$ (data of Schweers and Myer, 1970)

| Infected seedlings | Root system | 1132 |
|---|---|---|
| | Mycorrhizas | 1684 |
| | Uninfected roots | 875 |
| Uninfected seedlings | Root system | 863 |
| | Fine roots | 772 |

their growth rate, whereas that of the mycorrhizas is relatively constant. The equivalent comparison to that in *Pinus* would be between the mycorrhizas of *Fagus* and those uninfected apices of roots which are beginning to grow slowly and to differentiate close to the tip and which eventually become infected.

Although then on a weight basis we may conclude the different respiration rates of mycorrhizal and uninfected roots are predictable because of their different growth and function, there is no doubt that the fungal tissues of mycorrhizas consume significant amounts of carbohydrate in their respiration. Dissection of the fungal sheath from the mycorrhizas of *Fagus* led to the conclusion that the fungal tissue accounted for at least half the oxygen consumption and $CO_2$ emission, although it comprised only about 40% of the weight.

Relatively little is known about the characteristics of the respiration of ectomycorrhizas except for those of the mycorrhizas of *Fagus* of which an account was given by Harley (1969a), and only salient points of that account will be repeated. The respiratory quotient of excised mycorrhizas as measured in solution in Warburg flasks is approximately unity but it rises steeply when the atmosphere in the flasks is below an oxygen concentration of about 5%. It is at this level that deficiency of oxygen affects the rate of nutrient uptake. The release of $CO_2$ under anaerobic conditions can equal or exceed $CO_2$ production in air. The most striking feature of the respiration of *Fagus* mycorrhizas is that their oxygen uptake is not inhibited by substances which inhibit the cytochrome oxidase pathway. Sodium azide and potassium cyanide at millimolar concentrations or less, either greatly stimulate oxygen uptake or leave it relatively unaffected; moreover carbon monoxide in the dark does not inhibit respiration (Harley and McCready, 1953; Harley *et al.*, 1956; Harley and ap Rees, 1959). Harley and ap Rees concluded that the inhibitor-resistant respiration was a property of the fungal sheath but that the fungus also contained an inhibitor-sensitive cytochrome system. Coleman and Harley (1976) confirmed these

TABLE 50

Respiration rates of *Fagus* mycorrhizas and uninfected roots obtained over periods of years by sampling from beechwoods. $\mu lO_2$ 100 mg$^{-1}$ dry weight h$^{-1}$

| | Range | Mean | Authority |
|---|---|---|---|
| Mycorrhizas | 218–419 | 327 | Harley and McCready (1950–1952) |
| Mycorrhizas | 277–354 | 320 | ap Rees 1955–1957 |
| Uninfected roots | 134–696 | 269 | ap Rees 1955–1957 |
| Mycorrhizas | — | $284 \pm 56$ | Wedding and McCready (unpublished) |
| Uninfected roots | — | $328 \pm 78$ | |
| Mycorrhizas | — | 367 | Wedding and Harley (1976) |
| Uninfected roots | — | 239 | |

findings by extracting the mitochondria of *Fagus* mycorrhizas using polyvinylpyr-
rolidone, bovine serum and reducing agents to minimize the inhibitory effects of
the phenolic compounds in the tissue extracts. They were able to obtain
preparations of particles in which oxygen uptake was tightly coupled to
phosphorylation during the oxidation of Krebs cycle intermediates and they
obtained ADP/O ratios which indicated that the three sites of oxidative
phosphorylation were operative. The application of substrates and inhibitors to
these mitochondrial preparations allowed the construction of the scheme shown
in Fig. 18. The salient feature is the presence of the non-phosphorylating bypass
to oxygen which is inhibited by hydroxamates such as *m*-Clam. Cyanide-
resistant bypasses of this kind are not an uncommon feature of the respiration of
plants. Indeed, Antibus *et al.* (1980) have identified a similar bypass in the
ectomycorrhizal roots of *Salix rotundifolia* and the vesicular-arbuscular roots of *S.
nigra*. It is, however, the magnitude of the bypass that is remarkable in
ectomycorrhizas, for inhibition of the cytochrome pathway may result in a great
stimulation of oxygen uptake. This can be explained because cyanide and azide,
by inhibiting the cytochrome pathway, eliminate two of the sites of oxidative
phosphorylation, leaving only one in operation. This may allow a greater rate of
glycolysis by making ADP more available for the process.

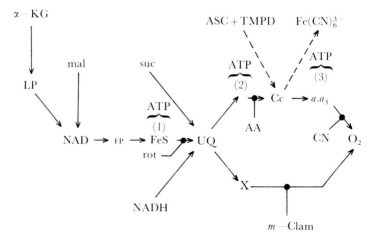

FIG. 18. Diagram of the suggested electron transport pathways to oxygen in isolated
mitochondria from beech mycorrhizas. αKG, αKetoglutarate; mal, malate; suc,
succinate; ASC + TMPD, ascorbate + tetramethyl-*p*-phenylene diamine; rot, rotenone;
AA, Antimycin A; CN, cyanide; LP, lipoate; FP, flavoprotein; FeS, iron sulphur protein;
UQ, ubiquinone; Cc, cytochrome c; a.a$_3$, cytochrome a.a$_3$; *m*-Clam, *m*-chloro-
benzhydroxamate; 1, 2, 3 are sites of oxidative phosphorylation. ●—, inhibition;
→, electron transport.

Neither the function of this oxidative pathway nor the means by which it is switched on and off in natural conditions is yet understood. Coleman and Harley observed that when succinate was being oxidized by *Fagus* mitochondria the addition of *m*-Clam caused a slight increase of ADP/O ratio and this might be an indication that both pathways operate when there is ADP limitation of the cytochrome system. Whether factors which cause such limitations switch the pathway on remains to be determined. Clearly the alternate pathway operates in particular when the cytochrome system is inhibited and then the glycolytic system remains active and may increase in rate. It was suggested by Coleman and Harley that the function of the alternate pathway might therefore be to allow production of metabolites such as organic acids which, if secreted in the soil, might have important chelating properties, but no investigation of this possibility has yet been made.

The respiration rate of mycorrhizas of *Fagus* depends on whether they are absorbing nutrients or not. If excised mycorrhizas are kept in aerated water or buffer for a period their respiration rate falls. As can be seen from Table 45 (and also Harley and Jennings, 1958; and Harley *et al.*, 1977), the fall is not at first dependent upon the exhaustion of available carbohydrates although after prolonged storage it may be so. After one or two days the respiration reaches a low, almost uniform, value which is usually maintained for several days. The application of nutrient salts such as those of ammonium or phosphate, and to a lesser extent chloride and sulphate, causes a rise in respiration rate, which in some cases is closely related to the quantity absorbed. Similar rises in respiratory rate are observed when carbohydrates are absorbed or when certain inhibitors are applied. Substances like 2,4-dinitrophenol which uncouple oxidation and phosphorylation, or azide and cyanide that inhibit the cytochrome system, also greatly increase oxygen uptake and $CO_2$ emission. All the effects are explicable in terms of the reduced or increased availability of phosphate acceptors like ADP. The work on the mitochondria of mycorrhizas demonstrates, if demonstration be needed, the operation of the TCA cycle in terms of electron transport in mycorrhizas. Carrodus and Harley (1968) showed that when [14]C-acetate was fed to *Fagus* mycorrhizas, rapid incorporation of [14]C occurred at first into organic acids, followed by incorporation into amino acids later. The order of labelling of citrate, succinate and malate was as one would expect for the operation of the TCA cycle.

All these features of respiration are clearly relevant to the operation of mycorrhizas as nutrient-absorbing organs.

## Conclusion

A great deal of work on the growth of seedlings in the mycorrhizal condition had until fairly recently been done using conifers symbiotic with relatively few fungal species. Now an increasing knowledge is accumulating about many kinds of ectomycorrhizal plants grown with a great variety of symbiotic fungi. Few of these, however, include full descriptions of the mycorrhizal organs and their development and senescence. It would seem important that in the future the extent of the external mycelium, the size, especially the mass, of the sheath and the degree of development of the Hartig net and the extent to which hyphae penetrate into the cells should be described. The external mycelium appears to be important in nutrient uptake, the mass of the fungus may have effects on the consumption of photosynthate, as may also the degree of penetration into the tissues. It might be that more extensive penetration results in a greater drain on the host, or a shorter life period of the joint organ or both. Using fungi like *Thelephora terrestris*, which, as Birraux and Fries (1981) pointed out, could be made to fruit on very young hosts, the carbon drain and indeed the whole efficiency of the symbiosis might be studied at different seasons and periods of development of a single kind of ectomycorrhiza composed of the same genetic strains of host and fungus.

*Chapter 10*

# Physiology of Ectomycorrhizas. Nutrient Uptake

## Introduction

In the late nineteenth century ectomycorrhizas were believed to be especially important in the absorption of nitrogen. In 1894 Frank put forward what came to be called the "Nitrogen Theory" of mycorrhiza which sought to explain the ability of mycorrhizal plants to absorb nitrogen compounds from soil lacking nitrates. It was believed that fungal infection allowed the host to use ammonium and organic nitrogen present in the organic layers of the soil. This nitrogen theory held prominence for over 40 years and much research was based upon it. The work of Hatch (1937), which included experiments on the growth and nutrient content of seedlings of *Pinus*, threw doubt upon it in its simple form. His analyses indicated that ectomycorrhizal seedlings of *Pinus* were not only larger in size but also contained greater quantities of the major elements, nitrogen, potassium and phosphorus per unit weight than their non-mycorrhizal controls. These results were soon confirmed by many others. Hatch put forward the view that the importance of mycorrhizas lay in their increase in efficiency of uptake of any nutrient in short supply. The reasons for greater efficiency lay, according to him, in the change in geometry of the absorbing surface of the root system following infection, coupled with an increased absorbing area due to the surfaces of the fungal hyphae. When root systems became infected the short roots continued to grow and branch to form complex mycorrhizal systems of branches. The host tissue of the infected rootlets increased in diameter and this was further increased by the fungal sheath.

As Bowen (1973) pointed out, estimates of the specific uptake rate of nutrients (quantity absorbed per mg of root) can be calculated from the results of Hatch and others (Table 51). Although they are clearly underestimates because the contents of nutrients in the seed are not known, many comparative analyses of mycorrhizal and non-mycorrhizal plants show a greater specific uptake rate in the former. Table 51 includes an example from the results of Finn (1942) which does not show this effect. However, all the experiments in the Table and many others show that a higher quantity of nitrogen, phosphorus and potassium per unit seedling weight is found in the mycorrhizal seedlings as a whole. Results such as these gave rise to the so-called "Mineral Salt" theory of mycorrhiza. All nutrients are absorbed through the fungus, and mycorrhizal infection tends to improve the absorption of whatever major nutrient is most deficient. Hatch emphasized the dominant extent to which the formation of mycorrhizal branch systems contributes to the increased absorbing surface of the root systems of both seedling and adult ectomycorrhizal trees over uninfected ones. Since they are covered by a fungal sheath, ectomycorrhizal infection, by its very structure, must affect the uptake of all nutrients and he emphasized that it was illogical to stress only its effect on the absorption of nitrogen, or on that of any other particular nutrient. Certain experimental results showed, however, that phosphate absorption was more stimulated than that of other nutrients by mycorrhizal infection of tree seedlings (McComb, 1938; McComb and Griffith, 1946; Stone, 1950). Such findings together with the facts that phosphate deficiency is common in natural soils (and especially in the soils available for reafforestation), and also that the isotope $^{32}P$ was extremely suitable for experimental purposes, led Kramer and Wilbur (1949), Harley and McCready (1950) and Melin and Nilsson (1950) to study phosphate uptake by mycorrhizal roots. Then, as so often happens, a fashion was set, and work on phosphate uptake and nutrition of mycorrhizas increased almost to the exclusion of work on the uptake of other elements. It must therefore be emphasized very strongly that the possession of the fungal sheath sets ectomycorrhizas apart from those mycorrhizas that do not possess a sheath. Sheathless mycorrhizas may absorb significant amounts of nutrients through the surfaces of the cells of the host in addition to absorbing them through the fungal hyphae. Ectomycorrhizas and others with a complete sheath cannot. All materials must pass through the fungus which therefore dominates all uptake by the mycorrhizas and often the greater part of the surface of the absorbing root system is ensheathed by it. It is therefore illogical to assume that ectomycorrhiza is important in phosphate absorption alone, for infection must *affect* the absorption of all substances.

## The process of absorption

The quantity of hyphae which join the sheath of ectomycorrhizas to the soil is variable. It might be expected that the more extensive and copious the connexion the greater the exploitation of the soil. Similar considerations must apply to those discussed above concerning the extramatrical hyphae of vesicular-arbuscular mycorrhiza. The hyphae from the mycorrhizal surface not only cross

TABLE 51

Growth and specific nutrient uptake of nitrogen, phosphorus and potassium by *Pinus strobus* seedlings. (1) Hatch (1937); (2) Mitchell *et al.* (1937); (3) Finn (1942)

| Degree of mycorrhizal infection | Dry weight mg | Root weight mg | Nitrogen | | Phosphorus | | Potassium | |
|---|---|---|---|---|---|---|---|---|
| | | | T | SA | T | SA | T | SA |
| (1) Mycorrhizal (+) | 448 | 180 | 5·39 | 0·030 | 0·849 | 0·0047 | 3·47 | 0·019 |
| | 361 | 170 | 4·62 | 0·027 | 0·729 | 0·0042 | 2·57 | 0·015 |
| Uninfected (0) | 300 | 174 | 3·16 | 0·013 | 0·229 | 0·0013 | 1·04 | 0·006 |
| | 361 | 182 | 2·51 | 0·018 | 0·268 | 0·0015 | 1·94 | 0·011 |
| | 301 | 152 | 2·40 | 0·016 | 0·211 | 0·0014 | 1·17 | 0·008 |
| (2) Mycorrhizal (70·2%) | 337 | 127 | 5·4 | 0·042 | 0·72 | 0·006 | 2·12 | 0·016 |
| Uninfected (9·5%) | 181 | 75 | 2·2 | 0·029 | 0·13 | 0·002 | 0·81 | 0·011 |
| (3) Mycorrhizal (87%) | 223 | 120 | 2·05 | 0·017 | 0·30 | 0·003 | 1·52 | 0·013 |
| Uninfected (10%) | 93 | 56 | 1·17 | 0·021 | 0·24 | 0·004 | 0·98 | 0·018 |

T, total absorbed mg.
SA, specific absorption mg mg$^{-1}$ root dry weight.

any zone depleted of nutrients near the sheath surface, but also may readily be extended or replaced with a small expense of carbon and nutrients per unit area of absorbing surface formed. Stone (1950) compared two samples of seedlings of *Pinus radiata* with very different extents of extramatrical hyphae. He found those with the more extensive system absorbed phosphate from the soil faster and translocated a greater quantity to their needles. By use of an elegant experimental method, Melin and Nilsson (1950, 1952, 1953a,b, 1955, 1958) and Melin *et al.* (1958) showed that nutrients labelled with isotopes, fed to the extramatrical hyphae of *Pinus sylvestris*, were translocated to the root by the hyphae and thence through the host to the needles. The nutrients translocated in this way were phosphorus from soluble orthophosphate, nitrogen from ammonium and glutamic acid, calcium, sodium and other cations. Skinner and Bowen (1974a,b) using *Pinus radiata* and *Rhizopogon luteolus* confirmed the transport of phosphate in mycelial strands. Phosphate was absorbed into the strand by a mechanism which was inhibited by $10^{-3}$ M KCN and temperature dependent. The translocation might occur over distances of up to 12 cm but they observed that there were large differences in extent of growth in the soil between strains of fungus and between samples of the same fungal isolate in different soil conditions (Skinner and Bowen, 1974b). Experiments such as these emphasize the need for the extent of production of extramatrical hyphae and rhizomorphs to be fully described in experiments on the efficacy of different combinations of fungus strain and host genotype.

Much of the detailed experimental work on the mechanism of uptake of nutrients by ectomycorrhizas has been done with excised mycorrhizas in which both the outgoing hyphae and the throughput of water in the system are eliminated. The effect of those factors has to be examined in mycorrhizas normally sited in soil and attached to their parent root. The overwhelming reason for using excised mycorrhizas for investigating certain aspects of mycorrhizal physiology is that uniform samples of them can be obtained for studying specific stages of the uptake process. Whole root systems are composed of mycorrhizas, uninfected primary roots and secondarily thickened axes in various proportions so complicated that it is almost impossible to conclude anything about detailed mechanisms from them. The mycorrhizas are covered with a sheath, the surface the properties of which can be assumed to resemble, qualitatively, those of the extramatrical hyphae closely and the tissues can be expected to react metabolically to applied conditions in much the same way. It is easy to obtain large numbers of similar mycorrhizal roots from the surface layers of forest soil, wash and prepare them with no great effort. Of course excised mycorrhizas and uninfected roots can also be obtained from aseptically grown plants in the laboratory, but the labour of providing them in sufficient quantity for experimental work on any large scale is daunting. The criticisms levelled at the use of excised mycorrhizas are those applicable to all experiments with excised roots, i.e. that the transpiration stream is eliminated and the tissue may suffer starvation of respiratory substrates during the experiment. The transpiration stream is not, however, of direct influence on nutrient absorption except in certain conditions and its effect can be examined by other means. As has been shown in Table 45, carbohydrate deficiency does not occur during short experiments with excised *Fagus* mycorrhizas and this probably applies to mycorrhizas sampled from natural conditions but not perhaps to those produced in the laboratory, unless high light intensities are provided.

The factors which affect the rate of absorption of nutrients by mycorrhizas are similar to those which affect the rate of absorption of substances by most plant material. When washed mycorrhizas are placed in nutrient solutions their apparent free space comes into equilibrium with the solution. Thereafter absorption of solutes takes place at rates which are metabolically dependent. The apparent free space for cations is, as usual, greater than that for anions so that with them there is a clear initial phase where uptake is apparently more rapid and less metabolically dependent than later. If account is taken of the equilibration phase absorption is found to be dependent upon temperature and oxygen supply, related to respiratory rate, decreased by inhibitors of the cytochrome system and by those that uncouple oxidative phosphorylation from oxygen uptake. The rate of absorption shows a tendency to reach a maximum with increasing concentration but the curve is not truly hyperbolic and no limiting value is reached over reasonable ranges of concentration. With phosphate uptake by *Fagus* mycorrhizas goes on increasing up to high values (Fig. 19) even in excess of 100 mM. The rate of absorption also depends upon the

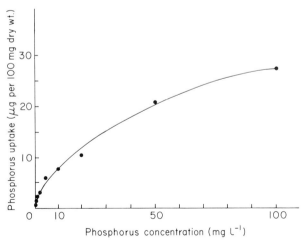

FIG. 19. Absorption of phosphate by excised mycorrhizas of *Fagus sylvatica* from phosphate solutions containing $KH_2PO_4$ up to $3{\cdot}2$ mM $\equiv 100$ mg P $l^{-1}$. Uptake in $\mu$g per 100 mg dry weight of mycorrhiza in 45 min. (Harley and McCready, 1952.)

availability of respiratory substrate but, as has been mentioned, the mycorrhizas of naturally grown adult trees contain large reserves of carbohydrate especially in the sheath. However, uptake of ammonium which is rapidly assimilated into organic compounds in the tissues may be accelerated by the application of carbohydrates externally.

These features of nutrient absorption lead to the conclusion that the process in mycorrhizas does not differ in essentials from that of other absorbing organs in its relation to metabolic processes. In the ensuing account where there are special points about the effect of external factors on the uptake of particular nutrients, they will be discussed, for its aim will be to consider the manner in which the absorption physiology of ectomycorrhizas differs from that of uninfected roots. The papers upon which the above account is mainly based are from researches on *Fagus* mycorrhiza (e.g. Harley and McCready, 1952a,b; Harley *et al.*, 1953; Harley *et al.*, 1954; Harley *et al.*, 1956; Harley and Jennings, 1958; Harley and Wilson, 1959; Carrodus, 1966). It is to be hoped that in time the results will be confirmed and widened by use of mycorrhizas of other kinds.

## Uptake into uninfected and mycorrhizal roots

As has been discussed, the most active uninfected root apices differ from mycorrhizas by the fact that they are dividing and growing. It is well known that rate of uptake in the tip region of a growing root is much greater than that in the region behind it. This is not generally true of mycorrhizal apices. McCready (unpublished) found that the uptake of phosphate by *Fagus* mycorrhizas did not change greatly over distances of 1–2 cm. However, the rate of absorption for centimetre lengths of mycorrhizal and non-mycorrhizal roots may be very

TABLE 52

Comparative uptake rates of phosphate by excised mycorrhizas and uninfected roots

| Authors and host | Fully infected | Uninfected | Sheath poorly developed |
|---|---|---|---|
| Harley and McCready (1950) | 5·18 | 0·88 | 4·76 |
| *Fagus sylvatica* | 6·68 | 0·75 | 1·20 |
| | 1·97 | 0·42 | 0·62 |
| | 2·72 | 0·61 | — |
| | 1·69 | 0·72 | 2·13 |
| Bowen and Theodorou (1967) | 7·5 | 3·5 | — |
| *Pinus radiata* | 15·5 | | |
| | 15·0 | 5·5 | 5·5 |

The values for each host are relative to one another.

different. Table 52 shows comparisons made by Harley and McCready (1950) with *F. sylvatica* and Bowen and Theodorou (1967) with *Pinus radiata*. This feature was emphasized by the autoradiographs of Kramer and Wilbur (1949) and also of Harley and McCready (1950) which show intense phosphate accumulation in the mycorrhizas and in the extreme apices of uninfected roots but not elsewhere. Bowen (1968) by an ingenious method scanned the long roots of *P. radiata* with a Geiger counter and showed that whereas the most active region of uptake in uninfected roots was at the apex and in the positions of the apices of developing short roots, that of a long root bearing mycorrhizas was at its apex and more particularly at the positions of the mycorrhizal rootlets.

These results were obtained with phosphate the uptake of which, as Harley and McCready (1952a) showed, resulted in a great accumulation in the fungal sheath. By dissecting the fungal layer from the host tissue they were able to estimate the relative quantities accumulated in the two tissue systems. This method has since been used for more detailed analyses, in particular of carbon compounds, phosphate fractions and other nutrients. In the case of phosphate, about 90% of the absorbed phosphate was found in the sheath after absorption from low concentrations. This observation has since then often been confirmed. Experiments with whole mycorrhizal seedlings of oak and pine (Lobanov, 1960; Clode, 1956; Morrison, 1962a) have shown that a higher proportion of the phosphate absorbed by mycorrhizal seedlings is found in their roots than in those of uninfected plants. Harley and McCready (1952a) verified that the accumulation was not dependent on or influenced by excision. They compared the distribution of phosphate between the sheath tissue and the subjacent core in mycorrhizas attached to adult trees in the forest and those detached. Comparisons were made on three occasions: when the trees were leafless, developing their leaves and in full leaf. On all occasions there was a great accumulation in the sheath (Table 53). In spite of this, their results have often been criticized as due to excision. Melin and Nilsson (1958) compared the accumulation in the fungal sheath of normal seedlings with that in the sheaths of seedlings whose shoots had been removed. They compared uptake per unit dry weight in the sheath and in the host tissues of the roots. Assuming that their dissections were satisfactory and a complete separation of the two tissue systems was achieved, a recalculation shows that there was a greater accumulation in the fungal sheath of decapitated plants (71–91%) than in intact plants (28–63%). This discrepancy seems to be due to decapitation and therefore might be caused by cessation of transpiration or by carbohydrate deficiency which is not unusual in laboratory-grown seedlings; the variability of the results of dissection leads to doubts about the successful recovery of all the sheath material. Indeed, the bulk of the available evidence shows that the primary location of the accumulation of nutrients, especially of phosphate, is in the fungal sheath. A similar but somewhat lesser accumulation of other ions and of carbohydrates in the sheath has been observed and examples are given in Table 54. These are results of dissection following a period of absorption when of course two features must be noted. First, the external concentration is applied to the sheath and since what is absorbed passes through it, an appearance of accumulation will occur unless that movement is very rapid. Secondly, there may be a real accumulation depending upon the activity of the fungus. Garrec and Gay (1978) have analysed the mycorrhizas of *Pinus halepensis* by means of electron probe and concluded that phosphate is mainly accumulated in the fungal sheath and Hartig net region and is lower in the host tissue; magnesium is found especially in the inner sheath and Hartig net region; calcium follows phosphate to a considerable extent and is particularly abundant in the sheath and Hartig net region. This parallel distribution of calcium and phosphorus is explicable since it has been shown that the phosphatic granules (polyphosphate) which are present in the fungal tissues also contain calcium (see below). By contrast potassium is fairly uniformly distributed. Sihanonth and Todd (1977) examined the distribution of elements in the tissues of mycorrhizal *Pinus taeda*. They record that the metals magnesium, potassium and calcium and the non-metals phosphorus and sulphur were all at higher concentrations in the fungal tissues of the sheath and Hartig net. The cells of the host in the mycorrhizas contained greater quantities than those in uninfected roots. Since the sheath, whether it accumulates nutrients permanently or semi-permanently, separates the host tissue from the soil, the mechanism of the passage of substances through it and factors affecting their rate of passage require investigation.

TABLE 53

Estimates of the proportion of phosphorus which accumulates in the fungal sheath of *Fagus* mycorrhizas when attached to the parent root system or when excised

*Experiments in Bagley Wood with roots of adult trees at three seasons. Mycorrhizas in aerated phosphate solution pH 5·5 at ambient temperature*

| Condition of mycorrhizas | Attached | | | | Detached | | | |
|---|---|---|---|---|---|---|---|---|
| Date | 31 March | 11 May | 23 July | | 31 March | 11 May | 23 July | |
| mm $KH_2$ $^{32}PO_4$ | 0·074 | 0·32 | 0·16 | 1·6 | 0·074 | 0·32 | 0·16 | 1·6 |
| Mean percentage in sheath | 88 | 88 | 90 | 87 | 91 | 89 | 85 | 91 |
| Range | 74–96 | 83–94 | 89–94 | 86–93 | 85–96 | 79–93 | 83–89 | 91–93 |

Harley (1978a,b) has stressed the probability that the sheath of ectomycorrhizas acts as an important storage organ both for inorganic nutrients derived from the soil and carbon compounds derived from the host. This property, which might have important ecological implications, will be discussed later.

TABLE 54

Accumulation of substances in the fungal sheath immediately following uptake

*The sheath is 39% of the total dry weight in* Fagus *and 34% in* Nothofagus

| | Concentration applied mM | Percentage in sheath | |
|---|---|---|---|
| $H_2PO_4^+$ (as $KH_2PO_4$) | 0·016 | 93 | ⎫ Harley, Brierley and |
| | 0·16 | 92 | ⎬    McCready (1954) |
| | 1·0 | 83 | F. A. Smith (1972) |
| | 30·0 | 65 | Brierley (1953) |
| $Cl^-$ (as KCl) | 0·2 | 45 | F. A. Smith (1972) |
| $SO_4^{2-}$ (as $MgSO_4$) | 0·31 | 53 | Morrison (1962a) |
| $K^+$ (as KCl) | 0·2 | 67 | Edmonds and Harley (unpublished) |
| $Rb^+$ (as Rbcl) | 0·1 | 62 | Harley and Wilson (unpublished) |
| $NH_4^+$ (as $NH_4Cl$) | 10·0 | 65 | Carrodus (1967) |
| | 10·0 | 75 | Harley (unpublished) |
| $NO_3^-$ (as $NaNO_3$) | 0·3 | 71[a] | Edmonds (unpublished) |

[a] *Nothofagus.*

## Absorption of nitrogen compounds

There have been several reviews in recent years on the nitrogen nutrition of mycorrhizal plants (Bowen and S. E. Smith, 1981; Alexander, 1982; Raven *et al.*, 1978), but there has been very little recent experimental work on it with most kinds of mycorrhiza, except in those of Ericaceae by Read and his colleagues which will be discussed later. The ectomycorrhizal fungi, as has been shown, are all able to use ammonium and some amino acids and other simple organic nitrogen compounds for their growth in culture, but species and even strains are very variable in their ability to use nitrate. The habitats of most ectotrophic trees are unlikely to contain nitrate in quantity in the soil. This is due to the low pH and the presence of phenolic compounds which inhibit organisms which might oxidize ammonium in nitrate in their surface horizons. It is in the very acid humic layers that the ectomycorrhizal roots are concentrated. There, the sources of nitrogen which might be available to them are ammonium, simple soluble organic nitrogenous substances, and complex, often insoluble, nitrogen compounds. Lundeberg (1970) showed by $^{15}$N-labelling experiments that complex nitrogenous compounds in the humus are not a satisfactory source of nitrogen for the growth of any of the mycorrhizal fungi that he tested. Melin and Nilsson (1952, 1953a) showed that the extramatrical hyphae of the fungal symbiont of *Pinus sylvestris* mycorrhiza absorbed $^{15}$N-labelled ammonium and glutamine and translocated nitrogen compounds through the fungus to the root and the needles. In considering this matter it is important to bear in mind that the competition for available nitrogen in the humic layers of the soil is great but that the availability of carbon compounds suitable as carbon sources for fungi is low, so that a mycorrhizal fungus using host-derived carbohydrates is at an advantage.

The process of ammonium absorption into excised mycorrhizas was examined by Carrodus (1965, 1966, 1967) and found to be metabolically dependent and influenced by the supply of carbohydrate which is utilized in part in the production of amino acids from the absorbed ammonium. Only a small quantity of ammonium (Budd, *see* Harley and Wilson, 1963) remains in the tissue, for nearly 90% of it is very rapidly converted to organic compounds and internal ammonium concentration rises very little. Exogenous carbohydrates, which in nature would come from the host, are essential for rapid uptake of ammonium by excised mycorrhizas, especially in samples taken in the spring when root growth is active and internal carbohydrates are presumably low. The carbohydrates, glucose and fructose, whether applied and absorbed before ammonium or simultaneously with it, both increase its rate of absorption. Since, if applied before $NH_4$ uptake, glucose would be converted to trehalose and glycogen, whilst fructose would mainly be converted to mannitol (Lewis and Harley, 1965b), Carrodus believed that all these storage carbohydrates might be used as sources of carbon for synthesis of organic nitrogen compounds during the absorption and assimilation of ammonium. Ammonium uptake, he observed, was also increased by application of some organic acids (notably succinate and malate) and by bicarbonate up to about 7·5 mM. He confirmed the observation of Harley (1964) that it was coupled with dark $CO_2$ fixation, which greatly increased during ammonium absorption and assimilation. The main compound into which the fixed carbon was synthesized was glutamine, but other amino acids and organic acids although labelled by dark fixation of $^{14}CO_2$ in the absence of ammonium did not increase their radioactivity very significantly when it was present (Table 55). The importance of glutamine as an early product of ammonium assimilation confirmed the unpublished results of Budd who by differential hydrolysis had concluded that the amide of glutamic acid, glutamine, increased rather than asparagine.

Ammonium salts greatly stimulate respiration of beech mycorrhizas, especially when they have been kept for a period in nutrient-free solutions. During storage the respiration of the mycorrhizas falls to a low steady rate controlled by the availability of phosphate acceptors such as ADP (Harley *et al.*, 1956; Harley, 1981). The application of ammonium causes an increase of oxygen uptake, carbon dioxide emission and an increased rate of turnover of phosphate. The increase in oxygen uptake is greater the higher the concentration applied. This behaviour is explicable, perhaps, by the need for energy during the actual absorption process, but certainly for the production of the carbon radicals for synthesis of amino acids and their amides, which are derived from glycolytic products and $CO_2$ fixation.

Carrodus found that the mycorrhizas of *Fagus* that he used did not take up nitrate significantly and did not reduce it. Their inability to absorb nitrate rapidly was confirmed by F. A. Smith (1972) who showed that the rate of entry of nitrate into *Fagus* mycorrhizas was only of the same order as that of chloride and much lower in rate than that recorded by Carrodus for ammonium. This agrees with the results of Harley *et al.* (1954) who showed that nitrate had a much lesser effect on the uptake of oxygen than ammonium or phosphate. The application of nitrate to *Fagus* mycorrhizas moreover did not increase dark $CO_2$ fixation as would be expected if it were reduced in the tissues and incorporated into organic compounds (Table 55). Indeed Carrodus showed that fixation in the presence of potassium phosphate or chloride was similar to that in the presence of potassium nitrate. Although from all these experiments it appears that the *Fagus* mycorrhizas investigated do not use nitrate, it does not imply that no ectomycorrhizas do so. The experiments of France and Reid (1978) appear to show a degree of nitrate absorption by excised mycorrhizas and non-mycorrhizal roots of *Pinus taeda*. The samples were grown in aseptic culture, excised and exposed to 50 mM ammonium chloride or potassium nitrate in nutrient solution. Uptake was measured by disappearance of ammonium or nitrate from these solutions. There were therefore no corrections for uptake into apparent free space. The curves did not, as France and Reid suggested, show saturation kinetics for they were time progresses of change of concentration in the solutions. It would appear that the uptake of nitrate was little more than that observed by Carrodus or F. A. Smith in *Fagus* mycorrhizas. Ammonium absorption clearly occurred for there was continued uptake following the initial rapid equilibration period.

TABLE 55

Incorporation of [14]C in glutamine during uptake of nitrogen in the presence of [14]$CO_2$

*Time: 4 h at 20°C. (After Carrodus, 1967)*

| Nutrient in medium | Control | $NH_4Cl$ | $KNO_3$ | KCl |
|---|---|---|---|---|
| CPM $\times 10^{-3}$ | 76·3 | 107·8 | 88·4 | 93·9 |
| % in glutamine | 12·3 | 38·8 | 10·7 | 8·9 |

It is possible of course that if nitrate were absorbed into the sheath it might be translocated to and reduced by the host tissue. Indeed Li *et al.* (1972) have shown that *Alnus rubra* tissues, but not those of *Pseudotsuga douglasii*, could reduce nitrate rapidly. Yet for this activity to be important nitrate would have to be absorbed much faster than it has been observed to be in any of the experiments quoted. Li *et al.* note that *P. douglasii* reacts by increased growth to the application of nitrate to forest soil, but that could be explained either by nitrate transformation to other compounds or by stimulation of organisms that then mineralize more nitrogen, rather than by some of its mycorrhizal fungi absorbing and reducing nitrate. Richards (1965) and others have shown that nitrate fertilization of soil diminished the mycorrhizal development in *Pinus*. This has always been held to indicate that nitrate was absorbed. Indeed Björkman believed that nitrate was absorbed, reduced and built into organic compounds using carbohydrate in the process so reducing mycorrhizal infection. Recently Bigg (1981) has shown that nitrate may have a direct inhibitory effect on some mycorrhizal fungi whether they can reduce it or no, and by that means nitrate in culture or the soil may diminish mycorrhizal development.

There have been a few experiments on the absorption of soluble organic compounds of nitrogen in addition to those of Melin and Nilsson showing uptake and translocation of nitrogen from [$^{15}$N]-glutamate. Carrodus (1966) showed that glutamate and aspartate and their amides, glutamine and asparagine, were absorbed from solution by *Fagus* mycorrhizas. The uptake of the amides was about three times as fast as that of the anions. Since Carrodus did not find ammonium in the medium containing the amides he assumed that the molecules were either absorbed whole, or if they were externally hydrolysed, the rate of ammonium release was much smaller than its uptake rate. It would be of interest to know whether the mycorrhizal fungi have surface-attached amidases as some of the fungi do. Alexander (1982) records that the uptake of aspartate and glutamate and their amides by *Picea abies* mycorrhizas had been observed by Marchenko in 1967. In his own experiments with seedlings of *Picea sichensis*, mycorrhizal with *Lactarius rufus*, he found that 0·5 mM aspartate or serine diminished their growth in axenic conditions even though they were already provided with ammonium as a nitrogen source. Mycorrhizal seedlings were much less affected than non-mycorrhizal seedlings, especially by aspartate. The different effects of aspartate and serine on them might have been due to aspartate being absorbed by the mycorrhizas.

In Carrodus' (1967) experiments and in those of Harley (1964 and unpublished), the greatest increase in radioactivy on feeding [$^{14}$C]-bicarbonate appeared to be in the sheath tissue, and it was a reasonable supposition that the labelled nitrogen compounds, especially glutamine, formed when ammonium was absorbed, were accumulated in the fungus particularly. Carrodus did not obtain any evidence of subsequent movement of labelled compounds from the fungal layer to the host tissue when the mycorrhizas were kept for a period in water. However, Reid and Lewis (as reported by Lewis, 1976), in an experiment based on the inhibition technique of Drew and D. C. Smith (1967), found that *Fagus* mycorrhizas during or after the fixation of $^{14}CO_2$ in the presence of ammonium might translocate radioactive carbon compounds in certain conditions (Table 56). The appearance of radioactivity in the medium in the presence of amino acids, especially glutamine, is evidence of their transfer to the host in normal conditions. The chromatographic analysis of the medium indicated that glutamine was the main radioactive compound released whatever amino acid was present in the external solution.

TABLE 56

Showing that glutamine is probably translocated from the sheath to the host tissue in *Fagus* mycorrhizas

*Estimates of the $^{14}C$ released into the medium as a proportion of the $^{14}C$ fixed into soluble compounds. "Inhibiting" compounds added at 1% (w/v) of external solution (NB. The principal compound released with every amino acid tested was glutamine)*

Experiments A   $^{14}C$-bicarbonate and ammonium applied to the mycorrhizas followed by the "inhibiting" compound

Experiments B   $^{14}C$-bicarbonate and ammonium applied to the mycorrhizas simultaneously with the "inhibiting" compound

*Results of Reid and Lewis (in Lewis, 1976)*

| Experiment type | "Inhibiting" amino acid in medium | Time h | $^{14}C$ in medium as percentage of total soluble $^{14}C$ | |
|---|---|---|---|---|
| | | | 1st series | 2nd series |
| A | Nil (water) | 4 | 0·33 | 0·13 |
| | Aspartate | 4 | 1·72 | — |
| | Glutamate | 4 | 2·33 | 1·92 |
| | Glutamine | 4 | 1·36 | 1·34 |
| B | Nil (water) | 4 | 0·08 | 0·04 |
| | Aspartate | 4 | 0·62 | — |
| | Alanine | 4 | 0·87 | — |
| | Glutamine | 4 | 1·07 | 0·75 |
| | Nil (water) | 5 | 0·14 | — |
| | Glutamate | 5 | 1·07 | — |
| | Glutamine | 5 | 1·18 | — |

## Absorption of phosphate

The absorption of phosphate into excised mycorrhizas of *Fagus* results in its immediate incorporation into phosphate compounds. As Harley and Loughman

(1963) showed, within a few minutes of placing washed mycorrhizas into dilute $KH_2{}^{32}PO_4$ solutions, nuceotides and sugar phosphate become labelled. Both the constituent tissues, separated sheath and host core, exhibited rapid incorporation of applied phosphate, but in intact mycorrhizas the core tissue received less phosphate and appeared more sluggish in incorporating inorganic phosphate into other compounds. The process of phosphate incorporation is similar to that found in uninfected roots of plants where it has been concluded that inorganic phosphate is incorporated via the nucelotide pools into the other fractions. The process, as we have seen, results in a great accumulation of the absorbed phosphate in the sheath tissue and as absorption from dilute (up to 0·3 mM) phosphate solution continues, about 10–20% only of that absorbed passes steadily into the core tissues through the sheath. By studying time progresses of esterification of the phosphate entering the host core Harley and Loughman were able to show that the labelling of inorganic orthophosphate represented 100% of the radioactivity in that tissue initially and that it fell to lower proportions later as nucelotides, sugar phosphates and other fractions became labelled. Since both the sheath tissue and the host tissue showed the same labelling pattern of soluble phosphates if allowed separately to absorb phosphate from radioactive solutions, it was concluded that in the intact excised mycorrhizas inorganic orthophosphate was the form which passed from the sheath to the host when low concentrations were applied.

Harley and McCready (1952b) and Harley *et al.* (1958) studied the possible routes by which phosphate might pass through the sheath from the external solution to the host. They showed first that the sheath prevented the host from absorbing phosphate at its maximum possible rate except when very high concentrations were applied. In 32 mM phosphate solutions, removal of the sheath had relatively little effect and the core seemed to be supplied by diffusive movement through the sheath, almost as if it were free of the sheath. In these conditions low temperatures and inhibitors reduced the incorporation of phosphate both into the sheath and into the host. At very low concentrations the removal of the sheath greatly increased the potential of the host to absorb phosphate, but in this state any inhibitor applied to intact mycorrhizas which inhibited uptake into the sheath also inhibited uptake into the core. By contrast, when mycorrhizas were placed in intermediate concentrations (1–3 mM) factors which inhibited uptake into the sheath tended to increase or leave unaffected the uptake into the host. The conclusion reached was therefore that from low phosphate concentrations such as might be expected in the soil, diffusive movement through the sheath did not take place at a significant rate. It might, however, occur at the ecologically impossibly high concentrations above about 1 mM. At the excessive concentrations of 32 mM or more the diffusive movement was such that the host was supplied with phosphate almost as if it were free in solution.

The phosphate passing through the sheath to the host does not equilibrate with a large part of the phosphate in the fungus. If it did so, as Harley *et al.* (1954) showed, an evident lag phase in the arrival of phosphate in the core tissue would be expected, and the quantity of phosphate in the pathway to the core would be related to the length of the lag phase. In an experiment using low concentrations it was found that the lag phase was exceedingly short, so that the quantity of phosphate in the sheath with which the passing inorganic phosphate equilibrated was very small—of the order of 0·017 μg phosphorus per 100 mg dry weight of mycorrhizas. This is a very small quantity compared with the amount of phosphate present in the sheath and even if it were underestimated by a factor of 10 or even 100 it would still be a small part of the 245 μg found in the sheath of 100 mg dry weight of mycorrhizas. Later work by Jennings (1964a) suggested that this small pool of phosphate with which that passing to the host tissue equilibrates might increase or decrease slightly according to the external supply of phosphate. It may be concluded therefore that inorganic phosphate, as it is absorbed by ectomycorrhizas, is incorporated first into the metabolic pools of the protoplast of the sheath and these constitute a small proportion of the total phosphate as they do in other roots (Crossett and Loughman, 1966).

In mycorrhizas it has been found that some of the phosphate in the sheath is present as polyphosphate. This was pointed out by Ashford *et al.* (1975) using *Eucalyptus* and subsequently shown to be true of *Pinus radiata*, of the arbutoid mycorrhizas of *Arbutus unedo* and of the vesicular-arbuscular mycorrhizas of *Liquidambar styraciflua* by Ling Lee *et al.* (1975). Using the same method as Ling Lee *et al.*, cytochemical staining of granules in the fungal tissues, Chilvers and Harley (1980) described particles believed to be polyphosphate in the sheath of *Fagus* mycorrhizas. The number and size of particles increased during phosphate absorption at rates similar to the rate of absorption of phosphate from similar concentrations, and similar factors affected their formation as affected phosphate uptake. Ultrastructural studies of these granules especially by Strullu and his colleagues (Strullu, 1976a; Garrec and Gay, 1978; Strullu *et al.*, 1981, 1982) have shown that calcium is an important constituent of them in many kinds of mycorrhiza including those of *Pinus*, *Pseudotsuga* and *Fagus* as well as in vesicular-arbuscular mycorrhizas. The formation of polyphosphate in *Fagus* mycorrhizas was further examined by Harley and McCready (1981) using the method of extraction and precipitation described by Aitchison and Butt (1973). Assuming that there was little hydrolysis in the extraction and that the precipitation with $BaCl_2$ was complete, a large amount of the phosphate which is accumulated in the sheath tissue is polyphosphate. This seems to be the form in which a great part of the phosphate in the sheath is separated from the mobile phosphate which can move to the host tissue. It may be that all the polyphosphate present is not in the form of granules. Loughman (personal communication) has obtained evidence by n.m.r. analysis that a considerable part may be soluble in *Fagus* mycorrhizas and both Loughman and Dodwell (personal communications) have evidence that the polyphosphate present in *Fagus* mycorrhizas is readily broken down by damage to the tissues. It is for these reasons that the results of Harley and McCready (1981) must be treated at the moment as somewhat preliminary in a quantitative sense.

Phosphate which is accumulated in the sheath may be translocated to the host tissue by a mechanism which is dependent on oxygen supply and deficiency of phosphate, inhibited by metabolic inhibitors and has a high temperature coefficient (Harley and Brierley, 1954, 1955; Harley *et al.*, 1958). These properties led to the conclusion that the process was under metabolic control. It now appears that this transfer is, in great probability, a movement of phosphate stored as polyphosphate and immobilized. It is possible that a diminution of the small phosphate pool in the protoplast of the sheath, which was observed to vary by Jennings (1964a), might be a trigger to this process. The uptake of phosphate into the sheath and its movement into the host may be represented by a modification of the scheme given by Harley (1969a) (Fig. 21) where the sites of accumulation are possibly largely of polyphosphate.

The movement of phosphate from the soil to the shoots of *Pinus radiata* was examined by Morrison (1957a, 1962a). He grew seedlings at two levels of phosphate in pots in the greenhouse for 17 weeks. At the high phosphate level the non-mycorrhizal plants grew somewhat better than the mycorrhizal ones, but the reverse was true in low phosphate. Plants of each kind were then grown with fresh supplies of phosphate labelled with $^{32}P$ in a variety of soils. In all experiments the movement of $^{32}P$ to the shoot tip of the non-mycorrhizal plants was rapid at first but later decreased in rate and almost ceased but could be increased again by further addition of phosphate to the soil. Movement to the shoot tips of mycorrhizal plants, although much slower than the initial rate in non-mycorrhizal plants, continued steadily for weeks and was little affected by further additions of labelled phosphate to the soil. The accumulated radioactivity of the shoot tips of mycorrhizal plants eventually exceeded that of the non-mycorrhizal controls. If mycorrhizal and non-mycorrhizal plants were deprived of phosphate after a period of uptake of $^{32}P$-labelled phosphate, radioactivity continued to pass to the shoots of mycorrhizal plants for three weeks but ceased to move to those of non-mycorrhizal plants after a short period. This behaviour is explicable in terms of the accumulation of phosphate in the fungal sheath coupled with a steady rate of transfer to the host when phosphate is available. When phosphate supplies are deficient movement from the stored phosphate in the sheath occurs.

The demonstration of the probable importance of polyphosphate in the phosphate metabolism of ectomycorrhizas helps to explain a number of puzzling physiological features which they exhibit. First, it may explain the great potential of the sheath to accumulate phosphate. For instance, the sheath of

*Fagus* mycorrhizas can almost double its internal phosphate concentration in 24 h by absorption from a solution of 0·16 mM (Fig. 20). Such an accumulation, if it were as inorganic orthophosphate, would be expected to have effects upon the osmotic potential and buffering system of the sheath cells. The formation of large polymeric molecules of phosphate would seem to obviate this. Secondly, the absorption of phosphate is coupled with an increase of oxygen consumption, so that in some conditions the quantity of phosphate absorbed is exactly proportional to the oxygen consumed (Harley *et al.*, 1954, 1956; Harley and McCready, 1981). This relationship might be explained by the operation of polyphosphate kinase in the synthesis of polyphosphate with the consumption of ATP. Such a reaction would result in a rapid turnover of the nucleotides and a proportionate change in oxygen consumption. Harley and McCready (1981) pointed out that the ratio of atoms of phosphorus absorbed to atoms of oxygen consumed was frequently between 0·6 and 1·0 and indicated that under the experimental conditions the efficiency of the coupling of phosphate absorption and oxidative phosphorylation was high and of the order of 25–30%. Since in their analysis only a portion, up to 40%, of the absorbed phosphate was incorporated as polyphosphate they suggested that this high efficiency indicated that the actual process of phosphate uptake as opposed to the formation of polyphosphate was also dependent on oxidative phosphorylation.

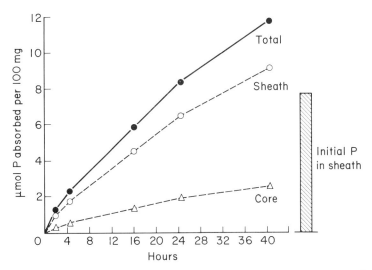

FIG. 20. Absorption of phosphate in $\mu$mol per 100 mg dry weight into the tissues of excised beech mycorrhizas from 0·16 mM $KH_2PO_4$ during 40 h, pH 5·5, 18°C. At the right the initial quantity in the fungal sheath. (Harley, 1978a.)

Jennings (1964b) showed that the application of strong solutions of monovalent cations ($K^+$, $Na^+$ and $NH_4^+$) as chlorides, to *Fagus* mycorrhiza led to the loss of phosphate from the tissues, whereas the divalent $Ca^{2+}$ tended to prevent this. Jennings pointed out that the loss of phosphate was associated with a decrease of bound phosphate in the tissue, especially that hydrolysed in 7 minutes by dilute acid. The latter fraction is of course often assumed to be approximately the polyphosphate fraction in tissues. The effect of calcium in preventing loss is particularly interesting because of the reports that phosphate and calcium are distributed together in ectomycorrhizal tissues, and that polyphosphate granules in several kinds of mycorrhiza contain calcium. It is possible that the granules are rendered insoluble by calcium and its replacement by monovalent ions may lead to their dissolution and the leakage of phosphate. This matter clearly needs further examination.

Lastly, as Harley (1978a) pointed out, it is not impossible that the polyphosphate present in the sheath can act as a temporary store of energy because the bond energy of the phosphate group of polyphosphate is said to be of the order of 9 kcal mol $^{-1}$. Obviously the quantities stored will not provide the energy for all the metabolic processes of the sheath for more than a short time. However, the inability to prevent uptake of some ions completely in short-term experiments by anaerobiosis or the application of KCN might be partially explicable in terms of the utilization of polyphosphate-bond energy.

The possibility that poorly soluble inorganic phosphate might be absorbed from the soil by ectomycorrhizal plants was examined by Stone (1950) and by Bowen and Theodorou (1967). The latter showed that four mycorrhizal fungi, *Boletus (Suillus) granulatus*, *B. (S.) luteus*, *Rhizopogon roseolus* and *Cenococcum graniforme (geophilum)*, could bring phosphate into solution from rock phosphate in culture. They ascribed this as possibly due to acid secretion and expressed doubt whether it could occur in conditions prevalent in soil. They also, like Stone, observed that mycorrhizal plants of *Pinus radiata*, inoculated with the same fungi, absorbed more phosphate from rock phosphate than uninfected plants when grown in soil. They queried, however, whether in these experiments and in those of Stone part or all of the effect might have been due to contaminating microorganisms on the mycorrhizal surface. The matter clearly needs further investigation in view of the possibility that mycorrhiza might excrete organic acids capable of chelating metals.

Since a large part of the phosphate in the soil horizons where most ectomycorrhizas occur is present as organic phosphates, these have been studied as possible nutrient sources. Of course uninfected root surfaces as well as mycorrhizal surfaces may possess phytase activity (Woolhouse, 1969). Theodorou (1968) showed that the group of mycorrhizal fungi mentioned above were able to grow when supplied in culture with phytates of calcium and sodium rather than inorganic phosphate. He also showed that *Rhizopogon roseolus* produced two phytase enzymes. Since then there have been a number of investigations of this matter. Bartlett and Lewis (1973) examined the surface phosphatase activity of mycorrhizas of *Fagus*. They showed that there was more than one phosphatase present because the activity had a double pH optimum and hydrolysed a range of phosphate compounds including inorganic pyrophosphates and organic compounds especially inositol phosphates. They emphasized that the possession of such an activity on the surface of the fungal component of ectomycorrhizas might result in the immediate recycling of the phosphates present in the fallen litter back into the mycorrhizal system. Williamson and Alexander (1975) also examined *Fagus* mycorrhiza. They found that acid phosphatase was present throughout the fungal tissue and was not associated with contaminating microflora to any significant extent. They agreed with Bartlett and Lewis that more than one phosphatase enzyme was present, and that each had different characteristics. The total activity, as Woolhouse had described, was much greater (eight-fold) than that of uninfected roots.

Alexander and Hardy (1981) also showed that mycorrhizas of *Pinus sichensis* possessed surface phosphatases inversely correlated in their activity with the concentration of extractable inorganic phosphate in the soil. In this respect the work is extremely reminiscent of that of Calleja *et al.* (1980) who showed that the phosphatase activities of four species of ectomycorrhizal fungi were more strongly developed in the absence of soluble phosphate from the culture medium. Alexander and Hardy sound a deep note of caution against concluding from these results alone that the surface phosphatase activity has any function in ecological situations. They suggest first that the activity may be derived from senescing cells and that phosphate deficiency might encourage senescence. This hardly seems tenable since Williamson and Alexander had found phosphatase activity throughout the sheath, and the materials used by Woolhouse and by Bartlett and Lewis were of young active mycorrhizas, and Calleja *et al.* were working with active mycelia. Secondly, Alexander and Hardy doubt whether there can be organic phosphates in solution in sufficient quantity to be important. They suggested that these compounds would either be adsorbed or present as insoluble salts of calcium, magnesium, iron or aluminium in mineral soil. However, the surface nature of the activity on the fungal hyphae and the sheath indicates that the close proximity of these surfaces would be necessary for and result in hydrolysis. Clearly there is a reasonable presumption that in the organic horizons these phosphatases will be active in releasing inorganic phosphate in the immediate environs of the fungal sheath or its hyphae. Indeed the sheath and hyphae would be positioned so as to compete efficiently for the phosphate produced. The subject clearly needs more attention together with an investigation of the possibility that the sheath and fungal hyphae secrete substances that chelate metals and may bring into solution organic and inorganic phosphate (Johnston and Miller, 1959; Cromack *et al.*, 1979).

The uptake of phosphate by mycorrhizas of different fungi on a single species of tree may vary much in rate. Mejstrik (1970) showed that two different kinds of excised mycorrhiza of *Pinus radiata*, classified according to the system of Dominik (1955) as Fa and Ca, both absorbed phosphate faster than excised uninfected roots over a range of temperatures from 5 to 30°C but differed from one another in rate of absorption and in the effect of temperature upon it. Similarly Mejstrik and Krause (1973) showed that seedlings, mycorrhizal with *Suillus (Boletus) luteus* or *Cenococcum graniforme (geophilum)*, differed both in the rate of phosphate absorption and in the extent to which they could exploit humic matter. Using *Abies balsamea*, Langlois and Fortin (1978) (Table 57) found that there were very significant differences in the uptake rates of four kinds of mycorrhizas obtained from adult trees in the forest. The difference in rate was about ten-fold between the highest and the lowest. As they pointed out, results such as these are extremely important in considering the ecology and practical importance of mycorrhizal symbiosis.

TABLE 57

Relative absorption rates of phosphate from dilute ($0.18 \times 10^{-9}$ M) phosphate solutions pH 4.5 at 25°C during 2.5 h by different types of mycorrhizas on *Abies balsamea* (data of Langlois and Fortin, 1978)

| Type of mycorrhiza | CPM mg$^{-1}$ $\pm \sigma$ |
|---|---|
| Brown | $217.2 \pm 68.6$ |
| Black | $129.9 \pm 25.1$ |
| White | $661.4 \pm 344.6$ |
| Yellow | $2118.6 \pm 467.5$ |

## Absorption of other ions

There has been little study of the absorption of substances other than nitrogen and phosphorus compounds by ectomycorrhiza either from the soil or from solutions. One exception, however, is in the uptake of potassium and some other metallic elements. Wilson (1957), Harley and Wilson (1959, 1963) and Edmonds *et al.* (1976) have published the results of investigations using excised mycorrhizas of *Fagus*, but the subject cannot be considered to be in a satisfactory state.

Potassium and sodium are absorbed from pure KCl or NaCl solutions at rates which are affected by temperature, oxygen supply, and the presence of metabolic inhibitors. Low concentrations of calcium in the solutions have little effect on their uptake rates. The rates of absorption of potassium and sodium reach high approximately level values between 0.1 and 0.2 mM but potassium is absorbed more than twice as fast as sodium from pure solutions. There is great selectivity for potassium from mixed solutions, so that its uptake from 0·1 mM solutions is unaffected by the presence of sodium at 0·6 mM. There is an interaction between ammonium and potassium. For example the uptake of potassium is decreased by about 40% when both are present in solution at 0·1 mM. Potassium at ten times the concentration of ammonium however, only decreases ammonium uptake by about 5%. Ammonium is absorbed at about ten times the rate of potassium and is rapidly incorporated into organic compounds. It is likely that the interactions in uptake are the result of direct competition for a common carrier, or of changes in the electrical properties of the plasma membranes associated with the uptake process, or of rapid incorporation of organic anions into amino compounds during ammonium uptake. These features are not very different from those of uninfected roots but certain properties of the mycorrhizas with regard to the loss of potassium from the tissues need further investigation.

Potassium is readily lost from mycorrhizal roots. It is present in the tissues at about 70 $\mu$mol per 100 mg dry weight or if assumed to be distributed in the total tissue water at about 0·1 M. Losses occur when the tissue is kept at low oxygen concentration where respiratory quotient is above unity, or when kept at temperatures above 20°C in normal oxygen supply (with the solutions in equilibrium with air). The most probable explanation of these losses is that there is a decrease of anions in the tissue in both these circumstances, and a loss of bicarbonate and potassium from it. Two kinds of anions may well decrease, organic anions and phosphate. As has been described, polyphosphates are formed and readily hydrolysed in the fungal tissue and their state of polymerization may alter the concentration of the counter ion which must be present. It seems probable that a decrease of organic anions is the primary factor correlated with loss of potassium at low oxygen concentrations. However, when mycorrhizas are kept in aerated water they rapidly decrease in respiratory rate to a low value although their carbohydrate reserves remain high. The limitation of the respiratory rate is (according to Harley, 1981) associated with an increase of ATP:ADP ratio. This is a condition inducive to the synthesis of polyphosphate by polyphosphate kinase. The rate of the processes leading to the respiratory decrease is enhanced by increase of temperature so that conditions for the loss of potassium would be expected to develop more rapidly at higher and more slowly at lower temperatures and this indeed they do. Addition of glucose to beech mycorrhizas leaking potassium at temperatures above 20°C caused a rapid cessation of the leak (Edmonds *et al.*, 1976) and this would be expected since glucose is rapidly phosphorylated on uptake and its absorption leads to a reduction of ATP:ADP ratio and hydrolysis of polyphosphate.

Clearly the processes of potassium uptake and loss and of the linkage of the processes with phosphate polymerization require much further investigation.

Clement *et al.* (1977) have examined the growth of *Pinus nigra* (ssp. *nigricans*) on calcareous soil. This pine, in common with some other conifers, sometimes becomes chlorotic and stunted in culture or in natural soils in which there is a large quantity of calcium carbonate and a high pH value. They pointed out that this was not a genetic feature of any particular variety of *P. nigra* and it did not become evident when the plants were mycorrhizal. The symptoms on calcareous soil, besides chlorosis, were a high rate of absorption of cations particularly of calcium and potassium. Table 58 shows the results of one of their experiments on natural calcareous soil which indicate that mycorrhizal development increased growth and internal phosphate concentration of seedlings but reduced the excessive intake of potassium and calcium. This effect is reminiscent of the exclusion of toxic metals by ericoid mycorrhizal infection, and its causation needs to be examined to explain the success of the growth of mycorrhizal plants of many types not only on calcareous soils, but on toxic mine spoil and diverse man-made sites.

TABLE 58

Growth of seedlings of *Pinus nigra* (ssp. *nigricans*) in 6 months on calcareous soil when inoculated with either *Boletus (Suillus) granulatus* or *Tuber melanosporum* (results of Clement *et al.*, 1977)

| Measure | Not sterilized mycorrhizal | Sterilized non-mycorrhizal | Sterilized and inoculated | |
|---|---|---|---|---|
| | | | *B. granulatus* | *T. melanosporum* |
| Fresh wt. (g) | 1·95 | 0·53 | 1·37 | 1·00 |
| Phosphate % dry wt. leaves | 0·17 | 0·09 | — | — |
| Potassium % dry wt. leaves | 1·00 | 1·35 | — | — |
| Calcium % dry wt. leaves | 0·56 | 0·76 | — | — |

Sulphate absorption by ectomycorrhizal *Pinus radiata* and *Fagus sylvatica* was examined by Morrison (1962b, 1963). He showed that over a period of three hours the rates of absorption of $^{35}SO_4$ into mycorrhizal and non-mycorrhizal *P. radiata* per unit weight of root were similar, whereas that of phosphate, in parallel experiments, was between 2 and 5 times as fast into mycorrhizal than non-mycorrhizal plants. Using *Fagus* mycorrhizas he showed that although 80–90% of the phosphate absorbed was, as expected, accumulated in the fungal sheath, only 50–60% of the absorbed sulphate was accumulated there. Moreover, although after a few minutes the percentage of sulphate accumulated in the sheath might be about 75%, it dropped during 2 hours of uptake to about 50%. This figure of 50% does of course represent almost twice the accumulation per unit weight in the sheath as in the host. Indeed the avidity of the sheath tissue for sulphate was nearly four times that of the host core, as shown by allowing both tissues to take up sulphate in the separated condition. Nevertheless the uptake rates of sulphate by both tissue systems were low, so that the rate of absorption by the host core was not reduced by the presence of the sheath even during absorption from 0·32 mM sulphate. Clearly movement of sulphate by diffusion through the interhyphal spaces of the sheath from this low concentration was enough to satisfy both the cells of the sheath in depth and the host core. This contrasts strongly with the process of uptake of phosphate. For that ion is so rapidly accumulated by the sheath from 0·32 mM solutions as to prevent diffusion through the interhyphal spaces being effective in delivering phosphate to the host at all.

## Movement of nutrients through the fungal layer

Mycorrhizas absorb nutrients which nourish both the fungus, its hyphae, sheath and Hartig net, and also the root and shoot tissues of the host. In ectomycorrhizas the sheath tissue is a barrier to direct uptake into the host and the movement of nutrients from it into the host is a process which competes with the use and storage of nutrients within the fungus itself. It is important to realize that the sheath is not a rapidly growing organ and in that respect it differs from fungal hyphae growing actively on a substrate. We have shown that nutrients, both carbon compounds from the host and inorganic substances from the soil, are accumulated in its cells, but at the same time some of the soil-derived nutrients pass through it to the host. The process may be represented by a diagram (Fig. 21) which is a modification of that given by Harley (1969a) to represent phosphate absorption and movement.

Nutrients may be absorbed metabolically (process $A_1$) from the substrate by hyphae which emanate from the sheath surface and be translocated within them to the sheath of the mycorrhiza. In that case movement from the very beginning is within the living fungus. Similar factors will affect the rate of uptake of nutrients into the hyphae as affect the rate of metabolic uptake into the cells of the sheath ($A_1$). Where external hyphae occur the exploitation by them of a large volume of soil may be envisaged and a close contact made with potential substrates in the solution, adsorbed or solid phases of the soil. Hydrolysis of insoluble compounds such as those of phosphorus or nitrogen might take place at points of contact between the hyphae and the fungal sheath with soil particles if the fungus possesses surface-attached enzymes of a suitable kind, as some undoubtedly do. When available nutrients become exhausted in their immediate neighbourhood, the hyphae as they senesce would be replaced by the growth of hyphal branches in new locations.

Diffusive or mass movement of nutrients in solution to the surface of the sheath will also take place. It would be expected to vary in rate with the nature of the substance concerned, its diffusion coefficient, its concentration in solution, its fixation on soil particles, with the moisture content of the soil and with the rate of transpiration of the host. Many of the important ions are relatively immobile because they are adsorbed to soil particles; examples are phosphate, ammonium and potassium. The relative immobility of phosphate and its low concentration in the soil solution are well known. However, ammonium, an ion required in large amounts per unit biomass and absorbed as we have seen very avidly by ectomycorrhizal roots, has a low mobility relative to demand, as has also potassium. Hence in a mineral soil of stable structure the same considerations apply to ectomycorrhizas as to vesicular-arbuscular mycorrhizas. The mycorrhiza itself with low surface to biomass ratio and slow growth rate would not be expected to be an efficient absorbing organ but its outgoing hyphae would be essential to its efficiency. The extent of the development of hyphae which penetrate the soil from the sheath is very variable indeed. Many ectomycorrhizas are almost smooth with no extramatrical hyphae. The kind and extent of extramatrical hyphal development in its many variations can be seen in papers which aim to classify mycorrhizas on the structural characteristics (e.g. Dominik, 1955). Mousain (1971) (Fig. 15) has illustrated the mycorrhizas of *Pinus maritima* showing many variations of hyphal connections with the soil, from smooth hyphal-free surfaces through hair-like or cystidium-like extensions, to

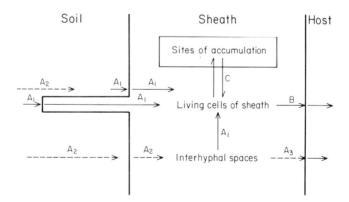

FIG. 21. Diagram showing the movement of nutrients through the sheath of ectomycorrhiza. $---$, diffusive or mass movement; $—$, metabolically dependent movement. Nutrients move by diffusive or mass movement in the soil. They are absorbed by metabolically dependent reactions $A_1$ into the hyphae or sheath surface and may either be released by metabolic reaction B to the host, or accumulated by the reversible reaction C. Nutrients may also pass non-metabolically from the external medium through the interhyphal spaces of the sheath, from where they may be accumulated by $A_1$ into the living fungal cells or into the host cells ($A_3$).

extensive hyphae or ropes of hyphae. It is therefore important to consider the possibility of efficient nutrient uptake by mycorrhizas lacking extensive outgoing hyphae. Smooth-surfaced mycorrhizal roots are common in all adult species in the L, F and H layers of the forest floor which have a structure quite distinct from that of the mineral soil. Here the soil is not stable, as the mineral soil horizons are, and there is no fixed structure of solids, spaces and liquid films. It is a zone usually avoided by soil scientists when considering the movement of nutrients. The mycorrhizas are found lying along and between leaf material and fragments of it, and surrounded by organic matter which can be moved by physical factors of wind and rain and by biological means. The organic matter is continually being comminuted, moved and mixed by the soil animals. Its voids are variable in position and extent, but much larger than those of the mineral layers. In the litter layers the release of nutrients from newly fallen or senescing leaves may be rapid and periodic, subject to changing conditions and leaf-fall. The movement of nutrients in solution may be rapid as rain percolates downwards and solution may actually flow over the surfaces including those of mycorrhizal roots. In such a situation the intense exploitation by mycorrhizas will lead to efficient trapping of nutrients without extensive extramatrical mycelium.

Metabolic uptake into the sheath ($A_1$) leads to accumulation of nutrients in its living cells by reaction C and a steady passage into the host tissue by reaction B. As we have seen, there is evidence of both these movements during absorption of phosphate and ammonium. With phosphate the competition between reaction C, accumulation, and reaction B, movement into the host is, during absorption from ecologically probable concentrations, in favour of C; for 80–90% of the phosphate which is absorbed is retained in the sheath and 10–20% passed by reaction B into the host. The accumulation of phosphate is as polyphosphate and soluble inorganic phosphate and that of ammonium seems to be as amino acids, particularly glutamine. In both cases the accumulated nutrient may be mobilized and passed to the host. Phosphate moves as inorganic phosphate by a metabolically dependent process as reversal of C combined with B when external phosphate supply is low (Harley and Brierley, 1955). The accumulated nitrogen passes to the host as glutamine. In the latter case since an external concentration of amino acid can elute radioactive glutamine from the tissue the nitrogenous substances passing to the core must also be released by the fungus, before being absorbed by the host tissue in reaction B.

A diffusive movement ($A_2$) from the soil solution through the interhyphal spaces may theoretically take place. In experiments with phosphate it has been shown to occur to an increasing extent when excised *Fagus* mycorrhizas are kept in phosphate concentrations above 1·0 mM, which is of course ecologically a very unlikely concentration indeed. It occurred in Morrison's experiments with sulphate at 0·32 mM in the external solution. As penetration occurs by this means, uptake takes place by reaction $A_1$ from the interhyphal spaces into the living hyphae. The depth or rate of penetration of any substance through the sheath by diffusion will therefore depend inversely upon the avidity with which the cells of the sheath absorb it and directly upon the factors of concentration and diffusion coefficient which affect its rate of movement. As pointed out by Harley (1969b), the form of the uptake of phosphate against concentration (Fig. 19) can be explained in these terms. It is a combination of a hyperbolic relation between the uptake of phosphate and concentration by individual cells and a diffusive penetration of the apoplast of the tissues which, as external concentration is increased, allows more and more cells deeper in the tissues to absorb. The ease of uptake and loss of potassium and of the penetration of sulphate leads to the view that the diffusion of these ions through the interhyphal spaces might occur more readily than that of phosphate or ammonium. Nevertheless it may be accepted that at all expected ecological concentrations absorption is into the extramatrical hyphae and the cells at or close below the surface of the sheath, and that movement through the sheath is within the fungal cells. The fungal sheath therefore seems to have functions analogous to the cortex of uninfected roots. In the cortex nutrients are accumulated at the same time as they also pass to the xylem. It seems likely that the sheath is also an efficient storage organ for carbohydrates for which there are especially great demands during the fruiting of the mycorrhizal fungi.

From what has been discussed it is clear that the processes of nutrient uptake into ectomycorrhizal roots do not differ in any fundamental way from that into other roots or absorbing organs. They are linked with respiratory metabolism and depend therefore upon adequate supplies of carbohydrates which provide both the respiratory substrate and the radicals for synthesis of organic compounds. The processes are complicated, however, by the presence of the fungal sheath through which the greater part of all substances must pass on their way to the host tissue, for most of the feeding root system is covered. Nutrient substances are accumulated in the sheath whilst uptake is proceeding. It is as if the accumulating processes of the sheath compete with the movement through the sheath to the host, just as those of the cortex do in uninfected roots. When the external supply is deficient the stored substances in the sheath are mobilized and move to the host. This has been verified by experiment in the case of phosphorus and nitrogen compounds. The magnitude of phosphate accumulation is great and the sheath may retain at first 90% of the phosphate absorbed. To put this stage into its ecological context Table 59 uses estimates of phosphate uptake by beech forest, weights of mycorrhiza per unit area, and experimentally estimated rates of phosphate absorption and accumulation from concentrations of phosphate over a range that might be expected in soils. The estimates of weight of mycorrhizas per decimetre are halved for this calculation because McQueen, whose figures are used, estimated absorbing roots. Although these would have been mostly mycorrhizal—more than 90%—his estimates are halved to allow both for uninfected and moribund roots. The calculated values might therefore be underestimates. At the concentrations of phosphate used in the uptake experiments about 90% of that absorbed would have been retained in the fungal sheath and about 10% passed to the host. As can be seen from the Table, the amount absorbed per day from $0.3 \times 10^{-6}$ M phosphate is greater than the average requirement of beech forest for phosphate, and the total absorbed from $3.2 \times 10^{-6}$ M per day is about ten times greater. Indeed at the latter concentration the amount passing to the host is equivalent to the daily requirement, and an amount almost equal to 10 days' supply is accumulated in the fungal sheath. It would be expected in natural conditions that the availability of phosphate in the soil would vary with season, but these figures indicate that the ectomycorrhizal roots can adequately cope with a range of phosphate concentrations over which the available phosphate in the soil might be expected to fluctuate. In times of phosphate flushes accumulation in the sheath is capable of storing a very significant amount of phosphate in terms of the demand of the tree.

TABLE 59

The uptake and storage of phosphate in the mycorrhizas of *Fagus sylvatica*.

1. *Duvigneaud and Denaeyer de Smet (1971) estimate that one ha of beech forest absorbs 14·0–37·6 µg dm$^{-2}$ in 24 h (mean 25·8 µg dm$^{-2}$ per 24 h)*
2. *McQueen (1968) estimates 2400 mg absorbing roots of* Fagus *dm$^{-2}$ of forest of 105 years' age. 1200 mg are taken as a safe estimate to allow for uninfected roots and moribund roots.*
3. *The experimentally determined phosphate absorption is derived from Harley and McCready (1952a).*

*Results expressed as µg phosphate dm$^{-2}$*

| Concentration of phosphate (M) | $0.03 \times 10^{-6}$ | $0.3 \times 10^{-6}$ | $3.2 \times 10^{-6}$ | $32 \times 10^{-6}$ |
|---|---|---|---|---|
| Uptake by 100 mg in 24 h (µg) | 0·275 | 2·70 | 22·2 | 80·6 |
| Uptake by 1200 mg in 24 h (µg) | 3·3 | 32·4 | 266 | 967 |
| Amount stored in sheath (µg) | 3·0 | 29·2 | 240 | 870 |
| Amount absorbed into core (µg) | 0·3 | 3·2 | 26 | 97 |

If we were to use the estimates of the amount of fungal sheath per unit area given by Vogt *et al.* (1982) of 450 and 740 mg dm$^{-2}$, higher quantities (nearly double) would be expected to be stored in the sheath. There seems no reason to doubt, as some have done, that the capacity for storage in the sheath is of ecological significance especially in seasonal climates with seasonal growth activity and phosphate mineralization.

## Water relations of ectomycorrhiza

The humidity of the soil may affect the activity of mycorrhizas either at high water potentials by its effect on aeration, or more directly where the soil has a low water potential by water stress. The physiological activities of the ectomycor-rhizal fungi in culture and especially the process of nutrient uptake by the mycorrhizas themselves are greatly dependent on adequate oxygen supply which may be reduced in some soil conditions. Skinner and Bowen (1974b) suggested that compaction of soil might reduce aeration and certainly waterlog-ging may do so in extreme cases. The surface horizons of forest soils where the main accumulations of mycorrhizas occur are well aerated, as Brierley (1955) showed. The levels of oxygen and carbon dioxide were never such, in the soils he examined, as to reduce the absorption by *Fagus* roots significantly even after heavy rain. *Pinus radiata* and *Fagus sylvatica* were shown by Theodorou (1978) and Harley (1940) to fail to form mycorrhizas in saturated waterlogged conditions. In such places the absence or deficiency of oxygen might be expected to reduce the ability of the fungus to form a sheath, for a good oxygen supply was found by Read and Armstrong (1972) to be essential to that process, and indeed the fungi themselves require good aeration for growth in culture. Even intermittent waterlogging has a deleterious influence on the activities of mycorrhizas, as Gadgil (1972) showed. He estimated the vitality of mycorrhizas of *Pinus radiata* and *Pseudotsuga douglasii* by estimating the activity of succinic dehydrogenase in them. At the same time he examined their capacity for absorbing $^{32}$P-phosphate from solution. The rate of phosphate absorption by the mycorrhizas of *P. douglasii* was about halved after two weeks' waterlogging, but that of *Pinus radiata* only insignificantly reduced. The vitality of both was significantly reduced after four weeks but it was clear that the *P. douglasii* was somewhat the more easily damaged by waterlogging.

It is generally agreed that a moderate moisture content of the soil encourages mycorrhiza formation and that dry conditions or drought discourage mycor-rhiza. This subject has been reviewed in detail by Reid (1979). As he showed, there is a good deal of cultural evidence that the mycorrhizal fungi vary very much in their ability to withstand low water potential in soil (Theodorou, 1978; Uhlig, 1972; Mexal and Reid, 1973; see Reid, 1979) and there is often a considerable variation in the behaviour of different strains of one species in this regard. As a result the ability of different mycorrhizas to withstand drought may be different, but there is general agreement that strains of *Cenococcum geophilum* are especially drought resistant. In the early experiments of Worley and Hacskaylo (1959) the total percentage of short roots which were mycorrhizal diminished with increasing drought, but this was due to a decrease of white mycorrhizas. The black mycorrhizas, presumably formed by *Cenococcum*, were much fewer in number but much less reduced by drought so that their proportion increased with increasing water stress. Read *et al.* (1977) suggested that the formation of ectomycorrhiza by *Helianthemum chamaecistus* with *Cenococ-cum* was an adaptation to dry soil conditions, for Mexal and Reid (1973) had shown that strains of this fungus will grow at very low water potentials in culture.

However, most ectomycorrhizas seem to be more resistant to desiccation than uninfected roots, as Cromer (1935) and Harley (1940) have shown. The mycorrhizas of *Pinus radiata* appeared to be unaffected by drought in conditions where the cortex of uninfected roots shrivelled and collapsed. Cromer stated that after a prolonged drought, three days' rain caused mycorrhizas to begin to grow again whereas uninfected roots did not recover for about 14 days. The experiments that Goss (1960) made with *Pinus ponderosa* seedlings under extreme drought did not show that mycorrhizas conferred resistance to desiccation. However, after limited drought mycorrhizal seedlings recovered sooner than non-mycorrhizal ones. He ascribed this more to their ability to renew the growth of their roots after drought than to the actual resistance of the mycorrhizas themselves, and this appears to agree with the views of Cromer. He drew attention to the observation by Robertson (1954) that the long roots of *Pinus sylvestris* which contained a Hartig net were more resistant than those without. It seemed more probable that the hyphae connecting the roots, mycorrhizal or long roots, to the soil constituted a more important factor in recovery than the resistance of the sheath to water loss. Hu (1977) showed that these connecting hyphae remained functional in soil with a water potential of $-1$ bar ($-0.1$ mPa) for they continued to absorb phosphate, nitrogen and potassium from the soil. Uninfected roots were much affected in their uptake rates of these substances at a much higher water potential, $-0.2$ bar ($-0.02$ mPa). By contrast Reid and Bowen (1979) found that there was no difference in uptake by excised, uninfected and mycorrhizal roots from solutions of lowered water potential. Solutions down to $-1.2$ mPa ($-12$ bar) water potential did not affect phosphate uptake significantly. Hence the difference between mycorrhizal and non-mycorrhizal roots in dry soil must be due, it seemed, to the shrinkage of the soil away from the root surface, which greatly increases the resistance to water uptake into uninfected roots. The hyphae and hyphal strands of mycorrhizas, however, keep their contact with the soil. This problem was also approached by Sands and Theodorou (1978). They made a direct comparison between mycorrhizal and non-mycorrhizal seedlings of similar weight and height. There were 72% of the short roots mycorrhizal in the inoculated and 6% in the uninoculated seedlings. As the water potential of the soil became more negative, the transpiration rates and the water potentials of the leaves of both samples fell. It was therefore concluded that the resistances to water movement of the whole soil and plant system increased. Since the transpiration rate of both kinds of seedlings was similar and the water potential of the leaves of the mycorrhizal plants was more negative, it followed that their plant/soil resistance was higher than that of non-mycorrhizal seedlings. It was judged that the origin of the difference in resistance lay in the passage of water from soil to plant due to a difference in geometry of the root systems between infected and uninfected plants. No suggestion was made that the fungal sheath itself might constitute a significant resistance to water movement which offset any effect of the hyphae. Against this view, however, must be put the observations of Sands *et al.* (1981) that in a pressure chamber the rate of exudation, i.e. the reciprocal of the resistance to the passage of water, is similar in the mycorrhizal and non-mycorrhizal roots of *Pinus taeda*.

The subject obviously needs more investigation especially in terms of the movement of water through hyphae and fungal sheath and the resistances involved. Read and Malabari (1979) used mycorrhizas synthesized with *Pinus sylvestris* to show that hyphal strands were capable of conducting $^3H_2O$ to the host. The number of strands per unit volume with seedlings in peat was very great and this might have given a lower resistance to absorption than that of the uninfected roots of the control seedlings. Duddridge *et al.* (1980) have confirmed that these mycorrhizal strands, like so many others, are differentiated and have a structure which might allow for rapid transport of water through them.

There have been rather few researches having a bearing on the transport of water or of nutrients within the hyphae of the fungi in culture and upon the relationship of the two processes. Nor have there been any satisfactory experiments on the effects of transpiration rate and water through-put on rates of nutrient absorption either from soils or from solutions.

## Conclusion

Ectomycorrhizal plants are mainly woody perennials and trees whose root systems are infected with fungi, mostly Ascomycetes and Basidiomycetes but also a few species of *Endogone*. On the root systems, although all or almost all roots are infected, the ultimate laterals are heavily infected with hyphae which form a tough fungal tissue on their surfaces. This sheath connects with a Hartig net of hyphae in the walls of the epidermis and cortex and with mycelium in the soil. The mass of fungus associated with each infected mycorrhizal rootlet is very large—40% or more of the weight of the mycorrhizas, probably much more if the hyphae and its fruit bodies are included. This mycelium and sheath are nourished by carbon compounds from the host and soil-derived nutrients pass through them into the host. The mass of the fungal tissue seems to provide some explanation why woody perennials almost exclusively develop this kind of mycorrhiza. The sheath stores considerable quantities of both soil-derived nutrients and carbon compounds, so that the selective advantage of the association may well lie in this property. This view is made more probable since ectomycorrhizal associations are almost exclusive to habitats where the growth period is seasonal or intermittent. The absence of close specificity of the fungi to their hosts and vice versa would appear to be of ecological importance.

## Chapter 11

# Ectendomycorrhizas in Conifers

Mycorrhizas with many of the characteristics of ectomycorrhizas, but also exhibiting a high degree of intracellular penetration, have been described at various times over the last 80 years in various species of tree. These ectendomyc-orrhizal structures seem to be common over and above those examples of ectomycorrhizas where a few cells are penetrated by the fungus, or where the senescent cortex is, as already described, fully colonized by hyphae in the late Hartig net zone. Melin (1917 and later) distinguished "pseudomycorrhizas" in pine as forms of infection by septate fungi, which did not form sheath and Hartig net, nor were the usual changes in the histology of the epidermal and cortical cells present in them, yet the cells were penetrated by hyphae. However, at various times he and others implied that a thin sheath and Hartig net might be present. These pseudomycorrhizas were believed to be formed by dark sterile mycelia which have been found to be commonly present in the root region, and variously called *Rhizoctonia sylvestris*, *Mycelium radicis atrovirens* (*Phialocephala dimorphospora*, Richards and Fortin, 1973) or some other non-committal name. They may belong to the constellation of dark sterile forms found by Robertson (1954) and Harley and Waid (1955) on the roots and mycorrhizas of pine and beech, and more recently by Read and Haselwandtner (1981) on alpine plants and have been described frequently in earlier researches on mycorrhiza (see Harley, 1950). The implication of the name pseudomycorrhiza was that the fungi looked as if they were behaving as weak parasites and since they were found often on unthrifty seedlings they were blamed for the unthriftiness. Levisohn (1954) described "aberrant" root infections as opposed to "true" mycorrhizas. "Pseudomycorrhiza" is a term which might well be abolished, as Mikola (1965) and Harley (see 1959 and 1969a) have suggested, because of its lack of precision. Mikola (1965) pointed out that various authors have also used the term "ectendotrophic" for a variety of kinds of structure including those called pseudomycorrhiza by many people. The term "ectendomycorrhiza" should be used, however, as a purely descriptive name for those forms which exhibit some of the structural characteristics of both ectomycorrhizas and endomycorrhizas, and it implies no functional significance. The intracellular penetration in many ectomycorrhizas is a state to which they progress as they age, but there are normal active forms of mycorrhizal infection which possess a Hartig net, and sometimes a sheath in which a considerable degree or a specialized kind of intracellular penetration occurs. These are the mycorrhizas of some Ericales, e.g. the Pyrolaceae, the Arbutoideae and the Monotropaceae, as well as sometimes those of the Cistaceae. Here we concern ourselves with a kind of mycorrhiza which occurs on trees that are ectomycorrhizal during most of their life. It has been most observed in the conifers but certainly also occurs in others.

Laiho and Mikola (1964) examined the initiation of mycorrhizal infection in pine (*Pinus sylvestris*) and spruce (*Picea abies*) seedlings. The spruce developed a normal ectomycorrhiza, but on pine a kind of mycorrhiza having a coarse Hartig net, intracellular infection and a thin or absent fungal sheath was formed. These ectendomycorrhizas were initially very abundant and invariably present on pine in nursery beds of soil of agricultural origin but very infrequent indeed on spruce.

The earliest sign of infection was the formation of a Hartig net which followed closely behind the apical meristem as the root grew. Behind this, intracellular penetration increased in intensity towards the older basal part of the root, so that the cells became almost filled with coils of septate hyphae which were up to 15 $\mu$m thick. Mycorrhizal roots infected in this way persisted for at least a year with few signs of hyphal degeneration or lysis of the fungus. As Mikola (1965) wrote: "Intracellular hyphae do not injure the cortical cells; both host cells and intracellular hyphae were observed to live at least one year after commencement of infection; even the nuclei of such heavily colonized cortical cells were clearly visible in stained preparations." This description shows that the contention sometimes made that the cortical cells are dead and devoid of protoplast in ectendomycorrhiza is not generally true. Mikola saw no signs either of hyphal "digestion" even in mycorrhizas two years old. He believed that the hyphae and cortical cells died simultaneously. The longevity of this kind of mycorrhiza therefore equalled that of mycorrhizas with a more usual ectomycorrhizal structure and no intracellular infection (1 year or more), and was many times longer than uninfected roots of similar position in the root system.

Mikola (1965) showed that these ectendomycorrhizas of *Pinus sylvestris* were extremely similar to others that have been described, such as those figured by Goss (1960) from *P. ponderosa* from grassland soils in Nebraska, those illustrated by Björkman (1942) on *P. sylvestris* from pine forest in Sweden, and those described by Rayner in England. He isolated endophytes from nursery-grown ectendomycorrhizal pine and from ectomycorrhizas on young spruce which, when back-inoculated, successfully reformed ectendomycorrhizas of the expected structure on pine. One hundred and fifty strains, called by him "E type" were isolated and possibly, so Mikola believed, belonged to a single species. The aerial hyphae were brown, septate and coarse, 4–9 μm thick, without clamp connexions, conidia or reproductive bodies. The substrate hyphae were hyaline, often producing swellings like chlamydospores which were up to 30 μm in diameter.

Mikola studied the behaviour in culture of the strains which he isolated. He showed that the pH optimum for growth was between 5 and 6 and this differentiated them clearly from *Mycelium radicis atrovirens* which was little affected in growth rate by pH values between 3·5 and 8·0 on the same media. All the strains that he tested grew well on glucose and sucrose, some also on starch. Cellulose was not used by any strain, indeed their carbon requirements resembled those of ectomycorrhizal fungi in general. Again in nitrogen requirements they were also similar to ectomycorrhizal fungi. They made good growth on ammonium tartrate, but the pH change in the medium with other ammonium salts resulted in a lesser growth. One strain was able to grow well on nitrate but most grew much less well on it than on ammonium. Most were able to make moderate growth on peptone and casein hydrolysate but not on aspartic acid or glycine. Taken all in all, these fungi behave as ectomycorrhizal fungi do, and there is no evidence from their use of carbon polymers as to how they penetrate the cortical cells so readily. The strains of fungus isolated into culture were tested by back-inoculation in hosts. Several of the isolates formed ectendomycorrhiza with pine which had a Hartig net, a thin mantle, and different amounts of intracellular penetration. All the four strains tested on spruce formed only typical ectomycorrhizas with well-developed sheath and Hartig net but no intracellular penetration. Strain E57 isolated from nursery-grown ectomycorrhizal spruce formed, on back inoculation, ectendomy-corrhizas with pine and ectomycorrhizas with spruce.

Mikola concluded that the kind of ectendomycorrhiza formed by E-type strains was almost confined to young 1–3 year pine in nursery soil over a wide range of soil fertility, acidity and humus content and that its intensity of formation was not greatly dependent on light intensity. He also concluded that although these mycorrhizas had much in common structurally with the so-called "haustorial" mycorrhizas of Levisohn (1954) they were not detrimental to seedling growth.

Laiho (1965) made a wide survey of ectendomycorrhiza in nurseries and forests in Europe and America and experimentally synthesized mycorrhizas on a number of species of tree, using the isolated fungal E-strains. He encountered one type of ectendomycorrhiza only and in this the roots were inhabited by coarse septate mycelium which formed a strong Hartig net and intracellular infection. He concluded that E-strain fungi were present in all ectendomycorrhizas that he examined in detail, but that there must be other causative fungi, for E-type would only infect species of *Pinus* and *Larix*, whereas some ectendomycorrhizas undoubtedly occurred on species of *Picea* and of other genera. He stressed, however, that there was a possibility of error because senescent mycorrhizas often had intracellular infection and he quoted Mikola's (1948) observation that *Cenococcum* could penetrate the cortical cell walls of hosts growing in unsuitable conditions. This point is even more strongly reinforced by the observation that it is quite usual for the senescent cortical cells to be invaded (Harley, 1936; Atkinson, 1975; Nylund, 1981). Laiho was impressed by the absence of a deleterious effect of ectendomycorrhizal infection on *Pinus* seedlings and concluded that it was a "balanced symbiosis" and that it gave place to ectomycorrhiza as the seedling aged, especially in woodland conditions.

Wilcox (1968b, 1971) described similar ectendomycorrhizas on young plants of *Pinus resinosa* in which coarse hyphae form the Hartig net in both long and short roots. These hyphae tend to spiral round in the cell-wall following the angle of the cellulose fibrils. They penetrate the cell-wall through pit areas and also by means of rather complicated appressoria and grow into the lumen of the cell. The hyphae seem to be stimulated to spread inter-cellularly behind the apex of the long roots and around developing laterals, but do not invade the meristems. They may be active around the lateral initials and infect the young roots as they pass through the cortex. Although ectendomycorrhizas are often present on most of the short roots of young seedlings at first in nursery soils, they are later replaced by ectomycorrhizas formed by slender hyphae probably of another fungus.

Wilcox (1971), Wilcox and Ganmore-Neumann (1974) and Wilcox *et al.* (1975) isolated a number of fungi from *P. resinosa*: a black imperfect fungus (BDG22) (possibly a *Chloridium* sp. from its conidia) and a second (BDG58), probably identical with an E-type strain isolated by Mikola. The fungi formed ectendomycorrhiza with *P. resinosa* seedlings but they had different morphological characters; BDG22 formed, in contrast to E-type and BDG58, fine intracellular hyphae and Hartig net and masses of spherical bodies in some cells.

The detail of infection by BDG22 is given by Wilcox and Ganmore-Neumann (1974). On *Pinus resinosa* it grew longitudinally along the roots, encircling root hairs and penetrating them, or the surface cells, to invade the cortex where it became both intracellular and intercellular. Its hyphae were very frequently septate and produced chlamydospores both on the root surface and within the tissues. Soon a Hartig net was formed in the cortex beginning near the passage cells of the endodermis in the inner cortex and eventually throughout the whole tissue. Rhizomorphs or hyphal strands which develop on the root surface formed a sheath similar to that of typical ectomycorrhizal roots. Wilcox and Ganmore-Neumann were of the opinion that this fungus, and the ectendomycorrhizas that it formed, were clearly different from those formed by the E-type strains of Mikola or by BDG58. BDG58 ($\equiv$ E-type?) was thought by Wilcox to be the most widespread of the ectendomycorrhizal fungi. He and his colleagues described it as having ascomycetous affinities by its septation and simple pores. From their descriptions of the chlamydospores found on the root surface and in the tissues, Hall (1977) tentatively equated BDG58 with the "endogonaceous" fungus described as "crenulate spores" by Mosse and Bowen (1968a,b) from Australia. He believed that this type of spore was associated in the areas that he sampled, with *Pinus radiata* rather than with vesicular-arbuscular mycorrhizal roots. Thomas and Jackson (1979) isolated fungi from the roots of *Picea sichensis* in nursery soil. Amongst them were two fungi with septate mycelia which formed ectendomycorrhiza when back-inoculated on seedlings of *P. sichensis*. They equated these mycelia with the E-strains of Mikola and Laiho (although those usually formed ectomycorrhizas with *Picea*) and showed that they might give place to other fungi, especially *Thelephora terrestris* which formed ectomycorrhizas on the same roots during the first years of seedling growth. Walker (1979) described a new genus and species, *Complexipes moniliformis* which was believed to belong to the same group of fungi as the E strains, and was tentatively included in the Endogonaceae but this now seems to be an error. R. M. Danielson (1982) has further considered the taxonomy of ectendomycorrhizal fungi. He found that *Pinus banksiana* and *Picea glauca* growing in mine spoils were heavily infected by E-strain fungi which were the dominant mycorrhiza formers in that habitat. The fungus, as isolated on 2% malt or Melin–Norkrans medium, formed septate mycelium with simple pores, Woronin bodies and septal plugs. Since it was also inhibited by Benomyl it was believed to be ascomycetous. Two types of chlamydospores were formed by it: terminal ones like those of so-called *Complexipes*, and spores in chains. Back-inoculation on *Pinus* produced ectendo-mycorrhizas with thin or absent sheath, Hartig net and intracellular infection. With *Picea* monopodial ectomycorrhizas were formed—a behaviour like that of E-strain fungi. One can conclude therefore that ectendomycorrhiza of pine is formed often, perhaps always, by an ascomycete.

The BDG58 isolate which seems to be the most widespread of the ectendomyc-orrhizal fungi, far from being deleterious to the growth of its host, as ectendomycorrhizal infection was suggested to be by earlier workers, was shown by Wilcox and Ganmore-Neumann (1975) and Wilcox (1978) to increase that of *Pinus resinosa* (see Table 60). It seems clear that ectendomycorrhizas are not "aberrant" or minor parasitic infections which occur in the absence of ectomycorrhizal fungi. They are symbiotic associations which may be mutualis-tic and are frequently the early infection state of seedlings. Perhaps comparative estimates of the drain of carbon from the host caused by them and by ectomycorrhizal fungi should be made to see if the least carbohydrate-demanding fungi are selected by seedlings before full photosynthetic potential is reached.

One very interesting question raised by Wilcox concerns the consistent formation of ectendomycorrhiza on pine and ectomycorrhiza on spruce by the same strains of E-type fungus. This is not a problem unique to these strains; it is met again in arbutoid mycorrhiza, where one and the same strain infects species of *Arbutus* or *Arctostaphylos* with extensive cell penetration to form ectendomycorrhiza, and forms typical ectomycorrhizas with conifers (see Chapter 13). A related problem is the formation by one fungal mycelium of monotropoid mycorrhiza on *Monotropa* with intracellular haustoria, and ectomycorrhiza of typical structure on forest trees such as spruce or beech.

TABLE 60

Effect of ectendomycorrhizal infection with fungus BDG58 ( ≡ E-type strain of Mikola and Laiho) on the growth of seedlings of *P. resinosa* at different temperatures

*Growth period 3 months. Weights in mg (results of Wilcox and Ganmore-Neumann, 1975)*

| Temperature (°C) | Infection | Fresh weight mg | | | Root weight |
| | | Root | Shoot | Total | Shoot weight |
|---|---|---|---|---|---|
| 15 | − | 1096 | 1597 | 2693 | 0·69 |
|    | + | 1787 | 2015 | 3802 | 0·89 |
| 20 | − | 983 | 1770 | 2753 | 0·56 |
|    | + | 2619 | 2058 | 4677 | 1·27 |
| 30 | − | 1142 | 1991 | 3133 | 0·57 |
|    | + | 1949 | 1948 | 3897 | 1·00 |

*Chapter 12*

# Ericoid Mycorrhizas

## Introduction

The order Ericales includes families whose members range in habit from the trees and shrubs in the Ericaceae to the herbaceous partial "saprophytes" and complete "saprophytes" of the Pyrolaceae and Monotropaceae. The families have often been regrouped in various ways in attempts to separate taxa having common habit and structure as well as common floral characters. In any event many of the tendencies in habit are paralleled by variation in mycorrhiza, for the whole order is mycorrhizal, although the kinds vary. Indeed the Ericales are extremely interesting in this regard for forms with sheath and Hartig net and specialized "haustoria", forms resembling the ectendomycorrhizas of conifers, and endomycorrhizas with no sheath or Hartig net but extensive hyphal colonization of the cells of the cortex, are all found in various species. Recently ectomycorrhizas have also been found on some trees. We now have evidence that, on the one hand, the physiology of some resembles that of the ectomycorrhizas of forest trees, while the physiology of the totally achlorophyllous or partly achlorophyllous species had something in common with mycorrhizal physiology of some of the Orchidaceae.

The kinds of mycorrhizas in the Ericales were classified by Harley (1959, 1969a) into ericoid and arbutoid mycorrhizas. The first is found on the very fine root-systems of most species of the Ericoideae, Vaccinioideae, and Rhododendroideae of the Ericaceae and of the Epacridaceae and Empetraceae. Hyphal development (of septate hyphae) around the root is copious and forms a loose weft, but that on the root surface itself is relatively slight. Within the cells very extensive hyphal coils develop so that large quantities of fungal tissue are associated with the roots. Although the fungus eventually degenerates, so does the host cell—a different behaviour from that seen in vesicular-arbuscular mycorrhizas and from that in arbutoid or monotropoid or orchid mycorrhizas.

The host plants, which are fully autotrophic shrubs and small trees, are mostly restricted to or characteristic of acid or peaty soils. These mycorrhizas are called "ericoid" (or sometimes "ericaceous").

The second type of mycorrhiza found in Ericales was called "arbutoid" because it was described in *Arbutus unedo* by Rivett in 1924. Similar kinds of mycorrhizal infection have since been found in other Arbutoideae and in many Pyrolaceae, e.g. in *Arctostaphylos* and *Pyrola*. In some Monotropaceae the structure is different again and the term "monotropoid" introduced by Duddridge (1980) is used for them. In many of the arborescent arbutoid mycorrhizal species, the root system is recognizably divided into long and short roots as in ectomycorrhizal trees. Moreover the short roots ("tubercles" Rivett called them) are encased in a pseudoparenchymatous sheath or mantle which connects not only with hyphae in the soil but also with a well-developed Hartig net between the epidermal and cortical cells of the host. A characteristic feature is, however, the penetration of the epidermal cells, at first by simple hyphae but later hyphal coils may develop in them. The coils eventually disintegrate and the cells may be recolonized by active hyphae. The resemblance of this kind of mycorrhiza to ecto- and ectendomycorrhizas is obvious and it is therefore perhaps not surprising to learn that ectomycorrhiza has also been described as occurring in some Arbutoideae (see Table 61).

In contrast to most of the species having ericoid or arbutoid infection which are woody, those in the Pyrolaceae and Monotropaceae are usually herbaceous. In Pyrolaceae some species lack a sheath and their mycorrhizas resemble the ericoid type, but in others a Hartig net and a sheath of some kind are present. The cells are penetrated by the fungus which may, as in ericoid mycorrhizas, form copious coils in them. The achlorophyllous Monotropaceae have mycorrhizas with well-developed sheaths and Hartig nets. Simple haustoria penetrate into the outer cells of the host and pass through a characteristic series of changes (see later). These monotropoid mycorrhizas are distinct from other kinds in the Ericales (Duddridge, 1980).

An idea of the variability observable in some species of Ericales may be obtained from Table 61 which is derived from Largent *et al.* (1980b). The table gives, perhaps, a general idea of frequency of different types within a species, but it is clearly more reliable for those species where a sufficient number were observed.

In view of the variability in mycorrhizal equipment in the order Ericales, it may well be asked why they are described together in two chapters. It is not for the reason of tradition, but because the very range of variation which occurs within an obviously related monophyletic taxon emphasizes some of the problems of the origin and evolution of mycorrhizal symbiosis, as will appear later. Moreover, much more observation is needed before some of them can be linked with confidence to other types.

TABLE 61
Kinds of mycorrhiza observed in some species of Ericales (from Largent *et al.*, 1980b)

| Species | Number of plants observed | Kind of mycorrhiza | | | | |
|---|---|---|---|---|---|---|
| | | Ecto | Arbutoid | Arb./Ecto | Ericoid | Zero |
| *Arbutus menziesii* | 55 | 23 | 24 | 8 | 0 | 0 |
| *Arctostaphylos canescens* | 13 | 7 | 2 | 4 | 0 | 1 |
| *A. columbriosa* | 47 | 10 | 34 | 7 | 0 | 0 |
| *A. glandulosa* | 1 | 0 | 0 | 1 | 0 | 0 |
| *A. intricata* | 21 | 12 | 11 | 1 | 0 | 0 |
| *A. manzanita* | 18 | 8 | 6 | 7 | 0 | 0 |
| *A. nevadensis* | 44 | 12 | 16 | 16 | 1 | 1 |
| *A. nummularia* | 1 | 1 | 0 | 0 | 0 | 0 |
| *A. patula* | 52 | 20 | 30 | 13 | 3 | 2 |
| *A. uva-ursi* | 6 | 1 | 4 | 1 | 0 | 0 |
| *A. viscida* | 10 | 4 | 4 | 3 | 0 | 1 |
| *Cassiope mertensiana* | 2 | 0 | 0 | 0 | 2 | 0 |
| *Gaultheria ovatifolia* | 2 | 1 | 0 | 0 | 1 | 1 |
| *G. shallon* | 29 | 3 | 1 | 2 | 5 | 18 |
| *Kalmia polifolia* | 9 | 2 | 0 | 0 | 4 | 3 |
| *Ledum glandulosum* | 8 | 6 | 0 | 0 | 1 | 2 |
| *L.* var. *californicum* | 5 | 0 | 0 | 0 | 3 | 2 |
| *Leucothoë davisiae* | 21 | 6 | 3 | 0 | 6 | 6 |
| *Phyllodoce empetriformis* | 14 | 0 | 0 | 0 | 4 | 10 |
| *Rhododendron occidentale* | 2 | 0 | 0 | 1 | 1 | 0 |
| *R. macrophyllum* | 6 | 3 | 0 | 2 | 0 | 1 |
| *Vaccinium arboreum* | 17 | 2 | 0 | 0 | 6 | 9 |
| *V. membranaceum* | 4 | 0 | 0 | 0 | 2 | 2 |
| *V. occidentale* | 3 | 0 | 0 | 0 | 2 | 1 |
| *V. ovatum* | 37 | 8 | 1 | 3 | 4 | 21 |
| *V. parvifolium* | 8 | 1 | 0 | 0 | 1 | 6 |
| *V. scoparium* | 17 | 2 | 0 | 1 | 0 | 14 |
| *V. menziesii* | 2 | 0 | 1 | 1 | 0 | 0 |
| *Chimaphila umbellata* | 15 | 2 | 0 | 0 | 1 | 12 |
| *Pyrola picta* var. *picta* | 8 | 0 | 1 | 0 | 1 | 6 |
| *P. aphylla* | 1 | 0 | 0 | 0 | 1 | 0 |
| *P. secunda* | 3 | 0 | 0 | 0 | 0 | 3 |

*Note:* Ecto: Sheath, Hartig net, no intracellular hyphae.
　　　Arbutoid: Sheath, Hartig net, and intracellular hyphae.
　　　Arbutoid/ecto: Sheath and either Hartig net or intracellular hyphae deficient.
　　　Ericoid: No sheath, nor Hartig net, intracellular hyphae.

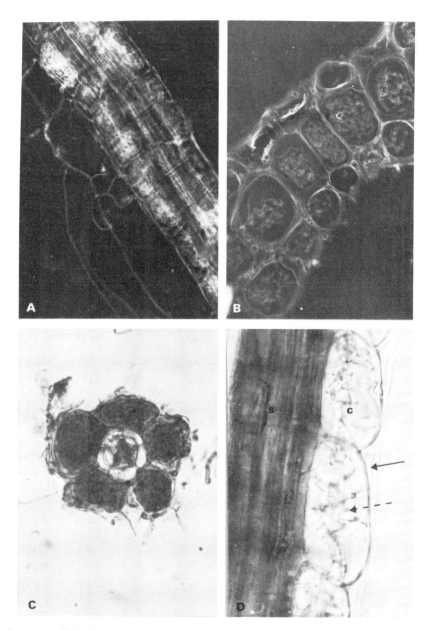

Plate 11. Ericoid mycorrhiza.

A. Surface view of ericoid mycorrhiza of *Rhododendron ponticum* showing intracellular infection and extramatrical hyphae. ×145. (Photo: J. Duddridge.)

B. LS mycorrhizal root of *R. ponticum* showing intracellular hyphal coils. ×145. (Photo: J. Duddridge.)

## Ericoid mycorrhizas

The root systems of most Ericaceae (excluding Arbutoideae), Empetraceae and Epacridaceae are very fine and have been called "hair roots" by Brook (1952) who worked with *Pernettya macrostigma*. This plant provides a good example of these roots for its small monarch stele is surrounded by a simple pericycle and endodermis (often tannin-filled), outside which is a single layer of larger cortical cells. The roots lack epidermis and root hairs and have only a reduced root-cap which indeed may be absent. The diameter in *Pernettya* is 50–100 $\mu$m, in *Calluna* it is about 40 $\mu$m, but in *Rhododendron* and *Vaccinium* it is, although fine, somewhat more robust than these (see Plate 11). The sequence of apical growth of the hair roots of *Calluna* is well described by Burgeff (1961). He figured both longitudinal and transverse sections of the apical region showing an apical cell giving rise distally to a loose root-cap of some eight cells, a ring of lateral cells which give rise to the cortical (epidermal) region 1 cell thick of some 9–10 cells in section, and a tiny proximal group of initials which form the endodermis and stelar region. Mycorrhizal infection is typically confined to the outermost or sole layer of cortical cells and in any case it never penetrates the endodermis or stele. Infection is initiated by growth of the fungus from the soil to form a sparse network of septate hyphae spreading over the surface of the root and between and through the walls of the cells of the root without the formation of appressoria. Nieuwdorp (1969) gives an account of the process in *Calluna*, *Erica* and *Vaccinium*, Bonfante-Fasolo and Gianinazzi-Pearson (1979) describe it in *Calluna*, Pearson *et al.* (1980) and Duddridge (1980) in *Rhododendron* and Bonfante-Fasolo, Berta *et al.* (1981a) in *Vaccinium myrtillus*. As the root matures the cells on its surface tend to separate, allowing some hyphae to grow between them. The surface of the root is surrounded by a thick mucigel layer (Burgeff, 1961; Leiser, 1968) and it is there that the hyphae spread. Presumably the pectolytic enzymes that they produce aid this process (Pearson and Read, 1975). The fine, septate hyphae that penetrate the wall become surrounded by cell-wall material within the lumen. These hyphae appear to originate from slightly swollen hyphae but no appressoria are formed. As the infective hypha enters the cell a "collar" of material is laid down around it (see Plate 12). This collar, on comparative grounds may be an "extrahaustorial matrix" or "papilla" of host origin (Bonfante-Fasolo and Gianinazzi-Pearson, 1979; Duddridge, 1980). Its presence indicates that the cell protoplast remains alive

C. TS root of *Calluna vulgaris* showing the cortex of one cell layer fully colonized by fungal hyphae. × 145. (Photo: J. Duddridge.)

D. Early stage of infection of *Calluna* root in aseptic culture. S, stele; C, cortex; →, extramatrical hyphae: − − →, internal hyphae. Note the developing intracellular hyphal coils. × 140 approx. (Pearson and Read, 1973a.)

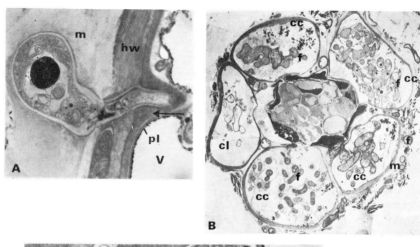

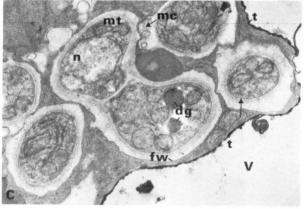

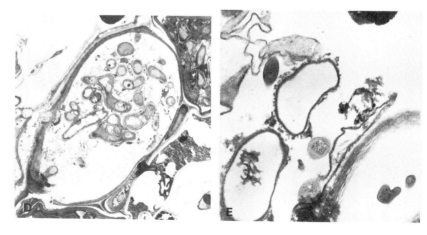

and active, as do the persistence and frequent enlargement of the nuclei of the host (Freisleben, 1934; Brook, 1952). Under the low power electron microscope it may be seen that the intracellular hyphae are surrounded by host cytoplasm (Nieuwdorp, 1969) and as Bonfante-Fasolo and Gianinazzi-Pearson (1979) and Duddridge (1980) have shown, they are enclosed by host plasmalemma. Presumably the "collars" may be what were earlier described as "röhrentüpfel" by Burgeff (1961) and others using the light microscope.

Intracellular hyphae, by growth and branching, form the characteristic coils, clearly described by Rayner (see 1927), which come to fill the cells. As the fungus invades the cortical cells to produce the intracellular coils, it appears to attract the host cytoplasm to surround it. The cytoplasm increases in volume and is separated from the hyphae by the host plasmalemma and a fibrillar interfacial matrix, which appears to be of pectic material as shown by its staining reaction (Bonfante-Fasolo and Gianinazzi-Pearson, 1979; Duddridge, 1980; see Plate 12). When the infection is mature the cell is almost filled by fungal hyphae and the remainder of the volume of the cell includes many small phenol-containing vacuoles, but at first both host and fungus possess apparently active organelles. The fungus spreads along the surface of the root and produces branches which infect the cortical cells. A spread from cell to cell is infrequent. Mapping of the occurrence and intensity of fungal colonization of entire root systems of seedlings of *Calluna vulgaris* has shown that infection can be extremely extensive. Read and Stribley (1975) give examples in their paper; almost 65% of the root length was

PLATE 12. Electron micrographs of mycorrhizal infection in *Calluna vulgaris* and *Vaccinium myrtillus*.

A. Penetration of a cortical cell of *C. vulgaris* by a hypha. A swollen hypha in the mucigel layer (m) has produced a narrow branch which penetrates the host wall (hw) and invaginates the host plasmalemma (pl) and vacuole (v). A layer of material derived from the host forms a "collar" around the penetrating hypha (arrowed). × 9500.

B. Reconstruction of a transverse section of a single hair root of *C. vulgaris* showing general structure of the infection. Mucigel (m) surrounds the single layer of cortical cells and contains fungal hyphae (f). Four of the cortical cells (cc) contain heavy fungal infection by apparently active hyphae, while a fifth (cl) has collapsed and contains empty hyphae. × 1275 approx.

C. A single cortical cell of *C. vulgaris* containing fungal hyphae. Both host and fungus have clearly distinguishable and apparently active cytoplasm. The fungal wall (fw) is electron-translucent. The fungal cytoplasm can be seen to contain nuclei (n), mitochondria (mt) and electron dense granules (dg). The interfacial zone between the symbionts contains fibrillar material (arrowed) and membraneous configurations (mc). The host vacuole (v) is surrounded by the tonoplast (t). × 18 900.

D. Living hyphae surrounded by disorganized cytoplasm in a cell of *V. myrtillus*. × 1760.

E. Empty and collapsed hyphae in an empty cell of *V. myrtillus*. × 3745.

A, B and C from Bonfante-Fasolo and Gianinazzi-Pearson, 1979; D and E from Bonfante-Fasolo *et al.* (1981).

infected, and of this some 48% fell into the intensity categories of 75–100% infected. In *Calluna* penetration of the cells of the roots was first observed 0·5 mm behind the root-cap. In *Vaccinium* the apical 2–3 mm is often uninfected, but such differences may well depend on the activity of the meristem and maturation zones of the root, that is upon the growth rate of the root. Read and Stribley calculated from their mapping that as much as 42% of the volume of the root was occupied by fungal hyphae (see Table 62), an extremely high proportion. Each infected cell had at least one entry point (frequently more) connecting intracellular with extra-matrical mycelium in the soil. Hence a *Calluna* root with 2000 infected cells per cm of root length would have at least 2000 connexions with the soil—a much greater number than in vesicular-arbuscular mycorrhiza (Table 63).

TABLE 62

*Calluna vulgaris.* The calculated mean amounts of fungal tissue in each category of fungal infection and in the whole root system (from D. J. Read and D. P. Stribley, 1975)

|  | Infection category | | | | |
|---|---|---|---|---|---|
|  | 1 | 2 | 3 | 4 | 5 |
| Proportion of roots falling into the category (mean of 6 seedlings) (%) | 28 | 20 | 9 | 4 | 4 |
| Mean level of infection assumed for category (%) | 100 | 87·5 | 62·5 | 37·5 | 12·5 |
| Calculated volume occupied by fungus (%) | 80 | 70 | 50 | 30 | 10 |
| Calculated volume as proportion of whole root system (%) | 22·4 | 14·0 | 4·5 | 1·2 | 0·3 |
| Mean total volume occupied by fungus (6 seedlings) (%) | | | 42·4% | | |

TABLE 63

*Calluna vulgaris.* Calculated number of entry points in the various categories of infection and in the entire root system (from D. J. Read and D.P. Stribley, 1975)

|  | Infection category | | | | |
|---|---|---|---|---|---|
|  | 1 | 2 | 3 | 4 | 5 |
| Number of entry points per cm | 2000 | 1750 | 1250 | 750 | 250 |
| Root length per category (cm) | 28 | 20 | 9 | 4 | 4 |
| Number of entry points per category | $56 \times 10^3$ | $35 \times 10^3$ | $11 \times 10^3$ | $3 \times 10^3$ | $1 \times 10^3$ |
| Mean total number of entry points for whole seedlings assuming, per cell, | | | | | |
| a.  1 entry point, | | | $106 \times 10^3$ | | |
| b.  6 entry points, | | | $636 \times 10^3$ | | |
| c.  12 entry points, | | | $127 \times 10^4$ | | |

The anatomy and histology of *Rhododendron* mycorrhiza as described recently by Peterson *et al.* (1980) is similar to that of *Calluna* and *Vaccinium*. The epidermis and cortex are both infected, often very intensely. Infection usually begins about 1·5 mm behind the apex; the closest observed was 0·54 mm from it in a root with an active growing point. Some roots lose their meristem, cease growing and become infected right up to the apex.

After a period of active interaction between the hyphae and the cytoplasm of the host, degeneration takes place. Until recently, this process was believed to be similar to that in vesicular-arbuscular mycorrhiza and in orchid mycorrhizas. Recent work, however, by Bonfante-Fasolo and Gianinazzi-Pearson (1979) and Duddridge (1980) (see also Read, 1982, in press) has shown their behaviour to be different. The first signs are not a degeneration of the hyphae within an active cell, but the degeneration of the contents of the cell of the host (see Plate 12D). The plasmalemma of the host eventually loses its integrity and living hyphae may exist in a degenerate host cell for a time. However, soon the vacuoles of the fungus increase in size and the hyphae degenerate. Bonfante-Fasolo and Gianinazzi-Pearson (1981) reported living hyphae in senescent cells of the host and suggested that they might be important in the recycling of material within the symbiotic system in a manner also suggested for ectomycorrhizas (see also Scannerini, 1982, in press). They also showed that adjacent cells of *Calluna* might be in different stages of senescence, not because there were repeated colonizations of cells as in Orchidaceae and other plants, but because in neighbouring cells both host and fungal development were at different stages (see Plate 12). Read (1982) reports that degeneration occurred at 4 and was complete at 7 weeks after penetration. Peterson *et al.* (1980) also stated that the "active physiological stage [of *Rhododendron* symbiosis] seems to be restricted to a small period of time" in spite of there being a permanent root axis in the relatively long-lived roots. The active period was from the initiation of infection in the epidermal and cortical cells after they had achieved full size to the beginning of the sloughing-off following the onset of secondary thickening of the stele.

The degree of infection of the root system is affected by external factors and there is a seasonal variation. Infection is sparse in early spring and becomes more intense later in the season. This is partly explained by the seasonal production, growth and colonization of new roots, but the rapidity with which the hyphae grow and with which the intracellular coils disintegrate will determine the sequence, pattern and intensity of infection. In a natural alpine vegetational system dominated by ericaceous plants, Haselwandter (1979) observed that the lowest intensities were at the highest altitudes where the growing season was short, the temperature lowest and the soils thin. It seems likely that the shortness of the growing season was the most important factor controlling root and fungal development. High levels of infection were found to be associated with high biomass and vigour of growth of the host plants (five species were observed) even in low nutrient soils. As with vesicular-arbuscular and ectomycorrhizas, fertilization of the plant with phosphate or ammonium nitrate in culture has been found to reduce infection (Morrison, 1957b; Stribley and Read, 1976).

There is no evidence that mycorrhizal infection extends beyond the root system into shoots, fruits and seeds. Rayner and her associates believed that infection was systemic and that the causal organism was a species of *Phoma*. Infection of shoots and roots was thought to originate from attenuated fungal hyphae in the seedcoat and embryo, and to be an absolute necessity for seed germination and normal root growth. It has been quite clear for several decades that sterile plants of ericaceous species may develop quite normally in suitable conditions, and that the fungi which have been isolated from shoot systems cannot form mycorrhizas with them. Some of the confusion may have arisen from a failure to distinguish between "attenuated" (real or imaginary) infection of tissue by hyphae of casual invaders, and mycorrhizal infection of the cells of the root by hyphal coils. It is essential to emphasize this point because systemic mycorrhizal infection was once reported in the Cistaceae (see pp. 115–116) and does not seem to occur there either. This disposes of the only two cases in which some belief in systemic infection by a mycorrhizal fungus has persisted.

## The fungi of ericoid mycorrhiza

It was for a long time believed that species of the genus *Phoma* were mycorrhizal with Ericaceae. Isolations of them had been made from roots by Ternetz (1907), and Rayner and Levisohn (1940), and from the seeds of *Calluna* by Rayner (see 1927). These fungi have now been shown, in so far as a negative can be proved, not to form mycorrhiza with Ericaceae. All recent investigations, which have involved isolation of fungi from the cells of the roots of many species of Ericaceae and re-synthesis of typical mycorrhizas, have indicated that slow-growing fungi giving dark-coloured mycelia in culture are the causal organisms of mycorrhizal infection (Singh, 1974; Pearson and Read, 1973a; and see Harley, 1969a). Similar sterile, slow-growing, dark-coloured mycelia were isolated long ago from *Vaccinium* by Doak (1928), Freisleben (1933, 1936) and from species of several genera by Bain (1937). Many of these were proved to be mycorrhiza-formers by back inoculation into ericaceous hosts. Subsequently Burgeff (1961), McNabb (1961), Singh (1964, 1974) and Pearson and Read (1973a) confirmed these findings. Pearson and Read used a similar method to that of Bain. They plated out individual or small groups of cortical cells from washed roots which could be observed to contain mycorrhizal hyphae. The hyphae that grew out from them were sub-cultured. Of these isolates 97% were "slow-growing dark fungi" which could infect and form typical mycorrhizas on roots of the species of origin and other Ericaceae. The non-specificity observed by Pearson and Read (1973a), that the endophytes obtained from *Calluna vulgaris*, *Vaccinium myrtillus*, *V.*

*oxycoccus,*[1] *V. macrocarpum,*[1] *Erica cinerea* and *Rhododendron ponticum* could each form mycorrhiza with every one of these hosts, agrees with the conclusions of Freisleben (1936). His *Mycelium radicis myrtillis-α* formed mycorrhiza with species of *Vaccinium* tested, and with fifteen species from ten other genera.

The lack of specificity of the fungal isolates which vary somewhat in cultural characterisitics is forcibly emphasized by the fact that of 41 similar dark mycelia obtained from a range of soil types, 30 were mycorrhizal with *Calluna* whether ericaceous plants had been growing in the soil or no. Somewhat similar dark sterile forms have been isolated from the rhizosphere of many plants including *Pinus* and *Fagus*. They have been categorized by some observers as pseudomycorrhizal fungi. Pearson and Read tested one of these from *Pinus* roots and found it unable to infect any of the tested species of Ericaceae in their conditions.

One of the isolates from *Calluna vulgaris* made by Pearson and Read and shown to be mycorrhizal with a range of species of Ericaceae produced apothecia in association with seedlings grown aseptically. It was described by Read (1974) as *Pezizella ericae* sp. nov. This species was shown to be homothallic by Webster (1976). Duddridge (1980) isolated *P. ericae* from *Rhododendron ponticum* and showed it to be mycorrhizal with that species. Vegh *et al.* (1979) isolated closely similar mycelia to those obtained by Pearson and Read from a *Rhododendron* hybrid and from *Erica gracilis* in France. Both isolates formed mycorrhizas under sterile conditions with the *Rhododendron* and they fruited in culture on Melin's medium (1959) by the method used by Webster to give *P. ericae* fruit bodies.

The similarity of the isolates obtained by several workers has led to the tacit assumption that they all may belong to the one species of *Pezizella* or at least to a group of very similar species, and this seems very likely. It should, however, be noted in passing, in case it becomes important, that the actual isolate used by Pearson and Read for most of their work appears to have been obtained from *Vaccinium macrocarpum*; that used in the study of the physiology (1975) was from *Calluna vulgaris*. No criticism is implied by this statement of the enlightening work by these workers; it simply is made because there has developed an unfortunate tendency to refer to the plexus of dark sterile mycorrhizal mycelia of Ericaceae as "the endophyte", and it is still essential to keep an open mind about the matter in view of the wide range of mycorrhizal fungi which may be associated with a single kind of mycorrhiza. Indeed, species of the basidiomycetous genus *Clavaria* which grow in close association with species of Ericaceae have also been suggested as possibly mycorrhizal with them (Seviour *et al.*, 1973; Englander and Hull, 1980; Hudson, personal communication, 1980). According to Hudson, mycorrhizal hyphae in "Clavaria" habitats may have a paler colour than elsewhere, but since all his attempts to isolate *Clavaria* sp. from infected roots and from its fruit bodies failed, proof of its mycorrhizal status is still lacking. Moreover, Dr Read has told

---

[1] *V. oxycoccus = Oxycoccus palustris* Pers. E. B. *V. macrocarpum = O. macrocarpus* (Ait.) Pers.

us that his attempts to infect *Calluna* with spores of *Clavaria argillacea*, a common species of heathland, have not been successful. On the other hand, Seviour *et al.* (1973) used a fluorescent antibody technique with the roots of *Azalea indica* and suggested that the hyphae within them were immunologically related to those of fruit bodies of a *Clavaria* of sub-genus *Syncoryne* (cf. *C. vermicularis* but possibly undescribed) fruiting nearby. Indirect evidence was also obtained by Englander and Hull (1980), namely that photosynthetically derived $^{14}$C could pass from *Rhododendron* sp. and *Pieris japonica* to expanding fruit bodies of *C. argillacea*, but not to mature ones. $^{32}$P was also observed to move from $NaH_2{}^{32}PO_4$ applied to the fruit bodies into the roots. It may be pertinent to recall, however, that both species *C. argillacea* and *C. vermicularis* are very variable (Corner, 1950). It will be noted later that similar evidence of the identity of a mycorrhizal associate of *Monotropa* was later confirmed by inoculation experiments (Björkman, 1960).

Some further evidence that both ascomycetous and basidiomycetous fungi may be involved in ericoid mycorrhiza is given by the descriptions of the septal pores of hyphae within the cells. Bonfante-Fasolo and Gianinazzi-Pearson (1979) described the pores seen under the electron microscope in the hyphae infecting *Calluna vulgaris* as simple, sometimes associated with Woronin bodies, as would be expected of an Ascomycete. Later, Bonfante-Fasolo (1980) showed that both an Ascomycete and a Basidiomycete might be present in the cells of the roots of *Calluna*. Both behaved similarly within the cells. Peterson *et al.* (1980) describe the infection of *Rhododendron* as variable, and that dolipore septa as well as septa with simple pores and Woronin bodies were observed in hyphae even in the same cell. They also recorded the presence of dolipore septa in *Pieris* and Woronin bodies in *Kalmia*. It seems probable that, as is the case with ectomycorrhizas, both Ascomycetes and Basidiomycetes are involved in ericoid mycorrhiza formation.

## Physiology of the fungi

Pearson and Read (1975) made a physiological study of an isolate, presumably *Pezizella ericae*, from *Calluna* and Vegh *et al.* (1979) have observed *Pezizella* isolated in France from *Rhododendron*. The optimum temperature for growth of the isolates lay between 20 and 30°C, growth being very slow below 5°C and above 30–32°C. The pH was optimal between pH 6 and pH 7 and the lower limit was pH 4 on the artificial medium used by Pearson and Read. There was a much less marked effect of pH between 4 and 8 in the malt medium used by Vegh *et al.* The fungus grew on amino nitrogen or ammonium nitrate, but least well on glycine. A wide range of carbon compounds supported growth: monosaccharides including the pentose xylose, disaccharides maltose, sucrose and cellobiose, and complex carbohydrates starch and pectin, but not cellulose. Phosphate was absorbed both as inorganic orthophosphate and from inositolhexaphosphate. It was shown that the phosphatase activity per unit dry weight was highest in mycelium grown in a low inorganic phosphate medium, a feature reminiscent of ectomycorrhizal fungi.

Nieuwdorp (1969) studied a group of isolates from the Ericaceae made by Freisleben, two from *Vaccinium myrtillus* (*Mycelium radicis myrtillis* α and β), and by Burgeff, one from *Vaccinium ligniosum* and three from *Vaccinium oxycoccus*. He tested their ability to decrease the viscosity of solutions of sodium pectate and carboxymethyl cellulose. The viscosity fell rapidly in 2–5 hours in sodium pectate, and fell significantly in 21 hours in carboxymethyl cellulose. This indicates that the strains produce pectinase which may assist in the utilization of the mucilage layer of the roots (see Leiser, 1968) and perhaps cellulase which might bring about some hydrolysis of cellulose. However, the cellulolytic activity does not seem to be such as to disagree with the findings of Pearson and Read that this carbohydrate will not act as a sufficient source of carbon for growth. The presence of the cellulolytic activity may, however, be of significance in the penetration of cell walls.

Translocation in the hyphae has been demonstrated in culture by Pearson and Read (1973b). The distances across which $^{14}$C-glucose or $^{32}$P-phosphate were translocated were small (6 mm) in the fungal hyphae alone. However, when the fungi were associated with seedlings it was up to 25 mm. It is possible that both cytoplasmic streaming and mass flow are involved as mechanisms.

## Physiology of ericoid mycorrhizas

### *Growth of the autotroph*

Some experiments by Brook (1952) and Morrison (1957b) on the physiology of mycorrhizal ericaceous plants gave an indication that improved mineral nutrition of infected plants was the cause of their increased growth on soils low in nutrients. However, other experiments did not show any improved growth of mycorrhizal plants; indeed, they often grew less in weight than the uninfected control plants (Bannister and Norton, 1974; Singh, 1974). It now seems that although differences in soil, and in methods used to sterilize it, might explain much of the variability of these results, real growth stimulations or inhibitions may arise according to whether the growth of the host is limited by soil-derived nutrients or by carbon assimilation. Most methods of soil sterilization (particularly those involving heat) increase soil nutrient levels and may also release toxins. These problems are particularly severe in soils high in organic matter, such as those which typically bear ericaceous vegetation. Hence, differences in growth between mycorrhizal plants grown on natural soil and non-mycorrhizal plants grown on sterilized soil could often be interpreted both in terms of nutrient availability or toxins produced by sterilization. It should be noted too that in some cases the toxins produced by heat sterilization have been found to be destroyed by fungi, whether mycorrhizal or not (see Harley, 1969a).

More recent work has taken these problems into account and a series of experiments by D. J. Read and his colleagues on *Calluna vulgaris* and *Vaccinium macrocarpum* have provided much essential information about the physiology of the association. If pre-inoculated mycorrhizal plants and uninfected plants are grown in soil sterilized by γ-irradiation (which results owing to death of living organisms in changes in nutrient availability, but in no great generation of toxicity), then improved growth of the mycorrhizal plants is regularly observed at low but not very low soil nutrient levels, (Stribley *et al.*, 1975; Stribley and Read, 1974a, 1976). In more fertile soils there were no such differences. It is worth noting that natural (unsterilized) soil treatments were also included in some of these experiments and that growth of mycorrhizal plants in them was almost always lower than in either of the treatments in sterilized soil. Stribley *et al.* (1975) estimated the changes in ammonium, nitrogen and phosphate arising immediately from γ-irradiation and at each harvest during their experiments. There is no doubt that the availability of these nutrients was maintained at a sufficiently high level after sterilization to afford some explanation of the increased growth of the plants.

## Nitrogen absorption

Most ericaceous species characteristically grow on nutrient-poor, acid soils. In these soils, nitrogen mineralization rates are low and since the pH is usually below 4·0, ammonium predominates over nitrate. Analysis by Read and Stribley (1973) and Stribley and Read (1974a, 1976) of nutrient content and concentrations in *Calluna vulgaris* and *Vaccinium macrocarpum* has shown that mycorrhizas have an effect on the nitrogen uptake as well as upon the phosphate uptake of these plants (Fig. 22). It is important not to stress too much the effect of infection on nitrogen absorption and to ignore that upon phosphate or indeed any other ion which is relatively immobile in soil solution, although quantitatively of course more nitrogen is required than any other soil-derived element. The importance of mycorrhizas in nitrogen absorption is of greater ecological interest, however, because the habitat of most Ericaceae is acid or peat soils. In *Vaccinium macrocarpum* the improved nitrogen nutrition has been shown to be the result of greater efficiency of uptake, i.e. specific absorption rate (nutrient absorbed per unit dry weight of root per unit time). Two factors probably contribute to this (Stribley *et al.*, 1975; Stribley and Read, 1976). First, the uptake of ammonium ions from the soil-solution in culture has been shown to be greater in mycorrhizal than in non-mycorrhizal plants growing in sand supplied with 2·7, 7·5 and 20·5 mg nitrogen per litre, applied as $(NH_4)_2SO_4$. Ammonium ions are relatively immobile in soil (though not as firmly bound as phosphate), so that even in sand-culture uptake may become diffusion-limited. Under these conditions extramatrical hyphae growing out from mycorrhizas may cross depletion zones round the roots, increase the volume of soil effectively colonized by the absorptive system, and so increase ammonium uptake in a manner comparable to phosphate uptake by vesicular-arbuscular mycorrhizas. The absorption systems of fungi have an extremely high affinity for ammonium (apparent $K_m$ of $NH_4^+$ uptake in *Neurospora* is 1–2 $\mu$M, i.e. lower by a factor of 10 than that of roots), so that rapid uptake of this ion from very dilute solutions by mycorrhizal hyphae could certainly take place. Uptake of nitrate is unimportant in the very acid soils where plants with ericoid mycorrhizas usually grow.

There was no increase of growth in weight nor increase of internal nitrogen content as a result of infection at either the very low (1 mg N l$^{-1}$) or very high (56 mg N l$^{-1}$) nitrogen concentration in the soil in the experiments of Stribley and Read. They suggest that at low concentrations, when the plants were very heavily infected, carbon drain to the fungi could have outweighed any advantages to the autotroph of improved nutrition. This may be compared with the similar reactions of vesicular-arbuscular and ectomycorrhizal plants. At the highest concentration all plants showed signs of ammonium toxicity and the degree of mycorrhizal infection was reduced. Increased concentrations undoubtedly would reduce the importance of diffusion across depletion zones in soil and also of the high affinity fungal uptake sites, so that mycorrhiza-dependent increases of ammonium uptake would not be expected, and other factors such as the availability of other soil-derived nutrients may have been limiting growth.

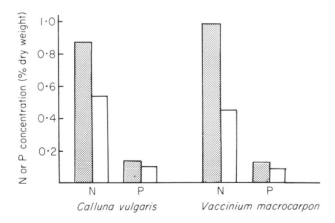

FIG. 22. The concentrations of nitrogen and phosphorus in the tissues of seedlings of *Calluna vulgaris* and *Vaccinium macrocarpon* after 6 months' growth in soil agar. Mycorrhizal plants, stippled; non-mycorrhizal plants, unshaded. (Redrawn from Read and Stribley, 1973.)

The second factor contributing to increased uptake of nitrogen by infected plants in natural conditions is the use of sources in the soil not available to non-mycorrhizal plants. Soil incubated with $(^{15}NH_4)_2SO_4$ and glucose was expected to become labelled in both the organic and inorganic nitrogen fractions, but the organic fraction showed a very much lower $^{15}N$ enrichment than the latter (see Lundeberg, 1970; Stribley and Read, 1974a). Acid washing reduced but did not eliminate the $^{15}NH_4^+$ content. Mycorrhizal plants grown on such labelled soil were found by Stribley and Read (1974a) to have lower $^{15}N$ enrichment than non-mycorrhizal plants, a result which indicates that mycorrhizas may be able to utilize nitrogen from the organic soil fractions, but which does not provide any information about the form of organic nitrogen absorbed.

In more recent experiments Stribley and Read (1980) have shown that mycorrhizal plants of *Vaccinium* may, under axenic conditions, utilize several amino acids as sole nitrogen sources; an ability not so well developed in non-mycorrhizal plants grown in the presence or absence of soil microflora (Table 64). The fungi probably do not utilize sources of the soil nitrogen more complex than amino acids. They showed no ability to degrade humic or fulvic acids in culture. In some highly organic soils active proteolysis can result in accumulation of amino acids, especially if the ammonification rate is low (e.g. at low temperatures). Amino acids are also available in freshly fallen litter in which successful competition by mycorrhizal fungi for these simple organic nitrogen compounds would result in efficient uptake by the mycorrhiza.

Translocation of $^{15}N$ (supplied as glutamine) by the endophyte occurs in axenic culture conditions and results in increased enrichment of associated mycorrhizal *Vaccinium* seedlings. Although it is unlikely that so readily

TABLE 64

Absorption of amino acids by mycorrhizal plants of *Vaccinium macrocarpum* as estimated by growth of the shoots (from Stribley and Read, 1980)

| Nitrogen source | Shoot weight (mg) | | |
| --- | --- | --- | --- |
| | Mycorrhizal | Sterile non-mycorrhizal | Non-sterile non-mycorrhizal |
| Ammonium | 43·4 | 41·4 | |
| Glycine | 43·9[a] | 11·6 | 11·4 |
| Alanine | 47·6[a] | 22·5 | 9·4 |
| Aspartic acid | 29·5[a] | 7·0 | 8·1 |
| Glutamic acid | 39·4[a] | 8·2 | 9·7 |
| Glutamine | 44·6[a] | 18·2 | 23·3 |
| No nitrogen | 6·0 | 8·3 | 8·7 |

[a] Significant difference between M and NM treatments.

metabolizable an organic nitrogen compound as glutamine would be available for absorption under natural conditions, this result does indicate the ability of the endophytes of Ericaceae to translocate $^{15}$N derived from organic nitrogen sources and also demonstrates transfer of nitrogen to the tissues of the autotroph.

## Phosphorus nutrition

Although the habitats of *Calluna* and *Vaccinium* on which most research has been done may be likely to be more nitrogen-deficient than phosphate-deficient, the experiments of Read and Stribley (1973) and Pearson and Read (1973b) clearly indicated the possible importance of mycorrhizal infection to phosphorus nutrition. The relatively immobile phosphate ion was shown to be absorbed into and translocated by the hyphae of an endophyte derived from *Calluna* (*Pezizella ericae*) from a source to seedlings of *Calluna vulgaris* and *Vaccinium oxycoccum* in aseptic culture. Strains of *P. ericae* from *Vaccinium macrocarpum* and *Rhododendron ponticum* were shown by Mitchell and Read (1981) to use sodium, aluminium, and iron salts of inositol phosphate in culture. Their growth rate with the sodium salt was similar to that with orthophosphate at 100 mg P l$^{-1}$. We can conclude therefore that if growth is limited by phosphate supply from the soil, it may be increased by mycorrhizal infection. Moreover, Read (1982, in press) reports that phosphate is stored as polyphosphate granules in the fungal hyphae.

## Carbon nutrition

The tested endophytes of ericaceous mycorrhizas, as has been mentioned, have some pectolytic but minimal cellulolytic ability. Their distribution in soil, wider than that of their autotrophic associates, also suggests that they may have some ability to survive as soil saprophytes or as resting structures in the root rgion of non-host plants, as ectomycorrhizal and even some vesicular-arbuscular fungi may possibly do. Studies with $^{14}$C-labelled compounds have revealed aspects of interdependence between the symbionts, and indeed show great similarities between ericoid and ectotrophic mycorrhizas.

When *Vaccinium* plants are supplied with $^{14}CO_2$, rapid photosynthesis is followed by incorporation of label into glucose, sucrose and fructose (Stribley and Read, 1974a). These sugars are the major soluble carbohydrates in shoots and uninfected roots. When mycorrhizal, $^{14}$C is incorporated in the root, into the fungal carbohydrates, mannitol and trehalose (see Table 65), and into glucose polymers (glycogen) and mannose polymers. All these indicate a carbohydrate transfer from autotroph to heterotroph. The labelling pattern observed in experiments is similar to that found by Lewis and Harley (1965b,c) in ectomycorrhizas and in many biotrophic symbioses between autotrophic and heterotrophic organisms (see D. C. Smith et al., 1969). It can be interpreted in the same way. If the carbohydrate derived from the autotroph is converted by the heterotroph to a compound (such as mannitol), which is not readily re-utilized by the autotroph, then a concentration gradient in favour of transport to the heterotroph will be maintained. Rapid incorporation of label derived from $^{14}CO_2$ into insoluble mannose polymers in the fungus also takes place in ericaceous mycorrhizas, and would further help to maintain net carbohydrate transfer. Higher concentrations of sucrose (four-fold) were observed in mycorrhizal roots of *Vaccinium* by Stribley and Read, although in the labelling experiments sucrose did not become more heavily labelled in them than in uninfected roots (Table 65). Greater sucrose concentrations have been also observed in mycorrhizal orchid roots (Purves and Hadley, 1975) and after pathogenic infection of leaves by rusts. It has been suggested that alterations in host enzyme levels or activities which maintain higher pools of soluble carbohydrate in infected organs would favour transport in the direction of the heterotroph. They might give evidence for fungal alteration of host metabolism on the lines of that suggested for ecto-mycorrhizas of beech by Wedding and Harley (1976), or by hormonal activity.

Pearson and Read (1973b) showed that $^{14}C$ glucose fed to the fungal hyphae did not readily appear in the tissues of the host plant, *Calluna*. Indeed it seems likely that net transfer of carbohydrate is in favour of the heterotroph. Nevertheless there is the expectation, and some evidence for it, that carbon (e.g. organic nitrogen compounds) must move from heterotroph to autotroph. Stribley and Read (1975) showed in addition that, in the absence of photosynthesis, $^{14}C$ from glucose and amino acids applied to the fungus appeared to pass to the host plant, *Vaccinium*, so that the radioactivity (cpm per mg dry weight) of the shoots increased. The increase was negligible in non-mycorrhizal controls, and when active photosynthesis was taking place. The result is of interest particularly because the movement of nitrogen from fungus to host is likely to be as an amino acid or amide (e.g. glutamic acid or glutamine) because absorption of ammonium into fungal hyphae and into ectomycorrhizas is followed by immediate synthesis of the amides, and ammonium does not

TABLE 65

Movement of $^{14}C$-photosynthate to the roots of mycorrhizal and non-mycorrhizal plants of *Vaccinium macrocarpum*.

*Plants fed 72 h with $^{14}CO_2$ in the light. Cpm mg $^{-1}$ dry weight of tissue (results of Stribley and Read, 1974a)*

| Type of root | Glucose | Fructose | Sucrose | Mannitol | Trehalose | Myoinositol |
|---|---|---|---|---|---|---|
| Non-mycorrhizal | $41 \pm 12$ | $9 \pm 3$ | $747 \pm 141$ | — | — | 5 |
| Mycorrhizal | $80 \pm 15$ | $26 \pm 9$ | $370 \pm 99$ | $108 \pm 42$ | $37 \pm 13$ | 3 |

Means of 7 estimates $\pm 95\%$ confidence levels.

accumulate in them (Budd and Harley, 1962a,b; and see Harley and Wilson, 1963). Since carbohydrates are required for the formation of organic acids, which provide the carbon radicals of amino acids and amides, a cycling of carbon from the autotrophic host as sugar, returning as nitrogen compounds must be envisaged in ericoid mycorrhizas, as in leguminous nodules and ectomycorrhizas (see S. E. Smith, 1980).

The reason why assimilation of $CO_2$ by the autotroph should reduce $^{14}C$ movement from the fungus as amino compounds is not easy to understand unless it results from the dilution of the labelled carbon compounds with unlabelled ones.

## Heavy metal tolerance

Read has pointed out that the heathland soils and mine spoil heaps on which Ericaceae such as *Calluna* in the northern hemisphere and species of Epacridaceae in the southern hemisphere dominate the vegetation have a high content of available heavy metals. In these copper, aluminium and zinc, for instance, may be in solution at toxic levels owing to high acidity. Bradley *et al.* (1981, 1982) grew *Calluna*, *Vaccinium* and *Rhododendron* in sand with a range of concentrations of copper and zinc added as sulphates to the culture solutions. Although non-mycorrhizal plants grew little and suffered high mortality, especially in the higher concentrations of both metals, mycorrhizal plants were able to grow even up to 50 mg $l^{-1}$ of copper and 100 mg $l^{-1}$ of zinc. The shoots of the non-mycorrhizal plants contained higher metal levels than the mycorrhizal plants, for the metals were accumulated in the roots of the latter. Read (1982) put forward the suggestion that the heavy metals became bound to the carboxyl groups in the pectic interfacial matrix between fungus and host of the mycorrhizal roots and hence did not affect the activity of the cells. This behaviour is extremely interesting especially in view of the increasing amount of information on the colonization of mine spoil heaps and derelict areas by mycorrhizal plants of all kinds, and of their ability to grow better than non-mycorrhizal plants in a variety of habitats with high contents of potentially toxic ions.

## Conclusion

The work with ericoid mycorrhizas has shown that they have much in common with other kinds of mycorrhiza in function and ecological importance. They differ from vesicular-arbuscular mycorrhiza most particularly in the behaviour of the cells of the host during the degeneration of the fungus, because there would seem to be no question of a lytic action by a living host cell. However, in the action of the extramatrical hyphae and the movement of nutrients through them into the host, as well as the movement of carbon compounds from the host into the fungus, there are close parallels with them and with ectomycorrhizas, and probably with ectendomycorrhiza of *Pinus*. The latter also have the common feature of penetration into the cells of the hosts as coiled hyphae, but differ in the possession of a Hartig net.

The emphasis on the importance of the mycorrhizal infection in the processes of nitrogen absorption is especially relevant to the interpretation of their ecological dominance on acid and peat soils. The fungus, however, also intervenes in the uptake of phosphate and presumably of other essential nutrients. A special interest also attaches to the experiment of Read on *Erica bauera*, one of the species from South Africa which flower in the dry season. He was able to show that mycorrhizal plants store nitrogenous substances absorbed in the moist period, and use them in the reproductive period. The storage in non-mycorrhizal plants was much less. This, he points out, is analogous to the storage in the fungal sheath of ectomycorrhizas (Read, 1978).

Read's conclusions about the importance of ericoid mycorrhizal infection in ecology are that it is particularly adapted to acid soils of high organic and polyphenol content and low in available nitrogen and sometimes in phosphorus. Here the ability of mycorrhizal plants to absorb organic nitrogen and phosphorus compounds as well as inorganic ones and their resistance to toxicity particularly fits them to achieve dominance.

## Chapter 13

# Arbutoid and Monotropoid Mycorrhizas

## Introduction

The structure of the roots of Arbutoideae, which are more sturdy than those of the Ericaceae and Epacridaceae, was described by Rivett (1924). According to him and to later authors such as Zak (1974, 1976a,b) and Mejstrík and Hadač (1975) the root system of Arbutoideae like that of many mycorrhizal trees is heterorhizic, and is clearly differentiated into long and short roots. The long roots are described as being sparsely infected in an intercellular fashion and to contain a Hartig net as in *Pinus*, and they may also have a sparse weft of hyphae on their surfaces. The short roots are clothed in a sheath which on different kinds and on different species of root varies in thickness from a few μm to 85 μm or even more; in *Arbutus unedo* for instance it was figured by Rivett as about 20 μm thick. The Hartig net is well developed between the outer cells of the cortex of which only one tier is usually affected, and into that tier there is extensive fungal penetration to form coils of hyphae. Mejstrík and Hadač, however, describe occasional penetration of roots of *Arctostaphylos uva-ursi* by short haustorial hyphae more in the manner of *Monotropa*.

Scannerini and Bonfante-Fasolo (1982) describe the fine structure of the mycorrhizal roots of *A. uva-ursi*. The fungal sheath is cemented together by fibrous electron-lucent material and it gives rise to a Hartig net from which hyphae penetrate the cells to form coils. These are enclosed by an electron-dense interfacial matrix and by the host plasmalemma. Both the Hartig net and the sheath, as well as the infected cells of the host, appear active in autumn according to them, and contain cytoplasm and organelles. The fungus contains glycogen rosettes and the host contains amyloplasts. Ling Lee *et al.* (1975) have also observed polyphosphate granules in the fungal tissues of *Arbutus*. Scannerini records that in spring the fungal and host cells may degenerate. However, no extensive examination of the ultrastructure throughout a year has yet been made, but, according to Read (1982), Duddridge has also described the mycorrhiza of *A. uva-ursi* in similar terms to Scannerini and Bonfante-Fasolo.

As shown in Table 61 Largent and his colleagues (1980) described considerable variability in the mycorrhizas of various Arbutoideae. Besides arbutoid mycorrhizas, ectomycorrhizas and intermediate forms reminiscent in many cases of the ectendomycorrhizas described by Wilcox and by Mikola and Laiho on conifers were observed. Mejstrík and Hadač agree also that *Arctostaphylos uva-ursi* is usually ectomycorrhizal. In their experience, ectendomycorrhizas with intracellular penetration comprised only 1–2% of the mycorrhizas which they observed in that species. Most were ectomycorrhizas of the types A, F, and G of Dominik (1955). On the other hand, Zak (1974, 1976a,b), who observed mycorrhizas in the field on Arbutoideae and synthesized them in pure culture, did not report ectomycorrhizas to occur on any of the species that he studied. Indeed he stressed the superficial morphological similarity of the ectomycorrhizas formed on *Pseudotsuga* and the arbutoid mycorrhizas formed on *Arbutus* and *Arctostaphylos* by the same species of fungi. These fungi exploited the cortical cells of different species in strikingly different ways. This behaviour is reminiscent of that of the E-type ectendomycorrhizal fungi of *Pinus* described by Mikola, which formed ectomycorrhizas with *Pseudotsuga*.

## Arbutoid mycorrhiza

The fungi of *Arbutus* and other arbutoid mycorrhizas were long believed to be Basidiomycetes by analogy with ectomycorrhizas. Recently this has been confirmed both by synthesis experiments and by the description of dolipore septa in *Arctostaphylos* (Duddridge, 1980; Scannerini and Bonfante-Fasolo, 1982; Read, 1982). Zak (1974) showed, by tracing the mycelium of *Cortinarius zakii* that it forms mycorrhiza of the arbutoid sort with *Arbutus menziesii*. At the same time he recognized at least 6 other mycorrhizas, one formed by *Cenococcum graniforme* (*geophilum*) and three of the others by unknown Basidiomycetes. *Cortinarius zakii* also forms ectomycorrhiza with *Pseudotsuga douglasii* and *Abies grandis* and of course *Cenococcum* forms ectomycorrhizas with many species. *Arctostaphylos uva-ursi* was observed (Zak, 1973) to form arbutoid mycorrhizas with *Cortinarius aurei-folius* var. *hesperus* which also formed ectomycorrhiza with *Pinus contorta* and *Pseudotsuga menziesii* and ectendomycorrhiza with *Polygonum paronychia*. Zak (1976a) also produced mycorrhizas in culture between *A. uva-ursi* and 8 fungi all known to form ectomycorrhizas with other hosts (*Hebeloma crustuliniforme, Laccaria laccata, Lactarius sanguifluus, Poria terrestris* var. *subluteus, Thelephora terrestris, Rhizopogon vinicolor, Pisolithus tinctorius* and *Poria terrestris* var. *cyaneus*). The first 5 associated excellently with *A. uva-ursi* as the host and the last one only fairly well. Zak failed to form mycorrhiza on *A. uva-ursi* with *Corticium* (*Piloderma*) *bicolor* and surprisingly with *Cenococcum graniforme* (*geophilum*). He remarks that the ectendomycorrhizas formed on *A. uva-ursi* by these 8 fungi had not been observed by him in nature. Zak obtained somewhat similar results (1976b) with *Arbutus menziesii* which was shown to form arbutoid mycorrhiza with *Thelephora terrestris*, *Corticium* (*Piloderma*) *bicolor*, *Cenococcum graniforme* (*geophilum*) and *Pisolithus tinctorius*, all of strains known to form ectomycorrhiza with forest trees. Molina and Trappe (1982a) tested the ability of 28 ectomycorrhizal fungi to form mycorrhizas with *Arbutus menziesii* and *Arctostaphylos uva-ursi* in pure culture. All but three produced arbutoid mycorrhizas with both the hosts. This emphasizes the similarity between these arbutoid mycorrhizas and ectomycorrhizas and raises two additional points of great importance. First the description given in earlier chapters of the properties of ectomycorrhizal fungi apply equally to those of arbutoid mycorrhizas in every respect. Secondly, since Molina and Trappe did not observe ectomycorrhizas in their synthesis experiments, the records of Largent and of Mejstrík and Hadač of their formation in the wild need further examination to determine in what conditions they may be formed. Of the fungi that failed to form arbutoid mycorrhizas with the ericaceous hosts, *Alpova diplophloeus*, which has been described as specific to the genus *Alnus*, spread along the surfaces of the roots but penetrated them only rarely and sparsely. *Cortinarius pistorius* from *Pinus* and *Zelleromyces gilkyae* from *Pseudotsuga* failed to form any kind of association with the *Arbutus* and *Arctostaphylos*.

The formation of ectomycorrhizas with some hosts and arbutoid mycorrhizas with others poses interesting questions reminiscent of those raised about E-strain ectendomycorrhizal fungi. It may be presumed from the papers quoted that the same strain was capable of penetrating through the cell-walls and colonizing the cells of the ericaceous plants and of forming ectomycorrhizas with minimum cell-wall penetration in the coniferous hosts. We must assume that the difference might lie in the fact that one host either must be resistant to or must inhibit cellulolytic enzymes or the wall-modifying system of the fungus. Alternatively, perhaps the ericaceous host might be caused to digest or not to form its own walls by the fungus which then enters the cells. This contrasts with the report of Largent *et al.* (1980b) that a single fungus, *Amanita geminata*, formed ectomycorrhizas with both *Arctostaphyllos manzanita* and *Pinus contorta*.

*Physiology of arbutoid mycorrhiza*

No experimental results are available on the physiological properties of arbutoid mycorrhiza. Since it is the common form of absorbing organ of an important and ecologically significant or dominant group of species, it must be assumed to be of selective advantage. This must be even more strongly accepted because the sheath on the roots, as in ectomycorrhiza, may not only have a storage function (polyphosphate is stored), but also insulates the host from the soil. Hence the fungus calls the tune in absorption by the short roots, and everything absorbed by them must pass through it.

It would be exceedingly interesting to determine the physiological properties of the strains of fungi used and their potential for digesting pectin, cellulose and lignin, for some appear to belong to species in which cellulose digestion has not been suggested. That knowledge would provide a basis for thought and experiment on the different effects of different hosts on these fungal properties.

## Mycorrhiza in Pyrolaceae

The mycorrhizas of the green Pyrolaceae were described by Christoph (1921) and by Lück (1940, 1941). They may well be classified as arbutoid mycorrhizas. A fungal sheath forms on the root surface and from it fungal hyphae connect with the soil and also penetrate between and into the cells of the cortex of the root. A recent investigation by Duddridge (1980) of *Pyrola rotundifolia*, a plant of bogs, fens and damp places, has shown that the roots are divisible into long mother roots and short roots. She observed two kinds of mycorrhizal short root, an abundant black mycorrhiza and a non-pigmented one. The fungal mycelium of the black mycorrhiza possesses clamp connexions and dolipore septa. It produces sparse wefts on the root surface on which the hyphae spread at first along the lines of the cell walls before they penetrate to form the richly lobed Hartig net. Plentiful intracellular hyphae also develop in the cortical cells and are surrounded by the host plasmalemma. The young infected cell has a large vacuole and the penetrating hyphae at first pass round it. Later there is a great increase of hyphal colonization coupled with a comparable increase of host cytoplasm, and small vacuoles replace the large one. The hyphal coils within the cells are surrounded by host plasmalemma, and their walls are surrounded by a lightly stained material, the interfacial matrix, which is distinct from the hyphal wall. The nucleus becomes enlarged and lobed, as in other intracellular infections, and the cisternae and dictyosomes may become hypertrophied. Glycogen is laid down in both the Hartig net and intracellular hyphae, especially during the resting period October to March. Although there is no starch in the infected cells it is present in the stelar tissues and inner cortex. The short roots are of restricted life, and their death and breakdown involve changes in both host and fungus. The Hartig net develops strongly in thickness; and the sheath becomes a dark rind.

In the functional mycorrhiza, as in the case of the ericoid type, not all the cortical cells are exploited by the fungus simultaneously. As the infection of any cell ages, the hyphal walls thicken and become osmiophilic and degenerate, but alongside may be cells newly infected with the hyphae apparently more active. Eventually the short rootlet becomes totally senescent and fungal hyphae, not necessarily of the mycorrhizal fungus, are present in the stele and penetrate the sheath.

The seed germination and early growth of *Pyrola* have been studied because some of them have been described as difficult to germinate. The work of Lihnell (1942) is of great interest because on germination aseptic culture of *P. rotundifolia* and *P. secunda* formed colourless branched axes with root-like structures. These plant bodies continued to develop for $3\frac{1}{2}$ years without further differentiation. Lihnell isolated four fungi from the mycorrhizas of *P. rotundifolia* and one from *P. secunda* which were sterile septate fungi perhaps similar to the sterile forms isolated from other Ericales, but although they had some effect on seed germination if present in the medium, they did not form, in the conditions used, any association with the "procaulomes" produced by germination. It is interesting that Lück (1940) had isolated a clamp-bearing basidiomycete from *Pyrola*, so here as with ericoid mycorrhiza there may be several potentially symbiotic fungi.

## Monotropoid mycorrhiza

The Monotropaceae including only holosaprophytes are an extreme culmination of a series in the Ericales. They form a kind of mycorrhiza peculiar to themselves. The name "monotropoid mycorrhiza" was used by Duddridge because they are different from those of other Ericales in many particulars, especially from arbutoid mycorrhizas with which they were loosely included by Harley (1969a). *Monotropa* itself was recognized by Kamienski (1881) as having its roots completely invested by a fungal layer. He noted the similarity between them and those of *Fagus*, and commented that whatever passed into the tissue of the higher plant from the soil must pass through the sheath of fungus. He was therefore simultaneously writing and thinking in similar ways about mycorrhiza to Frank who published the name "Mykorrhizen" in 1885. Kamienski thought that *Monotropa* might be dependent for nourishment on the neighbouring trees because he believed that they were associated with a common mycelium. Indeed after nearly 100 years this view has been experimentally confirmed by Björkman (1960) and by others later.

The root system of *Monotropa* consists of main roots bearing secondary and tertiary branches. Hirce and Finocchio (1973) described them as being hexarch, tetrarch or triarch and protostelic respectively and all being infected mycorrhizally. The roots are enclosed in a fungal sheath several layers of hyphae thick, which is connected with Hartig net involving the outermost layer of host cells (Plate 13). In addition, hyphal pegs called haustoria penetrate into the epidermal cells. The root system of *M. hypopitys*, which has been the most studied of the Monotropaceae, forms a "root-ball" in which extensive fungal mycelium ramifies and encloses the roots and mycorrhizas of neighbouring trees in a complex weft (Plate 13). *M. hypopitys* frequently grows in woods of beech or pine and other conifers whose roots and mycorrhizas as well as *Monotropa* roots are involved in the "root-ball". The plants overwinter underground and from the

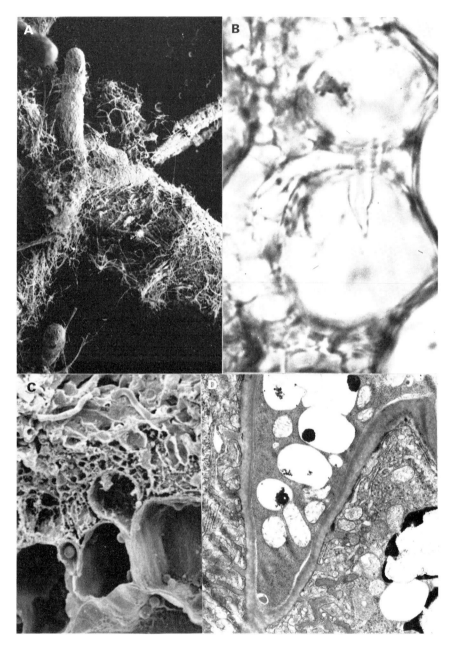

PLATE 13. A. Scanning electron microscope view of the association of the roots of *Pinus* and *Monotropa hypopitys*. The fungal hyphae and the sheaths of their roots are closely associated. ×75.

"root" system flowering scapes develop. Adventitious buds develop on the apices of some of the roots of first and second order, and the flowering shoots formed from them grow above the ground, mature and senesce over a period of months. As development takes place, the fungal peg-like haustoria in the epidermal or outer cortical cells of the host go through a sequence of changes, described by Lutz and Sjolund (1973) and Duddridge (1980). According to them, the cell-walls of the host are not at first actually penetrated by the fungal peg but the wall invaginates and encloses the peg. In the older literature (see Harley, 1969), it was suggested that a new wall secreted by the host enclosed the fungal peg, leaving the apex free as it grew into the host cell. However, the present interpretation, although similar in many respects, emphasizes the integrity of the host cell-wall and its complete enclosure of the peg at first. As the growth of the peg continues, glycogen tends to disappear from the fungal sheath and perhaps this carbohydrate may be used in the formation of extensive branching "ingrowths" of the wall of the peg which now emerge out into the cell (Plate 13). These make the surface of the peg reminiscent of the transfer cells described by Gunning *et al.* (see 1974) to occur in many kinds of cell systems in the plant kingdom through which active movement of nutrients is believed to take place. The ingrowths are of polysaccharide material deposited on the secondary wall. They are not of impermeable callose, as indicated by their staining reactions. During this period rough endoplasmic reticulum and mitochondria are plentiful in the cells, especially in the "transfer" regions. Duddridge (1980) observed an osmiophilic ring in the wall at the base of the peg which she thought might be related to function. She suggested that it prevented backward flow of nutrients through the walls of the cell at the point of the fungal entry, in a similar way to that believed to apply to *Erysiphe* by Gil and Gay (1977). The osmiophilic ring surrounding the haustorial neck in *Erysiphe* can be seen in Plate 15.

When mature, the tip of the peg as described by Francke (1934), undergoes modification, which he described as "digestion", but which is called "bursting" by Duddridge. She carefully distinguishes it from the process of "ptyophagy" described by Burgeff (1932, 1936; see Harley, 1969a) in orchids where the tips of the hyphae, penetrating host cells, are sealed off and "digested". The contents of the peg are released into a sac enclosed by host plasmalemma. The sac contains

B. Section of *Monotropa* root showing the fungal sheath, penetration of fungal tissue between the cells and penetration of the hyphal peg into a cell of *Monotropa*.   × 3750 approx.

C.  Scanning electron microscope view of a transversely fractured mycorrhizal lateral of *M. hypopitys* showing sheath and a hyphal peg.  × 1250.

D.  LS of mature fungal peg and extensive ingrowths from the invaginated cell wall of the host. Note the vacuoles containing polyphosphate granules and mitochondria within the fungal peg.  × 20 250.

Photos: J. Duddridge.

groups of organelles and later fungal protoplast as the peg collapses. This breakdown occurs when the flowering scape is fully developed and mature. At about the same time hyphae from the fungal sheath invade and colonize the cells of the cortex which become senescent.

This description of Duddridge emphasizes the changes in the structure of the mycorrhiza during the development of the plant. As vegetative activity and flower production take place the sheath and Hartig net are formed, followed by loss of glycogen from the sheath as the fungal pegs develop. These then burst and lose their contents to the cell. This is followed by the final senescence of the scape and branch roots.

In a paper yet to be published (*New Phytol* 1983) Robertson and Robertson describe an investigation of two other species of Monotropaceae, *Pterospora andromeda* and *Sarcodes sanguinea*. It is of great interest that in all important respects their findings agree with those of Duddridge (1980) and Duddridge and Read (1982) in the structure of the symbiotic organ and in the penetration and behaviour of the haustorial apparatus.

## *Physiology of the mycorrhiza of* Monotropa

The seed of *Monotropa* is small and has proved difficult to germinate, like the seeds of so many orchids which also have a period when they are dependent on an external supply of organic compounds. As in some of them, asymbiotic germination can be brought about by prolonged washing in water before the seed is set upon the germination medium (Francke, 1934). The embryo consists of a minute axis of very few cells set in an endosperm of about a dozen large cells. On germination it forms a small plant body, similar to the primary protocorm of orchids, which possesses an apical growing point and a broad base attached to the remainder of the endosperm. No further development occurred in Francke's cultures unless they were inoculated with a fungus which he had isolated from the roots of adult plants. This fungus, which he believed to be a species of *Boletus*, formed a hyphal sheath around the protocorm and penetrated into a few cells of the outer layer in a manner similar to the infection of the roots. Some further growth and cell division took place after infection but soon stagnation occurred. Francke was unable to obtain further development to produce plants that were differentiated into root and shoot. Clearly the embryo possesses some ability to absorb nutrients in the absence and in the presence of the fungus, but some factor or nutrient must be lacking because development comes to a halt. It is possibly derived under natural conditions via the fungus from a secondary host (see below).

As mentioned above, Kamienski (1881) thought that *Monotropa* possessed a common symbiotic fungus with the forest trees near which it grew. He also believed that it might be nourished, not saprophytically, but through the infecting mycelium from the neighbouring trees. Björkman (1960) tested this, pointing out that Francke had believed that its fungus was a *Boletus*, a genus well known as an ectomycorrhizal fungus of forest trees, and therefore Kamienski's hypothesis was tenable. He first separated *Monotropa* plants from the tree roots by metal sheets, and observed them to grow poorly compared with attached plants. Later, $^{14}$C-labelled glucose and $^{32}$P-labelled orthophosphate were found to be translocated in five days from the spruce and pine trees into which they had been injected to the tissues of *Monotropa* growing close by. The distance between the trees and *Monotropa* plants was 1–2 m and young developing plants became more radioactive than old mature plants. Björkman confirmed the view of Kamienski that the fungus infecting the tree roots and *Monotropa* was probably of the same mycelium. The isolates made from *Monotropa* roots produced mycorrhizas with pine, but with spruce the association produced a fungal sheath but no Hartig net. These results certainly suggest a hyphal connexion between the mycorrhizas of *Monotropa* and those of pine and spruce, although those with the latter were somewhat abnormal. Other plants in the neighbourhood, such as *Calluna vulgaris*, *Vaccinium vitis-idaea* and *V. myrtillus*, did not receive radioactivity from either glucose or phosphate injected into spruce and pine, and that was regarded as confirmation of the need for hyphal connexions. In a similar experiment briefly reported by Furman (1966), *Monotropa* plants were injected with $^{32}$P-labelled phosphate which was transported to neighbouring *Quercus* and other trees which were ectomycorrhizal and to an *Acer* which was endomycorrhizal. It seems probable from Björkman's work that an endophyte of *Monotropa* is mycorrhizal with pine but can also invade spruce. Whether or not the same or another species of endophyte was observed by Furman is unknown. The connexion of *Monotropa* with *Acer* complicates the picture.

Björkman called the behaviour which he observed "epiparasitism" and the habit has since been found to be quite common. The so-called "saprophytism" of some angiosperms has received comment from Furman and Trappe (1971) and Harley (1973). Both noted the similarity of the behaviour of *Monotropa* with that of orchids, like species of *Gastrodia*, in which the fungus *Armillaria* (*Armillariella*) *mellea* is parasitic on the neighbouring trees and mycorrhizal with the orchid. Campbell (1971) noted that *Monotropa uniflora* in the USA was also associated with *Armillaria mellea* in a similar fashion to the achlorophyllous orchid *Gastrodia elata*, although *Monotropa hypopitys* seemed to be associated, as Björkman observed, with the mycorrhizal fungi of neighbouring trees. The relationship of the *Armillaria* with the tree roots was described as less "harmonious" in the case of *M. uniflora*, as might indeed be expected from its parasitic behaviour.

This subject will be discussed in connection with "saprophytism" in orchids and elsewhere. Here let it be recorded that the supply of readily available carbon compounds even in the litter and humus layers of the soil of temperate forests does not seem to be quantitatively sufficient to support such large saprophytes even as *Monotropa* spp. and an indirect source of carbon from photosynthesis or via a large pool in a parasitized host seems necessary for achlorophyllous plants of any size.

## Summary of mycorrhizal symbioses in Ericales

The range of structure of the mycorrhizas of Ericales is wide, but it has similarity with ectotrophic and ectendotrophic mycorrhizas of trees and links with them all in a complex group of forms. Almost all the plants concerned in all these kinds of mycorrhizas are woody perennials, subshrubs, shrubs and trees. Almost all the fungal symbionts are Ascomycetes or Basidiomycetes with permanent rather than ephemeral mycelium.

Within the order Ericales all the fungal phases of mycorrhiza (extramatrical hyphae, fungal sheath, Hartig net, intracellular coils, and specialized haustoria) are found but they are differently combined and developed in different groups. It is an interesting exercise to speculate upon the selective value of the various combinations of host and fungus.

Those species with hair roots, the Ericoideae, Vaccinioideae, Epacridaceae and Empetraceae in particular, grow upon very acid peaty soils for the most part, but in any event upon very nutrient-poor soils. Their root systems consist of a mass of hair-like roots of small diameter which are heavily enmeshed in a weft of extramatrical hyphae and copiously infected in the cortical cells where fungal coils are formed. These mycorrhizas appear to be an outstanding exception to the generalization of Baylis (1972, 1975) that plants with fine roots are frequently not dependent upon mycorrhizal infection. Clearly plants with ericoid mycorrhizas are adapted to the poorest soils by the formation of short-lived hair-like roots, and exploitation of the soil is further improved by the extramatrical fungal mycelium. The Rhododendroideae resemble the taxa with hair roots although their root systems are more robust and have permanent axes, but they do not as a general rule inhabit such extreme soils. The Arbutoideae form mycorrhizas with a great many features in common with ecto- and ectendomyc-orrhizas. The well-developed sheath and Hartig net are formed in common with the former and the Hartig net and intracellular coils in active host cells with the latter. The reports of Largent *et al.* and of Mejstrík and Hadač that ectomycorrhizas also occur in Arbutoideae are of interest, especially so because they have not yet been synthesized in culture although arbutoid mycorrhizas have been, using fungi which form ectomycorrhizas with other hosts. A wider survey is required. The host plants of arbutoid mycorrhiza grow in seasonal climates in which the storage function of the sheath would be of selective advantage. This may explain the differences from ectendomycorrhizas of conifers which are apparently often infections of seedling stages where thin sheaths only can be supported by the developing seedling.

Until there is more knowledge of the relationship of the hyphal coils with the host cells in arbutoid mycorrhiza, little can be concluded of their selective advantage. To hazard a guess, however, these coils might provide an extensive interface for nutrient transfer in mycorrhizal roots which have a shorter life than most ectotrophic ones. It would therefore be interesting to determine whether the life period of the mycorrhizal rootlets of Arbutoideae is shorter than that of ectomycorrhizal plants.

The sheath and Hartig net of *Monotropa* require separate consideration. The fungal sheath in this case is supported by carbon from another host, not by *Monotropa*. It may therefore act as other sheaths do as a storage organ which releases its stores as the rapid growth of the scape and flower production takes place, an idea apparently supported by the work on ultrastructure. This kind of mycorrhiza might have evolved from the arbutoid type which might itself be considered to be "preadapted", in having a sheath and a Hartig net. There is clearly a wide field here for research to determine the factors which attract the fungus to invade *Monotropa* and the materials, if any, which it derives from the host.

*Chapter 14*

# Orchid Mycorrhizas

## Introduction

The Orchidaceae contains more species than almost any other flowering plant family, something between 12 000 and 30 000 species, being rivalled only by the Compositae. The family is cosmopolitan but with many more species in the tropics than in the temperate regions. It has a varied range of life forms, including both green terrestrial species such as are familiar in temperate regions, and in addition a large number of epiphytes of a great range of size in the tropics. Some orchids, like *Vanilla*, are lianes and many species of all life forms, including the large lianes, are achlorophyllous and are often regarded as saprophytic, although the carbon nutrition of some has been shown to be very complex. The smallest orchids, *Cryptoanthemis slateri* and *Rhizanthella gardneri*, are completely subterranean including their flower-buds and have a fresh weight of a few grams at most; the largest orchid is said to weigh over a ton. Apart from the beautiful, popular and sometimes bizarre flowers with complex pollination mechanisms, the orchids are remarkable for two characters: firstly, their seeds are extremely small, the largest is about $14\,\mu g$, and within them the embryo is little differentiated; secondly, they are all mycorrhizal, living throughout life in association with fungi. It is in research on seed germination, early growth and the part played in these processes by the fungi, that great efforts have been made, whereas by contrast the functions of the mycorrhizal association in adult green orchids has received much less attention. The importance of this work is partly because of the commercial value and beauty of orchid blooms and the need, on the part of growers, to raise new varieties including hybrids from seed. As a result much more is undoubtedly known in the trade about some aspects of orchid growth and physiology than has been published, and much information gained by amateurs is of extreme practical value but difficult of exact interpretation.

All orchids, whether they are chlorophyllous or achlorophyllous as adults, pass

through a prolonged seedling stage during which they are unable to photo-
synthesize. In most seed plants the homologous period is passed through at the
expense of reserves stored in the seed but orchid seeds contain very limited
reserves indeed in the form of starch or lipid (Arditti, 1979). Since glyoxysomes,
normally involved in lipid utilization through the glyoxylate cycle, are said to be
absent in *Cattleya* seeds, the possibility exists that lipids may not be used in
gluconeogenesis or for synthetic reactions, and that the lipid stores are of little
immediate significance in germination. If the seeds are spread on a moist
substratum the embryos, undifferentiated except for their apical meristematic
regions and perhaps the rudiment of a cotyledon, absorb water, swell slightly and
may burst the testa and produce epidermal hairs. The embryo does not develop
further unless it receives at least an exogenous supply of carbohydrate or is
infected by a compatible mycorrhizal fungus. In some species, besides carbohydrates, vitamins or growth factors are also required for the embryos to develop
asymbiotically. Discussions of this topic and its history are given by Harley
(1969a) and comprehensive reviews have been provided by Arditti (1967 and
1979), so selected information only will be presented here.

The soluble carbohydrates which have been found to be generally suitable for
the growth of orchid seedlings are those which are commonly metabolized by all
angiosperms. D-glucose, D-fructose and sucrose together with maltose give good
growth but L-sugars are unsuitable, as are most organic acids. The most
interesting carbohydrates from the point of view of the physiology of symbiosis
which have been tested are the common fungal metabolites trehalose and
mannitol.

The disaccharide trehalose, found in fungi, insects and some pteridophytes
such as *Selaginella*, is probably suitable for the growth of most orchid seedlings in
culture. Mannitol (linear hexapolyhydric alcohol) which, like other linear
polyols, is found in a number of higher plants as well as fungi (e.g. Oleaceae,
especially in the phloem transport system) is important in the carbon metabolism
of many symbiotic systems (D. C. Smith *et al.*, 1969), but is unsuitable for the
growth of many orchid seedlings. It does not, for instance, support the growth of
*Dactylorchis purpurella*, *Vanilla* sp., *Bletilla hyacintha* or *Goodyera repens* (Ernst, 1967;
S. E. Smith, 1973; Purves and Hadley, 1975, 1976) but does support the growth
of others such as *Phalaenopsis* sp. and *Dendrobium nobilis*, as shown by Quendow
(1930) and Ernst (1967). Tables 66 and 67 give examples of experiments with
*Dactylorchis purpurella*, *Bletilla hyacintha* and *Goodyera repens* which show the kinds of
differences between species that may be expected. Despite considerable
argument and discussion in the past, it must be now quite clear to all that
mycorrhizal infection is not an absolute requirement for the growth of orchid
seedlings in culture. Indeed the successful growth to flowering stage of *Laelia-
Cattleya* by Knudson (1930) and of *Miltonia* sp. by Bultel (1926) amply
demonstrated this many years ago. Further, the experience of tissue culture and

growth of "obligate" fungal parasites as well as orchid seeds in culture would lead one to believe that all species of plant can be grown asymbiotically in suitable conditions as soon as these have been discovered. Only recently Nakamura (1982) has successfully germinated seeds of *Galeola septentrionalis* and grown young plants in culture, although it has been sometimes held that seeds of achlorophyllous orchids usually resist such efforts. Nevertheless, under natural ecological conditions the necessary soluble sugars, vitamins, amino acids and growth factors are unlikely to be available in the substrate for more than very short periods of time in transitory microhabitats. Orchid seedlings can perhaps be likened to rust fungi, in that both can be grown in culture but in nature both are obligate symbionts, although their partners are not.

### TABLE 66

Sizes of protocorms of *Dactytorchis purpurella* and *Bletilla hyacintha* after growth in the dark for 17 and 9 weeks respectively in media containing various carbohydrates at 22 C (results of S. E. Smith, 1973)

| Orchid | | Carbohydrate supplied[a] | | | | |
|---|---|---|---|---|---|---|
| | | Minerals alone | Glucose | Sucrose | Trehalose | Mannitol |
| *D. purpurella* | Length ($\mu$m) | $362 \pm 8$ | $783 \pm 27$ | $703 \pm 33$ | $746 \pm 30$ | $321 \pm 5$ |
| | Width ($\mu$m) | $307 \pm 5$ | $516 \pm 16$ | $599 \pm 20$ | $497 \pm 15$ | $294 \pm 5$ |
| *B. hyacintha* | Length ($\mu$m) | $587 \pm 22$ | $2575 \pm 192$ | —[b] | $1750 \pm 123$ | $0^c$ |
| | Width ($\mu$m) | $387 \pm 20$ | $688 \pm 20$ | —[b] | $748 \pm 16$ | $0^c$ |

[a] 0·5% for *D. purpurella*, 1·5% for *B. hyacintha*.
[b] Not done.
[c] No germination.

### TABLE 67

Percentage germination and sizes of protocorms of *Goodyera repens* after growing in the dark for 8 weeks in media containing 1% of various carbohydrates at 20°C (results of Purves and Hadley, 1975)

| Dimension | Carbohydrate supplied | | | | | |
|---|---|---|---|---|---|---|
| | Minerals only | Glucose | Fructose | Sucrose | Trehalose | Mannitol |
| Length ($\mu$m) | $233 \pm 25^a$ | $423 \pm 66$ | $437 \pm 79$ | $417 \pm 64$ | $425 \pm 49$ | $0^b$ |
| Width ($\mu$m) | $121 \pm 12^a$ | $229 \pm 30$ | $242 \pm 36$ | $230 \pm 44$ | $231 \pm 24$ | $0^b$ |
| Percentage germination | 0 | 84 | 71 | 77 | 75 | 0 |

[a] Control size ungerminated seed.
[b] No germination.

The development of infection and the role of the fungi will be discussed later. Here let it be stated that one of their primary functions is the provision of carbon compounds to the seedling during its development which may be very prolonged, even in years, before green leaves are formed. However, the time period during which all carbon supplies must come from external sources is very variable and it is not yet clear how far most orchids, after they have produced green leaves, may or must derive carbohydrate from their fungal partners. It must therefore be stressed that the mycorrhizal infection of orchids differs from most of the others so far discussed, achlorophyllous plants like *Monotropa* excepted, in one particular fundamental respect. The net movement of carbon compounds is from the substrate through the fungus into the host. In other forms of mycorrhiza carbon compounds originate in photosynthesis by the host and their net movement is into the fungus. Adult orchids, whether or not they contain chlorophyll, usually have mycorrhizal roots and there is indirect evidence that some terrestrial orchids must continue to gain carbon through them. For instance, the interesting and carefully confirmed observations of Wells (1967) that *Spiranthes spiralis* may spend one or more years in subterranean existence and then produce a flowering scape in the following year argues for absorption of carbon via its mycorrhizal equipment. Summerhayes (1951) gives the examples of *Cephalanthera rubra* and *Goodyera repens* which, if heavily shaded in natural habitats, continue to exist underground for as long as twenty years. Similarly the experiments with epiphytes by Ruinen (1953) lead to the same conclusion. She found that many were connected by their mycorrhizal hyphae with the living tissues of their supports, which were improved in their growth when the epiphytes were removed from them. In this respect epiphytes resemble achlorophyllous orchids like *Gastrodia elata* (Kusano, 1911) and *Gastrodia minor* (Campbell, 1963) which, far from being saprophytic, were found to be attached by their fungal associates to living or dead trees which had been or were being parasitized indirectly. The degree, however, to which green chlorophyllous orchids are mycorrhizal as adults is likely to vary. In some species infection has been reported to be sporadic. There is, however, an infection cycle in many species and of course new roots that are produced by dormant tubers or rhizomes, especially in spring or after dormancy, are attacked and penetrated by fungi to form mature mycorrhizal roots in the ensuing months, so that infection is often most abundant in autumn. Hence many of the reports of small or nonexistent mycorrhizal infection are not always reliable unless observations through the year have been made. Nevertheless it seems possible that some, like *Cyprepedium calceolarius*, *Listera ovata*, *Cephalanthera oregana*, *Orchis globosa* and others, may often be little infected and perhaps virtually fungus-free (Ramsbottom, 1922; Knudson, 1927; Niewieczerzalowna, 1932; Summerhayes, 1951). On the other hand, the occurrence of almost or totally chlorophyll-free forms of green species, e.g. *Cephalanthera rubra* and *Epipactis purpurata*, shows that these are sufficiently mycorrhizal to be fully dependent upon their fungi for nutrition. The books of Hans Burgeff, especially *Saprophytismus und Symbiose* (1932), give many examples which can only be rationally interpreted in the sense that many adult orchids with green foliage are to a significant extent reliant on their fungi for carbon nutrition. Clearly this is a subject which needs much more research and as yet it has had too little attention (but see Hadley, 1975).

Besides carbohydrates, orchid seedlings require other substances for continued growth. Inorganic nutrients, including inorganic nitrogen sources, are required in the medium, and amino acids are sometimes stimulatory or even essential. In addition they require vitamins and growth factors, although the detail of their requirement varies with the species. Thiamin and nicotinic acid seem to be required by most. These requirements are provided by the activities of the mycorrhizal fungi after infection in nature (see Arditti, 1967, 1979).

## The fungi of orchid mycorrhizas

The fungi isolated from the roots of orchids and capable of stimulating the growth of their seedlings differ from those concerned with vesicular-arbuscular ectomycorrhiza and (most) mycorrhizas of Encales. Nutritionally, they are capable of rapid axenic growth upon soluble carbohydrates and upon insoluble resistant carbon polymers, such as cellulose. A very large number of the isolates were referred to the form genus *Rhizoctonia* (see Burgeff, 1936). These included *R. repens* which was much used in the pioneer experiments of Noël Bernard in the early years of this century and many of the species used by Burgeff. Others that have been much used in experiments are *R. goodyerae-repentis* isolated from *Goodyera repens* and *R. solani*, a very active parasite of herbaceous plants, from *Dactylorchis purpurella*. One *Rhizoctonia* isolate made by Catoni (1929) from Cyprepidium was observed to fruit in culture and was renamed *Corticium catonii*. Subsequently other *Rhizoctonia* isolates were encouraged to form perfect stages in culture, especially by Warcup and Talbot (1962–1970; Warcup, 1981). The basidiomycete genera to which they belong include *Thanatephorus (Corticium)*, *Ceratobasidium*, *Ypsilonidium*, *Sebacina* and *Tulasnella*. Warcup (1975b) and Jonsson and Nylund (1979) have shown that seeds of some green orchids may be stimulated by fungi not having a *Rhizoctonia* stage—*Oliveonia pauxilla* and *Favolaschia dybowskyana* respectively. Obvious basidiomycetous mycelia-bearing clamps were isolated from many achlorophyllus orchids and it was at one time believed that all their endophytes were of this kind, though that is now known not to be so. Some have been seen to produce fruit bodies: *Marazmius coniatus*, *Xerotus javanicus*, *Hymenochaete* sp., *Armillaria mellea* (agg.) and *Fomes* sp. (Kusano, 1911; Burgeff, 1932, 1936; Hamada, 1940; Hamada and Nakamura, 1963). Warcup (1981) tested 65 wood-rotting fungi with seeds of *Galeola foliata* an achlorophyllus liane. Seven white-rot species, including *Coriolus versicolor* and two species of *Fomes*, successfully encouraged germination as did three species isolated from *G. sesamoides*. Many of these fungi clearly have activities other than mycorrhiza formation with orchids. For instance, *Coriolus versicolor* is a very widely distributed saprophyte and *Rhizoctonia solani* and *Armillaria mellea* are very variable, extremely destructive parasites and the others must have a considerable capacity for an independent saprophytic or parasitic life. The fungus isolated from *Corybas macranthus* by Hall (1976) is reminiscent of the fungi of ectendomycorrhiza, ("*Complexipes*") which are septate and also form terminal and intercalary swellings.

All the mycorrhizal fungi from orchids which have been tested in culture can use soluble sugars as carbohydrate source. In addition most produce exoenzymes which hydrolyse complex polymers such as starch, pectin and cellulose and some destroy lignin (see Burgeff, 1936; see Harley, 1969a; Hadley and Ong, 1978). A comparison of the activity of the pectinases (endopolygalacturonase, endopolymethylgalacturonase and protopectinase) produced by pathogenic isolates of *R. solani (Thanatephorus cucumeris)*, *R. repens (Tulasnella calospora)* and *R. Goodyerae-repentis (Ceratobasidium cornigerum)* failed to reveal any association between pathogenicity to orchid seedlings and enzyme production. Indeed the highest polygalacuronase activity was found in cultures of *R. repens* which were symbiotic with a wide range of orchids. None of the fungi produced pectinases in the presence of glucose, which may be an example of metabolite or end-product repression (Pérombelon and Hadley, 1965). Cellulose decomposition by *R. repens* and *R. solani*, isolated from orchids, has also been demonstrated by S. E. Smith (1966), both in pure culture and in unsterilized soil, so that mycorrhizal strains of *R. solani* are similar in this respect to isolates from other sources. The cellulolytic and ligninoclastic properties of *Armillaria mellea* are well known and those of other fungi isolated from tropical orchids were demonstrated by Holländer (see Burgeff, 1936). The pathogenicity of the many strains of *Rhizoctonia solani* (*Thanatephorus cucumeris*) and *R. goodyerae-repentis (Ceratobasidium cornigerum)* to crop plants is not correlated with a pathogenicity to orchid seedlings. Some of the most active pathogens may be readily compatible with orchids (Warcup, 1981).

The requirement of orchid fungi for other nutrients beside carbon compounds is generally unspecialized also. Most can use a wide range of nitrogen sources both organic and inorganic (Holländer in Burgeff, 1936; see Arditti, 1979). Some, however, are stimulated in growth by B-vitamins, yeast extract and root exudates (Vermeulen, 1946; Silva and Wood, 1964; Pérombelon and Hadley, 1965; Hadley and Ong, 1978).

From this description it is clear that not only is the mycorrhizal association established at first between two heterotrophic organisms, but that the fungus and its "host" have changed roles as compared with the other mycorrhizal associations described. In Orchidaceae it is the angiosperm which has the more strict nutrient requirements and the demand for carbohydrate, and is unable to grow independently in natural surroundings at the seedling stage, while the fungi have a considerable capacity for saprophytic and parasitic existence.

## Primary infection and the development of symbiosis

There are two stages of infection in the life cycle of many, perhaps most, orchids; the primary infection of the germinating seedling, and the reinfection of the new roots of the adult. The latter is especially important in those forms which perennate as uninfected tubers or rhizomes and which form a new root system when dormancy is broken. The source of infection may be the soil or, according to Warcup (1971), sometimes the tuber, where fungi may persist on the surface or even in the tissues. Infection of the roots of adult orchids has been examined structurally, but the dynamics of the infection process has received little attention. Almost all the experimental work on the development of infection has been carried out on the primary infection stage of seedlings in agar culture. A study such as that made for vesicular-arbuscular infections on *Endymion non-scriptus*[1] by Daft *et al.* (1980) might be rewarding. Development of plants of *E. non-scriptus* and their mycorrhizas was studied by sampling plants from their natural habitats for one year at six-weekly intervals. As with many terrestrial orchids, the root systems of this species are renewed annually. Infections occur rapidly in the new root, arising from propagules in the soil or from the dead root system of the previous year, and are always confined, as in orchids, to the regions of the root in contact with the soil. The bulbs apparently contain an inhibitor which prevents mycorrhizal development in the roots growing through them, providing another parallel with mycorrhizal development in orchids.

In the absence of fungal infection (or exogenous sugars and other essentials) an orchid embryo takes up water, swells, bursts the testa and produces a few epidermal hairs. This process is here called germination as opposed to the growth of the seedling.[2] At this stage the cells may contain numbers of starch grains which are hydrolysed very slowly and do not support the growth of the protocorm (Hadley and Williamson, 1971; Purves and Hadley, 1975). If a suitable fungus is present, single fungal hyphae penetrate the wall of either the epidermal hairs or epidermal cells near the suspensor of the embryo. It is not known whether cellulolytic or pectolytic enzymes are important in this process, but if they are their action must be very localized because usually minimal disruption of the cells occurs. There are cases, however, where local disruption of tissues has been reported and these may merit further investigation. The tubers of both the achlorophyllous *Gastrodia elata* and *Galeola septentrionalis* when attacked by *Armillaria mellea* suffer local tissue breakdown before the symbiosis

---

[1] More safely called the English Bluebell.

[2] It should be noted that some workers (e.g. Warcup) call the growth of the seedling "germination".

becomes established and a flowering scape is formed (Kusano, 1911; Hamada, 1940).

The fungal hypha as it penetrates the cell of the embryo or root causes an invagination of the plasmalemma of the host and becomes surrounded by a layer of host cytoplasm which appears to remain healthy and protoplasmic streaming continues. The host cells appear physiologically active and contain numerous mitochondria, well-developed endoplasmic reticulum, dictyosomes and vacuoles of variable size, and few or no starch grains in their plastids. The nuclei of infected orchid cells are obviously hypertrophied as has been long known (see, for instance, Burgeff, 1932) and this is clearly apparent under the electron microscope. Williamson (1970), following Williamson and Hadley (1969), showed that the nuclei in the cells of infected protocorms had higher DNA contents than those of uninfected ones. He used the orchids *Dactylorchis purpurella* and *Spathiglottis plicata* in culture with *Tulasnella calospora* and examined the nuclei by microdensitometry after staining with Feulgen, and also by autoradiography after uptake of tritiated thymidine. In both species of orchid the nuclei of infected cells had 2–4 times the stable DNA of the cell of uninfected protocorms. The uninfected cells of infected protocorms often also contained hypertrophied nuclei as judged by their DNA content and hence some influence of infection must pass to them. It should be noted that Alvarez (1968) found that there was somatic doubling of DNA in the cells of uninfected roots of *Vanda*. Assuming the meristematic nuclei to be 2C, the parenchyma cells possessed 2C, 4C and 8C nuclei as well as sometimes more than 8C. He found that the increase of size up to 8C involved a parallel increase in DNA. This somatic polyploidy must occur in addition to the clear effects of infection seen in sections and measured by Williamson.

Infection spreads from cell to cell so that the basal region of the protocorm becomes extensively infected. Growth and anastomoses of the intracellular hyphae result in the formation of coils (pelotons) which very much increase the interfacial area between the symbionts (see Fig. 23). The fungal wall is described as being formed of two layers within the cell, an electron dense inner layer against its own plasmamembrane and an outer less dense more "flocculent" layer which is always present and of variable thickness in *Dactylorchis maculata* (Strullu and Gourret, 1974). Hadley *et al.* (1971) also described the fungal wall as appearing to be of two layers, one electron dense, and the other, outside, less dense, granular and thick. They thought, as did Strullu and Gourret, that two layers were also present in free-living hyphae on agar.

Strullu (1976a) discusses the origin of these two layers of the wall of the hyphae which he dubs C1 (inner) and C2 (outer). He tends to the view on comparative grounds, after consideration of the structure of other endomycorrhizas, that the C2 layer within the host cell is most likely to arise in whole or in part by the action of the host. It is perhaps homologous with the interfacial matrix of vesicular-

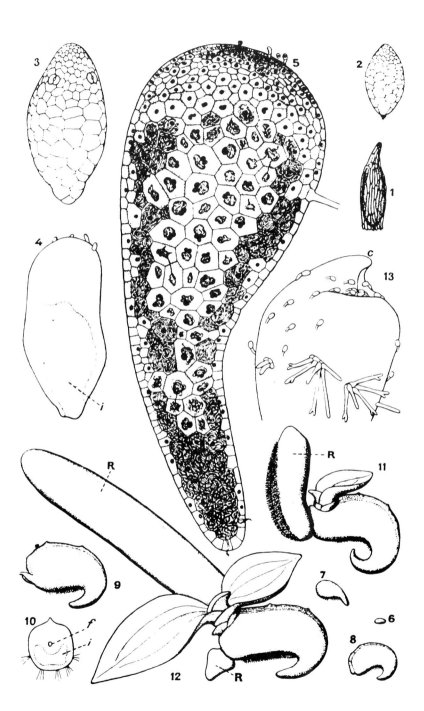

arbuscular mycorrhizas. The view of Nieuwdorp (1972), derived from a study of four chlorophyllous and four achlorophyllous orchids, appears at first sight somewhat different, but it basically agrees with Strullu. He suggests that the invaginated host plasmalemma continues to synthesize cell wall components, pectins and cellulose, but the young fungal hypha or peloton digests them. As the fungus ages the protoplasts within the hyphae disappear as a result of activity of the host (indicated by the formation of pinocytotic elaborations of the plasmamembrane) and the fungus becomes enveloped in a cellulosic slime layer continuous with the cell wall of the host. It can be seen that in this description the basis of the interpretation of when the fungal structure is young or aged is important (see below) and Hadley (1975) points out that the electron-lucent wall layer of free-living hyphae is very different from the thicker granular layer between hypha and host plasmalemma within the host cell. He therefore inclines to the view that the latter is formed by interaction of fungus and host.

Within the cell the hyphal coil takes up a large part of the volume so that it appears to lie within the vacuoles but surrounded not only by the invaginated plasmalemma but also by a parietal layer of cytoplasm which may in places be very thin, so that the tonoplast and the plasmalemma of the host are very close to one another. The cytoplasm of the fungus within the cells contains nuclei, vacuoles, mitochondria, lipid globules and glycogen rosettes but little ER, according to Strullu and Gourret. Hadley and his colleagues (1971, 1975) observed that the fungal plasmalemma was invaginated in places to form vesicles and tubules, structures which they believed might be associated with the transfer of substances from one symbiont to the other. They noted also that the fungal wall outside its plasmalemma bore protuberances on the outer wall to which the host plasmalemma was adpressed. They wondered whether these, which appeared to increase the contact area by about 15%, might play a part analogous to that believed to be played in transfer cells by their wall protuberances.

The intracellular hyphal coils have a limited life even in stable mycorrhizal associations of orchids. Mollison (1943) estimated that the collapse of the pelotons occurred in *Goodyera repens* eleven days after primary infection of a cell, but Hadley and Williamson (1971) showed that it could occur as soon as 30–40 hours after peloton or coil formation in *Dactylorhiza purpurella*. After this period of association with active host cells the hyphae degenerate. The stages were described in detail by Strullu and Gourret (1974). They classify the process in

FIG. 23. Seed germination of *Phalaenopsis amabilis* in association with *Rhizoctonia mucoroides*. 1, Seed; 2, 3, embryos a few days and 3 months after sowing without the fungus; 4, median section 6 days after infection with fungus; 5, 50 days after infection. All × 100. 6, Several months after germination without the fungus compared on the same scale; 7, 8, 9, 11, and 12, with plantlets formed 1, 2, 3, 5, and 12 months after infection. f, conducting strand; i, infected zone; R, root; c, young leaf. (From Bernard, 1909.)

four stages of which stage 1 is the association of living hyphae and cells already described. In stage 2 the hyphae, each still surrounded individually by the plasmalemma of the host, become flattened but the outer layer of the hyphal wall may be very thick (see Plate 14). Their content disappears except for some of the membranes and globules which break up more slowly. The flattened hyphae may now be enclosed in groups in the host plasmalemma rather than singly. Stage 3 consists of the formation of a complicated mass of associated hyphae in which both cell wall layers are apparent, but few hyphal contents exist. The whole is surrounded by host plasmalemma. In stage 4, the last stage, the identity of the hyphae themselves is lost as their walls become diffuse. During the course of this process reinfection of the cells may take place, as described by Burgeff (1934, 1936) and by Burges (1939). Strullu and Gourret (1974), who figure cells in which more than one stage exists, agree with this view. "A cell may be at the same time a digestion cell and a host cell." (Strullu, 1976a.) According to him, the reinfection may be a revitalization of hyphae within the cell, a hyphal penetration of the cell from a neighbouring cell, but only very improbably a new infection of the root.

This repeated invasion or spread of hyphae in the cells of the host is some evidence, as Strullu points out, that the degeneration of the fungus is generated by the host. For the host cell remains alive and active; and it receives anew a fungal penetration which ends again with the degeneration of the hyphae. Indeed a single cortical cell outlives several fungal colonizations. Hadley (1975) points out that elaborate membrane systems are present in the cells of the host during the phase of hyphal degeneration and that these are also to be seen in the preparations of Dorr and Kollman (1969) (see Plate 14).

The causes of the hyphal collapse are however not really known. Many workers, following the early view of Bernard, have believed it to be caused by the activity of the cell of the orchid and to be a manifestation of either a defence reaction against fungal invasion, or a means by which the orchid cell causes a release of nutrients from the fungus. Williamson (1973) showed that acid phosphatase activity, which is often thought to be a marker for lysosome activity, increases in cells where hyphal collapse is taking place. The localization of the activity was not, however, precise enough to determine whether the enzymes were of orchid or fungal origin, since autolysis of fungal hyphae would result in their own collapse. It is not impossible that orchinol, or one of the fungitoxic phenanthrenes which are synthesized by orchids (see below), might be important in fungal collapse. Increased activities of oxidase systems (polyphenol oxidase, catalase, ascorbic acid oxidase etc.) were found by Blakeman et al. (1976) to occur when hyphal collapse was proceeding, an observation again not very helpful in the interpretation of its role in this symbiosis.

When mycorrhizal symbiosis has been set up in a seedling (or root for that matter) there are therefore three phases or systems: an external hyphal system in

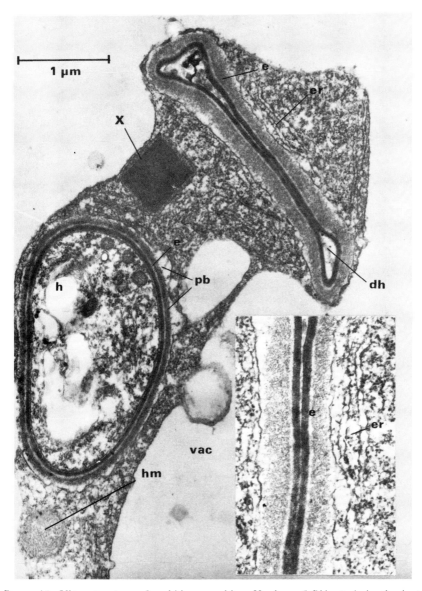

PLATE 14. Ultra structure of orchid mycorrhiza. Hyphae of *Rhizoctonia* in the host cytoplasm of a cell of *Dactylorhiza ( Dactylorchis) purpurella*. Both the living hyphae (h) and the dead collapsed hyphae (dh) are surrounded by an encasement layer (e), which is probably of host origin. Paramural bodies (pb) occur in the interfacial matrix between the host plasmalemma and the encasement layer. Host cytoplasm contains endoplasmic reticulum (er), mitochondria (hm) and a crystal (X). Inset: The host vacuole (vac) is visible, surrounded by the interfacial matrix in greater detail. (From Hadley, 1975.)

contact with the external substrate capable of absorbing carbonaceous and other nutrients, an internal system of cells in which active hyphae exist, and a second type of infected cell in which the hyphae are collapsing or being digested.

The functioning of this triple system is considered below.

## Growth of the seedling and nutrient supply

### Carbohydrates

As has been mentioned, the cells of the embryo often contain starch grains which remain there in the absence of suitable conditions for growth. It has been noted of many orchid embryos that, if soluble carbohydrate is provided to vitamin-dependent seedlings in the absence of suitable vitamins and growth factors, sugar is absorbed and the cells become glutted with starch grains. Clearly the embryos of many orchids are unable to form the enzymic systems required to utilize carbohydrate at a sufficient rate to maintain metabolism and to support growth. They require supplies of precursors of enzymes or coenzymes before they can do so (see Harley, 1969a). The fungus intervenes in this matter and one of the early results of infection is the disappearance of starch grains from the cells of the embryo. After this the continued growth depends upon the translocation of carbohydrates through the fungal hyphae into the cells of the embryo. The ability of orchid mycorrhizal fungi to hydrolyse complex carbohydrate polymers means that sources in the soil or in plant remains, or sometimes even in living plants, are available to the system. The translocatory ability of the fungi has been well established by Monson and Sudia (1963), S. E. Smith (1967), Purves and Hadley (1975) and by others, and is fully discussed in a later chapter (16). Hence the separation of the embryo from substrates in the soil poses no problems in this regard once the symbiosis is set up, for hyphae connect it to the substrate. Observations as early as 1920 by Beau showed that orchid seedlings only connected to a nutrient substrate by hyphae would grow; if the hyphae were

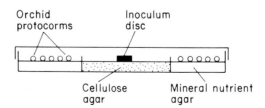

FIG. 24. "Schütte-dish" system used to study translocation to orchid seedlings. Seeds were sown on mineral nutrient agar in the outer dish. The fungus was inoculated onto the cellulose-containing medium in the inner dish; it grew over the barrier into the outer compartment and formed mycorrhizas with the seedlings.

severed growth ceased. More recently, S. E. Smith (1966) has used the experimental system illustrated in Fig. 24 to show that a fungus, *R. solani* (*Thanetephorus cucumeris*), was not only capable of hydrolysing cellulose and absorbing its products, but also of translocating them to orchid seedlings in sufficient quantities for growth to occur. The results are given in Table 68. It should be noted that the initiation of symbiosis is more certain if the fungus is provided with cellulose than if presented with a medium rich in soluble carbohydrate. In the latter case, destruction of the seedlings is more likely to occur (Harvais and Hadley, 1967; Hadley, 1969).

TABLE 68

Growth of seedlings of *Dactylorchis purpurella* in 14 weeks on substrates with or without cellulose in the presence or absence of *Thanetephorus cucumeris* at 22·5° C in the dark (results of S. E. Smith, 1966)

|  | Not inoculated | Inoculated | |
|  |  | + cellulose | − cellulose |
|---|---|---|---|
| Number of seedlings[a] | 14 | 226 | 324 |
| Length $(\mu m)$[b] | $248 \pm 9$ | $1170 \pm 119$ | $800 \pm 72$ |
| Width $(\mu m)$[b] | $206 \pm 10$ | $692 \pm 44$ | $519 \pm 39$ |

[a] Remaining healthy seedlings out of approximately 800.
[b] $\pm$ S.E.

These experimental results are a full confirmation of the work of Burgeff and Holländer (see Burgeff, 1936; Harley, 1969a) which showed that although the fungus "*Corticium catonii*" released no soluble reducing substances into the medium when grown on polypodium fibre it was capable of increasing the dry-weight gain of *Cymbidium* sp. ten-fold in three months in symbiotic culture.

In spite of the result of such experiments, the role of the fungus in discharging carbohydrate actually into the seedlings has been questioned, and indeed the results of most of the experiments cannot distinguish between direct transport of soluble carbohydrate into the seedlings and leakage of them from the fungus into the medium followed by uptake by the seedlings. However, Purves and Hadley (1975) point out that in any population of seedlings on cellulose only a proportion become infected and that only the infected ones show any increase in growth rate over the controls. Moreover, the direct transport of carbohydrate into the seedlings, which according to these experiments seems probable, has been confirmed by the use of radioactive substances (see Fig. 25).

The work of S. E. Smith (1967) and of Purves and Hadley (1975), using the tracers $^{14}C$ and $^{32}P$, has confirmed the ability of orchid mycorrhizal fungi to translocate and has given further information on the transfer of soluble carbohydrates to seedlings. The split-plate technique illustrated in Fig. 32

(Chapter 16) has been used so that labelled compounds could be supplied to the fungus alone, and diffusion between the two halves of the plate prevented. If [14]C-glucose is supplied at X, radioactivity appears in infected seedlings on the other side of the split within a few hours. Radioactivity continued to increase in the seedlings for 7 days in S. E. Smith's (1967) experiments with *Dactylorchis purpurella*, and for 18 days in Purves and Hadley's (1975) experiments with *Goodyera repens* (Fig. 25). Both alcohol soluble and alcohol insoluble fractions of the

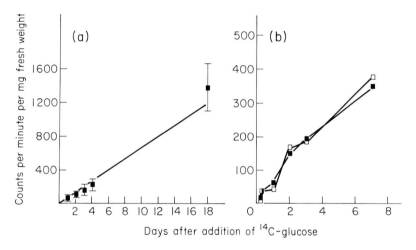

FIG. 25. Translocation of [14]C-labelled compounds by mycorrhizal fungi into orchid protocorms after the fungi had been supplied with [14]C-glucose on split-plates. (a) *Goodyera repens* infected by *Rhizoctonia goodyerae-repentis* (redrawn from Purves and Hadley, 1975). (b) *Dactylorchis purpurella* infected by *Rhizoctonia solani*. ■, soluble fraction; □, insoluble fraction. (Redrawn from S. E. Smith, 1967.)

seedlings became labelled in *Dactylorchis* and chromatographic analysis of the soluble fractions gave extremely informative results. The soluble carbohydrates of uninfected orchid tissues are sucrose, glucose and fructose, whereas those of the mycorrhizal fungi are predominantly trehalose accompanied by glucose and occasionally by mannitol, but not sucrose. Seedlings fed with [14]C-glucose via the split-plates became labelled not only in the fungal sugars but also in the orchid sugar, sucrose. Changes in the pattern of labelling with time in *D. purpurella* are shown in Fig. 26 and indicate that the fungal sugar, trehalose, is the most heavily labelled in the early samples but as time elapses sucrose becomes proportionately more heavily labelled as trehalose labelling declines. There is thus reasonably good evidence that carbohydrate is translocated in the fungus, and during or following transfer to the orchid cells it is converted to sucrose. Since trehalose acts as a source of carbohydrate for asymbiotic growth of the seedlings of several orchids it is possible that trehalose is absorbed directly from the fungus, but external hydrolysis to glucose before absorption by the action of a fungal or orchid trehalase cannot be ruled out. Trehalose is a suitable source of carbohydrate for *D. purpurella* (S. E. Smith, 1973), for *G. repens* (Purves and Hadley, 1975) and for *Bletilla hyacintha* and *Phalaenopsis* sp. (Ernst *et al.*, 1971; S. E. Smith, 1973). It can also be absorbed by the leaves of *B. hyacintha* and metabolized by them. The labelling patterns following absorption of [14]C-trehalose are identical with those following [14]C-glucose absorption (S. E. Smith and F. A. Smith, 1973). There is no good evidence in any of this work to indicate whether trehalose is absorbed intact or only after hydrolysis. The fact that no glucose was detected in the medium surrounding germinating *Phalaenopsis* seeds by Ernst *et al.* (1971) might suggest direct absorption, but a similar result would be obtained if the rate of glucose absorption equalled or could keep pace with that of trehalose hydrolysis.

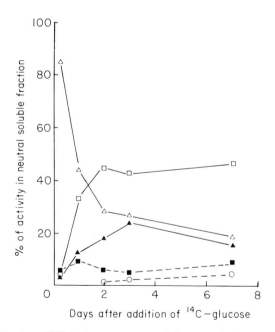

FIG. 26. Distribution of [14]C in the components of the neutral ethanol-soluble fraction of mycorrhizal protocorms of *Dactylorchis purpurella* after the fungus, *Rhizoctonia solani*, had been supplied with [14]C-glucose on split-plates. △, trehalose; ☐, sucrose; ▲, mannitol; ■, glucose; ○, fructose. (Redrawn from S. E. Smith, 1967.)

Mannitol, although present in some orchid mycorrhizal fungi, even in varieties of *R. solani*, is less likely to be important as a translocatory and transfer carbohydrate. It occurs in only a proportion of the fungi examined and is suitable for the asymbiotic germination of only a few species of orchid. The leaves of *Bletilla hyacintha* absorb and accumulate mannitol but do not seem to metabolize it, nor is it suitable for the germination of the seeds of this species (S. E. Smith, 1973; S. E. Smith and F. A. Smith, 1973). It would be most interesting to know whether an orchid such as *Phalaenopsis* sp. which can germinate on mannitol can use it in adult tissues. Metabolism of mannitol might possibly provide a basis of crude specificity between fungi producing it and orchids capable of using it.

Vitamins and growth factors synthesized by the fungi may also be important or essential for the growth of the seedlings of some orchids. Direct evidence of the transfer of such substances from fungus to host is lacking, but reasonable circumstantial evidence is available that the requirements of seedlings for vitamins in asymbiotic culture are provided by the fungus when a mycorrhizal association is formed.

## Translocation and transfer of mineral nutrients

It has been shown that orchid mycorrhizal fungi can translocate $^{32}$P supplied as orthophosphate both from hyphal tip to older mycelium and vice versa (S. E. Harley, 1965; S. E. Smith, 1966). Such translocation can lead to the accumulation of phosphate in associated seedlings, as shown by experiments with split-plates. It is also likely that mycorrhizal fungi play a part in mineral nutrition of both seedling and adult orchids in a manner comparable with other mycorrhizal systems. For instance, many orchids have thick "magnolioid" roots (*sensu* Baylis, 1975) with a few lateral rootlets and root hairs. The inefficiency of these roots in absorption of nutrients with low movement coefficients in soil would be offset by the fungal hyphae emanating from mycorrhizas.

The experiments of Holländer and Burgeff (Burgeff, 1936) indicate a great uptake of nitrogenous substances via the fungus, for the increase in weight of *Cymbidium* seedlings on polypodium fibre was about ten-fold in three months, whereas the increase in nitrogen content was twenty-five fold. The problem of whether the hyphae intervene in mineral absorption requires, however, experimental examination and is being considered by Clare Alexander in Hadley's laboratory. Inhibition by 1% thiobenzadol of the external mycelium of *Rhizoctonia goodyerae-repentis* attached to plantlets of *Goodyera repens* reduced nitrogen and phosphorus uptake and the growth rate of the plantlets. In *Dactylorchis purpurella* also, evidence was found of inorganic nutrient transport to the plantlets. Later experiments (personal communication) with plants of *Goodyera repens* grown in culture to leafy stage demonstrated that *Rhizoctonia goodyerae-repentis* does continue to intervene in the expected manner in inorganic nutrition. Clare Alexander compared mycorrhizal plants, mycorrhizal plants treated with the fungicide thiobenzadole, and non-mycorrhizal plants at the 4-leaf stage. The fungicide did not appear to affect the metabolism of the host as far as was ascertained. In a series of experiments the fungus absorbed and translocated nitrogen and phosphorus compounds from the medium to the orchid plants. The untreated mycorrhizal plants had higher relative growth rates, and nitrogen and phosphorus contents than the fungicide-treated or non-mycorrhizal plants. The problem is of ecological relevance because orchid roots under natural conditions, as has been described, are usually infected even in most green orchids, and, on comparative grounds, the fungus would be expected to operate as other mycorrhizal fungi do.

## Mechanisms of transfer

"Digestion" or "lysis" of the fungus in orchid mycorrhiza was first thought to be a mode of nutrient transfer but many have viewed it as a manifestation of defence of the host against invasion. They believed that nutrient transfer, which appeared to be a continuous process, was more likely to occur across the intact membranes of fungus and host. Under these circumstances control would be operated by the membranes of both symbionts in uptake, and natural or induced leakiness would be involved in the release of substances. If nutrient transfer occurred following hyphal "collapse", selectivity and control would operate at the orchid membrane only. If uptake were only in one direction the orchid might possibly be viewed as a necrotrophic parasite of the fungus (see Lewis, 1973), but of course it is equally possible that the fungus might be absorbing through its intact hyphal coils material leaking from the orchid whilst digestion or lysis was taking place elsewhere.

The fungal invasion clearly alters the internal metabolism of the orchid. The studies on *Dactylorchis purpurella* and *Goodyera repens* by Mollison (1943) and Hadley and Williamson (1971) suggest that increase of growth rate of protocorms precedes the onset of digestion, although other work in Hadley's laboratory (see Purves and Hadley, 1975) has questioned this. On the other hand, the hydrolysis of starch and prevention of excess starch formation in sugar media may be brought about either by infection or by the provision of yeast extract or nicotinic acid in the presence of other vitamins (Burgeff, 1934; Schafferstein, 1938, 1941). The fungus may in this case relieve an enzymic blockage by providing precursors of NAD and NADP. The different destination of $^{14}CO_2$ fixed by green protocorms of *Dactylorchis purpurella* into the hexoses glucose and fructose in infected protocorms, and into sucrose in uninfected ones, is another example of the effect of the fungus in generating conditions for normal metabolism of reserves (see Hadley and Purves, 1974). These results, the detailed explanation of which is not yet forthcoming, argue, as Hadley and Purves (1974) have pointed out, against a naïve interpretation of the orchid being a necrotrophic parasite on the fungus.

## Carbohydrate transfer from orchid to fungus

Authoritative statements have been made on fairly slender grounds that carbon transfer from fungus to orchid seedlings is reversed when the orchid becomes photosynthetic. These assertions were unsupported by experimental evidence (see Burgeff, 1959; and see Scott, 1969) and recent work has indicated that carbon movement from photosynthetic orchid to fungus is slight if it occurs at all, but almost all the experiments are open to more than one interpretation. For instance, growth of hyphae from infected protocorms onto an agar medium containing only mineral nutrients has been cited as evidence that "reversed" carbon movement into the fungus is occurring. However, *Rhizoctonia* species, including orchid fungi, can, as S. E. Smith (1967) pointed out, make considerable growth on "carbohydrate-free" agar even in the absence of orchid seedlings, so they may be able to use some carbon source from the agar itself (see also Hadley and Purves, 1974; Howard, 1978). More direct evidence of lack of transfer from host to fungus has been obtained using $^{14}CO_2$ as tracer. Early experiments by S. E. Smith and by Lewis and Hadley (unpublished) failed to detect any radioactivity in the fungal sugars after green infected seedlings of *D. purpurella* had assimilated $^{14}CO_2$ in the light. Nor was there any in the hyphae growing out onto carbohydrate-free medium. In later work with the same orchid, Lewis, Hadley and Purves (see Purves and Hadley, 1975) showed, however, that fungal metabolites sometimes became labelled in similar experiments, but since the control (uninfected plants) released $^{14}C$-labelled nutrients into the medium, carbohydrates released in a similar way by infected plants might have been the source of $^{14}C$ in the mycelium around infected protocorms; moreover, dark fixation of $^{14}CO_2$ by the fungus might also have occurred. In experiments with *Goodyera repens* Hadley and Purves (1974) showed that no $^{14}C$-labelled metabolites leaked into the medium and in this case when $^{14}CO_2$ was applied to the green top alone in the light little radioactivity passed to the rhizome and none into the growing hyphae in the medium. When the rhizomes alone were exposed to $^{14}CO_2$, small quantities of radioactivity appeared in the emerging mycelium. It was concluded that carbon movement from the seedlings to the fungus was at most very small in quantity. In contrast, movement into the mycelium from killed $^{14}CO_2$-labelled seedlings was fairly rapid (Table 69).

One can conclude, therefore, that a reciprocal movement of carbon compounds into the fungus does not readily occur, although there is no direct evidence of the presence of the kind of "biochemical valve" postulated by Lewis and Harley (1965c) for ectomycorrhizas of beech. This appears to be so, even in

## TABLE 69

Movement of $^{14}C$-labelled compounds from seedlings of *Goodyera repens* into mineral agar after exposure to $^{14}CO_2$ in the light. Radioactivity as cpm per sample of $5 \times 6$ mm discs of agar. (Results of Purves and Hadley, 1975)

| Treatment of seedlings | Distance from seedling source mm | Days after $^{14}CO_2$ exposure | | | | |
|---|---|---|---|---|---|---|
| | | 2 | 5 | 8 | 11 | 15 |
| Uninfected (living) | 5 | 0 | 2 | 1 | 15 | 11 |
| | 25 | 0 | 0 | 0 | 0 | 0 |
| | 50 | 0 | 0 | 0 | 0 | 0 |
| Infected (living) | 5 | $281^a$ | $74^a$ | $46^a$ | $55^a$ | $39^a$ |
| | 25 | 0 | $63^a$ | $1i^a$ | $16^a$ | $11^a$ |
| | 50 | 0 | 0 | 0 | $14^a$ | $0.1^a$ |
| Infected (killed) | 5 | — | 0 | $1688^a$ | $650^a$ | $205^a$ |
| | 25 | — | 0 | 0 | $143^a$ | $104^a$ |
| | 50 | — | 0 | 0 | 0 | $101^a$ |

$^a$ Fungus present in the discs sampled.

## TABLE 70

Translocation of carbohydrates through hyphae of *Rhizoctonia solani* (S10) into seedlings of *Dactylorchis purpurella* and incorporated into various organic fractions during 150 h

*Protocorms analysed after dissolving fractions in 80% alcohol, taking to dryness in reduced pressure and dissolution in water. Percentage radioactivity in the fractions separated on columns.* (Recalculated from S. E. Smith, 1967)

| Fraction | Percentage of $^{14}C$ |
|---|---|
| Total insoluble$^b$ | 50 |
| Soluble fractions: | |
| Anionic (organic acid) | 5 |
| Cationic (amino acid) | 15 |
| Neutral | |
|     Sucrose | 15 |
|     Glucose | |
|     Fructose | 5 |
|     Trehalose$^a$ | 6 |
|     Mannitol$^a$ | 4 |
| Total soluble | 50 |

$^a$ Uncharged fungal carbohydrates.
$^b$ Includes proteins and cell walls.

those experiments where no carbon source was available to the fungus outside the seedling. Harley (1975) has however expressed the opinion that since the orchid seedlings in the experiments of S. E. Smith (1967) were actually growing, the sink for carbohydrates is the conversion of them into amino acids, proteins and cell walls. In her experiments with *Rhizoctonia solani* and *Dactylorchis purpurella*, illustrated in Fig. 26 and Table 70, only 30% of the $^{14}$C-labelled compounds were soluble carbohydrate in the orchid seedlings; 50% was in insoluble compounds and 15% in amino acids.

One can only assume from the available evidence that in the adult state the green orchids so far examined do not provide a carbon source for the associated fungus but that the fungus continues to be self-sufficient in that respect.

## Seed germination and growth in natural conditions

The number of seeds produced by the individuals of a species of plant is related to the probability of successful germination and seedling establishment. The orchids represent an extreme case where the embryo in the seed is reduced to tiny proportions; it is unequipped with reserves except enough for a negligible growth, and the seeds are produced in millions per capsule. Since on an average each plant must produce only one successful reproducing progeny in its life-period, and most orchids are perennial, living for several years and usually producing many seed capsules each year, the odds on any given seed germinating successfully and producing a plantlet are extremely small indeed in natural conditions. It is not therefore surprising that the demonstration of the successful symbiotic germination of the seed of orchids with fungi isolated from their roots has proved to be a chancy business. Noel Bernard (1909) showed that sometimes typical stable mycorrhizas were formed and seedling growth was good; with other fungi growth ceased for one of many reasons. The fungus might invade the seedlings completely, killing them. Or it might be itself destroyed in the cells so that growth ceased but could be renewed by the introduction of a compatible fungal strain. Similar results have been obtained by later researchers. A scheme of possible pathways of fungus interaction with orchid seeds has been drawn up by Hadley (1970), after experiments with 32 strains of 10 species of fungus isolated from 5 north temperate and 3 tropical species of orchid with 10 species of protocorm in culture. This is presented, slightly modified, in Fig. 27.

Even if a compatible symbiosis (with some consequent growth stimulation) occurs at the beginning, it may break down at a later stage. A single culture vessel containing apparently compatible orchid seeds and fungus will normally contain a mixed population of uninfected and infected protocorms of all the kinds shown in the diagram. The proportion in each class depends upon the genetic strains of fungus and orchid and also upon the environmental conditions. Little is known of what controls the balance between the two organisms, but obviously several different mechanisms may be involved. Considering the large seed production by each individual in each season, a fine control might not be expected. On the other hand, it seems unlikely that all is left to chance. In the following sections a few important aspects of the control of the balance between the two symbionts are discussed.

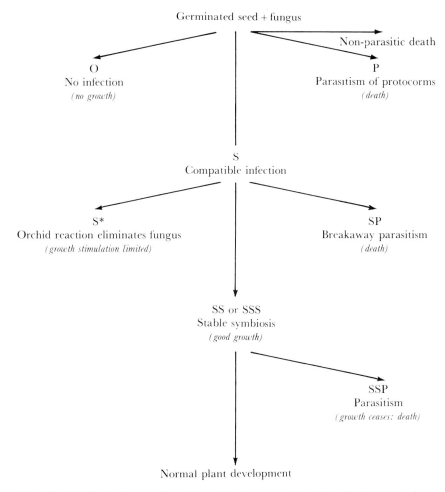

FIG. 27. Modified scheme of symbiotic development of orchids. O, Uninfected, no growth; S, SS, SSS, greater numbers of protocorms develop; S*, initial infection followed by hypersensitive reaction or complete fungal disintegration; P, fungus becomes an aggressive parasite killing cells; —, no infection but protocorms die for other reasons. (After Hadley, 1970.)

## Hydrolytic enzymes

Orchid fungi, like some unspecialized pathogens, produce cellulases and pectinases which may play a significant part in the saprophytic utilization of insoluble substrates and in obtaining sugars required by both fungus and orchid. The amounts of pectinases produced by different species of *Rhizoctonia* were found by Pérombelon and Hadley (1965) not to be related to their pathogenicity towards orchid seedlings. A considerable range of ability to decompose cellulose is also found among successful orchid endophytes (S. E. Smith, 1966; Hadley, 1969; see Burgeff, 1936). The activities of pectinases and cellulases must therefore be controlled in some way within the orchid, either by high levels of sugars such as glucose which may inhibit or repress pectinases and cellulases, or by some other means. In the first case starch hydrolysis which occurs in some newly invaded seedlings might bring about a temporary increase in glucose concentration which might be enough for immediate enzyme repression until photosynthesis starts, but this must be doubtful. Purves and Hadley (1975), however, observed that the infected roots of adult *Goodyera repens* have higher hexose levels than uninfected roots, so it might well be asked whether mechanisms exist in them for maintaining conditions which repress hydrolase activity by maintaining high soluble sugar concentrations; but it is difficult to see how such repression could operate in achlorophyllous orchids which had no other source of carbohydrate than the fungus.

On the other hand, mechanisms such as those favoured by Vanderplank (1978) for pathogens which depend on the co-polymerization of proteins of fungus and host, as discussed in a later chapter (18), might well operate as a mode of enzyme inhibition. Selection for such action in evolution would be strong, as it maintains the symbiosis.

## Phytotoxic substances and phytoalexins

The initial work on the formation of fungitoxic substances was due to Bernard who suggested that the absence of fungal infection from the tubers of some orchids and the resistance of some seeds to fungal attack was due to the presence within them of an antifungal principle. He showed in a paper published in 1911 that tubers of *Loroglossum* contained a substance which was toxic to many orchid endophytes including *Rhizoctonia repens* (*Tulasnella calospora*) but not to a strain of *R. solani*. Bernard believed, and Nobécourt (1923) agreed, that the toxic substance was formed after the fungal attack (i.e. that it was in modern terms a "phytoalexin"). Magrou (1924) did not agree and he believed that the fungitoxic principles in orchid tissues were there before attack. A series of papers by Gäumann and his colleagues (1956–1961), Nüesch (1963), Fisch *et al.* (1973), and Arditti *et al.* (1975) has followed up this work. They showed that orchinol, a dihydroxyphenanthrene, was formed by living tuber tissues of *Orchis militaris* when the tissue was kept in contact with *Rhizoctonia repens*. It was formed not only by the cells immediately in contact, but also, more slowly at first, by cells up to 12 mm from the contact. Subsequent work has shown that many terrestrial European orchids produce similar phytotoxic compounds and it is likely that the ability is much more widespread (see Arditti, 1979).

Three phytoalexins were initially identified—hircinol, loroglossol and orchinol—and others have since been discovered. The dihydroxyphenanthrenes all inhibit mycorrhizal fungi and many other fungi and bacteria also. They are indeed wide-spectrum toxins or antibiotics. They differ, as Wood has pointed out (1967), from other phytoalexins in not being formed only around the lesion or point of attack but at some distance away also. In any event their formation at a distance from the point of attack leads one to believe that their synthesis is by no means a simple matter, but must involve not only a chemical sequence of production but also a signal emanating from the point of attack.

It seems certain that these substances play a part, as the early workers believed, in controlling or restricting fungal invasion of orchid tissues. However, the problem arises as to how it is that mycorrhizal fungi can penetrate some orchid organs such as roots and not others such as tubers; and how it is that most other fungi seem to be excluded from both. Gäumann and his colleagues were of the view that the difference in susceptibility of the roots and the tubers was due mainly to a difference in the ability of the two kinds of tissue to produce the phytoalexins. Some mycorrhizal fungi, e.g. strains of *Rhizoctonia solani*, will destroy orchinol and it may be that the balance between phytoalexin production and its inactivation in any combination of orchid tissue and fungus might determine the stability of the symbiosis. Some other fungi have been found to be resistant to these phytoalexins, and others to attack orchid tissues so rapidly that time is not available for their formation before complete parasitization. It is clear that these compounds where they are produced in high concentrations, in the tubers for instance, may confer resistance that is nearly complete. At the lower concentrations in the root they may be only a coarse control of infection by alien fungi and of the extent of exploitation by mycorrhizal fungi. The subject is therefore in a somewhat unsatisfactory position and there is no doubt that a fresh impartial examination of it with selected genotypes of fungus and orchid would be worthwhile. Moreover, the effects of low concentrations of these compounds, not only on growth but also on the physiology, including the activity of the exoenzymes, of orchid fungi would bear examination.

## Phosphatases and other enzymes

Burges (1939) extracted by means of a micromanipulator the contents of cells of orchids in which hyphal disintegration was occurring. He showed that the contents of the "digestion cells" caused disintegration of fungal hyphae in culture. This was a practical indication that the disintegration was an enzymic event. Since then phosphatases have been shown to be important, but undoubtedly there must be other enzymes essential to the extensive disintegration of the hyphae which is observed. There is no point in speculating to what extent this "hyphal collapse" is a manifestation of the control of infection until it is known what generates it and whether the enzymes are produced by the host or the fungus, and from where comes the signal for their production and the extent to which their activity is affected by the phenanthrenes.

## Specificity and ecology of orchid fungi

Although it was at one time believed that symbiosis between orchids and their mycorrhizal fungi was always highly specific, it is clearly not so. Indeed it is becoming apparent that in these, as in all other kinds of mycorrhiza, the specificity of the symbiotic relationship is not very close. As long ago as 1936 Burgeff came to the conclusion that although not closely specific there was a tendency for certain fungal species to form more effective symbiotic unions with certain taxa of orchids. Curtis (1937, 1939) agreed with this conclusion which was based in both cases mainly upon surveys of the distribution of orchid "root fungi", that is the kinds of accepted *Rhizoctonia* species which were present in or could be isolated from the roots of different species in different habitats, different soils and so on. They gave surveys of information about the fungi present in the mycorrhiza of adult orchids only. They indicated in particular that there are ecological groupings of fungi and orchids related to species, habitat and ecological conditions, just as field observation has done in the case of ectomycorrhizas of trees.

It should be noted, however, that there is another test of compatibility of fungus and host which has been used, and most frequently used, in research on the mycorrhiza of orchids. That is, the test that a given fungus stimulates the early growth of the seedling of the orchid so that it passes successfully through all the stages shown in Fig. 27. It should be noted that such a test not only involves the successful completion of the histological interaction between the symbionts such as can be verified visually, but also a successful and efficient physiological interaction leading to growth and development.

The fungi found in and isolated from the roots of adult orchids do not always encourage the growth of the seedlings of that or indeed any orchid in tests in sterile media. There may be several reasons for this. First, roots may harbour a wide range of fungi and the isolation method used may yield surface inhabitants as well as fungi from within the tissues, and only some of the latter may form coils or pelotons within the cells (Warcup, 1975b). More than one fungus may be associated with the pelotons of a single root and even if the extracted pelotons alone are used for isolation more than one fungus may be obtained (Warcup, 1971). Indeed, sometimes a second fungus may intrude into the cells in which there is already a peloton. It is possible, too, that even if a fungus is symbiotically efficient with an adult root, it may not be equally efficient in improving growth of the seedling, especially if the adult is a green photosynthesizing organism. Indeed, the necessary properties of the fungus in the two cases may be different. In consequence the tests which examine what fungi are effective in seedling growth in particular sterile media and other conditions of nutrient supply and physical environment, may not be relevant to mycorrhiza formation and effectiveness in other growth phases.

The orchid fungi themselves may have a distribution in a vegetatively active state far wider than that of their hosts. Obvious examples of this are *Coriolus versicolor*, *Rhizoctonia solani* and *Armillaria mellea*. However, all these species are conspicuously variable and it remains to be clearly demonstrated how many of the variants are or are not potential mycorrhiza formers. Downie (1943) showed that strains of *Rhizoctonia goodyerae-repentis* (*Ceratobasidium cornigerum*), effective in seedling stimulation, were widespread in soil, humus, pine litter, and on senescent pine needles still attached to the tree. Although this observation may be explained by the presence of spores or sclerotia it is also possible that vegetative mycelium is present on some of these materials. The case is reminiscent of the distribution of the "dark sterile mycelia" of ericoid mycorrhiza. However, fungi with a great potential for the breakdown and use of cellulose would be expected to be able to pass a saprophytic existence more easily than those not similarly endowed.

In spite of a general lack of specificity, Warcup (1971, 1973, 1975b) pointed out that some fungi appear fairly specific to a given genus or species, although others are less so, as shown by seedling growth tests. Isolation of fungi from hyphal coils in the tissues showed that species of the genera *Pterostylis*, *Caladenia* and *Thelmitra* growing within a few centimetres of each other in dry sclerophyll forests of Australia had each its own endophytic species of fungus. Rarely were there "cross overs" and then only in roots infected by two fungi. Examples of fairly close specificity were also given by the observation (Warcup, 1973) that the seedlings of two species of *Pterostylis* were stimulated only by *Ceratobasidium cornigerum*, and those of two species of *Diuris* only by *Tulasnella calospora*, and that the seedlings of *Thelmitra* were stimulated by several species of *Tulasnella* and not by *C. cornigerum*. Again the species of the allied group of genera, *Caladenia*, *Glossodia*, *Elythanthera* and *Eriochilus* harboured *Sebacina vermifera* in their roots, although the fungus was not conspicuously successful in stimulating the growth of seedlings of that group of genera, yet in one test it greatly stimulated the growth of *Caladenia carnea*.

We are led to conclude from this that specificity is generally not close but is variable in the same sense as it is in ectomycorrhiza. Nevertheless we have no very good information on specificity in adult orchids except in respect of isolation from their roots, as risky a class of evidence as the ecological distribution of sporophores is of the specificity of ectomycorrhizas.

## Summary of mycorrhiza in Orchidaceae

The most outstanding point about research in the physiology of the symbiosis in Orchidaceae is that it has rarely been directly concerned with the adult stages of the host plant. Work has concentrated on the seedling stages when all orchids are heterotrophic for carbon. Such evidence as is available from observations on adult plants indicates that even when they contain chlorophyll, many may rely in part on additional carbohydrate from the fungus.

Again very little is known of the mineral nutrition of the orchids either in the seedling or adult stages. It seems very probable that the fungal hyphae intervene in nutrient absorption in orchid symbioses in a way similar to their interventions in other kinds of mycorrhiza and the experiments with protocorms in culture and more recently with young leafy plants seem to be the first clear demonstration of this. Indeed, if any credence can be put on the views of Baylis (1975) that plants with coarse roots are usually dependent on mycorrhizal infection for nutrient uptake, orchids must depend in that way on their fungal symbionts.

The fungi of orchid mycorrhiza, in contrast to most other mycorrhizal fungi, are active in using complex carbohydrate polymers for growth, and it is on this ability that their success as symbionts of the partially or wholly achlorophyllous orchids depends. Even if it were possible to believe that the small amounts of carbon compounds required for the germination and early growth of the seedlings of green terrestrial orchids could be obtained from the humus and litter layers of the soil, it seems very unlikely that these layers could be the source of the carbon supplies of the large totally achlorophyllous ones. The description of the achlorophyllous liane *Galeola hydra* given by Burgeff (1936) is an example which underlines this doubt, reinforced by the fact that active aggressive parasitic fungi like *Armillaria mellea* and species of *Fomes* have been found to be symbiotic with large achlorophyllous orchids. The work of Ruinen (1953) which showed that many tropical epiphytic orchids had hyphal connexions which penetrated the tissues of their supporting plants further emphasizes the common occurrence of this habit which has been called epiparasitism (see also Jonsson and Nylund, 1979). Björkman (1960) used the term to describe the habit of *Monotropa hypopitys* which was shown to be dependent for carbohydrate on neighbouring trees by having a common mycorrhizal fungus with them. Later Campbell (1963) showed that epiparasitism was probably the means whereby the achlorophyllous so-called "saprophytes" of several families, including Orchidaceae, obtained carbon compounds, either through fungi parasitic upon or mycorrhizal with green plants.

This common feature may relate the mycorrhiza of orchids with that of Monotropaceae, but the relationship is not necessarily very close. *Monotropa*, in contrast to orchids, possesses a well-developed fungal sheath, a Hartig net and very specialized fungal haustoria which the orchids lack. Similarly the achlorophyllous plants which have vesicular-arbuscular mycorrhizas may also be epiparasitic, but the transfer mechanism to the achlorophyllous host and the structure associated with it need investigation.

It may be recalled that Burgeff described several variations of the way the hyphae of the fungus were distributed and "digested" in the tissues of orchids. There is a similar range of behaviour in this respect in the Ericales which also includes chlorophyllous and partially and wholly achlorophyllous species. A comparative ultrastructural investigation of all these variations would be of great value because it might allow us to decide whether the diverse "digestion stages" or the intact hyphae which are common to all infections are more important in the transfer of metabolites between the symbionts.

The death of the fungus which precedes the death of the host cells in orchid mycorrhizas and in most other mycorrhizas where the fungus inhabits the living host cells might be a form of control of fungal exploitation, particularly if transfer occurs across the intact interface. In the hair roots of Ericaceae, which are ephemeral short-lived structures, there would seem to be no need for such a control and the partners cease activity almost simultaneously. It remains for future research to determine the role of phytoalexins like orchinol in the control of spread of the fungus through orchid tissues.

# PART 2
# ESSAYS ON MYCORRHIZAL SUBJECTS

*Chapter 15*

# The Causal Anatomy of Ectomycorrhizas

The apical regions of ectomycorrhizas differ in structure from those of uninfected roots of equivalent position in the root system. The differences include the formation of a fungal sheath which encloses the mycorrhizal organ in a layer of compact tissue, and modifications to the tissues of the host. The root-cap is reduced in extent, the meristem is small, and mature stelar, cortical and epidermal cells occur close behind the meristematic initials. The epidermal or outer cortical cells are radially elongated, and some distance behind the meristem their walls are penetrated by the Hartig net hyphae which do not, however, actually enter the cells. Root hairs are usually absent, but if present are distorted beneath the sheath. Within the cortex there is often a cylinder of cells which possess thickened walls and past these the Hartig net does not penetrate (Clowes, 1951; Chilvers and Pryor, 1965). The diameter of the host tissues of the mycorrhizas is greater than that of the equivalent uninfected roots, not because of increase in the number of cell layers but mainly because of the increase in radial diameter of the exterior layers of cells.

In this essay we discuss the origin of these differences of morphology and histology. It is clearly unsatisfying simply to conclude that they are due to hormones produced by the mycorrhizal fungus or host tissue. Such assumptions produce no explanation and no ideas that would lead to experiment except to look for the active substances and then to test them. There may be, indeed there is likely to be, a hormonal component in the processes of differentiation of both kinds of root, mycorrhizal and uninfected, and also of the fungal structures, but it is important first to define the problems as clearly as can be.

## The fungal sheath

The fungal sheath bears a resemblance in its structure to that of fruit bodies and other bulky fungal organs. The resemblance to fruit bodies was observed, for instance, by Luppi and Gautero (1967) in the differentiation of lactiferous vessels in the sheath of the mycorrhiza of *Quercus* formed in culture with *Lactarius flexuosus*, *L. piperatus* and *L. vellerius*. Other mycorrhizas synthesized with species of *Russula* bore characteristic cystidial ornamentation. Fassi and de Vecchi (1967), Fontana and Centrella (1967) and Scannerini and Palenzona (1967) all noted similarities between the peridium of mycorrhizal Tuberales and the fungal sheaths produced by them. Such similarities have also received comment by Trappe (1971). Since mycelium isolated into culture from fragments of the sheath tissue or from spores grows to produce vegetative mycelial colonies of branching hyphae, it seems certain that some particular set of conditions is necessary for the formation of sheath tissue. As a first suggestion, factors which encourage fruit-body initiation and formation might provide clues. The available information which seems relevant is however rather scanty, both about the initiation of primordia and their further development. Bevan and Kemp (1958) noted that, whereas the mycelium of *Collybia* (*Flammulina*) *velutipes* fruits in culture after many days, small lengths of the stipe of fruit bodies placed in a medium rapidly regenerate whole fruit bodies with little mycelial growth. They believed that both nutritional and hormonal factors might be operative.

This kind of behaviour was noted also by later observers using *Boletus edulis*, *Pleurotus florida* and *Agaricus bisporus* in addition to *C.* (*F.*) *velutipes*. Eger (1965a,b,c) showed that vegetative mycelium of *P. florida* and *C. velutipes* produced fruit bodies if macerated fragments of fruit bodies were placed on it. Heat sterilized or frozen and thawed fragments were active and fragments of each would encourage fruiting in the others. *A. bisporus*, although it regenerated from fruit body pieces, did not react to macerated fragments of its own fruit bodies placed on its mycelium, nor to those of the other species. It might be stimulated to fruit by contaminating microorganisms. These results suggest that a sporophore-inducing principle, which is stable to moderate heat and freezing, is formed by or present in fruit bodies. Regeneration is possible usually from pieces of all parts of the fruit body, but especially from lengths of stipe, provided they are placed in good contact with a nutrient medium. It is as if the tissue, being set on the path to form fruit bodies, continues to produce them, given adequate nutriment. It can be diverted from this course if the tissue is gently teased out before use as an inoculum. According to Giltrap (1981) the ability to form incomplete fruiting bodies in culture by two isolates each of three species of *Boletus* and one of *Suillus*, all of which had been found to form mycorrhizas with *Betula*, decreased with the time that they were held in culture. It was most frequent just after isolation, but was absent after 3–4 months. These observations may also suggest that the cultures were "set in the path" of fruit body formation on isolation from the flesh of the mature fruiting bodies but that the influence faded with time.

Watkinson (1979) has discussed the formation and growth of hyphal strands, rhizomorphs, coremia and sclerotia in a very valuable review. Certain specific substances such as nitrate, amino acids, auxin, ethanol, copper ions, etc. have been found to stimulate the formation of these complex structures in various organisms. Amongst the common conditions required for the formation of these organs is a high or critical ratio between carbon and nitrogen supplies. The process of morphogenesis includes differentiation of hyphae into core, cortex and binding hyphae, change or ordering of their growth and their adhesion together. The adhesion is often associated with the production of an extra-hyphal fibrillar material analogous to that found in fungal sheaths of ectomycorrhiza. The formation and growth of mycelial strands such as those of *Serpula* has been investigated (see Garrett, 1970). They develop in a mycelium growing from a food base over a non-nutrient medium. If unstranded mycelium is made to grow over a non-nutrient gap from one nutrient medium to another, strands will develop from the mycelium in the gap. It should be noted that this behaviour is relevant to the formation of the sheath in ectomycorrhiza. A branching system of hyphae colonizes the surface of the root. This has sometimes been called a sheath or an incipient sheath. Nylund (1981) insisted that a true sheath, i.e. the thickening and development of the sheath tissue, only occurred after the Hartig net had been formed. He was of the opinion that a sheath-forming factor was passed from the Hartig net to the incipient sheath. It may now be seen that this factor might be, in whole or in part, nutrient carbon compounds on which the morphogenesis of the complicated sheath structure depends. This idea also has a relevance to two other factors. First, excess nitrogenous nutrient in the soil decreases infection and sheath formation, and excess carbon in media used for axenic mycorrhizal inoculation results in the abnormal development of thick sheaths. Secondly, a sheath may form on the long roots of *Fagus*, *Nothofagus* and other plants with no Hartig net formation. It is easier to relate this behaviour to a supply of nutrient than to a hormonal stimulus of the kind that might be produced by a Hartig net.

Rhizomorph and sheath formation seem to require an oxygen supply or an oxygen gradient. Read and Armstrong (1972) showed that *Boletus* (*Suillus*) *variegatus* formed an incipient sheath or thick weft of hyphae on the surfaces of conifer roots in agar medium at points where oxygen was released from them. This effect only occurred in complete seedlings the tops of which were in air. Using silicone tubes, steeped in "seeded" nutrient agar, Read and Armstrong showed that *B.* (*S.*) *variegatus* formed a sheath closely resembling a mycorrhizal sheath on the tube if it was open at one end to the air. No sheath formed if the end was closed, nor in non-nutrient agar.

The process of the formation of hyphal strands in *Serpula lachrymans* was described by Watkinson (1971a,b; 1975). She showed that the oldest leading hyphae exude nutrients as they grow through the nutrient-deficient zone and that in consequence the laterals run parallel and adhere to them. Garrett (1970) applies this explanation of the formation of hyphal strands to the "ectotrophic spread" of sheath-like sheets of mycelium of pathogens like *Fomes annosus* on the surfaces of the roots of their host. The hyphae spreading from a point of infection over a resistant area, i.e. one deficient in nutrient, aggregate in response to exudation by the leading hyphae. The hypothesis seems sound as far as it goes. However, the aggregation might be expected to include the hyphae of many other species present in the habitat, for nutrients are unspecific in their action. Moreover, the differentiation of the hyphae of the strand to form binding hyphae and central core hyphae of large diameter might perhaps seem to demand some hormonal influence also. There are reports that small amounts of ethanol, butanol and propanol added to synthetic media encourage the formation of hyphal strands (Weinhold, 1963; Sortkjaer and Allermann, 1972; Bille-Hansen, 1973) and their biochemical intervention in carbon pathways is therefore probably worth following up. Moreover there are stimuli to form aggregates between the cells of Myxomycetes, Acrasiales, sponges etc. formed at various growth phases. These appear also to involve recognition between the cells which can aggregate and the results of the work on them may well be relevant to the formation of mycelial strands and of mycorrhizal sheaths.

From this discussion we can learn that much remains to be done in an experimental way on the effects of nutrition and other factors on the form of ectomycorrhizal fungi. Many of them produce hyphal strands and sclerotial structures as well as fruit bodies and fungal sheath. There seems to be some evidence of a morphogenic hormone in the experiments on fruit-body regeneration, but many other morphological effects are affected by nutrition. It would seem important as a first approach to determine at what stages and for what processes are we driven to assume the existence of hormones whose actions are relevant to mycorrhizal sheath formation.

## The host tissues

### Mycorrhizal infection zone

Marks and Foster (1973) in their discussion of structure and morphogenesis of ectomycorrhiza gave a diagram of a root apex in which a zone behind the root-cap and apical meristem was designated as the mycorrhizal infection zone. It extended up to the region where the primary cortex becomes moribund. They state that this is the zone in which mycorrhizal infection takes place and that the time during which its cells remain in the receptive condition is influenced by factors affecting root growth and tissue morphogenesis. Direct experimentation by Atkinson (1975) has corroborated the assumption made by Marks and Foster, and also tacitly by many others, that there is a subapical zone of "susceptibility" to mycorrhizal fungi. Using *Betula* seedlings in aseptic culture he was readily able to infect the roots of all denominations in the subapical region with a strain of *Amanita muscaria*, but using taproots or first order laterals up to 30–35 days old, he was unable, in 42 attempts, to cause them to form a sheath or Hartig net in any intercalary position, although loose wefts of mycelium might colonize their surfaces. We may conclude with Marks and Foster that at any rate some of the tissues and cells of the subapical regions have properties that allow or encourage the formation of mycorrhizal associations with them. In contrast to Marks and Foster, Chilvers (1974) has expressed the view that the morphogenesis of mycorrhizas of *Eucalyptus* depends on events at the very apical region. He bases this on the behaviour of the epidermal cells which he believes retain the "embryonic trait" of very elongated ("columnar") diameter in a radial direction in mycorrhizas. He therefore views the ensheathing of the apex by the fungus as the important primary stage of development. Chilvers and Gust (1982a,b) also stress that the growth and development of the mycorrhizal structure proceeds in a distal direction; both the root of the host and the fungal sheath extend distally and Hartig net formation within the tissues follows. There is no development backward, i.e. proximally, from the region where the initial sheath and Hartig net establishment takes place.

An inescapable corollary to these views is that the same properties of susceptibility may be possessed by the whole of each short root of *Pinus*, for they usually become wholly mycorrhizal (although sometimes there is a proximal length of uninfected axis). It necessarily follows that all parts of such short roots possess, with the exception of the extreme meristem in the apical region, the property of receptiveness for Hartig net formation. The structure of the short roots of *Pinus* has been well described. The root-cap is much reduced or absent, the apical meristem very small indeed, differentiation of mature tissues is close to the apex and the stelar structure is simple, often monarch. Recent investigation by Fortin *et al.* (1980) and Piché *et al.* (1981) has made use of growth pouches so that the initial development and infection of short roots of *Pinus strobus* was kept under continual microscopical observation. Five days after inoculation, sheath and Hartig net began to be formed on the existing short roots. These became mature by the 9th day when signs of dichotomy were evident, and well visible by the 12th day. At this time there was no dichotomy of the uninfected short roots. This might lead to the view that although dichotomy is not solely dependent upon infection, it is indeed encouraged by it. Moreover, Piché *et al.* observed that the infected rootlets appeared to continue to grow more than the equivalent uninfected ones. It is of interest that although a network of hyphae developed upon and parallel to the long roots, strands were not formed till the 21st day. It is attractive to conclude that at this time sufficient carbohydrate was being derived

from the host plant to allow this development. Before infection the short roots
contained phenolic compounds especially in the cortex and cap cells and lacked
starch. Both these features are often described as characteristic of mycorrhizal
short roots. However, Ling Lee *et al.* (1977a) showed that in *Eucalyptus*, although
there were phenolic compounds in uninfected roots, they were present in the
epidermal cells of the mycorrhizal roots, not in those of uninfected ones.

Characteristics often attributed solely to fungal influences may often develop
in the absence of infection. In contrast to *Pinus*, the apical regions of roots of any
order may become fully mycorrhizal in *Fagus*, *Betula*, *Eucalyptus* and probably in
most species with racemosely branched mycorrhizal systems and little fundamental differentiation of root orders. As mentioned before, Warren Wilson (1951)
and Warren Wilson and Harley (in preparation) observed that roots of any order
on *Fagus* seedlings might pass through a series of changes at their apices in which
they came to resemble the short roots of *Pinus*, as described above (except in the
monarchy of the stele). These changes occurred in the absence of mycorrhizal
fungi and in the presence of high or low nutrient. In a similar way Faye *et al.*
(1981) described how "without any symbiotic or parasitic fungus" short roots of
*Pinus pinaster* produced dichotomous branches in all the environmental
conditions used. Their experiments were particularly enlightening because they
grew explants of *P. pinaster* from axillary buds *in vitro* and compared the
behaviour of the roots of clones thus obtained. The root systems of different
clones each showed a similar tendency, strong or weak, to dichotomize,
indicating that this trait had a genetic basis and did not fundamentally depend
on the activities of a mycorrhizal fungus.

We conclude then that many of the characteristics of the host tissues of
mycorrhizal roots could arise from internal causes and need not be the result of
fungal influence, although it might be so. A closer definition of the changes
brought about by infection is clearly needed, in which the growth pouch
technique of Fortin *et al.* (1980) could be of great value.


## Hormonal theory of the changed anatomy of host tissues

In a series of papers between 1948 and 1973, Slankis examined the effects of the
products of mycorrhizal fungi on host histology and morphogenesis. In a first
group of papers (1948, 1949, 1951) he investigated the effects of mycorrhizal
fungi, of their metabolic products and of auxins on the growth of *Pinus* roots in
culture. Mycelia of *Boletus* (*Suillus*) *luteus* and *B.* (*S*) *variegatus* introduced into the
root cultures caused the short roots to swell and to produce dichotomous
branches reminiscent of mycorrhizas. Filtrates of aseptic cultures of the fungi
had similar properties, and on the assumption that they contained auxins their
effects on the roots were compared with those of synthetic $\beta$-indole acetic (IAA),
$\beta$-indole propionic (IPA) and $\beta$-indole butyric (IBA), and $\alpha$-naphthalene

acetic acids (NAA), at concentrations of 0·05–500 $\mu$g l$^{-1}$. These compounds had rather similar effects to the culture filtrates. IAA up to 50 $\mu$g l$^{-1}$ stimulated the formation and growth of long roots in culture; at higher concentrations it inhibited long roots but increased the development and dichotomy of short roots and decreased the number of root hairs upon them. Of these synthetic auxins, α-naphthalene acetic acid (NAA) was the most active followed by IAA, IPA and IBA. Amongst those who followed up Slankis' work, Ulrich (1960a,b) was unable to repeat his results with *Pinus lambertiana* using IAA, although indole acetic nitrile stimulated dichotomy of short roots. She noted, however, that *P. lambertiana* roots contained an oxidase of IAA which of course is likely to have affected the results.

It should be noted that in this work the actions of auxins in morphogenesis were judged by the swelling and dichotomy of the short roots, not by observation of anatomical structure in sections, in spite of the fact that even in 1925 Melin had noted that some uninfected short roots of pine might dichotomize. This has been repeatedly noticed since. Besides Faye *et al.* mentioned above, F. A. L. Clowes and L. Leyton observed them to dichotomize in aseptic culture (see Harley, 1959). Barnes and Naylor (1959a,b) showed that cultured roots of *P. serotina* produced dichotomous roots in the presence of the vitamins thiamin, pyridoxin, nicotinic acid and folic acid at concentrations greater than 10 mg l$^{-1}$; of the growth factors IAA, NAA, and kinetin at 1 mg l$^{-1}$; of the amino acids glycine, serine, ornithine, citrulline and arginine, at concentrations greater than 1 mg l$^{-1}$. R. W. Goss (1960) observed that the roots of *P. ponderosa* may dichotomize even when uninfected in natural conditions. Although these observations call into question the certainty of the assumption that fungal hormones are involved in histological changes associated with mycorrhiza formation, Slankis (1958, 1963) held the view that they were, and that probably IAA was the most important. He showed that IAA could be absorbed and translocated in pine root-systems and it appeared to exert its effect in regions which had not been directly supplied. He concluded (a) that it caused morphological deviations essential to mycorrhiza formation; (b) it influenced the whole root-system; (c) it determined the pattern of the root-system by affecting the frequency and sequence of long and short roots; (d) it not only increased dichotomy but caused histological modifications of differentiation near the promeristem and expansion of the cortical cells.

In subsequent work (1967, 1971 and 1973), Slankis elaborated the hypothesis that the structure of the host tissues can be attributed to the effects of fungal auxins. He states (1973): "The specific structure of ectomycorrhizae reflects a specific physiological and metabolic state that is induced and maintained by symbiotic fungus auxins and that this state is a prerequisite for the establishment and functioning of the symbiotic relationship." In reaching this conclusion he quotes results which show that the mycorrhiza-like structure of roots treated with auxins is only maintained if auxins are permanently present. In their absence renewed growth occurs. He draws the parallel that increased nitrogen supply causes renewed growth of the host tissue of mycorrhizal rootlets and supposes that this arises because high nitrogen supply diminishes auxin formation by the fungi. As mentioned, Moser (1959) indeed observed that extra nitrogen supply reduced auxin formation by the fungi in culture, but this was in media containing tryptophane which was present as sole source of nitrogen as well as being a precursor of auxin. Hence the addition of other nitrogen compounds would simply provide an alternative source of nitrogen for growth and diminish tryptophane breakdown. In his experiments (1971) Slankis showed that aseptic *P. strobus* seedlings in both low or high nitrogen supply (5 mg or 100 mg N l$^{-1}$ as $NH_4NO_3$) produced swollen dichotomies in the presence of 2·5–5·0 mg l$^{-1}$ NAA and hence concluded that the effect of nitrogen supply on mycorrhiza formation and maintenance is through effects on synthesis of auxins by the fungi, not directly on the host.

On the matter of the effect of light on mycorrhiza formation, Slankis concludes that this is not through its effects on carbohydrate synthesis. In an experiment (1963) he showed that IAA caused no swellings or dichotomies at 500 foot candles (5380 lx), but at 2500 foot candles (26 900 lx) the expected morphological effects occurred; but they did not occur when glucose, fructose or sucrose were supplied in the medium at the lower light intensity. He therefore suggests that some light-dependent hormone is produced by the host which is essential to the process.

As can be seen, Slankis has built up a comprehensive set of hypotheses based upon the known effects of external factors on mycorrhiza formation, upon the demonstration that many mycorrhizal fungi have been shown to form auxins in media especially when suitable precursors, such as tryptophane, are present, and upon his own experiments. The plexus of hypotheses lacks full credibility especially because hormone physiology does not provide adequate explanations. It is impossible to accept the suggestions with any satisfaction because there is as yet no real biochemical explanation of the action of any of the hormones, no real conception of what biochemical events occur between hormone application and effects observed in experimental work.

The question has been raised as to whether the criteria of external changes of form used by Slankis are acceptable indications of similarity in histology in the treated rootlets and mycorrhizas. He published in 1960 and 1967 photographs showing histological modification caused by 2·5 mg l$^{-1}$ IAA but no quantitative data were given of comparative cell sizes in auxin-treated and untreated material. This matter was however investigated, using cultured roots of *Eucalyptus*, by Chilvers and Pryor (1965). They treated roots of *E. grandis* with various concentrations of NAA. Concentrations between 0·19 and 1·9 mg l $(10^{-6}$–$10^{-5}$ M) decreased linear growth and caused proliferation of side branches and swellings on the main apices. Examination of the sizes of the epidermal cells of affected roots showed that there had been changes (Table 72). The axial diameters had been reduced and the radial and tangential diameters had been increased. Chilvers and Pryor point out that the treated roots had a decreased rate of linear growth and a 70–100% increase in the diameter of the growing portions. It is instructive to compare these figures with those for untreated roots of which they examined the distribution of diameter classes in mycorrhizal as compared with uninfected roots. The infected root has a larger diameter than the uninfected, an overall increase of perhaps 50%, but allowing for the presence of the fungal sheath about 25% increase in diameter of the host tissue. The comparison of epidermal cell sizes in mycorrhizal and uninfected roots is given in Table 71.

These changes are in the same direction in the two cases shown in Tables 71 and 72. A difference lies in the fact that in the NAA treated roots, although the epidermal cells were radially expanded, their radial axes were at right angles to the root axis. Chilvers and Pryor make rather much of this, but it would be easily

TABLE 71

Comparison of uninfected and mycorrhizal cells on different parts of a single axis of *Eucalyptus grandis* (from Chilvers and Pryor, 1965)

| Origin of epidermal cells | Mean radial diameter $\mu$m | Mean axial diameter $\mu$m | Mean tangential diameter $\mu$m | Volume $\mu$m$^3$ |
|---|---|---|---|---|
| Uninfected | 8·5 | 41·8 | 15·4 | 5472 |
| Mycorrhiza | 23·6 | 18·6 | 15·3 | 6701 |
| Difference | $+15·1^a$ | $-23·2^a$ | $-0·1$ (NS) | $\times 1·2$ |

[a] Significant at 0·1% level.

TABLE 72

Comparison of unaffected and naphthalene acetic ($10^{-5}$–$10^{-6}$ M) acid affected epidermal cells on different parts of a single axis of *Eucalyptus grandis* (from Chilvers and Pryor, 1965)

| Origin of epidermal cells | Mean radial diameter $\mu$m | Mean axial diameter $\mu$m | Mean tangential diameter $\mu$m | Volume $\mu$m$^3$ |
|---|---|---|---|---|
| Unaffected | 8·5 | 36·2 | 12·5 | 3846 |
| NAA treated | 11·6 | 23·6 | 18·3 | 5009 |
| Difference | $+3·1$ (NS) | $-12·6^a$ | $+5·8$ (NS) | $\times 1·3$ |

[a] Significant at 0·1% level.

explained by the fact that in mycorrhizas the outer face of the epidermal cells is carried forward by the sheath, as shown by Clowes (1951).

Clowes (1949) grew beech seedlings in the presence of IAA and showed that the lengths of the roots were increased in $0.01$ mg l$^{-1}$ but decreased in $0.1$ mg l$^{-1}$. At concentrations between these two the number of lateral roots increased. Application of IAA in lanoline (at the rate of 1/400 parts) caused local swellings due to increased radial and tangential size of the cortical cells, proliferation of the pericycle and initiation of many lateral primordia. Colchicine at 100 mg l$^{-1}$ caused vacuolation and maturation of the cells close to the promeristem and transverse enlargement but reduction of elongation of the cortical cells. These effects could, Clowes thought, be explained on the basis of a check on the rate of division of cells in the promeristem so that the cells immediately proximal to it had an enhanced supply of nutrients or hormones. He further suggested that the fungus either produced some factor or removed some factor, so that cell division was reduced in frequency. He suggested that in the latter case thiamine, which is required by all ectomycorrhizal fungi in culture, might possibly be the substance removed.

Harley and Lewis (1969) pointed out that the growth processes of higher plants involve the interaction of several classes of hormones, gibberellins, kinins and auxins. All these are produced by some, but by no means all, ectomycorrhizal fungi in some conditions. Ethylene is also produced by some. Little has so far been written about the possible effects of the compounds other than auxins on mycorrhizal form and structure. Gibberellic acid has been found to inhibit some mycorrhizal basidiomycetes (Levisohn, 1960; Santoro and Cassida, 1962; Gogala, 1967, 1970), as indeed has IAA (Fortin, 1970). Gogala found gibberellic acid in the root exudate of *Pinus sylvestris* but it is not certain that it was at a sufficient concentration to inhibit mycorrhizal fungi. In the same extracts a kinin was present in a sufficient quantity to stimulate *Boletus edulis* var. *pinicolus* and Gogala wondered whether it could be a constituent of the M factor of Melin. These results contrast with those of Fortin with IAA for he showed that *Pinus sylvestris* and *P. resinosa* roots absorbed IAA, so allowing *Suillus* (*Boletus*) *variegatus* to grow in culture to which IAA had been added. He believed that the roots encouraged the growth of the fungus by removing the auto-inhibition caused by the auxins and at the same time the root might be histologically modified by absorbing them. The importance of such an activity would of course depend upon the conditions, especially the presence of suitable precursors for IAA production and upon the actual strain of mycorrhizal fungus.

Since the ability to form auxins and growth factors in culture is not possessed by all ectomycorrhizal fungi and it is very variable in those that possess it, nothing is to be gained at this stage by further speculation in the absence of knowledge of the biochemical effects of these compounds. It is however profitable to draw parallels between the known effects of external factors on mycorrhiza formation and the work on the effects of external factors on the hormone physiology of roots in general.

## Hormones and the growth of roots

In the work of Slankis there is one single unifying theme which should be kept in mind and shorn of all complicating discussion. This is the contention that mycorrhizal fungi produce auxin (IAA is preferred) and that this auxin affects the growth of susceptible roots by modifying apices so that they assume the structure seen in the host tissues of mycorrhizas. All effects of other external factors on the roots, e.g. nitrogen supply or light intensity, operate through stimulation or otherwise of fungal growth and auxin production. It is of interest therefore to relate this theme to the results of experiments in which auxins were applied to root systems other than those of ectomycorrhizal hosts, using pure auxins in a variety of external conditions.

It is usually assumed that auxins at concentrations that stimulate the growth of shoots inhibit that of roots, but that concentrations several orders of magnitude lower stimulate root growth and are essential for it. Street (1968) classified variation in the need of an auxin for the growth of excised roots in culture; some roots soon cease to grow in the absence of auxins or their precursors; others grow in the absence and are only increased in growth rate by a supply of auxins; and yet others will only grow in the presence of external auxins in special conditions. Hence there seems to be a variable dependence on or even need of an external auxin, or conversely of an ability of roots to produce sufficient auxin themselves in culture.

In the third group one of the conditions which particularly affects growth is the external concentration of sucrose. Sucrose at 2% or 3% (w/v) allows rapid growth of tomato roots at first, but later growth diminishes and ceases and their apices may be said to have "aged" (see Fig. 28). Whereas 1% sucrose allows continued growth without ageing of the apices, the addition of auxins, e.g. NAA at concentrations between 0·1 and 0·5 $\mu$g l$^{-1}$ to roots in 1% sucrose, induces ageing and decreases the mean rate of growth (Fig. 29). It might be argued of mycorrhiza formation in susceptible roots that since intense infection is correlated with high internal sucrose content (Marx et al., 1977), the roots are made sensitive to the action of auxin derived from the fungi by the presence of sucrose in the tissue. It should be noted in passing that Slankis used 7% sucrose in his root cultures and they may well have been sensitized to auxin.

Street and his colleagues (see Street, 1967) showed that excised roots of different species reacted by ageing with different concentrations of sucrose: Senecio vulgaris 3%, tomato 2%, Lycopersicon pimpinellifolium 0·5%. The ageing process was coupled with an increased diameter near the apex and a decreased linear growth, and they ascribed the effects to an internal factor, an auxin, synthesized in older tissues behind the apex in high sucrose supply. They concluded that IAA was not the auxin involved because, although by itself in high concentrations it produced the ageing symptoms described, it did not, like NAA, act in concert with sucrose concentration. NAA had also a cumulative effect. It was without effect at $10^{-10}$ M ($0 \cdot 019$ mg $1^{-1}$) after the first sub-culture with $2\%$ sucrose but this concentration became increasingly active in producing ageing in subsequent subcultures.

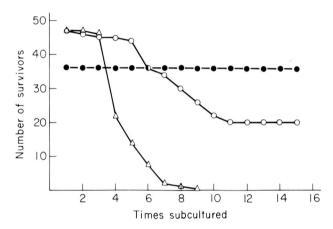

FIG. 28. Ageing of excised roots of tomato in culture solutions containing different concentrations of sucrose. Number of root tips continuing growth after repeated subculture, i.e. numbers surviving plotted against number of subculturings (which occurred each 7 days). ●, $1\%$; ○, $2\%$; △, $3\%$ sucrose. (Redrawn after Street, 1967.)

Again the anti-auxin 1-naphthoxyacetic acid (1-NOA) counteracted the effects of NAA in the presence of $3\%$ sucrose on tomato roots (see Fig. 30), so that $0 \cdot 2$ mg $1^{-1}$ almost offset the ageing effect of the sucrose and maintained the rate of axial extension.

### Internal factors and apical ageing

As has been described, work with *Fagus* (Warren Wilson, 1951; Warren Wilson and Harley, in preparation) and with *Pinus* (Wilcox, 1968a,b) has shown that all roots on a seedling go through a phase of rapid growth in length followed by a gradual cessation of growth. The work with *Fagus* demonstrates that as the growth rate of the root apices decreases they become differentiated into mature tissues and form root hairs closer to the apex. The meristem is greatly reduced in size (see Table 27, Fig. 16, Plate 10) and they come to resemble the apical tissues

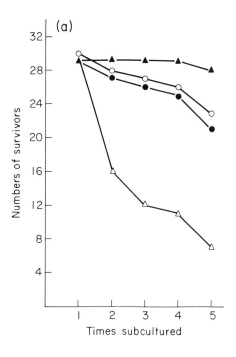

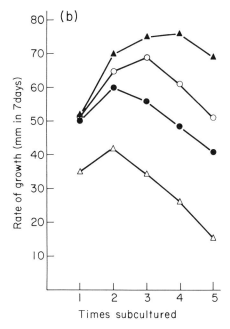

FIG. 29. Ageing of excised roots of tomato in culture solutions containing 2% sucrose with added NAA. (a) Number of root tips continuing growth after repeated subculture each 7 days. (b) Mean rate of growth of surviving apices (mm in 7 days). ▲, control; ○, 0·1 μg l⁻¹ NAA; ●, 0·2 μg l⁻¹ NAA; △, 0·5 μg l⁻¹ NAA. (Redrawn after Street, 1967.)

of ectomycorrhizas. These changes are not dependent on the presence of a mycorrhizal fungus because they occur equally in sterile or non-sterile surroundings. Nevertheless in forest soil the colonization of the root apices takes place in the later stages of the ageing of the apex, which therefore appears to

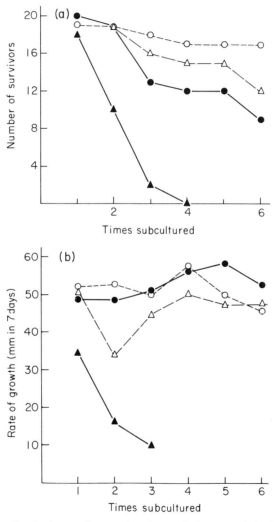

FIG. 30. Ageing of excised roots of tomato in culture solutions containing 5% sucrose with added l-NAO (l-naphthoxy acetic acid). (a) Number of root tips continuing growth after repeated subculture each 7 days. (b) Mean rate of growth of surviving apices (mm in 7 days). ▲, control; ●, 0·05 mg l⁻¹; △, 0·1 mg l⁻¹; ○, 0·2 mg l⁻¹ l-NOA. (Redrawn after Street, 1967.)

precede rather than follow fungal colonization in unsterile soil. One is therefore driven to conclude that the process of differentiation of the host tissue to the state in which it exists in the mycorrhizal root is at least in part caused by internal factors such as Street invokes in his work on the ageing of excised root apices. This conclusion agrees fully with the results of Piché *et al.* (1981) and of Faye *et al.* (1981) with pine and suggests that the fungus rather than causing or stimulating the changed histology and morphology of the host tissue is primarily active in promoting the continued growth, branching and physiological activity of the infected aged apex. The apex in *Pinus* would abort, or in other species would either stagnate or abort were it to remain uninfected.

## The fungal sheath and compact soil as morphogenic agents

Clowes (1951) commented on the radially elongated epidermis of mycorrhizal roots and suggested that the cells retained approximately the same orientation as they had at the very apex because of their attachment to the sheath, which was pulled forward by the elongation of the stelar tissue. The sheath can indeed withstand considerable tensions as shown by some unpublished experiments of Harley. Pieces of sheath dissected off leading roots of *Fagus* were clamped between perforated sheets of Perspex so that an increased atmospheric pressure could be applied to them on one side. They were shown to stretch but not to break when the pressure difference between the sides was in excess of 100 kPa (1 bar). Hence pressures of that order could be generated by extension of the host tissue within the sheath without it splitting. It is possible therefore that the rounded apex, the crushed cap-cells and epidermis and approximately radially oriented cells near the apex could depend for their state or shape upon the tension of the sheath. It is interesting that Chilvers and Pryor (1965) stated of uninfected roots of *Eucalyptus*: ". . . roots growing slowly through unfavourable media exhibited the spatially precocious differentiation of tissues associated with mycorrhizal development."

Atkinson and M. J. Goss (in Atkinson, 1975) noted that roots of *Betula* growing against a small mechanical impedance in the soil develop swellings similar to those that Goss (1977) observed in Barley. They bear some resemblance to the effects of mycorrhizal infection on the form of roots. Samples of sterile seedlings of *Betula verrucosa* were therefore grown in media of glass ballotini beads which were contained in plastic pots. The lids of the pots were perforated by a hole through which the stem of a seedling passed together with a tube through which the ballotini could be constantly or periodically injected with nutrient solution. Low levels of mechanical impedance could be developed in the ballotini by screwing down the lids more tightly. A pressure of approximately 30 kPa was used with pore diameters between the beads of 69 $\mu$m, i.e. less than the diameters of *Betula* tap roots or first order laterals. This low level of impedance resulted in the production of densely branched short lateral roots borne on a very short tap root, as compared with long tap roots bearing many long first-order laterals with second-order branches in the absence of impedance. Measurements are given in Tables 73 and 74 of a number of characteristics. Laterals developed in impeded conditions possessed many more root hairs per unit length of root reaching closer to the apex and steles which also matured closer to the apex than those of unimpeded roots. The increased diameter of these laterals was not due to increased numbers of cortical cells but to differences in dimensions of the cells. Cell lengths were decreased and cell diameters increased in impeded conditions, although in some cases only were the differences highly significant.

The impedance of the soil to the penetration of the root therefore provides another method by which some of the characteristics of the host tissue in mycorrhiza might arise. Moreover the pressure between the sheath and lateral branches of a mycorrhizal rootlet as they pass through the cortex may have an immediate morphogenic effect also since the sheath can withstand pressures which might be exerted by the host up to more than 100 kPa.

TABLE 73

The effect of low soil impedance of 30 kPa (approximately) on the growth of roots of *Betula* (Atkinson and M. J. Goss, see Atkinson, 1975)

| | Mean fresh weight of shoot g | Mean distance from tap-root apex of first laterals $\mu$m | Mean diameter of laterals ($\mu$m) | |
| --- | --- | --- | --- | --- |
| | | | 1st-order | 2nd-order |
| Unimpeded | $0\cdot22 \pm 0\cdot10$ | $7\cdot4 \pm 3\cdot1$ | $409\cdot3 \pm 25\cdot2$ | $226\cdot9 \pm 23\cdot5$ |
| Impeded | $0\cdot23 \pm 0\cdot14$ | $3\cdot0 \pm 0\cdot7$ | $571\cdot9 \pm 25\cdot2$ | $330\cdot3 \pm 23\cdot5$ |

TABLE 74

Dimensions of epidermal and cortical cells ($\mu$m) of impeded and unimpeded roots of *Betula* (Atkinson and M. J. Goss, see Atkinson, 1975)

| | | Tap root | | Lateral root | |
| --- | --- | --- | --- | --- | --- |
| | | Impeded | Unimpeded | Impeded | Unimpeded |
| Epidermal cells | Cell length | $57\cdot6^a$ | $83\cdot2$ | $44\cdot1$ | $52\cdot7$ |
| | Cell diameter | $25\cdot9$ | $19\cdot6$ | $20\cdot5$ | $17\cdot0$ |
| Cortical cells | Cell length | $66\cdot2^a$ | $89\cdot9$ | $67\cdot2$ | $69\cdot0$ |
| | Cell diameter | $51\cdot4$ | $45\cdot7$ | $41\cdot7^a$ | $28\cdot2^a$ |

[a] Significant at 1% level.

It is clear from the foregoing that a number of different factors are capable of changing the structure of the apex of the root from that typical of a growing root to a state resembling that within the fungal sheath of ectomycorrhiza. The important problem is to decide whether these changes are dependent solely on the influence of the fungus, i.e. on fungal auxins, as suggested by Slankis, or whether some of them are initiated by some of the other means, including internal factors, which have been observed to modify the morphology and histology of apices of roots in the same kinds of way.

There seems to be no doubt whatever, that the short roots of *Pinus* develop their structure as they are formed. This was described in the early work (e.g. Aldrich Blake, 1930; Hatch and Doak, 1933; Hatch, 1937). Moreover they are also described as aborting if not infected. Recent work (Faye *et al.*, 1981; Piché *et al.*, 1981) has also confirmed that short roots develop some of the characters of the host tissues of mycorrhizas in sterile conditions. Similarly the work of Wilcox and of Warren Wilson has amply shown that changes in the apex of roots precede mycorrhizal infection. An apex when susceptible to mycorrhizal infection is different from one which grows actively.

The concept of Foster and Marks of an infection zone must therefore be modified. The infection zone is not a zone behind the apex of a rapidly growing root, but a zone behind the apex of a root exhibiting slow divisions and becoming differentiated close to the apex. This condition which we have called the "ageing apex" may arise from any of a number of causes, internal, fungal or from soil impedance. It is in this "aged" condition that hyphae spread upon the surface of the root and penetrate between the cells to form the Hartig net. They penetrate in a region behind the dividing cells, in a zone where the cells are reaching or have reached their mature size. It is in this zone that the cell walls are being formed on the skeleton of the middle lamella, and the carbohydrate destined to be polymerized to form the secondary layers of the walls is secreted through the cell plasma membranes. It might be held that this is a condition of the cell-walls where penetration was easiest from a mechanical point of view. However not all fungi do penetrate the root even at this stage; it is only the mycorrhizal fungi which succeed, so that they must have some peculiar attributes. It is a reasonable hypothesis that the mycorrhizal fungi interfere with the polymerization of the walls of the host. The mechanism by which this may occur is that proteins in the fungal walls inhibit the wall polymerizing enzymes of the host either by complexing with them or by some other means so that a "modified contact layer" is formed between the two organisms and carbohydrate is released to the fungus.

## Conclusion

It is easy to build complex hypotheses concerning the intervention of fungus derived hormones to explain the morphogenesis of mycorrhiza. Such hypotheses, by appealing to factors whose physiological and biochemical functions are vague or unknown, simply obscure the problems and prevent rational experimentation. What are required first are careful descriptions of the process of infection. The behaviour of the fungus—germination of spores, growth of hyphae and its directional control if any, penetration of the tissues and sheath formation—all these processes are altered in conditions where external sugar is present in the medium so that experiments and observations on them will only give useful information relevant to natural conditions if the source of carbohydrate is entirely or mainly from the host. Descriptions of the Hartig net and its formation are greatly needed, especially in cultural conditions close to the natural. The method using growth pouches to observe the growth of hosts and fungi provides great possibilities for experiment. On the cultural side, quantitative information on the formation of auxins or other growth factors in ecologically probable conditions (i.e. in terms of carbohydrate, nitrogenous and other nutrient concentrations) by the fungi is essential. It is perfectly obvious that the infection of a root by spores germinating in the rhizosphere cannot involve the production of the kinds of concentration of auxins used in previous experiments. Auxin production depends during initial infection (as opposed to spreading of hyphae from other roots) upon the mass of the fungus and the availability of precursors. These must be estimated before full credence for auxin effects can be gained. It is most important that hypotheses about the initiation and development of mycorrhizal roots (i.e. their causal anatomy) should take into account a wide variety of factors and conditions both endogenous and exogenous. These factors may include hormones, but it is essential that any theory of hormone action should be viewed critically and in the proper context of development of fungus and root in ecological conditions.

## Chapter 16

# Translocation in Mycorrhizal Fungi

## Introduction

Of all the work on translocation in the fungi, and there is now a large number of published papers, the experiments on mycorrhizal systems have provided the most precise data on amounts of material translocated and on distances over which translocation is effective. They have also provided evidence about the mechanisms which may be important in translocation in fungal hyphae and fruit bodies. For this reason we intend to write about translocation in mycorrhizal fungi within a wider context than its role in higher plant nutrition. It is very important that mycorrhizasts should consider their experiments in relation to more generally discussed phenomena in plant and fungal physiology and, indeed, recognize the contributions they have the opportunity of making in this sphere.

### *Definition of translocation*

The term "translocation" has frequently been used without precision and before going further we must define what we mean and indicate how others have used this and related terms in the past. Hill (1965) uses the terms *translocation, transport, intracellular transport* and *streaming* more or less interchangeably in a discussion of the movements of nutrients, nuclei and cytoplasm along fungal hyphae. It is clear that cytoplasmic streaming and the movement of nuclei are cytological events which can be observed directly or by genetic changes in the mycelium (see Buller, 1933, for an early example). These phenomena *may* be related to nutrient transport processes, but the inclusion of *cytoplasmic streaming* within the term *translocation* (see Hill, 1965; Wilcoxson and Sudia, 1968; Jennings *et al.*, 1974) may lead to confusion between the nutrient transport process itself and one of the mechanisms by which it has been presumed to occur. If translocation of protoplasm is meant this should be made very clear, though the term has now been largely superseded by cytoplasmic streaming.

The distinction between nutrient *transport* and *translocation* is less clear. In this essay we use transport to describe the overall process of the movement of nutrients from one place to another. This may involve one or all of the following processes: uptake into cells from an external medium, long distance movement along (in this case) fungal hyphae, and transfer from one organism to another, which itself may comprise several constituent processes. Translocation is used quite specifically to describe the long distance intracellular transference of nutrients (usually of ions or small molecules) along fungal hyphae. A similar distinction between translocation and transfer in the fungi has been adopted by Cooper and Tinker (1978). A distinction between translocation, as defined here, and the movement of materials as a consequence of extension of mycelium during hyphal growth has also been made (Lucas, 1960, 1970). Translocation in the fungi defined in the above way can be directly compared with translocation in higher plants (i.e. translocation in the phloem).

## The pathway of translocation in fungi

Intracellular translocation in fungal hyphae has much in common with symplastic transport between cells of higher plants and algae, via plasmodesmata (Buller, 1933; Gunning and Robards, 1976; Spanswick, 1976). Translocation in vegetative fungal hyphae must be symplastic because apoplastic transport in non-cuticularized, unicellular filaments surrounded by a solution or nutrient medium is not possible; compounds would, as Marchant (1976) has stressed, simply leak to the medium. Dyes such as fluorescein have been used successfully to demonstrate symplastic connections in fungal hyphae (Roberts, 1950) as well as in plants (see Spanswick, 1976). In vegetative fungal hyphae it is possible to separate the sites of dye application from that of transport, so that any loss into the medium, following apoplastic transport, can be detected. Schütte (1956) showed that fluorescein appeared in the medium only at hyphal tips some distance from the site of application. Cytoplasmic continuity between hyphal segments has also been demonstrated repeatedly (see Buller, 1933; see Burnett 1976) and the movement of cytoplasmic contents between cells observed.

Mycorrhizal fungi come from many different taxonomic groups so that it is relevant to consider the differences between septal pores in Basidiomycetes, Ascomycetes and Phycomycetes, as these differ considerably (see Bracker, 1967; Burnett, 1976). In most Phycomycetes, including vesicular-arbuscular mycorrhizal fungi, the healthy vegetative hyphae are aseptate and cytoplasm is continuous throughout the thallus. Old parts of the mycelium may be separated by complete septa which are assumed to prevent not only cytoplasmic but also nutrient movements. However, apoplastic nutrient movement across septa might not be prevented.

In Ascomycetes (e.g. most ericoid and some ectomycorrhizal fungi), the frequent septa are perforated by simple pores with or without Woronin bodies, which allow the passage of cytoplasm and organelles including nuclei (see Burnett, 1976; Marchant, 1976). Simple pores in the septa of the hyphae of the Ascomycetes in ericoid and ectomycorrhizas are clearly visible between the fungal cells within the tissues or cells. The dolipore septum characteristic of most Basidiomycetes (and therefore most ectomycorrhizal and orchid mycorrhizal fungi) is much more complex in structure. A central pore is surrounded by an annular septum and capped by dome-shaped membranous vesicles which are connected to endoplasmic reticulum. The discontinuity of these vesicles means that cytoplasm is continuous between the fungal segments and that organelles may pass from one compartment to another. The dimensions of the dolipores given by Bracker and Butler (1963), Burnett (1976) and Marchant (1976) of 90–200 nm in diameter and 500 nm in length are similar to those of mycorrhizal fungi. For instance the dolipores of an intracellular Basidiomycete observed by Bonfante-Fasolo (1980) in *Calluna vulgaris* were 150 nm in diameter and 500 nm long. It is important to note that cytoplasm is continuous in all types of fungal hyphae, and that both types of perforated septum permit cytoplasmic streaming. The types of cytoplasmic streaming found in fungi are illustrated in Fig. 31.

## The occurrence and direction of translocation in mycorrhizal fungi

Here we need not be concerned with supposed evidence for fungal translocation obtained by growing fungi from nutrient onto non-nutrient medium. Such experiments are often very difficult to interpret and fortunately mycorrhizal fungi have rarely been used for them. Other fungi were used by Schütte (1956), Lucas (1960, 1970), Thrower and Thrower (1961, 1968) and Howard (1978).

Radioactive tracers can give unequivocal evidence for intracellular translocation in the fungi and have been used in most mycorrhizal studies. The first experiments were those of Melin and Nilsson in 1950 in which uptake of radioactive phosphate and its movement through the mycelium of an ectomycorrhizal fungus to the roots of a seedling of *Pinus* and through the *Pinus* tissues to the needles were demonstrated. This method was subsequently used to demonstrate movement involving translocation in hyphae of nitrogenous substances and cations (Melin and Nilsson, 1952, 1953a,b, 1955, 1958; Melin *et al.*, 1958). The method did not, however, lend itself to quantitative measurement of amounts or rates. The later experimental design has usually been similar to, or a modification of, the "split-plate" system of Lucas (1960) (see Fig. 32). A strip of agar is removed across the diameter of a petri dish to provide a barrier to limit or prevent diffusion between the two sections of agar. A fungus may then be inoculated on one side of the plate so that its hyphae grow across the diffusion barrier and colonize the agar on the opposite side. Tracer solutions can be applied to one (the donor) side, so that sampling of the agar on the opposite (receiver) side will give a measure of translocation which may be arranged to be either towards or away from growing hyphal tips. If mycorrhizal host plants are also included in the experimental design, then nutrient transfer in both directions between them and the associated hyphae can also be studied.

The details of the modifications to the "split-plate" method need not be described in detail. They have been made to overcome various problems and sources of error. Clearly with vesicular-arbuscular mycorrhizas host plants form an essential part of the experimental design because the fungi cannot be cultured without them. In some systems slight apoplastic movement of tracer has been detected in hyphae crossing the diffusion barrier. Lanoline has been used to minimize this (Pearson and Read, 1973; Stribley and Read, 1975; Howard, 1978). In work with $^{14}$C-labelled compounds care must be taken to exclude or

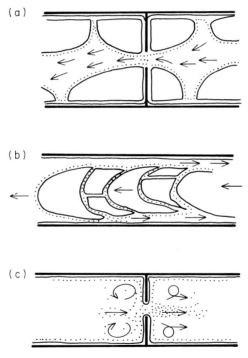

Fig. 31. Cytoplasmic streaming in fungi. (a) Unidirectional streaming in *Pyronema confluens* (redrawn from Buller, 1933). As in many Ascomycetes and Basidiomycetes streaming may be maintained in one direction for long periods of time and the vacuoles rarely move in the cytoplasm. (b) Bidirectional streaming in *Rhizopus nigricans* (redrawn diagrammatically from Arthur, 1897). In many Phycomycetes frequent reversals of flow of both cytoplasmic streams occur. Large vacuoles move in the main stream of cytoplasm but are absent from the peripheral streams. (c) Streaming accompanied by eddy formation (partial cyclosis) in *Humaria humicola* (after Jahn, 1934).

correct for the fixation of $^{14}CO_2$ generated by respiration of $^{14}$C-carbon sources (see Thrower and Thrower, 1968; Reid and Woods, 1969; S. E. Smith, 1967; Milne and Cooke, 1969; Howard, 1978).

Typical results for uptake and translocation of $^{32}$P-phosphate by a pure mycelial culture of *Rhizoctonia repens* are shown in Fig. 33. It is clear that $^{32}$P was translocated through previously established mycelium and that the amount crossing the diffusion barrier increased with increasing time. The results with this mycorrhizal fungus (isolated from orchid roots) do not differ in major respects from results obtained with *R. solani* and with some non-mycorrhizal fungi.

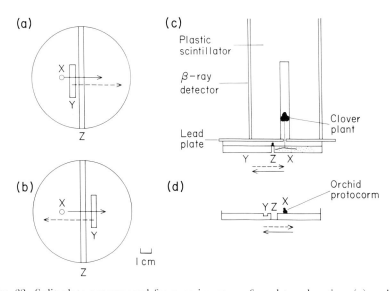

FIG. 32. Split plate systems used for experiments on fungal translocation. (a) and (b) Basic split-plates with a strip of agar removed across one diameter to provide a barrier to diffusion (Z) from one half of the plate to the other. The fungus was inoculated at X and grew in the direction of the solid arrow. Tracer was applied at Y and translocation in the direction of the broken arrow was measured by sampling and counting agar discs or squares. (a) Forward or acropetal translocation. (b) Backward or basipetal translocation. (c) Modification of the split-plate system, used by Pearson and Tinker (1975), Cooper and Tinker (1978, 1981). The diffusion barrier is a glass wall anointed with lanoline. Clover or onion plants were grown in soil or soil agar in one half of the plate and inoculated at X. Tracer was applied in sand or in water agar at Y. Fungal growth was in the direction of the solid arrow. Translocation in the direction of the broken arrow was measured by sampling whole plants or from detection of isotope ($^{32}$P) in shoots using a specially designed $\beta$-ray counter and plastic scintillator as shown. (Redrawn from Pearson and Tinker, 1975.) (d) Split-plate used to measure translocation to orchid protocorms. Mycorrhizal protocorms provided inoculum at X. Hyphae grew across the diffusion barrier (Z) in the direction of the solid arrow. Tracer was applied at Y and translocation in the direction of the broken arrow measured by sampling orchid protocorms.

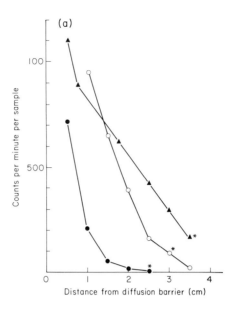

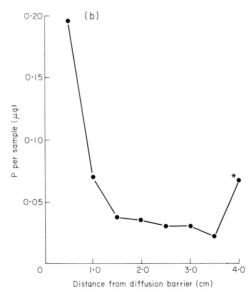

Fig. 33. Profiles of translocation of $^{32}$P in fungi growing on split plates. (a) Forward translocation in *Rhizoctonia repens* 24 h (●), 48 h (○) and 120 h (▲) after application of $^{32}$P-orthophosphate sampled as counts per minute per 0·5 cm diameter disc of agar. Margin of colony marked *. (Unpublished data of S. E. Smith.) (b) Translocation in *Rhizopus* sp., sampled as μg P per 0·5 cm cube of agar. Sporulation at the margin of the colony marked *. (Redrawn from Lucas, 1970.)

Results such as these are, however, difficult to explain in detail. The approximately logarithmic profile observed at 24 and 48 h is similar to that observed for translocation in phloem and makes measurement of rates of translocation very difficult (see Canny, 1960). Work with actively growing fungi on split plates is further complicated by several factors. These are: (1) continuous "loading" of the translocation system with tracer from the site of application; (2) the growth of the culture, which may increase the number of hyphae crossing the diffusion barrier during the period of the experiment, increase the amount of mycelium present in each sample disc and extend the margin of the colony; and (3) sporulation of the culture (Lucas, 1970; Littlefield, 1967) (see Fig. 33b). In addition, leakage from the hyphae into the agar (see the 48 h profile, Fig. 33a), while actually representing translocated material, could certainly influence the shape of the profile and add to difficulties of interpretation. It is not surprising that measurements of translocation rates, as opposed to simple demonstrations of its occurrence, have seldom been achieved for fungi growing in culture on split plates (but see Lyon and Lucas, 1969a). In several cases translocation has been shown to occur in either direction in the mycelium, although sometimes at different rates (see Table 75). We do not know whether bidirectional translocation occurs simultaneously in the same hypha or in different hyphae of a mycelium.

TABLE 75

Bidirectional translocation of $^{32}$P-phosphate by *Rhizoctonia solani*. Samples were counted 24 h after application of 10 $\mu$Ci of $^{32}$P-labelled ortho-phosphate to the donor side of 7-day-old cultures grown at 22·5 C, on split-plates (results of S. E. Harley, quoted in Burnett, 1976)

| Sample number | Distance from diffusion barrier cm | Counts min$^{-1}$ per sample[a] | |
|---|---|---|---|
| | | Base to apex | Apex to base |
| 1 | 0–0·5 | 425 ± 67 | 727 ± 92 |
| 2 | 0·5–1·0 | 122 ± 25 | 437 ± 84 |
| 3 | 1·0–1·5 | 53 ± 9 | 291 ± 83 |
| 4 | 1·5–2·0 | 34 ± 7 | 205 ± 62 |
| 5 | 2·0–2·5 | 29 ± 10 | 182 ± 70 |
| TOTAL | | 685 ± 102 | 1831 ± 302 |

[a] Values are means and standard errors of means of three replicate split-plates for each treatment.

In many of the investigations involving mycorrhizas, host plants were present in the experimental systems. Translocation of a variety of nutrients both towards and away from the plant has been clearly demonstrated (see Table 76). Sampling of the mycelium in agar outside the host is subject to the same

## TABLE 76
Examples of translocating fungi. A, mycorrhizal; B, non-mycorrhizal

| Fungus | Host (if present) | Isotope supplied | Direction of translocation | Distance (if known) cm | Reference |
|---|---|---|---|---|---|
| **A. Mycorrhizal fungi** | | | | | |
| **i. Vesicular-arbuscular mycorrhizal fungi** | | | | | |
| *Glomus mosseae* | *Allium cepa* | $^{32}PO_4$ | fungus to host | 7·5 | Hattingh *et al.* (1973) |
| *G. mosseae* | *Trifolium repens* | $^{32}PO_4$ | fungus to host | ~2·0 | Pearson and Tinker (1975) |
| *G. mosseae* | *T. repens* and *A. cepa* | $\left\{\begin{array}{l}^{32}PO_4 \\ ^{35}SO_4 \\ ^{65}Zn\end{array}\right.$ | fungus to host | ~2·0 | Cooper and Tinker (1978) |
| *G. fasciculatus* | *A. cepa* | $\left\{\begin{array}{l}^{32}PO_4 \\ ^{45}Ca \\ ^{35}SO_4\end{array}\right.$ | fungus to host | <7·0 | Rhodes and Gerdemann (1975; 1978a, b) |
| **ii. Ericoid mycorrhizal fungi** | | | | | |
| *Pezizella ericae* | in pure culture | $\left\{\begin{array}{l}^{32}PO_4 \\ ^{14}C\text{-glucose}\end{array}\right.$ | forwards[a] | 0·6 | Pearson and Read (1973b) |
| *P. ericae* | *Vaccinium* | $\left\{\begin{array}{l}^{32}PO_4 \\ ^{14}C\text{-glucose}\end{array}\right.$ | fungus to host | 2·5 | Pearson and Read (1973b) |
| *P. ericae* | *Vaccinium* | $\left\{\begin{array}{l}^{14}CO_2 \\ ^{14}C\text{-glutamine}\end{array}\right.$ | host to fungus / fungus to host | ~2·0 / ~2·0 | Stribley and Read (1975) |
| **iii. Ectomycorrhizal fungi** | | | | | |
| *Boletus* spp. | *Pinus sylvestris* | $\left\{\begin{array}{l}^{32}PO_4 \\ ^{15}NH_4NO_3 \\ ^{15}N\text{-glutamate} \\ ^{45}Ca\end{array}\right.$ fungus to host | | | Melin and Nilsson (1950, 1952, 1953a, 1955) |

| Fungus | Host / culture | Tracer | Direction | | Reference |
|---|---|---|---|---|---|
| *Boletus* spp. | *P. sylvestris* | $^{14}CO_2$ | host to fungus | | Melin and Nilsson (1957) |
| *Boletus* spp. | *Picea excelsa* | $^{14}C$-glucose | host to fungus | | Bjorkman (1960) |
| *Cortinarius glaucopus* | *P. sylvestris* | $^{32}PO_4$ | fungus to host | | Melin and Nilsson (1953b) |
| *Rhizopogon roseolus* | *P. virginiana* | $^{22}Na$ | fungus to host | | Melin *et al.* (1958) |
| *R. vinicolor* | *P. taeda* | $^{14}CO_2$ | host to fungus | | Ahrens and Reid (1973) |
| *Pisolithus tinctorius* | *P. taeda* | $^{14}C$-glucose | host to fungus | | Reid and Woods (1969) |
| *Cenococcum graniforme* | *P. taeda* | $^{14}CO_2$ | host to fungus | | Ahrens and Reid (1973) |

iv. Orchid mycorrhizal fungi

| Fungus | Host / culture | Tracer | Direction | | Reference |
|---|---|---|---|---|---|
| *Rhizoctonia repens* | pure culture | { $^{14}C$-glucose, $^{32}PO_4$ } | bidirectional[a] | ~4 | |
| *R. solani* | pure culture | { $^{14}C$-glucose, $^{32}PO_4$ } | bidirectional[a] | ~4 | Harley, S. E. (1965) Smith, S. E. (1966, 1967) |
| *R. solani* | *Dactylorchis purpurella* | { $^{14}C$-glucose, $^{32}PO_4$ } | fungus to host | ~1 | |
| *R. goodyerae-repentis* | *Goodyera repens* | $^{14}C$-glucose | fungus to host | | Purves and Hadley (1975) |

B. Non-mycorrhizal fungi in mycelial culture

| Fungus | Host / culture | Tracer | Direction | | Reference |
|---|---|---|---|---|---|
| *R. solani* | pure culture | $^{14}C$-glucose | forward[a] | up to 5 | Milne and Cooke (1969) |
| | pure culture | $^{32}PO_4$ | bidirectional[a] | up to 4 | Littlefield *et al.* (1965) |
| *Rhizopus stolonifer* | pure culture | $^{32}PO_4$ | backwards[a] | up to 4 | Lucas (1970) |
| *Cyathus stercorius* *R. stolonifer* *Aspergillus niger* *Sordaria fimicolor* *Pyronema domesticum* | pure culture | $^{14}C$-glucose | bidirectional[a] | ~1 | Howard (1978) |
| *Sclerotium rolfsii* | pure culture | $^{32}PO_4$ | to sclerotia | 4–6 | Wilcoxson and Subbarayadu (1968) |

[a]See Fig. 32.

problems outlined for pure culture studies in that accurate rates are impossible to
determine. Translocation to the host plant is less subject to these problems, and
the advantages of the system are described below. Translocation in this direction
is expected to occur in all examples of mycorrhiza, and a list of mycorrhizal fungi
which have been shown to have translocating ability for a variety of compounds
is given in Table 76. Lyon and Lucas (1969b) were of the opinion that ability to
translocate phosphate was associated with rapid turnover of phosphate
compounds and they showed that *Chaetomium* sp. with a slow rate of translocation
did not readily release orthophosphate from other phosphorylated compounds.
It is notable that there are no reports of absence of translocating ability amongst
mycorrhizal fungi, although effective distances over which compounds labelled
with $^{32}P$ and $^{14}C$ were translocated by *Pezizella ericae* in pure culture were very
small (6 mm). In association with a host plant the distances were increased to
25 mm, a point to be discussed later. In many experiments the maximum
distances over which translocation could be measured were limited by the
physical size of the experimental system and were usually a few centimetres (up to
12 cm for *Rhizopogon roseolus*, Skinner and Bowen, 1974). In all cases the distances
exceed the 40 $\mu$m over which diffusion can be shown by calculation to provide an
adequate and credible mechanism of symplastic solute transfer at reasonable
physiological concentrations (Tyree, 1970; Osmond and F. A. Smith, 1976). The
concentration gradients ($\Delta C$) necessary to explain translocation in terms of
diffusion are given in Table 77. Any mechanism of fungal translocation must
take these facts into account.

The capacity to absorb a variety of nutrients from the medium and translocate
them (or their metabolites) some distance through an established hyphal system
is a common, and probably a universal, attribute of mycorrhizal fungi. It is
clearly an essential and expected property, if the proposed role of mycorrhizas in
exploiting large volumes of soil at some distance from the root is to be upheld.
Similarly carbon compounds derived from autotrophic hosts must be translocated to growing hyphal apices and fruit bodies outside the root via
extramatrical hyphae. Hence translocation must occur in both directions in the
mycelium (and probably in each individual hypha) and this requires to be
explained when possible mechanisms of translocation are proposed.

## The rate of translocation

Measurements of the rate of translocation of nutrients in hyphae, expressed as a
flux, mol $cm^{-2} s^{-1}$, are essential before problems of the relative efficiency of
translocation in different directions by different fungi in various environmental
conditions can be resolved satisfactorily. Such flux rates are also important
in consideration of the role of mycorrhizal fungi in plant nutrition. Expression of
translocation rate as cm $h^{-1}$ as used by Burnett (1976) and Jennings *et al.* (1974)

for the fungi and many others for translocation in the phloem has only limited usefulness, as such measurements give no indication of the amounts of material being translocated nor the mechanism by which translocation occurs. Moreover estimates of flow rates by measurement of the time of first appearance of a dye or tracer at a particular point in a hypha are only useful for making comparisons if the translocation profile is sharp and continuous loading of tracer into the system does not occur (see Canny, 1960, for discussion). Neither of these criteria is met in any of the fungal systems used to date.

TABLE 77

Translocation fluxes through hyphae of mycorrhizal fungi and the concentration gradients ($\Delta C$) that would have to be maintained over a path length of 1 cm if such fluxes were to be by diffusion only.

A. Fluxes in *Glomus mosseae* measured using radioactive isotopes

| Isotope | Flux mol cm$^{-2}$s$^{-1}$ | $\Delta C$ mol l$^{-1}$ | Reference |
|---------|---------|---------|-----------|
| $^{32}PO_4$ | $3.0–10.0 \times 10^{-10}$ | $0.03–0.1$ | Pearson and Tinker (1975) |
| $^{32}PO_4$ | $2.0 \times 10^{-10}$ | $0.02$ | Cooper and Tinker (1978) |
| $^{32}PO_4$ | $2.0 \times 10^{-9}$ | $0.2$ | Cooper and Tinker (1981) |
| $^{65}Zn$ | $2.1 \times 10^{-12}$ | $2.8 \times 10^{-4}$ | Cooper and Tinker (1978) |
| $^{35}SO_4$ | $16.5 \times 10^{-12}$ | $13.7 \times 10^{-4}$ | Cooper and Tinker (1978) |

B. Fluxes in *Rhizoctonia* sp. measured from estimates of growth

| Molecule translocated | Flux mol cm$^{-2}$s$^{-1}$ | $\Delta C$ mol l$^{-1}$ | |
|---------|---------|---------|---|
| hexose | $2.3–10.3 \times 10^{-7}$ | $32.9–147.1$ | S. E. Smith (unpublished) |

[a] $\Delta C$ values calculated from the equation $\tilde{J} = D\,(\Delta C/l)$ where $\tilde{J}$ is the flux, $D$ the diffusion coefficient, $\Delta C$ the concentration difference between the two ends of the pathway and $l$ the path length (taken as 1 cm in these calculations).

Diffusion coefficients were taken as $1.0 \times 10^{-5}$ cm$^2$ s$^{-1}$ for phosphate, $0.75 \times 10^{-5}$ for Zn and ZnSO$_4$, $1.2 \times 10^{-5}$ for SO$_4$ as Na$_2$SO$_4$ and $0.7 \times 10^{-5}$ for hexose.

The best way of measuring rates of fungal translocation is to sample a discrete sink for translocated material at different times after application of tracer to the mycelium at some distant point. From this information, together with numbers and dimensions of connecting hyphae, a flux can be calculated. It is in making these measurements that mycorrhizal systems have proved so useful, and could yield considerably more information relevant to unravelling the mechanism of translocation in physiologically and taxonomically distinct fungi. The measurement of translocation in this way will not be confounded by the rate of transfer of tracer from fungus to host, for over short time periods the total estimated will be made up of that in the fungus plus that in the host. If a time course of appearance of isotope in the infected host is obtained (see Fig. 34) then linear parts of the curve may be used to determine the rate of translocation.

Measured rates of translocation of phosphate in *Glomus mosseae* to *Trifolium repens* and to *Allium cepa* are given in Table 77. The highest flux obtained was about $2 \times 10^{-9}$ mol P cm$^{-2}$ s$^{-1}$ (Cooper and Tinker, 1981) when *T. repens* plants were allowed to transpire freely at 20–25°C. Reduction in transpiration rate resulted in lower translocation rates. If we can assume that uptake of $^{32}$P-phosphate by the hyphae from agar was not affected by changed water relations, then the results indicate that mass flow of solution in the fungal hyphae can contribute to translocation. Such a result is in line with the experiments of Lucas (1970), who studied the effects of different water potentials generated by polyethylene glycol at opposite ends of a mycelium. The magnitude of any error in the experiments of Cooper and Tinker arising from an increased phosphate concentration around the absorbing hyphae by water absorption was minimized by periodic replacement of water in the experimental dishes. The effect of

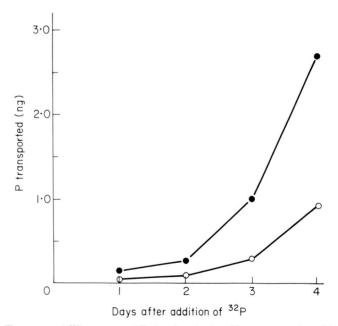

Fig. 34. Transport of $^{32}$P across a diffusion barrier by *Glomus mosseae* in soil/sand split plates. The host plant was *Allium cepa*. ●, Total phosphate transported detected in roots, shoots, soil and external mycelium; ○, phosphate transported to shoot. (Drawn from data presented as a table by Cooper and Tinker, 1978).

transpiration on translocation in *G. mosseae* highlights the finding that effective translocation distances from $^{32}$P-phosphate in *Pezizella ericae* were increased from 6 mm to 25 mm when the fungus was associated with presumably transpiring *Vaccinium macrocarpum* seedlings (Pearson and Read, 1973b). This Ascomycete may thus have a component of translocation which is dependent on mass flow.

An additional feature of experimental results, which requires to be explained by any mechanism of translocation, is the lag phase which frequently occurs after application of $^{32}$P to split plates and the achievement of a steady rate of appearance in the host (see Fig. 34; Cooper and Tinker, 1978, 1981). This is discussed below.

Carbohydrate translocation was not included in the studies with *G. mosseae*. Net carbon transport is presumed to take place from host to fungus and is therefore not susceptible of study in the same way as translocation in the opposite direction. Fluxes were not measured in the $^{14}$C studies with *P. ericae*, but results indicate that labelled photosynthate passes from host to fungus under normal circumstances. If the host plant is deprived of carbon dioxide then net carbon (but not necessarily carbohydrate) movement is in the reverse direction, possibly indicating changes in the source–sink relationships (Pearson and Read, 1973b; Stribley and Read, 1974b, 1975).

Orchid mycorrhizas do offer an opportunity of studying both mineral nutrient and carbon translocation into a host plant. Unfortunately the published studies do not include calculation of fluxes (S. E. Smith, 1967; Purves and Hadley, 1975). However, recalculation of data on the growth of orchid protocorms (S. E. Harley, 1965), making approximations of the shape, density and fresh weight : dry weight ratio of the protocorms, can give a rough indication of carbon translocation in hyphae of *Rhizoctonia* sp. When the fungus was supplied with cellulose and growth measured over several weeks, a mean translocation rate of about 8·1 $\mu$mol hexose s$^{-1}$ per seedling could be calculated. The number and dimensions of hyphal connexions were not recorded, but if we assume that there was no loss by respiration of $^{14}$C and 10 connexions per seedling of radius 5 $\mu$m, the calculated flux of hexose would be about $2\cdot3 \times 10^{-7}$ mol cm$^{-2}$ s$^{-1}$. If trehalose was actually the sugar translocated (see below) then the flux must be reduced by half (to $1\cdot2 \times 10^{-7}$ mol cm$^{-2}$ s$^{-1}$) and if there were 100 hyphal connexions per seedling the flux would be $1\cdot2 \times 10^{-8}$ mol cm$^{-2}$ s$^{-1}$. Such calculations can only be very approximate, but it is worth comparing them with other published calculations. For example, measurements of growth of strands in *Serpula lachrymans* have allowed calculations of translocation rates as $1\cdot27 \times 10^{-8}$ mol hexose cm$^{-2}$ s$^{-1}$ (Jennings *et al.*, 1974), while phloem translocation rates, also determined from dry weight increases in organs such as fruits ("specific mass transfer"), are in the range $4\cdot3 \times 10^{-7}$ to $3\cdot5 \times 10^{-6}$ mol sucrose cm$^{-2}$ s$^{-1}$ (recalculated from Canny, 1960). All these estimates must be minimum values as no account was taken of respiratory carbon losses.

*Effect of environmental conditions on rates of translocation*

The effect of changing transpiration rate in the host plant on phosphate translocation has already been mentioned and is in line with the previous observations (not based on determination of rates) that low relative humidity favours its movement and that of dyes (Plunkett, 1956; Schütte, 1956; Littlefield *et al.*, 1965a) in sporophores.

Temperature also affects the total amount of phosphate transported by *G. mosseae* into *T. repens*. Cooper and Tinker (1981) found that phosphate translocation in split plates was highest between 15–25 °C and that very little was translocated at 5 °C (see Fig. 35). In this, as in other investigations (e.g. Milne and Cooke, 1969; Littlefield *et al.*, 1965b), no attempt was made to distinguish between effects of temperature upon uptake and upon translocation itself. Phosphate uptake is an energy-requiring process in most organisms in most physiologically predictable circumstances and has been shown to be reduced at low temperatures. Results showing the sensitivity of the overall transport process to low temperatures do not directly indicate effects upon translocation itself.

A similar objection can be raised against the finding that cytochalasin B prevented translocation when applied extensively to a fungal culture. This compound certainly prevents cytoplasmic streaming, one of the mechanisms by

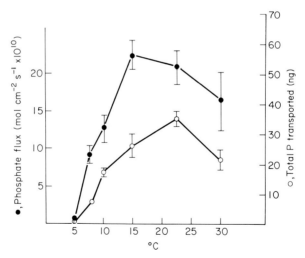

Fig. 35. The effect of temperature on translocation in *Glomus mosseae* in soil agar/agar plates. The host plant was *Trifolium repens*. Phosphate flux 14 days after addition of [32]P-orthophosphate detected in shoots, roots, soil agar and external mycelium. ●, Flux; ○, quantity transported. Drawn from data presented as tables by Cooper and Tinker (1981). Values given for 20–25 °C are plotted at 22·5 °C. Each point is the mean ± standard error of mean of five replicates.

which fungal translocation has been suggested to occur (see below). However, the effect of cytochalasins on cellular movements (including cell division, streaming, muscle contraction) by preventing actin polymerization may also be accompanied by inhibition of membrane transport of hexoses, purines and nucleosides across the plasmalemma, by quite a different mechanism (see Lin *et al.*, 1978; Tanenbaum, 1978; Baldwin, Leinhard and Baldwin, 1980). Use of cytochalasins to investigate sugar translocation must therefore be suspect and it is most important that their effects on phosphate uptake be investigated before unequivocal statements about the role of cytoplasmic streaming in translocation can be made.

## The identity of translocated compounds

There has been a general assumption that both organic and inorganic compounds are translocated in solution by fungi. Indeed, preliminary evidence indicates that the disaccharide trehalose is a major translocated carbohydrate in *Rhizoctonia* sp. (S. E. Smith, 1967), while glucose may be important in *Phytophthora cactorum* (Bokhary and Cooke, 1974). Recently it has been suggested that insoluble compounds may be translocated as granules propelled by cytoplasmic streaming. Cox and Tinker (1976) and Cox *et al.* (1980) calculated that polyphosphate granules in the vacuoles of *Glomus mosseae*, if transported by cytoplasmic streaming at measured rates ($12·6$ cm h$^{-1}$, Cox *et al.*, 1980) contain sufficient phosphate to account for measured fluxes into onion seedlings ($2·7 \times 10^{-8}$ mol P cm$^{-2}$ s$^{-1}$). Similarly, Milne and Cooke (1969) working with *Rhizoctonia solani* suggested that transfer of carbohydrate as glycogen granules (in the cytoplasm) would explain the $^{14}$C-labelling patterns obtained as cultures aged. Both of these suggestions remain to be confirmed by experiment. n.m.r. studies with *Fagus* mycorrhizas by B. C. Loughman (personal communication) have shown that polyphosphate need not be in granule form, for soluble polyphosphates of significant chain length are present in the vacuoles of the fungal sheath. Indeed the granular nature of vacuolar polyphosphate has been questioned and it has been suggested that formation of granules may in many cases be an artifact of staining methods using toliudine blue or lead nitrate (see Beever and Burns, 1980). Nevertheless polyphosphate, when not granular, occurs in large molecules of considerable chain length. Some carbohydrates also are believed to be enclosed in vesicles which presumably contain wall precursors. So migration of both substances in "packages" is possible even if not as granules.

## The mechanism of translocation

The mechanism of translocation is at present as uncertain in the fungi as in other plants, but it is subject to active investigation. For compounds in solution the mechanisms might include (i) diffusion, (ii) cytoplasmic streaming or cyclosis, important in mixing the cytoplasmic contents, and (iii) mass flow of solution driven by a pressure gradient. As with translocation in phloem, there are other possibilities, including contractile proteins which can cause mass flow of solutions. These latter are not discussed here, but Fensom (1975) can be consulted for a general discussion of the possible roles of contractile proteins in phloem. Diffusion alone is unlikely to be quantitatively significant in the translocation of carbohydrates or ions in solution by fungi. The distances over which translocation has been measured all exceed 40 $\mu$m and the concentration gradients which would have to exist in the hyphae if translocating at observed rates by diffusion alone are very large (see Table 77). Cowan *et al.* (1972) calculated from electrochemical potential gradients that diffusion could be the mechanism of translocation in sporangiophores of *Phycomyces blakesleanus*. However, Jennings *et al.* (1974) have pointed out that the calculations were based on appearance of $K^+$ in the bathing medium, which almost certainly occurs by passive diffusion across the fungal plasmalemma. Thus the results and calculations bear little relation to intracellular events.

The second mechanism, involving protoplasmic circulation or streaming superimposed upon difference in concentration between the ends of the system, may be important in intracellular translocation. There is no doubt that cytoplasmic streaming occurs in all groups of fungi. In Phycomycetes (e.g. *Rhizopus*, *Mucor*, *Phycomyes*) the stream can be bidirectional (Fig. 31) and frequent reversals of flow have been observed (see Buller, 1933). Bidirectional streaming certainly occurs in hyphae of *Glomus* spp. Rhodes (in Rhodes and Gerdemann, 1980) has described rapid bidirectional movement of cytoplasm containing vacuoles and observed the collision of bodies moving in opposite directions. Cox *et al.* (1980) measured the rate of cytoplasmic streaming in *Glomus mosseae*, but they did not record whether the streaming was bidirectional. In Ascomycetes and Basidiomycetes flow appears to be normally acropetal, through the septa. However, both acropetal and basipetal streaming have been observed in *Fimentaria fimicola* (Buller, 1933) and are presumed to occur in other fungal groups. Streaming is very difficult to observe in narrow hyphae and it is quite likely that reverse currents do exist. Certainly "eddies" in the neigh-bourhood of septa have been described in *Humaria leucoma* by Jahn (see Burnett, 1976) (Fig. 31).

Cytoplasmic streaming is almost certainly brought about by the activity of actomyosin-like filaments in the cytoplasm. It is an active process requiring ATP, affected by changes in temperature and inhibited by cytochalasin B (see Williamson, 1976; Tanenbaum, 1978; Lin *et al.*, 1978; Wagner, 1979).

Any form of cytoplasmic streaming involving partial cyclosis, eddies or reversal of flow would result in effective stirring of intracellular solution. Translocation of solutes could then result from the maintenance of a concentration gradient along the pathway, i.e. between "sources" and "sinks", and could be much faster than would occur by diffusion through stagnant solution. The concentration gradient itself would depend on loading from the source and removal into the sink and the rate of translocation would be determined by the rates of loading and removal as well as by the efficiency of stirring by cytoplasmic streaming along the pathway of translocation. By analogy with translocation in phloem, we would expect loading and removal to involve specific membrane transport processes for individual solutes. Translocation of solutes could then occur between sources and sinks. The driving force would be the difference in concentration between the source and sink for each individual solute. Translocation of different solutes at different rates in the same direction would occur if rates of loading and removal differed, giving rise to different concentration gradients. Translocation in opposite directions could occur as long as sources and sinks were opposed. How much mixing would occur if cytoplasmic streaming were unidirectional is hard to determine (but see below). However, we can see that under many conditions a mechanism of this type would explain bidirectional translocation and effects of temperature (see e.g. Subbarayadu *et al.*, 1966; Milne and Cooke, 1969; Lyon and Lucas, 1969a; Cooper and Tinker, 1981) and of cytochalasin B (Cooper and Tinker, 1981) on the rate of translocation (but see our earlier comments on this inhibitor). It is most important to emphasize bidirectional translocation (at least in the mycelium as a whole) in most mycorrhizal associations. In these, mineral nutrients such as phosphate, nitrogen compounds, potassium etc. have been shown to be translocated in the direction of the host, while host-derived carbohydrate has been demonstrated or is presumed to be translocated in the reverse direction towards hyphal apices in the soil (Table 76).

It is certainly worth considering the suggestion that material is moved in compartmented form (glycogen granules in the cytoplasm or polyphosphate in the vacuoles). This theory has been best developed for the movement of phosphate, as inorganic polyphosphate, by mycorrhizal hyphae of vesicular-arbuscular mycorrhizas (Cox and Tinker, 1976; Callow *et al.*, 1978; Cox *et al.*, 1980). There is no doubt that vacuoles (presumably containing polyphosphate) and granules do move within streaming cytoplasm and for the purposes of the following discussion it does not matter that the polyphosphate is bounded by the tonoplast, while the glycogen granules are possibly free in the cytoplasm. It is not a prerequisite of the theory that polyphosphate is insoluble, only that it is moved in "packets", although calculations are based on the amount of phosphate in "granules", made visible by staining methods which would result in pre-cipitation of the vacuolar phosphate (see Cox and Tinker, 1976; Cox *et al.*, 1980). *Net* directional movement (translocation) of phosphate in granule (or packet) form could occur as long as they are locally generated and locally dispersed or used. The same applies to granules of glycogen. Thus, unidirectional streaming could result in movement of granules along a hypha, but well organized "loading" and "unloading" systems must be envisaged, and the granules must be regarded as "tight" (i.e. not equilibrating rapidly with phosphate or soluble carbohydrate in the cells) otherwise net translocation in granules would not occur down a concentration gradient. If the granules do equilibrate rapidly with the bathing solution then they would "unload" continuously as they passed along a hypha down a concentration gradient, and no net transfer *in granule form* would occur. This would make some of the calculations based on granule size and rate of movement unreliable. If streaming is bidirectional, as in *Glomus* spp., it seems likely that this would tend to uniform distribution of "tight" granules in the cytoplasm in a similar way to stirring of solution. Net movement of granules along the pathway would again depend upon their generation near the source and unloading near the sink. The way in which granules would be redistributed by this form of streaming should be taken into account in calculations such as those of Cox *et al.* (1980). Rate of movement of material in granule form would depend upon the rates of formation and breakdown of the granules.

In most organisms in which polyphosphate metabolism has been studied, the formation of polyphosphate granules is regarded as a method of "mopping up" large amounts of inorganic phosphate from the cytoplasm (see Harold, 1966; Harley, 1978a; Beever and Burns, 1980). Hydrolysis of polyphosphate is envisaged as occurring when cytoplasmic inorganic phosphate levels fall; if this were so, polyphosphate granules would appear to equilibrate with cytoplasmic phosphate and probably could not be regarded as "tight". The breakdown of polyphosphate can be envisaged as occurring by two reactions: (1) by hydrolysis mediated by polyphosphatase, or (2) by a reversal of polyphosphate kinase with the re-formation of ATP in the presence of ADP. Enzymes of polyphosphate metabolism have been detected in vesicular-arbuscular mycorrhizas by Capaccio and Callow (1982). They found low activities of exopolyphosphatase in uninfected roots of onion. Infected roots contained higher activities of this enzyme and also of endopolyphosphatase, but no polyphosphatase activity was found in external hyphae. Polyphosphate kinase was detectable in mycorrhizal roots and external hyphae, but not in non-mycorrhizal roots, while polyphosphate hexokinase was only found in the external hyphae. The absence of polyphosphatases in external hyphae is interesting, as it suggests that polyphosphate turnover might be low and vacuoles therefore "tight", but much more work on control of polyphosphate metabolism is required before this could be regarded as certain or indeed likely. According to B. C. Loughman (personal communication) polyphosphate breakdown probably by polyphosphatase appears very active in damaged ectomycorrhizas, perhaps indicating a degree of separation of enzyme and substrate in the undamaged organ. If we accept that glycogen granules and vacuoles containing polyphosphate do equilibrate rapidly with solutions in the cytoplasm, then we must consider what role they may have in translocation of solutes in the fungi and how the existence of such "compartments" might be expected to affect the results of experiments. The ideas outlined below owe much to "the mechanism of translocation" suggested by Canny (1962). That the theory has not been upheld for phloem (on cytological grounds) does not concern us here; the theoretical arguments seem to be particularly pertinent for translocation in a fungal mycelium. The details of the hyphal system are presumed to be as follows:

1. Cytoplasmic streaming occurs, either bidirectionally in a single hypha or acropetally in one hypha and basipetally in another.

2. Rapidly equilibrating polyphosphate-containing vacuoles or glycogen granules are carried along in the cytoplasmic stream and are distributed at random. The granules equilibrate with a solution in the cytoplasm.

3. A concentration gradient of inorganic phosphate or glucose for instance exists from one end of the hypha to the other, due to loading from the source and removal into the sink.

The rapid equilibration of moving particles with the bathing solution results in "mixing" of the solution and the result is that net translocation of solute occurs down a concentration gradient. Bidirectional translocation can occur as long as cytoplasmic streaming also occurs in both directions, either in a single hypha or in different hyphae. In addition, the fact that some nutrients (inorganic phosphate and possibly carbohydrate) equilibrate with granules does not mean that all nutrients must be postulated to behave in the same way.

The existence of septa in most fungi, which may impede cytoplasmic streaming along the whole length of a hypha, does not seriously affect the arguments presented here, as long as the septa permit diffusion of ions or molecules in solution, and do not exceed about 40 $\mu$m in length.

The translocation pathway can then be regarded as a source and a sink (as before) linked by a chain of hyphal compartments in each of which cytoplasmic streaming and mixing of solution occur and between which diffusion over short distances within the septa takes place. The system is similar to a chain of internodal cells of *Chara*, linked via plasmodesmata in the nodal cells. It has been calculated (Walker, 1976) that cyclosis provides an adequate means of mixing the cytoplasmic contents of each internodal cell and of bringing ions to one side of a plasmodesma. Streaming in the adjacent internodal cell removes ions which arrive by diffusion through the plasmodesma. In this way a concentration gradient is maintained through the plasmodesma and net translocation can occur along a chain of cells, which is limited by the rate of streaming.

High concentrations of free inorganic phosphate may facilitate phosphorus translocation. Lyon and Lucas (1969a) have shown that in *Rhizopus* at 20 C not only was translocation at the highest rate ($1\cdot09 \times 10^{-3}$ $\mu$g P h$^{-1}$ mg$^{-1}$ fresh weight of the mycelial bridge) but that the relative level of orthophosphate was also highest at that temperature. Similarly apices of *Chaetomium*, a "non-translocating fungus", had much lower levels of free orthophosphate than *R. stolonifer*.

Mass flow of solution in vegetative hyphae may be important under some conditions, as shown by the effects of host transpiration on translocation in mycorrhizal fungi (Cooper and Tinker, 1981; Pearson and Read, 1973b) and of sporophore transpiration in *Polyporus* (Plunkett, 1956) and *Lentinus* (Littlefield *et al.*, 1965a). It has also been suggested for acropetal carbon translocation in *Serpula lachrymans* (Jennings *et al.*, 1974). If intracellular mass flow occurs then it will override any translocation for which the driving force is a concentration gradient, and all nutrients will move in the same direction down a pressure gradient, regardless of concentrations. The fluxes of each solute will depend upon their individual concentrations. Bidirectional translocation in the same hypha is not then theoretically possible, and it is hard to envisage small-scale differences which would allow mass flow in different directions in different hyphae of the same mycelium.

There are many reports that conditions which enhance evaporation or osmotic water loss by hyphae (e.g. air currents, increased temperature, application of osmotica) and are thus postulated to give rise to pressure differences and so to mass flow of solution, also result in an increase in the rate of cytoplasmic streaming (see Buller, 1933). If this is found to hold for streaming in mycorrhizal fungi then it is possible that results which seem to show the importance of mass flow may be equally well interpreted in terms of changed rates of streaming, and thus of concentration-dependent translocation (Cooper and Tinker, 1981). Further studies with cytochalasins under different "transpiration" conditions could prove most illuminating, as long as the possible effects of this compound on uptake as well as streaming are borne in mind.

### Translocation in mycorrhizas and plant nutrition—a summary

Translocation of phosphate and other nutrients over distances up to several centimetres in hyphae has been demonstrated in ericoid, vesicular-arbuscular, orchid and ectomycorrhizas. In all of these, transfer of some nutrients to the host cells has also been shown. Such translocation is clearly a prerequisite if these nutrients are absorbed by fungi some distance from infected roots. It is clear that the absorbing and translocating abilities of the fungi could account for the spatial aspects of mycorrhizal exploitation of nutrients in soil. It is less clear that measured rates of phosphate translocation (fluxes) in hyphae of vesicular-arbuscular mycorrhizal fungi can account for the amounts of phosphorus calculated to be transferred from fungus to root in plants growing in soil. The experiments of Pearson and Tinker (1975) and Cooper and Tinker (1978) gave values of the phosphate flux of $3–10 \times 10^{-10}$ and $2 \times 10^{-10}$ mol cm$^{-2}$ s$^{-1}$

respectively. More recently Cooper and Tinker (1981) have obtained maximum rates of $2 \times 10^{-9}$ mol cm$^{-2}$ s$^{-1}$ in transpiring plants but this is still considerably lower than fluxes calculated from the numbers of entry point hyphae and total P uptake $(3 \cdot 8 \times 10^{-8}$ mol P cm$^{-2}$ s$^{-1})$ by Sanders and Tinker (1973). The discrepancy may be partly accounted for by the different conditions under which the plants were grown that would have affected growth rates, plant transpiration rates, etc. In addition, calculations depend upon accurate measurement of the numbers and dimensions of hyphae, and upon the assumption that all hyphae included are both living and involved in translocation. This is very hard to determine. Overestimation of hyphal numbers could lead to considerable underestimation of fluxes.

Our information on carbohydrate translocation is much more fragmentary. We know that carbon is transferred from autotroph to heterotroph in ecto-, ericoid and vesicular-arbuscular mycorrhizas and that [14]C-labelled compounds can be translocated some distance away from the root. We do not know what compounds are translocated or at what rates. In orchids, mycorrhizal translocation of carbohydrate over considerable distances into chlorophyll-free protocorms and transfer to host cells have been demonstrated. We have given above calculated rates of transfer and there seems to be no technical reason why such calculations should not be subject to experimental verification, both in terms of actual (rather than calculated) dry weight accumulation and in terms of [14]C tracer fluxes into the protocorms.

A mechanism of translocation based on cytoplasmic streaming has been discussed which accommodates many of the observations that need to be explained by such a theory. In particular, bidirectional translocation of nutrients in hyphae can be explained if cytoplasmic streaming acts as a stirring mechanism, which results in increased rates of translocation down concentration gradients between loading and unloading sites in the hyphae. The mechanism also provides for different rates of translocation of different nutrients or ions in either direction. Superimposed on this basic hypothesis (which is not new and has been discussed by many others, particularly with respect to movement through plasmodesmata) is the suggestion that the existence of granules or vacuoles containing localized high concentrations of particular molecules may be involved in the stirring process and hence in translocation. Directional movement of granules or vacuoles from source to sink is not envisaged. It is possible that the involvement of vacuoles containing polyphosphate in the translocation process may also help to explain the lag phase in appearance of labelled phosphate in mycorrhizal plants when these are grown on split plates. Suggestions that this lag is due to a delay between arbuscule formation and degeneration (Cox and Tinker, 1976) can, as Cooper and Tinker (1978) point out, be discounted, for the lag occurs not only in appearance of [32]P in the shoots but also in whole plants. It seems likely that the lag phase is compounded of two factors: (1) a simple delay in time of appearance of $^{32}$P due to the distance between the source and the site of detection, and (2) a delay caused by the equilibration of $^{32}$P with $^{31}$P already present in the translocation pathway. Thus, although translocation of total phosphate [$^{31}$P + $^{32}$P] is at a constant rate this will not be apparent from measurements of radioactivity until the specific activity of $^{32}$P is uniform throughout the translocation pathway. This problem has been discussed by Harley *et al.* (1954) who concluded that the amounts of phosphate present in the translocation pathways in ectomycorrhizal roots of beech were small and that errors in estimation of the total phosphate passing through the fungal sheath to the root core were likely to be small. Although both ectomycorrhizas and vesicular-arbuscular mycorrhizas contain large amounts of polyphosphate the difference in extent of the lag phase for transfer of $^{32}$P from fungus to host tissue suggests that in vesicular-arbuscular mycorrhizas the polyphosphate may be present in and equilibrating with other forms of phosphate in the translocation pathway itself.

An extension of work on translocation in mycorrhizal systems could well yield information relevant to fungal translocation as a whole. In particular, measurement of fluxes of material through hyphae into mycorrhizal plants would give a sound basis for comparison of translocation of different molecules or ions. Use of orchid mycorrhizas would allow study of both mineral nutrient and carbohydrate translocation. Further biochemical work on the identity of the compounds translocated and their turnover in hyphae and mycorrhizal roots would also assist in elucidating details of the translocation process.

## Chapter 17

# Transfer of Metabolites between Symbionts in Mycorrhizal Associations

In our discussion it will be important to keep several questions in mind. Partial answers to some of them have already been obtained, others require considerable experimental investigation. One very important question is: What substances have been shown to pass across the interface? We also need to enquire how the structure of the interface may affect the transfer processes and whether the organisms are naturally leaky or whether juxtaposition of the organisms or substances produced by them induces leakiness.

There are several important generalizations about the contact zone between the symbionts in mycorrhizal associations that can be made. Some of these are obvious at the level of light microscopy, but the electron microscope has greatly improved knowledge of these matters. No mycorrhizal organ of any type has a static form or structure. They all have an initial infection stage, and then their constituent tissues, both of host and of fungus, undergo a sequence of development to a mature form before they senesce, either both components simultaneously or in sequence. Vesicular-arbuscular mycorrhizal roots grow apically and branch like uninfected roots. The new tissues become infected by the spreading of the hyphae within the root or from new infection points which form coils, arbuscules and vesicles in the cells. The fungus degenerates and then the host cortex ages and is sloughed away. Each arbuscule completes its life span in up to 15 days, but new ones are continually formed. Ericoid mycorrhizas are similar in that the intracellular hyphae are short-lived, but there the whole cortex of the hair-root is also a much more transitory structure than that of most roots, and the protoplasts of the cell and of the fungus seem to degenerate almost simultaneously, although the fungal hyphae may outlast the host. In the mycorrhizas of orchids the fungus forms physiologically active coils in the cells which later degenerate. The cell of the host outlives the fungus and may be reinfected again. In orchids with long-lived root systems, the roots may be differentiated into long and short axes. The latter are relatively heavily infected and shorter lived, although there may be many cycles of infection by and degeneration of the fungus in their cells. This differentiation into long and short roots is particularly developed in ectomycorrhizal and arbutoid mycorrhizal root systems and amongst them it finds its clearest development in the genus *Pinus* where the short roots persist for months only. However, in all mycorrhizas there is a period of development and persistence of the mature structure before senescence of the fungus and of the host occurs. An infected root system will exhibit all the phases of development and senescence in which the activities of the symbionts and the nature of the contact between them change. Every stage may be relevant to the process of transfer of nutrients.

## The mature interface

Whatever the developmental sequence there is always a condition, which we will call the "mature state", in which living and metabolically active cells of both symbionts are juxtaposed. Both organisms retain intact plasmalemmae which are separated by an interfacial zone. This normally includes the wall of the fungal hypha which may or may not show some structural modification especially in its outer region, and also material derived from the activity of the plasmalemma of the host. In ectomycorrhizas and in the intercellular hyphae of other kinds of mycorrhiza this may consist of a relatively unmodified host wall and a more or less modified middle lamella; or, it may be extremely modified to form what has been called an "involving layer" or "contact layer" and then the host cell-wall may appear thin (see Fig. 36, Plate 9). In the intracellular systems of endomycorrhizas and ectendomycorrhizas there is an interfacial matrix between the plasmalemma of the host and the fungal wall which is visually unlike the cell-wall of the host over much of its length. At the base, where the trunk of the arbuscule or penetrating haustorium or hypha passes through the wall of the host, it is usual to find that a layer of host wall, the collar, encloses the hyphae (Plates 5, 12, 15). This wall layer becomes thinner away from the cell-wall and is very thin or absent over the surface of most of the intracellular fungus (see Fig. 36). Instead, an interfacial layer often containing vesicles and fibres derived from the host plasmalemma is present. It is described as if the host maintains its ability to secrete carbohydrate and polymerize it into fibres to some extent, but is unable, especially close to the fungus, to organize the fibres. As the fungus senesces it becomes encapsulated in a thick fibrous layer, perhaps as the plasmalemma of the host recovers its ability to organize the fibres into a structure.

Not only are the plasmalemmae of both symbionts in a dynamic state, as shown by the presence of lomasomes, paramural bodies and vesicles, but also their cytoplasms contain typical organelles and the products of metabolism. There may however be, as we have seen in the arbuscules of vesicular-arbuscular mycorrhizas, modifications in the kind and distribution of enzymes of fungi and of infected cells. We cannot doubt that such changes must occur also in monotropoid mycorrhiza where the complex behaviour of the haustorium and its bursting during the growth and degeneration must be enzymically complex.

It is clear that simultaneous bidirectional transfer of molecules or ions can only occur when the interface separates two living cells. Such a state exists during "the mature stage" of all mycorrhizas. In stages where one or other symbiont is senescent or dead, monodirectional transfer only can be envisaged. For example, in the pre-Hartig net zone of ectomycorrhizas the cells of the root-cap and epidermis are crushed and destroyed so that the fungus might derive material from the dying cells, but transport in the reverse direction to the host would be

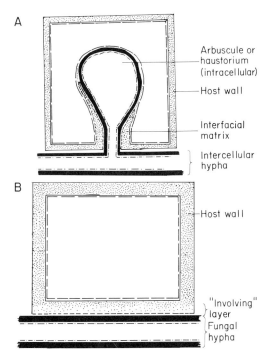

Fig. 36. Diagrammatic representations (not drawn to scale) of the membranes and wall components in intracellular (A) and intercellular (B) mycorrhizal associations. – – –, Host plasmalemma; –.–.–, fungal plasmalemma; stippled, host wall; black, fungal wall. Interfacial matrix derived from host wall in (A) stippled and bounded by a broken line. Modified host wall (e.g. in involving layer) in (B) indicated by less dense stippling and bounded by a broken line.

very unlikely. Similarly the endophytes of *Calluna* (which have some pectolytic ability) colonize the rich mucigel layer of the root-surface before penetrating the tissues and may derive carbon from this source. Conversely nutrient transfer from fungus to root cells following hyphal or arbuscular collapse might occur in orchid, ericoid and vesicular-arbuscular mycorrhizas, but it would be of amounts restricted to the contents of the collapsed hyphae.

The general similarity of the mature interface of different kinds of mycorrhiza and also applies to those of other biotrophic associations between fungi and their autotrophic symbionts. This is true whether the tissue inhabited by the fungus is of a root, a shoot or indeed of a unicellular alga. The interfaces of mycorrhizas, rusts, smuts, mildews and lichens have much in common, as illustrated in Fig. 36 and Plate 15, also Plates 4, 5, 9B, 12, 14.

These are morphological similarities and the common components may in reality have functional differences. For instance, the plasmalemma of the host which surrounds a mycorrhizal arbuscule is apparently not modified morphologically, whereas those of diseased hosts may change in thickness and staining properties, as Littlefield and Bracker (1972), Bracker and Littlefield (1973) and Gil and Gay (1977) have shown. However, the ATPase enzymes on the plasmalemma of the host in cells infected by mycorrhizal arbuscules or by mildew haustoria have both been shown to differ from those uninfected cells, indicating fundamental changes in the membranes of both infections (Spencer-Phillips and Gay, 1981; Marx *et al.*, 1982) as shown in Plate 15.

Broadly the interface is composed of the following:

1. Host plasmalemma.
2. A matrix, which may include host cell-wall and material derived from the host. Especially where hyphae grow intracellularly the amount of material derived from the host-wall present in the interface may vary with the development and age of the intracellular structure.

PLATE 15. The effect of fungal infection on the distribution of ATPase activity on peripheral and invaginated plasmalemma of host cells.

A. Uninfected cells of *Acer pseudoplatanus*. Dense staining due to lead deposition following activity of ATPase is chiefly confined to the uninvaginated plasmalemma (Pm) adjacent to the cell wall. × 13 000.

B. Mycorrhizal cell of *A. pseudoplatanus* invaginated by branches of an arbuscule. ATPase activity is much stronger on the invaginated plasmalemma (Em) than on the peripheral plasmalemma (Pm). × 13 000.

C. An epidermal cell of *Pisum sativum* invaginated by branches of an haustorium of *Erysiphe pisi*. ATPase activity is confined to (1) the peripheral host plasmalemma (Pm) adjacent to the host wall, (2) islands of plasmalemma within the host wall (W) and (3) the fungal plasmalemma of the haustorium (Hpm). Activity is absent from the invaginated host plasmalemma (Em). Transition occurs at the point of invagination of the host cell plasmalemma by the haustorial neck. × 15 000.

A and B from Marx *et al.* (1982). C from Spencer-Phillips and Gay (1981).

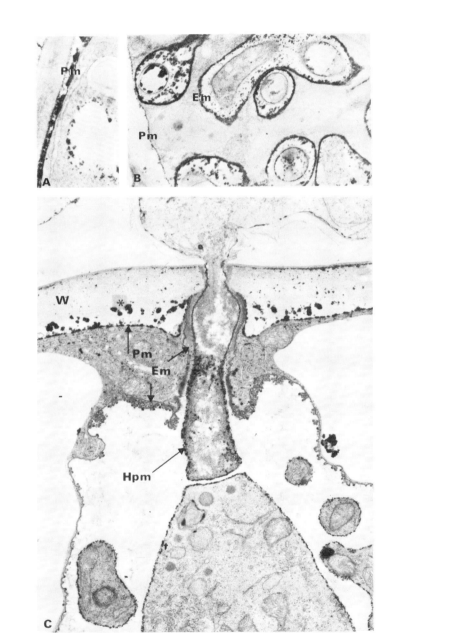

3. The wall of the fungal hypha or arbuscule, which often appears little modified.

4. The fungal plasmalemma.

It is important to note that an "apoplastic space" separates the fungal plasmalemma from the host plasmalemma and through this compounds and ions must pass as they are transferred from one organism to another. The existence of this space has been exploited experimentally in lichens (Drew and D. C. Smith, 1967) and also in ectomycorrhizas (Lewis, 1976). More recently it has received theoretical and speculative consideration by Woolhouse (1975). Some evidence for transfer enzymes associated with the membrane boundary has now been obtained (Marx *et al.*, 1982). Speculation apart, the morphology of the interface suggests that transfer is likely to consist of loss, passive or active, by one symbiont and uptake by the other, and that these two processes are in some sense adjusted to one another in rate in the functioning system. We are led to assume that it is in the region where healthy cells of both symbionts are juxtaposed (i.e. at the mature interface) that the important part of nutrient transfer is likely to take place.

## Substances transferred

The most complete studies of transfer between mycorrhizal symbionts have been carried out in ectomycorrhizas. The sheath of these organs so insulates the root tissue from direct access of substances from the soil that we may safely assume that everything absorbed passes in some way through the fungal tissue. Direct experimental data are available from the work of Melin and Nilsson (1950–1958) that phosphate compounds, cations including calcium and nitrogenous substances derived from ammonium or glutamic acid were transferred from the substrate to the host via the fungus; and that $^{14}C$-photosynthates passed from the host to the fungus. Other observers have shown that $^{14}C$-labelled substances may also pass back across the interface into the host. The nature of the actual compounds passing across the interface has been discovered in some instances. Harley and Loughman (1963) by short-term labelling experiments with *Fagus* showed that it was orthophosphate that passed from fungus to host in phosphorus transfer. Harley (1964) and Carrodus (1967) showed that $^{14}CO_2$ is fixed and incorporated largely into glutamine during ammonium uptake by the sheath, and Reid and Lewis (in Lewis, 1976) showed that after being labelled in this way this compound was transferred to the host. The dark $CO_2$ fixation that Harley (1964) showed to occur in beech mycorrhizas (recently shown to occur also in pine mycorrhiza by France and Reid, 1979) is of course an energy-requiring process. In all probability, by analogy with other plant materials, it depends on the reaction of phosphoenol pyruvate with $CO_2$ catalysed by phosphoenol pyruvate carboxylase, with the production of oxalacetic acid, which is rapidly converted to malate or to acids of the TCA cycle. Hence the carbon skeleton of glutamate which is transferred from fungus to host is probably derived from host photosynthate to the extent of three-quarters of the carbon involved, the remaining quarter coming from $CO_2$. There is thus undoubtedly some recycling of carbon derived from the host through the fungus and back to the root. The system can be compared with that operating in leguminous root-nodules where incorporation of ammonium into organic combinations requires carbohydrate (in addition to that powering nitrogen fixation) delivered via the phloem and subsequently re-exported from the nodule as amino acids, amides and ureides into the xylem (see Minchin and Pate, 1973; Sprent, 1980). A similar recycling of amino acids in corals as described by D. C. Smith (1978) is equally complex because the animal releases breakdown products of proteins which it cannot itself directly use, and these are synthesized into amino acids and proteins by the alga which are in fact passed back to the animal.

In mycorrhizas nitrogen accumulation may be from ammonium, nitrate or organic sources. The soil conditions will affect the availability of these three forms so that nitrate will be a significant nitrogen source only in neutral or alkaline soils and especially on disturbed sites where nitrification rates are high. In any event, since the demand for nitrogen is greater than that for other nutrients, there must often be some form of redress of ionic balance especially following the absorption of nitrate or ammonium. This may involve one or more of the processes of proton extrusion, cation uptake or loss, or organic acid synthesis via dark $CO_2$ fixation. Such of these activities as occur must be properties of the fungus in ectomycorrhiza and probably also in endomycorrhizas (Harley, 1964; see Raven and F. A. Smith, 1976; Raven, S. E. Smith and F. A. Smith, 1978; S. E. Smith, 1980; Bowen and S. E. Smith, 1981).

As has been described above, [14]C-compounds derived from photosynthesis were shown by Melin and Nilsson (1959) to be transferred from the leaves to the roots and thence to the fungus in ectomycorrhiza. Lewis and Harley (1965a–c, and see Harley and Lewis, 1969) have pointed out that the translocated sugar of the phloem in *Fagus* and probably in most ectomycorrhizal hosts is sucrose. It is, however, unlikely that the disaccharide passes across the interface into the fungus. Harley and Jennings (1958) showed that exogenous sucrose applied to mycorrhizal roots of *Fagus* was rapidly hydrolysed by an enzyme system on the fungal surface to its constituent glucose and fructose moieties, and that glucose was preferentially absorbed from the resulting glucose–fructose mixture. Lewis and Harley (1965b) showed that absorbed glucose was mainly converted to trehalose and glycogen in the fungal sheath, whereas mannitol was the main destination of fructose. Since in their experiments, in which [14]C-sucrose was applied to the host tissues of excised mycorrhizas, [14]C-mannitol was produced in large quantities in the sheath along with trehalose, it is highly probable that both glucose and fructose are transferred across the host–fungus interface.

The cation potassium is probably able to move readily to and fro across the interface. As shown by Harley and Wilson (1959) and Wilson (1957), monovalent cations are absorbed by *Fagus* mycorrhiza and part passes through the sheath to the host, but at first there is an accumulation in the fungal sheath. However, at temperatures above $20°C$, potassium is readily lost from mycorrhizal tissue and as Edmonds *et al.* (1976) showed, the cumulative loss may be very great—more than the $40 \mu mol$ present in the sheath of $100 mg$ of mycorrhiza. Hence potassium must have moved from the living host tissues through the fungus into the bathing medium.

Other kinds of mycorrhizas have not been studied to the same extent, although the facts that are known about them do fit the same general picture. For instance, the carbon nutrition of ericoid mycorrhiza is not greatly dissimilar from that described for ectomycorrhiza. If trehalose is the form in which carbohydrate is translocated in orchid mycorrhizal fungi, then its hydrolysis to glucose by trehalase (possibly of orchid origin) probably precedes transfer of hexose to the orchid and synthesis of sucrose (S. E. Smith, 1967; S. E. Smith and F. A. Smith, 1973). Speculative comparisons based on what experimental data are available and on similarities of general physiology have been made (see Lewis, 1975, 1976; Raven *et al.*, 1978; S. E. Smith, 1980). Other biotrophic associations also provide useful ideas. It is noteworthy, however, that the evidence available suggests that simple molecules or ions are transferred between the symbionts.

## Mechanisms involved in transfer

Exudation or loss of organic molecules from the cells of roots is quite normal. The compounds lost include soluble sugars, amino acids, organic acids and a great variety of other compounds in considerable quantities (Barber and Martin, 1976; see, for instance, Rovira, 1965; and Scott Russell, 1977). In addition, around the apex and primary active part of the root insoluble mucilage (mucigel) is secreted, which is composed largely of carbohydrate. This mucigel forms a layer up to $50 \mu m$ thick around the root surface (see Oades, 1978). The microorganisms of the rhizosphere and root surface are in a position to absorb nutrients from the soluble exudates and mucigel, given suitable enzyme systems. The quantities of organic substances released in this way are often very large even from root systems growing in sterile conditions and they may be increased by the activity of microorganisms around them, as well as by the soil conditions and the physiological state of the plant (see Scott Russell, 1977). Mycorrhizal fungi growing in close proximity to the cortical and epidermal cells of the root appear to be in a position to compete for exudates with advantage. Indeed, failing other information, it could be argued that on the grounds of the quantity of exudates often produced and of the advantageous position of mycorrhizal fungi, it should not be assumed that it is essential for them to have any influence, or produce any factor, that will increase exudation. However, not only is the loss of carbohydrates from roots into non-sterile soil often nearly double that into sterile soil, but there is evidence in other symbiotic systems that the autobiont becomes more leaky in association with the heterobiont than alone. A good example is the carbohydrate leakiness of the algae of lichens which cease to leak if cultured alone (Green and D. C. Smith, 1974; see D. C. Smith, 1975). Olsen (1971) has shown that triterpene glycosides such as avenacin and aescin inhibit the uptake of $K^+$, $Mg^{2+}$ and inorganic phosphate, and promote the leakage of these ions from the mycelia of a number of fungi including *Ophiobolus* (*Gäumannomyces*) *graminis* and *Pythium irregulare* amongst the plant parasites. There are many other examples of fungal toxins which alter the permeability of host plasmalemmae (see Wood, 1967).

Experiments of Pederson (1970) with the mycorrhizal fungus *Suillus* (*Boletus*) *variegatus* showed that fatty acids such as acetic at concentrations between 0·125 mM and 128 mM reduced respiratory rate. In a comparison of acids of different numbers of carbon atoms there was an increased effect with increasing chain length up to 8 carbon atoms, at 2 mM at pH 4·85. Methyl octanoate had the same effect as the acid on respiration, and like the fatty acids caused a leakage of u.v. absorbing substances. In subsequent papers (Lode and Pederson, 1970; Pederson and Lode, 1971) it was shown that the substances which leaked were relatively simple nucleotidic compounds with insignificant amounts of proteins and ribonucleic acids. The effects of the acids, like those of triterpene glycosides, appeared to be on the fungal membranes. With increasing chain length the lipophilic effect and the consequent membrane disruption would be expected to increase. Pederson took the view that the correlation of the reduction of oxygen uptake and leakage indicated an effect on internal membranes as well as upon the plasmalemma. In his comparison of the effects of acetic acid on various fungi in culture, Hintikka (1969) examined the reactions of four mycorrhizal basidiomycetes, *Suillus* (*Boletus*) *bovinus*, *S.* (*B.*) *variegatus* and *Rhizopogon* sp.; all were intolerant and only *S.(B.) elegans* and *S.(B.) variegatus* grew at the lowest concentration tested (0·8 mM). The relatively unspecific effect of these compounds, their apparent effect on internal membranes, and their lethality do not lead one at once to believe that they or compounds like them are likely to be operative in transfer between symbionts. Nor is it likely that an increase in general permeability of the associated membranes of the symbionts would provide a viable mechanism for transfer of material. The effect of decreased phosphate supply on the leakiness of host membranes described by Ratnayake *et al.* (1978) may have an effect on initial colonization, but it is hard to see how such a mechanism could operate in a mature mycorrhiza, although D. H. Lewis has suggested that the sequestration of phosphate in polyphosphate synthesis might engender a deficiency and affect permeability of plasma membranes (personal communication).

Cell-wall deposition is also a normal means by which carbohydrates are transferred (probably via golgi vesicles) from an intracellular to an extra-cellular location. Such deposition is clearly altered in the interfacial matrix of endomycorrhizas and will be discussed below as a possible source of carbon compounds to the fungus.

Changes in the membrane structure in the bacterioid cells of soybean nodules have been revealed by freeze fracture techniques (Tu, 1979). As the bacteroids develop, intramembrane protein particles (IMP) increase in frequency in the host membrane. The IMP might be enzyme systems concerned with the transfer of carbohydrate into, and organic nitrogen out of the cells. Concurrent changes in the distribution of aparticulate areas of the membrane were interpreted to be the result of vesicular fusions, possibly associated with membrane transport. Recently the membrane-bound ATPase enzymes have been studied cytochemically in both vesicular-arbuscular mycorrhizas (Marx *et al.*, 1982) and in the association of *Erysiphe* with its host (Spencer-Phillips and Gay, 1981). ATPases are associated with the normal uninvaginated plasmalemma and there is good reason to suppose that they are involved in membrane transport of ions and other substances. These ATPases become redistributed during arbuscule formation, so that the activity is concentrated on the invaginated membrane adjacent to the mature arbuscules (which themselves possess membrane-bound ATPase). Plasmalemma surrounding young arbuscules or degenerating arbuscular branches does not exhibit ATPase activity. The peripheral uninvaginated plasmalemma of infected cells shows reduced activity. Marx *et al.* (1982) conclude that the redistribution of enzyme activity may be significant in the transfer process. In cells invaginated by haustoria of *Erysiphe* no such redistribution of ATPase activity occurs (see Plate 15).

Infection of roots by vesicular-arbuscular mycorrhizal fungi is also associated with the development of phosphatase enzymes of alkaline pH optimum (Gianinazzi-Pearson and Gianinazzi, 1976; 1978). The enzyme activity has been found to be closely associated both with the arbuscular phase of mycorrhizal development and with the occurrence of a mycorrhizal growth stimulation in the host plant. The alkaline phosphatases are present in hyphae separated from the roots by digestion techniques (V. Gianinazzi-Pearson, personal communication) and similar (possibly identical) enzymes have been shown to be localized in the fungal vacuoles by electronmicroscopical cytochemistry (Gianinazzi *et al.*, 1979). The occurrence of enzymes involved in polyphosphate turnover were discussed elsewhere and it is clear that infection of roots results in considerable changes in the character and distribution of enzymes involved in phosphate metabolism, some of which may well be concerned in the transfer of phosphate from fungus to host.

Concentration differences across the membranes or interface may be very important in transfer mechanisms. Movement of ions would be affected not only by concentration but also by electrical potential gradients. Nothing is known about such gradients across host/fungus interfaces. However there is some information about factors which may affect concentration differences. Changes in amounts of soluble carbohydrates in plant cells are frequently associated with fungal infection. Decreases in starch are often observed in the cells in mycorrhizas infected by fungi or enclosed by the Hartig net. The most commonly detected changes, apart from synthesis of specific fungal carbohydrates, have been the rise in hexose concentration often associated with increased invertase activity and a decline in sucrose. In *Puccinia poarum* (Yuen quoted in Long *et al.*, 1975) hexose transfer to the fungus was confirmed. In the lichen *Peltigera polydactyla* repression of glutamine synthetase results in raised levels of both glucose and $NH_4^+$ in the alga and these compounds pass readily to the fungal symbiont (via the apoplastic space). If the algae are isolated from the fungus the repression of the synthetase is removed and the $NH_4^+$ and glucose are incorporated into glutamine in the alga (see D. C. Smith, 1978). Increased $NH_4^+$ levels are also associated with loss of photosynthate by the coral alga *Gymnodinium microadriaticum* (Taylor, 1978).

An increase of hexose concentration coupled with a decrease of sucrose concentration has been observed following mycorrhizal infection in *Goodyera repens* (Purves and Hadley, 1975) but not in *Vaccinium macrocarpum* (Stribley and Read, 1975). This kind of comparison, however, would certainly be unsuitable with ectomycorrhiza because of the presence of the large mass of fungal sheath tissue—which contains trehalose rather than sucrose together with the hexoses. It is not clear how far the presence of the fungus in orchid or ericoid mycorrhizas can vitiate the comparison with them. In any event failure to detect elevation of hexose levels does not necessarily mean that enzyme changes have not occurred, for the actual concentrations depend on relative rates of formation of hexoses, their uptake rates by the fungus and their conversion to fungal carbohydrates.

The directed transport of photosynthate to the fungal layer of ectomycorrhiza is frequently explained (*see* Slankis, 1973) by hormonal substances released by the fungus which act upon the host cells. There is no doubt that such an effect is possible but its operation has not been confirmed by direct experiment or explained in detail. The same applies to the mechanism outlined by Wedding and Harley (1976) who examined the possibility that a fungal metabolite produced in quantity might have an effect on the enzymes concerned with carbohydrate metabolism of the host. Mannitol was found so to affect the enzyme systems as to result, if it were to penetrate the host, in an increase in hexose concentration (see Fig. 17). It is possible to argue that since all root systems exude carbohydrates, the net effect of the suggested activity of mannitol would be to increase exudation quantitatively without any additional effect on membrane permeability.

It was suggested earlier (Harley, 1969a) that in ectomycorrhiza the fungus could interfere with the normal synthesis of cell walls in the root-cap and Hartig net regions, so diverting the precursor carbohydrates into the hyphae. The mechanism might be that envisaged by Vanderplank (1978). The proteins of the hyphal surface (the presence of carbohydrases and phosphatases there has been demonstrated by Harley and Jennings, 1958; Bartlett and Lewis, 1973; Williamson and Alexander, 1975; Calleja *et al.*, 1980) might copolymerize with the cell-wall condensing enzymes of the root so that the formation of the cell-wall polymers was inhibited or altered and soluble carbohydrate, released through the root cell-membranes, made available to the fungus. The suggestion by the Italian workers that glycoproteins may be present in the interfacial zone is relevant, for these compounds have been shown to have considerable and sometimes specific binding capabilities. Duddridge (1980) showed that the fungus in *Pinus* mycorrhizas distorted and softened the walls of the epidermis and cortex to produce the "contact layer" or "involving layer" in which the Hartig net lay (Plate 9). Similarly the fungal infection in *Eucalyptus* was observed by Ling Lee *et al.* (1977) to reduce the formation of starch and the release of slime in the root-cap, while intracellular hyphae of *Glomus tenuis* in raspberry roots apparently spread by destruction of the host middle lamella and separation of the walls of adjacent cortical cells (Gianinazzi-Pearson *et al.*, 1981). Evidence for alteration of wall deposition is provided by electron microscopical study of vesicular-arbuscular mycorrhiza, where invaginated plasmalemmae of root cells are apparently very active in producing carbohydrate, but the aggregation of this into cell-wall material is modified in the presence of the fungus (Dexheimer *et al.*, 1979; Bonfante-Fasolo *et al.*, 1981). Active synthesis and deposition of primary cell-wall material is typical of cells where the plasmalemma is under tension (normally in the elongating region of a root). It has been suggested (V. Gianinazzi-Pearson, personal communication) that extension of the plasmalemma during invagination by biotrophic fungal hyphae may act as a trigger to carbohydrate movement through the host plasmalemma in infected cells behind the elongating region. However, we must clearly bear in mind that the encapsulation of collapsing fungal structures also involves carbohydrate deposition under circumstances where the extension of the plasmalemma is unlikely. Possible interactions between fungus and root resulting in carbohydrate transfer between them are illustrated in Fig. 37.

We have less information on the possible means of transfer of soil-derived nutrients from the fungus to the host. Much has been made of the possible translocation of polyphosphate within the hyphae of vesicular-arbuscular mycorrhiza. But it is unlikely that polyphosphate is actually transferred across the interfacial apoplast. It has, however, been demonstrated that inorganic orthophosphate passed from the fungus to the host in *Fagus* mycorrhiza and that the movement through the sheath tissue from the soil to the host tissue is through the symplast of the sheath. Polyphosphate synthesis and degradation seem most likely to play a part in controlling the levels of inorganic phosphate within the fungal cytoplasm, as these processes have been shown to do in other microorganisms. Localization of polyphosphate in vacuoles may well serve to isolate soluble polyphosphate osmotically from the cytoplasm, as there is now some doubt about the existence of insoluble polyphosphate granules to the extent previously believed. Continuing hydrolysis of polyphosphate and transfer of orthophosphate across the tonoplast would certainly provide a means of maintaining a high cytoplasmic concentration of inorganic orthophosphate adjacent to the interface, which would (assuming appropriate electrochemical potential gradients or transport systems) facilitate phosphate transfer into the apoplast and hence across the interface (see Fig. 38). There is no need to involve mechanisms peculiar to mycorrhizas, although future work may conceivably show that some exist.

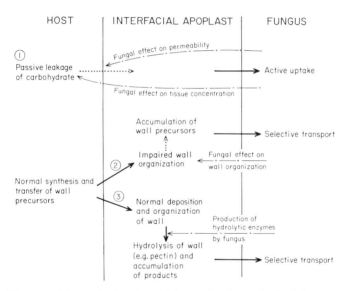

FIG. 37. Three possible mechanisms by which transfer of carbohydrate from autotrophic host to fungus might occur in mycorrhizas of different types and at different stages of development. Passive transport, ········▸ ; active or selective transport, ───────▸ ; fungal influences, ──·──·──▸ . (1) Passive leakage from host to apoplast, followed by active transport into fungus. Fungal influence could be on membrane permeability and/or concentrations of molecules in the host cytoplasm. (2) Normal synthesis of wall material. Wall deposition and organization affected by fungus so that wall precursors accumulate in the interfacial apoplast and are selectively absorbed by fungus. (3) Both wall synthesis and deposition are normal. Fungus produces hydrolytic enzymes which degrade the wall. Products accumulate in the interfacial apoplast and are selectively absorbed by fungus.

*Uptake from the apoplastic space*

Roots, fungi, the fungal sheath and the host core of ectomycorrhizas can absorb, by metabolically dependent means, a wide range of nutrients. Net transfer between the symbionts must depend upon the relative rates of passive (or active) release and active uptake by the opposed membrane (see Fig. 37). The suggestion that transpiratory water flow might affect the process (Woolhouse, 1975) seems at first sight credible, until one realizes that exudation from a root surface of multifarious compounds and of ions is not prevented and perhaps not significantly affected by inwards water movement. In any event it would be expected that if active release into the apoplastic space occurred, water throughput in the opposite direction would maintain or increase the concentration of the released material in that space, by preventing loss by passive leakage from it into the surroundings. The osmotic relations of root and fungal cells must be complicated, particularly when intracellular penetration and plasmalemma invagination occur.

The suggestion has been made by Woolhouse (1975) and by others that carbohydrate uptake into the fungus from the host might be coupled to breakdown of polyphosphate in its hyphae. No adequate details have been described. It might be conceived that polyphosphate kinase in the presence of

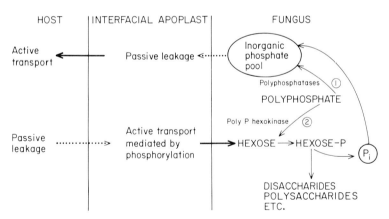

FIG. 38. Phosphate and hexose transfer between fungus and host. Passive leakage from one organism ( ·······► ) and active uptake by the other ( ────► ) is illustrated. Possible pathways by which polyphosphate metabolism might be linked to phosphate transfer are also shown ( ────► ). (1) Polyphosphate hydrolysis by polyphosphatases might raise or maintain the concentration of inorganic phosphate within the fungal pool, leading to increased passive efflux. (2) Polyphosphate breakdown could be linked to hexose uptake into the fungus during synthesis of hexose-phosphate, catalysed by polyphosphate hexokinase. Subsequent metabolism of hexose-phosphate could result in release of inorganic phosphate, which would raise the inorganic phosphate concentration.

excess ADP would generate ATP at the site of transfer. From the ATP, with hexokinases, hexose phosphates might be formed. Alternatively the hexose phosphate might arise from the action of polyphosphate hexokinase in the presence of hexose. The phosphorylated hexoses might be used in metabolism or in the formation of disaccharides, polysaccharides or other fungal storage substances and the phosphate released. Both the kinases have been detected in vesicular-arbuscular mycorrhizas and their extramatrical hyphae by Capaccio and Callow (1982). These reactions might reduce the cytoplasmic concentration of hexose in the fungus and so maintain a concentration gradient across its plasmalemma. If the quantity of polyphosphate in the fungus were limited, the maximum hexose that could be absorbed by this mechanism would also be limited and would be one mol per mol of phosphate transferred. A mechanism hitched to oxidative phosphorylation might then provide a more effective and continuous driving force. However, it is possible that polyphosphate is generated continuously during absorption of phosphate from the soil by the fungus. Polyphosphate concentration at the transfer surface might then be maintained by translocation and low hexose and high inorganic orthophosphate continuously generated there. As a result gradients allowing their movement across the apoplastic space would be maintained. It is to be noted that by this mechanism hexose and phosphate transfer would be reciprocal and quantitatively related, and so the mechanism would be readily testable. In the form described the membranes are assumed to leak into the apoplastic space and perhaps subsidiary assumptions have to be made to explain this and also to explain what cation if any is associated with the passage of phosphate.

The mechanisms which would favour the maintenance of a concentration gradient of (uncharged) carbohydrate molecules in the direction of the heterotroph have been extensively discussed (D. C. Smith et al., 1969). In particular the conversion of hexose molecules to mannitol, trehalose and glycogen in the fungal sheath of ectomycorrhizas would favour hexose transfer and a net carbohydrate transfer, for trehalose and mannitol are not readily utilized by the root cells. A similar system could operate in ericoid mycorrhizas, which apparently have the same array of carbohydrates (Stribley and Read, 1975). It should be noted that this mechanism depends upon phosphorylation of incoming hexose. It might therefore be associated also with phosphate absorption and polyphosphate turnover in the fungus.

## Lysis of hyphae

The collapse of hyphae is frequently observed in vesicular-arbuscular, ericoid, arbutoid and orchid mycorrhiza and was mentioned earlier in this chapter. There is little evidence which would help us to determine why such collapse takes place and whether it is mediated by the host or is a result of senescence of the fungus. Hyphal collapse was well known to the earliest workers on mycorrhizal symbiosis and, as Rayner (1927) mentioned, Frank and others viewed it as a mechanism whereby the host obtained a supply of food, especially nitrogenous compounds. Noel Bernard by contrast was unable to accept their hypothesis and compared "digestion" of the fungus with phagocytosis, the defence reaction of animal cells. It is implicit in these views, which are both canvassed to this day, that the host cells are active in the destruction or disorganization of the fungal hyphae. Indeed three views are taken. Fungal collapse is (i) the result of a defence reaction of the host; (ii) has been evolved to facilitate nutrient transfer from fungus to host; (iii) is a consequence of hyphal age and senescence. It is quite clear that it is not essential to fungal symbiosis for it is absent from pathogenic associations and from ectomycorrhizas and lichens, but of course the fungus is not usually intracellular in the last two examples.

Whatever the underlying reason for fungal collapse, the nutrients present in the intracellular hyphae must either be reabsorbed by the fungus or pass into the apoplastic space and be available for absorption by the host cells; any selectivity or even specificity in the transport process would then be exerted by the host plasmalemma and absorptive processes. It has frequently been assumed that such a selectivity, almost amounting to specificity to the uptake of one element, does exist and emphasis has been placed on the transfer of phosphorus in vesicular-arbuscular mycorrhizas, of nitrogen in ericoid mycorrhizas and of carbon compounds in orchid mycorrhizas. Experimental evidence which bears on this problem is scanty, but reports that the absorption rates of trace elements like Zn, Mn, etc. may be increased, or following Hatch's (1937) views on ectomycorrhiza that infection results in the increased absorption of those elements in short supply, suggest that the selectivity depends upon "host demand". In any event all nutrients passing into ectomycorrhizas pass via the sheath cells, so selectivity in them is the equivalent simply of selectivity of a root surface. Some nutrients are absorbed at different rates and in different quantities from others and some compete with others in their uptake mechanisms.

Suggestive evidence that carbohydrates move steadily from fungus to orchid seedlings (rather than by spasmodic "digestion" processes) is provided by the results of S. E. Smith (1967) and Purves and Hadley (1975) (see Fig. 25). The importance of neutral carbohydrates rather than ionized amino and organic acids in the transfer process also indicates some degree of specificity, or there must exist complex enzyme interconversions in the apoplast. These are extremely unlikely to occur. Similar steady transfer of phosphorus to clover and onion seedlings by *Glomus* spp. and through the sheath of ectomycorrhiza have been demonstrated (Pearson and Tinker, 1975; Cooper and Tinker, 1978, 1981; and *see* Harley, 1969a). Cox and Tinker (1976) have calculated, on the basis of arbuscule volume in the root, arbuscule life-span and phosphate content of arbuscules, that "digestion" of arbuscules could only contribute about 1% of the measured phosphate inflow into onions. The temporal association of mycor-rhizal growth response, development of mycorrhiza-specific phosphatases and active arbuscular phase of mycorrhizal development—together with localiz-ation of membrane-bound ATPases on the plasmalemma associated with active (but not young or collapsed) arbuscules—also lends weight to the role of mature arbuscules in phosphate transfer. Moreover, Mollison (1943) and Hadley and Williamson (1971) obtained evidence that growth of orchid protocorms was increased after penetration by the mycorrhizal fungus but before any digestion of intracellular hyphal coils could be observed.

Despite the rather scanty evidence against the occurrence of a non-specific "digestion" process, it is worth noting that a collapsed arbuscule or hyphal coil surrounded by the plasmalemma of the root cell can be likened to a simple protozoan food vacuole, as indeed Noel Bernard noticed at the turn of the century. But these do not occur in those ericoid mycorrhizas where collapsed hyphae are not observed in living cells. Moreover, *active* phagocytosis by the root cell cannot take place because of its cell-wall, so continued digestion depends upon the continual growth of the fungal hyphae into the lumina of the cells, and requires an agent or factor attracting the fungus to invade again and again. Isolated compartments in the cell into which the fungi could grow, and into which enzymes could be secreted, might exist and have been postulated to be important in the transfer of organic nitrogen compounds to the roots of *Erica bauera* (Read, 1978). The process of hyphal collapse occurs seasonally in this species of *Erica* as it does in many orchids (Burges, 1939; Vermeulen, 1946; and *see* Harley, 1969a). It may however occur continuously.

In ectomycorrhizas collapse of fungal hyphae does not occur in a comparable manner, and indeed the hyphae may outlast the cortical cells. In the late-Hartig net region, the autolysis or senescence of the primary cortical cells is encouraged by their being cut off by the cambial activity in the pericycle and by the phellogen in the cortex. The cortical cells become colonized by fungal hyphae which appear to absorb their contents. This recycling of materials back into the fungus is reminiscent of the close recycling described in corals. It is evidence that the interface and mode of transfer of substances change with age even in ectomycorrhizas. Nevertheless, the direct and immediate appearance of [14]C-labelled compounds in the fungal tissues in photosynthesizing seedlings of *Pinus* in experiments of Melin and Nilsson (1957) and the movement of [14]C-carbohydrate from host to fungus and the movement of phosphate from fungus to host in mycorrhizas of *Fagus* lacking the late-Hartig net zones (*see* Harley, 1969a) shows that a movement between living cells is extremely likely to occur and to be much more important.

## Outstanding problems

This chapter has essentially been a rehearsal of outstanding problems, but there are certain general points that may be emphasized. It must be at once clear that we need to know much more (and techniques are available) of the structure of, and the location of enzymes on the interface between the symbionts and the way these are modified as mycorrhizas develop. Then we need to know, above all, more about the nature of the compounds passing between the associated organisms, for then we would be in a stronger position to ask experimentally soluble questions about factors affecting transfer and the mechanism involved. The methods so far used have been few, but have by no means been fully exploited. The "inhibition" technique of D. C. Smith and his colleagues, despite theoretical problems, can certainly be used in ectomycorrhiza in conjunction with the dissection apart of the symbionts. But it is also possible for it to be more widely used in investigations where the symbionts cannot be separated. Progressive labelling techniques have been used by S. E. Smith (1967) in the investigation of transfer of carbohydrates between fungus and orchid protocorms and the further metabolism of the transferred material. They have also been used by Harley and Loughman (1963) in determining the nature of the phosphate passing to the sheath in ectomycorrhiza.

Even these methods and variants of them will give us more knowledge of the substances passing; and following this we need to investigate factors affecting release of specific substances into the external medium by mycorrhizal fungal hyphae or uninfected roots of mycorrhizal plants. Non-specific damage or change to the membrane is not relevant no matter how brought about. There is certainly a case for examining such mycorrhizal fungi as can be successfully grown in culture and determining in the first place what substances they release into the medium and the circumstances of and factors affecting release. Implicit in such research is the determination of the effects of symbiotic union on the internal concentrations of carbohydrates, amino acids, phosphates and other compounds, and of the enzymes associated with these changes of concentration. It would also be important to determine whether there is any relationship between the quantities of phosphate and nitrogen passing to the autobiont and of the carbohydrate passing from it into the fungus.

These suggestions for the future have their roots in fundamental work and have an equal bearing on research on biotrophic plant pathogens and other symbiotic associations such as rhizobial and actinorhizal nodules. It is in such biochemical, physiological and ultrastructural research that advances in understanding symbiosis are likely to be made. They may be usefully associated with experiments on growth of plants, as in that way the mechanisms underlying growth responses may be elucidated. This could lead to informed exploitation of mycorrhizal symbiosis, which empirical description of growth responses alone is unlikely to do.

## Chapter 18

# Specificity and Recognition in Symbiotic Systems

## Introduction

Recent writers (e.g. Lewis, 1973; Heslop-Harrison, 1978) have to some extent failed, by omission or commission, to make clear the degree of specificity which exists between mycorrhizal fungi and their hosts, and how specificity differs from that of other biotrophic fungal symbionts. For instance, Heslop-Harrison presents a table in which it is clearly stated that the angiosperm root systems in association with mycorrhizal fungi exhibit a "high or very high" degree of specificity. It will be clear to those who have read the rest of this book that at best this statement requires qualification. Indeed, there is little evidence for close specificity between hosts and fungi at the level of species, let alone at the level of different host biotypes (at the sub-specific level) or at the level of physiological races or strains of fungi. It is this latter degree of specificity which we have come to expect in associations involving biotrophic pathogens, and which has been frequently discussed (see reviews by Brian, 1967; Flor, 1971; Lewis, 1973; Day, 1974). P. W. Brian recognized the gulf between mycorrhizal fungi and plant pathogens in this respect for he stated (1976): "But there are exceptions to this generalisation [the strong correlation between obligate biotrophy and high specificity]. . . . I have in mind the very wide host range of the fungal partners in vesicular-arbuscular mycorrhizae . . . these exceptions need explaining." Unfortunately mycorrhizas were not mentioned again in the multi-author volume entitled "Specificity in Plant Disease" (edited by Wood and Graniti, 1976), but an explanation has been put forward briefly in very general terms by Vanderplank.

Vanderplank recognized that the selection pressures on mutualistic and antagonistic biotrophic associations are very different. The last paragraph of his book (1978) is extremely enlightening and reads:

> Opposite selection pressures are clearly involved. In parasitic symbiosis the host plant benefits by mutation to resistance because this ends, for the host, an unwanted

symbiosis. In mutualistic symbiosis the host loses by mutation to resistance because this ends the symbiosis. Mutations to resistance in mycorrhizal plants are eliminated by selection because they are disadvantageous; and the elimination also eliminates a major source of specificity.

It is the role of this essay to compare specificity in different biotrophic associations, and to put forward suggestions which may help further to explain the differences or at least to generate thought and experiment about them. Mycorrhizas, which are normally mutualistic, will be compared with the *Rhizobium*/legume symbiosis which is also mutualistic, and with symbioses between hosts and biotrophic parasites which are antagonistic.

## Degrees of specificity

There are several levels of specificity between host and fungus which must be clearly distinguished and understood before there can be any meaningful discussion of the problems. At the taxonomic level we may ask the question whether a given species of fungus forms a mycorrhizal relationship with more than one species of host, and in the contrary case whether a given species of host associates mycorrhizally with more than one species of fungus.

As we have seen, there is very considerable strain variation in many of the mycorrhizal fungi especially in those of ectomycorrhiza, so there may be a yet finer level of specificity. We therefore may ask whether the genotypes of a species (fungus or host) which forms a relationship with several taxa of symbionts can be distinguished so that there are recognizable genotypes peculiar to each associated taxon or genetic strain of that taxon. That is, are the subspecific genetic strains of fungus attuned in some way to the species or subspecific genetic strains of the host?

At a still finer level we ask whether there are in mycorrhizal symbionts anything resembling the genes for "resistance" or "susceptibility" or "virulence" that have been recognized in some kinds of antagonisitic symbioses (e.g. Uredinales).

Finally we must consider ecological specificity. By this we imply that although in cultural conditions a symbiotic association between two component organisms may be set up, the kind of mycorrhiza so formed may not occur in natural or ecological conditions. That is, specificity is closer in competition or in the available natural habitats than in axenic culture. This level of specificity is comparable to ecologically determined interspecific infertility between closely related species that can be shown by experiment to be interfertile (see Briggs and Walters, 1969).

In our discussion there seems to us no need to take any cognizance of the theoretical arguments as to whether those kinds of mycorrhiza in which lysis or death of cells occurs are biotrophic. This is particularly because in every kind of mycorrhiza there is a phase, sometimes prolonged, where active undamaged partners are in association. Harley (1969a) gave attention to the problem of specificity in ectomycorrhizas, ericoid mycorrhizas, orchid mycorrhizas and considered in a minor way, vesicular-arbuscular mycorrhizas. There was, however, at that time, not enough collected information of adequate reliability to allow a general discussion. Now that the taxonomy of mycorrhizal fungi is on a much sounder basis, relevant experiments and theoretical considerations can be discussed with a little more satisfaction and with some hope of engendering hypotheses for experimental appraisal.

## Specificity in mycorrhizas

Most plants are susceptible to mycorrhizal infection of some sort. Although the members of a few genera or families are not usually infected, even some of these may become inhabited by known mycorrhizal fungi if grown in juxtaposition with suitable heavily infected host plants. The "non-mycorrhizal" taxa present a problem for future investigation to determine their mechanism of "resistance" or "avoidance" of infection.

The ability of a single species of host to form mycorrhizas of more than one sort is widespread and will be discussed below. Here it is necessary to recall that, under natural conditions, species of many other genera may form either ectomycorrhiza or vesicular-arbuscular mycorrhiza; species of *Pinus* may form ecto- or ectendomycorrhiza; species of *Arctostaphylos* may form ecto- or arbutoid mycorrhiza (see Table 61). Perhaps it is not surprising that this degree of toleration of taxonomically diverse symbionts should exist. The widespread occurrence in and around all roots of many kinds of "dark sterile" mycelia is widely recognized. These merge almost imperceptibly into the "pseudomycorrhizal" fungi of ectomycorrhizal hosts and into the constellation of sterile endophytes isolated from Ericales and so on. Perhaps we should keep an open mind at this stage, that there may be a very great many fungi which higher plants tolerate as denizens of their root surface, senescing cells, and even as cell endophytes (see, for instance, Harley, 1950; Warcup, 1971). This amounts to asking the question whether tolerance is general and intolerance, resistance or exclusion are specific for certain kinds or classes of fungi.

The fungi which form mycorrhiza, of whatever sort, have at least one characteristic in common. Their hyphae and mycelia are not evanescent but long-lived and they belong to groups of which some members form fruit bodies with sterile tissues. They are in a real sense perennial fungi and in contrast to ephemerals, need a permanent food base which they can exploit and from which they translocate materials to their growing apices. There are two different groups of perennial fungi in the soil: (i) fungi capable of using the resistant substrates of wood, litter and the like, i.e. using, as it were, a large but limited expendable source; (ii) fungi inhabiting roots and using the products of current or recent photosynthesis. For the latter non-specificity to particular hosts is of selective advantage and ensures that there will be minimal spatial and seasonal discontinuites in their habitat. Moreover, as has been mentioned earlier, the existence of cellulolytic and lignoclastic genotypes in their populations (provided their mycelium is dikaryotic or diploid) would add an extra dimension to the factors promoting persistence even if these genotypes are not themselves mycorrhizal.

## Ectomycorrhiza

The number of kinds of ectomycorrhiza of essentially similar structure, differing only in detail, which may be formed by a single species of host, may be very large. Each is usually formed in association with a different kind of fungus. The early work of Melin (1927) and Dominik (1955) on the classification of ectomycor-rhizas emphasizes this, and it has been re-emphasized by Zak (1973) and Trappe (1967). Indeed the description of the distribution of types of mycorrhizas has become a valuable tool in ecological investigation. One example is to be found in the work of Boullard and Dominik (1960, 1966) who described ectomycorrhizas in the *Fagus sylvatica* forests of Poland and France, and showed a range of variation which emphasized the lack of close specificity in ectomycorrhizal relations. A further more direct kind of evidence is illustrated in Table 78 which shows some of the species of fungi which have been demonstrated to form ectomycorrhizas with *Pinus sylvestris* and *P. strobus*. Although it is not complete, for the information dates from 1962 with a few later additions, it clearly emphasizes the lack of specificity at the taxonomic level. There is no point in labouring such evidence for ectomycorrhizas for further examples could be legion. There is no doubt that this particular sort of mycorrhiza, with complex structure, with sheath, Hartig net, infrequent intracellular penetration and characteristic host histology, may be formed by fungi of very diverse taxonomic origin in a single host. By contrast, Table 79 shows the potential for a given species of fungus to form similar mycorrhiza with many hosts. To underline this Trappe (1962) gives 8 species of fungi proved in culture to form ectomycorrhizas with *Betula pendula* and *B. pubescens*; but recently Giltrap (1979), in the course of work on other aspects of ectomycorrhizal symbiosis, added some 40 species of Basidiomycetes to the list of fungi capable of forming ectomycorrhizas with the species of *Betula*.

Although many of the hosts of ectomycorrhiza each form associations with many taxa of fungi there is some variability in this respect and in the extent of the host range of the fungi. At the extreme, *Cenococcum geophilum* (*graniforme*) forms ectomycorrhiza with the species of numerous taxa—well over 20 genera from all vascular plant phyla—and *Pisolithus tinctorius* has a similar and increasing list of host trees, gymnosperm and angiosperm. At the other extreme a few fungi are reported to be very restricted in their hosts. The most quoted is *Suillus grevillii* (*Boletus elegans*) which is widely said to form mycorrhiza with species of *Larix* only, but that has been found not to be true. Indeed, Linnemann (1971) reports it as forming mycorrhiza with *Pseudotsuga menziesii* in culture and Trappe mentions it being reported on field evidence to form mycorrhizas with *Pinus elliottii*, *P. sylvestris* and *P. taeda*. Trappe also reports that species of *Larix* may form mycorrhiza in axenic culture with other fungi, e.g. *L. decidua*, with *Amanita muscaria*, *Suillus* (*Boletus*) *luteus*, *S.* (*B.*)*variegatus* and *Tricholoma psammopodium*.

TABLE 78

List of fungi (by no means exhaustive) proved to form ectomycorrhiza with *Pinus sylvestris* and *P. strobus* in culture

| Genus | Species |
|-------|---------|
| **A. *P. SYLVESTRIS*** | |
| *Amanita* | *muscaria, pantharina* |
| *Cenococcum* | *graniforme (geophilum)* |
| *Clitopilus* | *prunulus* |
| *Cortinarius* | *glaucopus, mucosus* |
| *Lactarius* | *deliciosus, helvus* |
| *Lyophyllum* | *immundum (Tricholoma fumosum)* |
| *Rhizopogon* | *roseolus, luteolus* |
| *Rhodophyllus* | *rhodopolius* |
| *Russula* | *emetica* |
| *Scleroderma* | *aurantium* |
| *Suillus* | *bovinus, flavidus, granulatus, luteus variegatus* |
| *Tricholoma* | *flavobrunneum, flavovirens, imbricatum, pessundatum, saponaceum, vaccinium* |
| | |
| **B. *P. STROBUS*** | |
| *Amanita* | *muscaria* |
| *Boletinus* | *pictus* |
| *Boletus* | *rubellus* |
| *Cantharellus* | *cibarius* |
| *Cenococcum* | *geophilum (graniforme)* |
| *Gyrodon* | *merulioides* |
| *Gyroporus* | *castaneus* |
| *Lactarius* | *chrysorrheus, deliciosus* |
| *Russula* | *lepida* |
| *Scleroderma* | *aurantium* |
| *Suillus* | *granulatus, luteus* |
| *Tuber* | *maculatum,[a] albidum[b]* |
| *Endogone* | *lactiflua[c]* |

*Source*: Trappe (1962, 1964).
[a] Fassi and Fontana (1967). [b] Fontana and Palenzona (1969). [c] Fassi and Palenzona (1969).

*Suillus grevillii* also forms arbutoid mycorrhiza with *Arctostaphylos uva-ursi* and *Arbutus menziesii*, a final demonstration that it is not as specific to *Larix* on a world scale as is believed in Europe.

The species of *Alnus* are examples of host plants which appear to form mycorrhiza with a small number of symbiotic fungi only. Neal *et al.* (1968) found by field observation that probably as few as 5 of the 30 terrestrial species that he found in *Alnus rubra* communities were mycorrhiza formers, including a species of *Alpova* and *Lactarius obscuratus*. In 1973 Froidevaux showed by chemical testing of sheath tissue that *L. obscuratus* was probably an important mycorrhiza-former of *Alnus rubra*. In 1979 Molina tested 28 known mycorrhizal fungi on *A. rubra* in culture, finding only four of them formed ectomycorrhizas with it. These were *Alpova diplophloeus, Paxillus involutus, Astraeus pteroides* and *Scleroderma hypogaeus*. *A. diplophloeus* did not form mycorrhizas with *Pseudotsuga menziesii, Pinus ponderosa,*

TABLE 79

Some of the hosts which have been proved to form ectomycorrhizas in culture with *Amanita muscaria*

| Genus | Species |
|-------|---------|
| *Eucalyptus* | *camaldulensis, calophylla, dalrympleana, diversicola, maculata, marginata, obliqua, regnans, st johnii, sieberi* |
| *Betula* | *pendula* |
| *Larix* | *decidua, occidentalis* |
| *Picea* | *abies, sichensis* |
| *Pinus* | *contorta, echinata, monticola, mugo, ponderosa, radiata, strobus, sylvestris, taeda, virginiana* |
| *Pseudotsuga* | *menziesii* |

*Sources*: Trappe (1962); Molina and Trappe (1982b); Malajczuk *et al.* (1982).

*Pinus contorta, Tsuga heterophylla* or *Larix occidentalis* and he therefore suggested that it might be specific to *Alnus* spp. Laiho (1970) had tested two of his strains of *Paxillus involutus* on *Alnus* with somewhat equivocal results, obtaining weak, often imperfect, mycorrhiza formation on *A. glutinosa* and *A. incana* with one strain and no infection with the other, nor was he able to confirm that *P. involutus* formed mycorrhiza with any species of *Alnus* in the field. These then seem to be the extremes of ectomycorrhizal specificity at the species level, *Suillus grevillii* with *Larix* spp. and few other trees, and *Alnus* spp. with a few fungal partners, particularly *Alpova* spp. which seems almost restricted to it.

Molina and Trappe (1982b) in their work on specificity of ectomycorrhizal associations, recognize from field experience what they call sporocarp-specific host associations. In these a fungus is only known to fruit either in association with a single genus, a single species, or with a wider taxon, e.g. Pinaceae. This is a facet of what we have called ecological specificity. They recognize that the mycelium of a fungal species may exist ecologically in other associations than those in which the fruit bodies are found. In laboratory tests they showed conclusively that fungal species were not necessarily restricted in the formation of mycorrhiza to the species of host with which they formed sporocarps. Molina and Trappe therefore recognize "broad host range" fungi forming mycorrhizas widely with coniferous and broadleaf trees, e.g. *Lactarius deliciosus*, *Laccaria laccata*, *Pisolithus tinctorius*, *Paxillus involutus* etc., conifer-specific fungi, e.g. those of the genus *Suillus*, and genus-specific fungi, e.g. *Hymenogaster albellus* with *Eucalyptus*, and *Alpova diplophloeus* with *Alnus*. In a detailed investigation of a single fungal genus *Rhizopogon* they showed that species of that genus only fruit in nature with Pinaceae. They tested 23 species with *Pseudotsuga douglasii*, *Tsuga heterophylla* and *Pinus contorta* using 29 isolates. Some (11 species) only formed an association with the host with which their fruit bodies were associated in nature, but 12 species formed good mycorrhizas with one or both of the other hosts. In further work (Malajczuk *et al.*, 1982) they studied the fungi capable of forming mycorrhizas with *Pinus radiata* and *Eucalyptus* spp. both often grown in plantations well outside their normal range. Of 22 species of fungus tested, 20 formed mycorrhiza with *Pinus radiata* and 9 or 10 formed them with each of 11 *Eucalyptus* spp. As they remark, the success of *Eucalyptus* and *Pinus radiata* in the establishment of exotic plantations and as invaders of indigenous communities can be attributed in part to their compatibility with "broad host ranging fungi". On the other hand, each possesses genus-specific associates with which they can form mycorrhiza. Malajczuk *et al.* point out that there appears to be a change from broad host range fungi to more specific fungi as plantations age and that this may have implications in reafforestation, especially in establishment of a second rotation of the crop. A more fundamental point, however, is that when seedlings of the host are confronted with incompatible fungi an early reaction may be the production of phenolic compounds in the epidermis and cortex. This kind of reaction was previously noted in the experiments of Molina (1981) on *Alnus* and *Paxillus involutus* and those of Grand (1968) on the host range of *Suillus* spp. This reaction clearly needs further investigation in respect of the compounds formed and their precursors, in respect of their detailed effect on the fungi and in respect of the genetics of its formation and effect.

We can conclude that close specificity is not a common characteristic of either ectomycorrhizal hosts or fungi. The specificity of ectendomycorrhizal fungi, in so far as there is evidence, is also not close. It has been shown for instance that the same fungal strain may produce ectendomycorrhiza in *Pinus* seedlings and ectomycorrhiza in those of *Picea*.

*Vesicular-arbuscular mycorrhiza*

A similar lack of specificity is expressed in vesicular-arbuscular mycorrhizal relationships. In considering the information available we must set aside for the present any quantitative differences in the response of the host plant to infection by different fungi with respect to growth or yield. In some articles (e.g. Mosse, 1975) specificity is considered in those terms—here we must first consider qualitative presence or absence of mycorrhizal infection. There is no clear evidence that any absolute specificity exists between taxa of vesicular-arbuscular mycorrhizal fungi and taxa of potential host plants and it is indeed of interest that even before modern methods were available investigators like Magrou (1936), Stahl (1949), Gerdemann (1955) and Koch (1961) had reached this conclusion. In general (accepting that a few plant families do not form mycorrhizas or usually form another type of mycorrhiza), a vesicular-arbuscular fungus isolated from one species of host plant can be expected, with reasonable confidence, to infect any other species which has been shown to be capable of forming vesicular-arbuscular mycorrhizas. Baylis (1962) refers to this type of fungus as the "Catholic Symbiont", while Nicolson (1967) entitled another excellent review article "VA mycorrhiza. A universal plant symbiosis". Although these reviews were both written before the modern reappraisal of the taxonomy of Endogonaceae (see Gerdemann and Trappe, 1974), their conclusions are still valid. Vesicular-arbuscular mycorrhizal fungi combine wide host range with permanence of association—a condition which (if we extend Garrett's analogy) can be likened to the sailor with a wife in every port rather than the Don Juan whose associations were more fleeting. This is illustrated in Table 80 which lists several species of vesicular-arbuscular mycorrhizal fungi together with the species of host plants which they have been shown by Gerdemann and Trappe (1974) to infect. The number of potential hosts could be considerably extended by listing the host/fungus combinations used in the very large number of experiments that have been carried out with taxonomically well-defined species of symbiont. In a few investigations, however, some of the fungi are reported to have a more restricted host range. For instance, in a survey of 19 species of host plant *Glomus gerdemanni* formed mycorrhizas only with *Eupatorium odoratum* (Graw *et al.*, 1979). By contrast, in field investigation of the mycorrhizal associates of a single genus, *Festuca*, Molina *et al.* (1978) found that in the Western United States and Canada species of *Festuca* were usually extensively infected with vesicular-arbuscular mycorrhizal fungi, as indeed they are elsewhere in the world (Crush, 1973; Sparling and Tinker, 1975a,b). There was no specificity amongst the eleven species of mycorrhizal fungi recognized, which belonged to the genera *Glomus*, *Gigaspora*, *Acaulospora* and *Sclerocystis*, although they showed some differences of frequency and variation in intensity of infection. Similar differences in intensity of infection of different species of plant in a sand-dune ecosystem have been observed by Koske (1981b).

## Mycorrhiza of Ericales

As long ago as 1936 Freisleben isolated a number of dark sterile mycelia from various Ericaceae. He tested them in culture, successfully producing ericoid mycorrhizal infection with several hosts, and concluded that they possessed low specificity. For instance, *Mycelium radicis myrtilli* α would form mycorrhizas with 15 species of Ericaceae. He failed to obtain mycorrhizas with it on 7 other species, including *Arbutus unedo* and *Arctostaphylos uva-ursi* which in any event form arbutoid mycorrhizas. The species *Pezizella ericae*, isolated by Read from ericoid mycorrhizas, similarly shows no close specificity. Isolates from *Calluna vulgaris*, *Vaccinium myrtillus*, *V. oxycoccum*, *Erica cinerea* and *Rhododendron ponticum* all formed mycorrhiza with all the hosts mentioned (Pearson and Read, 1973a).

TABLE 80

Mycorrhiza formation by species of *Glomus*

*The following species of* Glomus *are categorically stated to form mycorrhiza with the host mentioned.*
*(Gerdemann and Trappe, 1974)*

| Species of *Glomus* | Hosts proved in pot culture |
|---|---|
| *mosseae* | *Allium cepa, Fragaria vesca, Sambucus caerulea, Triticum aestivum, Zea mays* |
| *monosporus* | *Bellis perennis, Lycopersicum esculentum, Maianthemum dilatatum, Trillium ovatum, Zea mays* |
| *microcarpus* | *Fragaria* spp., *Geum* sp., *Phleum pratense, Rubus spectabilis, Taxus brevifolius, Thuja plicata, Zea mays, Juniperus communis* var. *siberica* |
| *fasciculatus* | *Allium cepa, Clintonia uniflora, Crataegus douglasii, Deschampsia danthioides, Epilobium watsonii, Fragaria vesca, F. chiloensis, Geum* sp., *Hypochaeris radicata, Maianthemum dilatatum, Malus* sp., *Mentha arvensis, Plantago lanceolata, Potentilla* sp., *Rubus spectabilis, R. ursinus, Sitanion lystrix, Stachys mexicana, Taxus brevifolius, Thuja plicata, Zea mays* |
| *macrocarpus* var. *macrocarpus* | *Allium cepa, Epilobium glandulosum, Fragaria chiloensis, Galium aparine, Stachys mexicana, Trifolium repens, Triticum aestivum, Zea mays* |
| *macrocarpus* var. *geosporus* | *Lycopersicum esculentum, Zea mays, Fragaria* sp. |
| *caledonius* | *Zea mays, Triticum* |

Zak (1974) described an arbutoid mycorrhiza of *Arbutus menziesii* and showed it to be formed by *Cortinarius zakii*, a species which he also verified by field observation to form typical ectomycorrhizas with *Pseudotsuga menziesii* and *Abies grandis*. In a later paper (1976a) he showed, by testing the infection of sterile seedlings in culture with 10 species of fungi, that *Arctostaphylos uva-ursi* could form arbutoid mycorrhizas with *Hebeloma crustuliniforme*, *Laccaria laccata*,

*Rhizopogon vinicolor, Thelephora terrestris, Lactarius sanguifluus* and *Pisolithus tinctorius*, all well-known ectomycorrhizal fungi of trees. He could not confirm that any of the mycorrhizas of *A. uva-ursi* were formed by these fungi in the wild.

Recently Molina and Trappe (1982a) have made an extensive investigation of the specificity of *Arbutus menziesii* and *Arctostaphylos uva-ursi* using isolates of 28 known ectomycorrhizal fungi (including *Cenococcum*) both of observed broad host-range and of ranges restricted to one or few coniferous or broad-leaf hosts. Of the 28, three failed to form arbutoid mycorrhizas: *Alpova diplophloeus* from *Alnus* spp., *Cortinarius pistorius* from *Pinus* spp. and *Zelleromyces gilkyae* from *Pseudotsuga menziesii* and *Tsuga heterophylla*. All these have a host range possibly restricted to the genera or species named. There was some variation in the frequency and intensity of mycorrhiza formation by the other 25 species tested but all formed arbutoid mycorrhizas, including some like *Suillus grevillii* which have also been reported to be of very restricted host range. No evidence was found that any of these fungi formed ectomycorrhiza rather than arbutoid mycorrhiza with either host in spite of reports to the contrary by Largent *et al.* (1980a) and Mejstřík and Hadač (1975) and others. Molina and Trappe also report that fungi like *Leccinum manzanitae*, restricted in distribution of its sporophores to habitats of these Arbutoideae, also forms ectomycorrhiza of typical structure in culture with conifers. These results amply confirm Zak's findings, especially that a single strain may form ectomycorrhiza with one host and arbutoid mycorrhiza with another. This shows that the actual structure of the mycorrhiza, especially as regards the extent of penetration of the fungus into active living cells, is determined by factors derived from the host plant.

The mycorrhiza of the achlorophyllous *Monotropa* was reported by Björkman (1960) to be formed by a fungus believed to be a *Boletus* sp. which he proved to be simultaneously ectomycorrhizal with *Pinus*. Linnemann (1969) also showed that a similar isolate from *Monotropa* produced ectomycorrhiza with *Picea abies*. These fungi are therefore non-specific and the structure of the mycorrhiza is again dependent on the host plant.

## Orchid mycorrhiza

Specificity in orchid mycorrhiza has been the subject of some controversy dating from 1909 to the present time (see Harley, 1969a; Hadley, 1970; Warcup, 1981). There is no doubt, as has been emphasized above, that both host and fungus must be subject to similar selection processes as in other mycorrhizal associations. As with ectomycorrhizas, there appears to be a range of specificities from the close association of *Caladenia* species with *Sebacina vermifera*, to host plants such as *Lyperanthus nigricans* or *Dactylorchis purpurella* which can accept a large number of different fungal species (see Hadley, 1970; Warcup, 1971, 1981). Similarly the fungi can usually infect a range of host plants, so that even though *Caladenia* spp.

are apparently normally dependent upon *S. vermifera* this fungus has been shown to associate frequently with species of *Glossodia*, *Elythranthera* and *Eriochilus* as well as *Acianthus* and *Microtis*. Warcup (1981) makes it clear that even when specificity normally appears to be quite close it is never absolute in the range of Australian orchids and their symbionts that he has investigated.

There are three features of the behaviour of orchid endophytes that require special comment. As pointed out by Warcup and Talbot (1967) and subsequently re-emphasized in the recent work of Warcup, orchid tissues may be inhabited by several fungi. In a single root, the species of *Rhizoctonia* on the surface may differ from those growing intercellularly and again from those forming active coils in living cells. Warcup (1971) states that coils formed by orchid endophytes may occasionally be colonized by other fungi. This feature is not found solely in orchid roots for there may be many fungi inhabiting a single root in other plants. In orchids it is the multiplicity of *Rhizoctonia* fungi that emphasizes this point. The second feature is that all the known orchid fungi have ecological niches other than as symbionts with orchids. Fungi like *Rhizoctonia solani* and *Armillaria mellea* may be extremely complex taxa or even composed of several separate species or genera, but one and the same strain may be symbiotic with orchids and parasitic with other plants. *Coriolus versicolor* similarly is a common wood-destroying saprophyte or weak parasite as well as symbiotic with *Galeola foliata*. These fungi are not specific either to their orchid symbionts or the the host that they parasitize. The third important feature is that specificity in Orchidaceae has been usually judged on the effect of the fungi on early seedling growth rather than on their ability to form mycorrhizas with the roots of adult orchids.

## Strain variation in mycorrhizal fungi and specificity

Nothing is known of the genetics of host/fungus interaction in mycorrhizas and indeed investigation has so far been minimal. There have been experiments in which different strains of a given mycorrhizal fungus have been tested against a series of "provenances" of host, that is, host grown from seed from different locations or of different cultivars. Mason (1975), using a number of different isolates of *Amanita muscaria*, showed that each produced a similar extent of mycorrhizal development on the progeny of two samples of seed of *Betula verrucosa* (*pendula*), one from Scotland and one from Latvia, but each of the isolates produced a different percentage infection on them. The interest lay in one isolate from pine in USA which failed to form mycorrhiza with both "provenances" of *Betula* under the conditions tested. Neither this example nor any other which might be quoted (Modess, 1941; Theodorou and Bowen, 1970; Laiho, 1970; Molina, 1979; etc.) has shown that the strain variation of mycorrhizal fungi involves *specific genes* that confer virulence or avirulence in respect of mycorrhiza formation. The same applies to the results of Marx and Bryan (1971) who made use of control-pollinated half-sib progenies of *Pinus elliottii* var. *elliottii* and one isolate each of *Thelephora terrestris* and *Pisolithus tinctorius*. This work indeed showed that the genotype of the host could affect intensity of infection but not that specific genes were involved.

The same conclusion can be drawn from the work on vesicular-arbuscular mycorrhiza of Hall (1978a) on *Zea mays*, of Bertheau *et al.* (1980) on wheat, and numerous investigations with *Citrus* spp. with respect to the host varieties, and that of Daft and Nicolson (1966), Mosse (1972a,b), Sanders *et al.* (1977), Daniels *et al.* (1981) and others in respect of endophyte variation. In most cases the work with endophytes has been concerned with differences between species and not between strains of a single species.

We can conclude, therefore, that we have as yet no evidence whatever of specific genes that affect virulence or avirulence in mycorrhizal fungi or concern susceptibility or resistance in the hosts, but we do have evidence that often a single strain of fungus may infect a very wide range of hosts and that a single provenance or cultivar of host may accept a wide range of species or strains.

## Ecological specificity

Evidence that a given pair, of host genotype and fungal strain, can form mycorrhiza in culture is no proof that this combination occurs in nature. This contention recurs in the writings of Zak, Trappe and others who have contributed so much in recent years to our knowledge of mycorrhizal fungi, their taxonomy and ecology. Nevertheless, successful synthesis does indicate that in suitable conditions such mycorrhizal combinations might exist, and that there may be an ecological component in the specificity. It is paralleled, perhaps, with the fact that ecologically non-mycorrhizal hosts can be made to accept vesicular-arbuscular endophytes in experimental conditions. It probably explains the common experience that mycorrhizal Basidiomycetes may appear by the distribution of their sporocarps to be usually associated with particular trees in nature, although in experiments they are found to form mycorrhizas much more widely. On the other hand, the ecological restriction is only very rarely, as far as is known, so pronounced as to ensure that a single host/fungus combination only is found on a particular site. Indeed there is good reason to expect selection to operate for multiplicity of combinations especially when one considers, on a single site, the diversity of soil conditions including soil microorganisms, the diversity of neighbouring plants and the diversity of light intensity etc. throughout a growing season.

Facultative mycorrhizal fungi are found amongst the endophytes of Orchidaceae many of which seem to have habitats away from their orchid hosts, as shown by Downie and by Mollison for *Rhizoctonia* spp. It is also true of the parasitic and wood-destroying species like *Fomes*, *Armillaria* and *Coriolus*. There is yet no convincing evidence of facultative mycorrhizal fungi associated with other types of mycorrhiza although there may be non-mycorrhizal strains of some mycorrhizal species.

## Summary of mycorrhizal specificity

There is no known case of specificity in any mycorrhizal association such that a single strain of a fungus is restricted to a single genotype, cultivar or provenance of host. There is indeed no described case of absolute species-to-species specificity. There are a few examples where a single species of fungus has been described as restricted to the species of a single genus of host plant, but on further examination the association has usually proved wider than it was first thought to be. Similarly there are host species which appear to accept few mycorrhizal fungi. The experience of synthesis of mycorrhizas in pure culture has shown that both fungi and hosts appear to be more selective of their partners in ecological conditions than in culture. Nevertheless, amongst ectomycorrhizal fungi, which, because of their conspicuous fruit bodies, have received most attention, restriction of genera like *Rhizopogon* to the Pinaceae has been borne out in culture. At the ecological level "sporocarp specific" fungi, only fruiting with particular hosts, as well as "broad host range" fungi, have been recognized.

The most interesting and theoretically important species of broad host range are those where a single fungal species forms a different kind of mycorrhiza with one taxon of host from that formed with another, and where a single host forms different kinds of mycorrhiza with widely different taxa of fungi. There is no case whatever for believing that in any kind of mycorrhiza there is a close specificity of host and fungus.

## *Rhizobium*/legume symbioses

The association between legume roots and *Rhizobium* is mutualistic under most conditions. Gaseous $N_2$ is fixed by the bacteroids in root nodules and becomes available to the host plants. At the same time carbohydrate derived from the host plant is used by the bacteria. It is therefore worth comparing this symbiosis both with mycorrhizas and with pathogenic associations.

It is generally considered that a considerable degree of specificity exists between host plants and *Rhizobium*. For many years the classification of the genus *Rhizobium* into different species has been based on the existence of apparent "cross-inoculation groups", although this approach is now being questioned with greater and greater frequency (e.g. Lange, 1961; Dart, 1974; Graham, 1976; Napoli *et al.*, 1980), chiefly on the grounds that many isolates of *Rhizobium* are symbiotically promiscuous, capable of nodulating plants outside their generally recognized cross-inoculation groups. Use of biochemical reactions in culture as well as host affinities have led to some modification of the species concept in *Rhizobium* (see Graham, 1976) as shown in Table 81. The changes are important for our present discussion, since it is difficult to consider host/symbiont specificity when the chief criterion delimiting the taxa within *Rhizobium* is ability to nodulate particular hosts.

Table 81 makes it clear that, at the broad level of species of *Rhizobium*, nodules can be formed not simply with single species of host plant but usually with all the members of a genus or group of genera. Specificity at this level is not at all close, and is comparable with that in mycorrhizal associations.

Host/*Rhizobium* interactions become much more complicated when strains of a particular bacterial species are considered. There is no doubt that there can be large differences in "symbiotic effectiveness" (ability to fix nitrogen) between strains of a bacterium in association with particular species or cultivars of a host. Such "specificity" can find expression at different stages of the infection process (see Table 82 from Sprent, 1979). Infection may be controlled by the genotype of *Rhizobium*, and mutants are known which can stop the infection process at several different stages.

The genotype of the host can also affect the development of symbiosis. Here resistance to infection appears to be superficially similar to that found in pathogenic associations (see below) but it should be noted that the resistance genes are recessive, not dominant. Non-nodulating lines of *Trifolium pratense* and *Glycine max* have received considerable attention. In *T. pratense* resistance is determined by a recessive gene (r) which acts in conjunction with a cytoplasmic factor. Thus, plants possessing double recessive (rr) genotype fail to nodulate in the presence of the cytoplasmic factor ($\sigma$). The block to infection is probably at infection-thread formation, since bacteria multiply in the rhizosphere and cause normal root-hair curling (Nutman, 1969).

Interpretation of nodulating characterisitics in terms of host/bacterium specificity and evolution of the symbiosis is much confused by the fact that most experimental work has been carried out on cultivated crop plants inoculated with strains of *Rhizobium* which may well have arisen in completely different geographical regions, rather than co-evolved with the host plant in question. An example of such a problem is provided by the work of Lie and his associates on *Pisum sativum* (Lie *et al.*, 1976; Lie, 1978; Lie *et al.*, 1981). A survey of the nodulating and nitrogen-fixing ability of a large number of genetically uniform cultivars of *P. sativum* indicated that varieties from the Middle East (Kabul, Isfahan and Afghanistan) nodulated poorly or not at all with *Rhizobium* strains from the culture collection at Wageningen (none of which actually came from the Middle East). Lie *et al.* (1976) thought that this finding supported the idea (discussed below) that host plants resistant to attack by pathogens are most likely to be found in the gene centres of cultivated plants. This for *Pisum* is the Middle

TABLE 81

Species of *Rhizobium* and *Agrobacterium* and their characteristics (from Graham, 1976)

| Species | Relation to species of Breed et al. (1957) | Flagellation | DNA base ratio % G + C | Serum zone litmus milk | Acid reaction litmus milk | Growth rate | Nodule forming characteristics, special features |
|---|---|---|---|---|---|---|---|
| R. leguminosarum | R. phaseoli + R. trifolii + R. leguminosarum | Peritrichous | 59·0–63·5 | + | − | Fast | Forms nodules on one or more of *Trifolium. Phaseolus vulgaris, Vicia, Pisum, Lathyrus, Lens* |
| R. meliloti | Unchanged | Peritrichous | 62·0–63·5 | + | + | Fast | Forms nodules with *Melilotus, Medicago, Trigonella* |
| R. rhizogenes | A. rhizogenes | Peritrichous | 61·0–63·0 | + | − | Fast | Causes hairy root disease of apples and other plants |
| R. radiobacter | A. tumefaciens + A. radiobacter + A. rubi | Peritrichous | 59·5–63·0 | + | − | Fast | Frequently produce galls on angiosperms. Produce 3-ketoglycosides |
| R. japonicum | R. japonicum + R. lupini + Cowpea miscellany | Subpolar | 59·5–65·5 | − | − | Slow | Nodulates many different legumes including one or more of *Vigna, Glycine, Lupinus, Ornithopus, Centrosema*, etc. |

East. The results seemed to imply that the evolution of the mutualistic *Rhizobium*/legume symbiosis was similar to the evolution of pathogens with their hosts. However, a subsequent study, using differently chosen bacteria and host plants, gave contradictory results (Lie *et al.*, 1981). It was found that twelve lines of the primitive and genetically very variable *Pisum sativum* ecotype *fulvum* from Israel would nodulate readily and fix nitrogen with a *Rhizobium* strain isolated from the same region. A bacterial isolate from Turkey produced effective nodules with only two lines, while an isolate from Holland was partially effective

TABLE 82

Host–rhizobium interactions during legume nodule formation (from Sprent, 1979)

| | "Developmental" stage | Genetical/physiological requirements | Comments |
|---|---|---|---|
| *Increasing specificity and interaction between host and bacterial genomes* | *Rhizobium* multiplication in soil | Host secretions may stimulate or inhibit | Degree of specificity variable |
| | Root hair curling and branching | Compatible host and *Rhizobium* | Degree of specificity variable |
| | Attraction between compatible host/rhizobial cells | ? Matching of cross-reacting groups. ? Bridging by lectins | Found by some to be highly specific: others less so |
| | Entry of bacteria | Dissolution or stretching of host cell wall | Temperature affects strain of *Rhizobium* entering and thread development. Plants may have genetically controlled anti-nodulating factors |
| | Growth of infection thread | Matched development of rhizobia with host cell wall material | |
| | Formation of nodules | Correct balance of growth factors from both partners | *Rhizobium* mutants known which cause development to stop at any stage. Correct matching of host and rhizobial genotypes essential |
| | Formation of mature infected cells | Release of rhizobia from threads. Matched growth of rhizobia and host membranes | |
| | Formation of bacteroids, development of nitrogenase and haemoglobin, etc. | Specific interactions involving both partners | Host may act by providing correct environment |
| | Maintenance of bacteroid tissue | Correct interchange of materials between symbionts | Not well understood. Strongly conditioned by environment |

with only one line. This result shows that while specificity certainly occurs between strains and cultivars there may well be a selection towards nodulation and effectiveness within a particular geographical area. There does not seem to be any evidence of selection for resistant lines, which would be expected in genetically variable hosts at the centre of origin of the genus if resistance had any selective advantage.

## Host/pathogen symbiosis

Concepts of host/fungus specificity and of recognition between the two organisms forming a symbiotic association are, in the literature of plant pathology, almost inextricably linked to considerations of host resistance to infection and pathogenicity of the fungus. The emphasis has been upon host resistance to disease rather than tolerance of infection, which may be more important in the present context. Most plant pathologists would agree that resistance can be either general (non-specific)—of the kind that can be effective against a wide range of fungal biotypes—or specific, that is effective against some fungal biotypes but not others.

From a genetic standpoint there are three ways by which resistance can be inherited. Oligogenic (or major gene) resistance is determined by one or a few genes, which are usually dominant (in contrast to non-nodulating legumes). The effects of such major genes are usually highly specific. Selection of resistant individuals having such genes and breeding from them has led to the existence of resistant lines or cultivars of particular crops. Very close specificity of this type is usually found in associations of plants with pathogenic biotrophic fungi. Polygenic resistance is determined by a large number of genes and confers resistance to a wide variety of fungal biotypes. It does not usually therefore confer specificity on the relationship between a host and a pathogen and may confer resistance to a range of both biotrophic and necrotrophic parasites. Cytoplasmic inheritance of resistance to disease is rare, but we have already seen that it is implicated in non-nodulating strains of legumes. Specificity in the relationship between pathogens and their hosts is most likely to be mediated via the operation of major, dominant genes for resistance and their recessive alleles, which result in susceptibility, together with complementary genes in the fungal genotype.

### Evolution of resistance and the gene-for-gene hypothesis

It is a realistic starting point to envisage that the state of a population of plants which had not been challenged by a potential symbiont (mutualistic or pathogenic) would be susceptibility to infection or non-specific resistance. There seems to be no need to invoke the existence of a specific resistance mechanism to invasion by an organism which, as far as the population of plants is concerned, does not exist. Vanderplank (1976) would suggest that genes for, that is allowing, susceptibility must have functions other than those that are concerned with their role in relation to invading pathogens. Alleles of them which confer resistance to infection must, he contends, have a two-fold role: that of the susceptible allele *plus* the role conferring resistance. This contention rests on the axiom that if a gene can mutate it must exist and if it exists it has a role. Alleles for resistance arise by mutation and their frequency in a population is determined by their selective advantage. Interactions between populations of hosts and symbionts which cause disease result in the selection of individual resistant plants (arising by mutation), whose subsequent successful reproduction leads to the increase in numbers of resistant individuals and biotypes. The allele for resistance, conferring increased fitness on the plant *in the presence of the pathogen*, would become more frequent in the population, while the allele for susceptibility would become relatively less frequent. For the pathogen, the existence of resistant biotypes of its host-plant provides the impetus which results in selection of new pathotypes in which alleles for virulence once again confer upon the fungus an ability to infect the host plant. These alleles, like those for resistance in the host, must retain their original functions as well as conferring virulence on the fungus. Resistance genes (R) in the host and avirulence genes (A) in the parasite are normally dominant. Co-evolution of host and parasite may thus result in the existence of a range of biotypes of the host plant carrying major genes for resistance and a number of pathotypes with complementary genes for avir-ulence. Resistance is expressed with Rr or RR associated with Aa or AA. Such relationships have been formalized in the gene-for-gene hypothesis (see Flor, 1971), which has been shown to operate in several symbioses between plants and their pathogens (see Table 83 after Day, 1974), and apparently forms the basis of the very close specificity existing in the obligate biotrophic pathogens such as rusts, mildews, etc.

Two further lines of evidence support the idea of co-evolution of host and symbiont in pathogenic systems. First, the widest variety of host biotypes and fungal pathotypes normally occurs in the geographical centre of origin of the host species as long as the pathogen is also present (see Vavilov, 1949; Leppik, 1970; Eshed and Wahl, 1970). It is at such centres that co-evolution would have been operating longest and resistance genes would be most likely to be found. The second line of evidence is that ancient species of pathogen (such as *Uredinopsis* on fossil *Osmunda* spp. from the Triassic) seem to have a much narrower host range than those associated with more recently evolved hosts. *Puccinia graminis*, for instance, has physiological races which can infect a wide range of genera and species (see Anikster and Wahl, 1979).

TABLE 83

Host-parasite systems in which a gene-for-gene relationship has been demonstrated (d) or suggested (s) (after Day, 1974)

| System | Root/shoot | Reference |
|---|---|---|
| **Rusts** | | |
| *Linum-Melampsora lini* (d) | Shoot | Flor (1942) |
| *Zea-Puccinia sorghi* (s) | Shoot | Flangas and Dickson (1969) |
| | | Hooker and Russell (1962) |
| *Triticum-P. graminis tritici* (d) | Shoot | Green (1964, 1966) |
| | | Kao and Knott (1969) |
| | | Loegering and Powers (1962) |
| | | Williams *et al.* (1966) |
| | | Luig and Watson (1961) |
| *Triticum-P. striiformis* (s) | Shoot | Zadoks (1961) |
| | | Lewellen *et al.* (1967) |
| | | Line *et al.* (1970) |
| *Triticum-P. recondita* (d) | Shoot | Samborski and Dyck (1968) |
| | | Bartos *et al.* (1969) |
| *Avena-P. g. avenae* (d) | Shoot | Martens *et al.* (1970) |
| *Helianthus-P. helianthi* (s) | Shoot | Sackston (1962) |
| | | Miah and Sackston (1970) |
| *Coffea-Hemileia vastatrix* (s) | Shoot | Noronha-Wagner and Bettencourt (1967) |
| **Smuts** | | |
| *Triticum-Ustilago tritici* (s) | Shoot | Oort (1963) |
| *Hordeum-U. hordei* (d) | Shoot | Sidhu and Person (1971, 1972) |
| **Bunts** | | |
| *Triticum-Tilletia caries* (d) | Shoot | Metzger and Trione (1962) |
| **Mildews** | | |
| *Hordeum-Erysiphe graminis hordei* (d) | Shoot | Moseman (1957, 1959) |
| *Triticum-E. graminis tritici* (d) | Shoot | Powers and Sando (1957) |
| **Other parasitic fungi** | | |
| *Malus-Venturia inaequalis* (d) | Shoot | Boone and Keitt (1957), Day (1960) |
| *Solanum-Phytophthora infestans* (s) | Shoot/tuber | Black *et al.* (1953) |
| | | Toxopeus (1956) |
| *Lycopersicum-Cladosporium fulvum* (s) | Shoot | Day (1956) |
| *Solanum-Synchytrium endobioticum* (s) | Root | Howard (1968) |

## Biotrophy in root and shoot symbionts

It is apparent from Table 83 that the majority of the fungi which have so far been proved to be involved in specific gene-for-gene relationships with their hosts are shoot pathogens. Lewis has pointed out (1973) that such fungi, having no access to nutrients outside the plant, have no opportunity to evolve to a mutualistic relationship with their hosts and this seems to be acceptable, with minor exceptions such as the production of hormones (see also Harley, 1950). In contrast, fungal biotrophic root-infecting *pathogens* are rare (parasitic *Mastigomycotina* and *Myxomycota* excepted) but a very wide taxonomic range of hyphal fungi form mutualistic mycorrhizal associations with the roots of almost all species of plants. These root-inhabiting fungi have access to mineral nutrients in the soil which are efficiently absorbed and a proportion of which passes to the host plant. Any comparison of biotrophic parasites with mycorrhizal fungi must therefore take into account the fact that not only are the two groups of fungi distinguished by their parasitic versus mutualistic modes of nutrition, but also that the biotrophic parasites are generally characteristic of shoot tissue, while mycorrhizal fungi infect the roots. This second distinction may have been almost as important as the former in determining ways in which interactions between host and symbiont may have evolved.

## Dispersal and spread of infection

Shoot- and root-inhabiting fungi have quite different problems in the dispersal of propagules from one host to another and in the spread of infection within a population of host plants. Moreover, the conditions for dispersal which pertain in natural vegetation systems are different from those in crop monoculture for reasons of distance between and variability of the targets.

Garrett (1970) has considered the ways by which pathogenic root-infecting fungi survive in the absence of host roots. Many of these ways are equally applicable to mutualistic symbionts. For example, mycorrhizal fungi can survive on living root systems of any *susceptible* plant, and they possess a wide host range. Many root pathogens also have wide host ranges. If a fungus depends on vegetative mycelial growth through the soil from one root to another to effect spread and survival, it is clear that a wide host range will be more advantageous than any close host/symbiont specificity especially in a species-rich plant community. All types of mycorrhizal fungi spread from root to root in this manner, as do many root pathogens. Root-infecting pathogenic fungi also survive as resting propagules (sclerotia or spores) and in this way they achieve persistence in time, rather than dispersal in space. Such propagules may require relatively large storage reserves and may be few in number. The time for which they need to survive may be considerably reduced if the fungus has a wide host range. Most mycorrhizal fungi can survive as spores, sclerotia or other resting structures and may also persist as mycelium in previously active roots. Orchid endophytes can persist on organic matter, in soil or as plant pathogens, until an orchid seed appears in their vicinity. In this particular association the dispersal of the symbiosis is presumably largely achieved by the orchids, which have numerous, very small, wind-dispersed seeds. It has also been suggested that fungi which form ericoid mycorrhizas and ectomycorrhizas may have some limited saprophytic capabilities which allow them to persist as members of the root surface flora. Evidence for this is very slender and the subject requires further investigation, but the existence of active saprophytic strains of a few species has been verified which, although they may not form mycorrhizas, may represent a state in which the dicaryon can persist and give rise to non-saprophytic mycorrhizal progeny.

## Conclusions

There appear to be two possible and interrelated reasons why the mycorrhizal symbioses and also *Rhizobium*/legume symbioses have not evolved the same degree of specificity with respect to host plants as have biotrophic pathogens.

The first and probably more important reason is that mycorrhizas and leguminous root nodules are advantageous to both autotroph and heterotroph. Hence, assuming that susceptibility to infection is primitive and oligogenic resistance highly evolved, there seems to be no reason why specific resistance mechanisms should occur in the host plant. As Vanderplank (1978) pointed out, there would have been no selection in favour of oligogenic resistance and a gene-for-gene relationship which might have led to host/symbiont specificity. Some legumes are described as having major gene resistance to *Rhizobium* infection, but as we have already observed the resistance genes are recessive and not dominant as is usual in resistance to plant pathogens. This is to be expected in a mutualistic association and might, unlike the situation in pathogens, be associated with an absent or defective gene product, and with the requirement for a "resistance" gene to affect some other activity not concerned with resistance.

Since mutualistic associations have a selective advantage to both symbionts, factors leading to "epidemics", that is the universal presence of such associations, will have selective advantage. It is instructive to read articles on spread of disease epidemics in natural vegetation systems with the mental posture of either a host plant or a fungus potentially capable of forming a mutualistic symbiosis. For instance, Burdon (1978) has pointed out that the incidence of a specific pathogen will decrease in an area containing several species of host if the number of species per unit area increases. The reasons include decrease in the amount of infectible tissue per unit area and increase in distance and difficulty of dispersal. Any fungus which is unspecific will not be restricted in this way. *Almost all mycorrhizal fungi come into this category.*

A second possible reason why specificity is rare in mycorrhizal associations is that many root-inhabiting fungi, including a large number of mycorrhizal fungi, depend upon the spread of perennial mycelium and persistence as resting propagules much more than upon widespread dissemination of spores. Chances of survival are clearly improved by the existence of a wide range of susceptible species of host plants, so there appear to be selection pressures on the fungi against the evolution of close host/fungus specificity. This argument applies to mycorrhizal fungi as much as it does to many pathogenic root-infecting fungi which are usually considered in this context.

## Symbiotic effectiveness in mycorrhiza

As has been mentioned in connexion with the discussion of *Rhizobium*, genetic variation of both host and *Rhizobium* can alter symbiotic effectiveness. This may operate both through processes concerned in the initiation and maturation of the nodule, and also through direct biochemical alteration. The complexity of the process of nodule initiation is such that it is undoubtedly affected by many genes in both participants and the same applies to the biochemical fixation of $N_2$, as well as the translocation of carbon compounds to, and the products of fixation from, the nodule. It seems extremely likely that similar considerations apply to mycorrhizas of all types. We know that there is considerable strain variation of the fungi and a complex pattern of translocation and transfer of compounds to and from the participants, and each type of mycorrhiza is complex in morphology, anatomy and histology. The symbionts must therefore be genetically attuned in a number of respects so that the whole infection process unfolds in an integrated orderly fashion and the biochemical interaction occurs efficiently at each stage.

It is extremely enlightening at this point to consider the classic work of Chen and Thornton (1940) on morphology and effectiveness in legume symbiosis. They demonstrated that the effectiveness of a nodule might depend upon the volume and longevity of the leghaemoglobin-containing zone. These were affected by the rate of growth and maturation of the nodule and by the rate of the senescence of its tissues. Hence various combinations of these rates might result in prolonged or transient effectiveness as measured by rate of fixation. Of course the effectiveness of legume nodules is readily measured by $N_2$ fixation rate whereas the effectiveness of mycorrhizas is more difficult to assess. So far, except in orchids where initiation of seedling growth has been used, there have only been estimates of comparative growth rates of given hosts with a range of fungal symbionts of one or a number of species. The results usually measure effectiveness in terms of dry weight increment, which is of course the resultant of complicated antecedent causes. For instance, increase of dry weight depends fundamentally on the release of the host from factors limiting carbon dioxide fixation by the provision of soil-derived components by the fungus. It is offset by the consumption of carbon compounds by the fungus. The gain in dry weight is essentially the difference between these two competing activities and is therefore a very coarse measure of effectiveness. Moreover its actual value as a measure of effectiveness in any symbiosis will be very sensitive to the availability of nutrients in the soil, and the nature of the actual nutrient, the supply of which is limiting the growth in dry weight of the host. It may not bear a close relationship to reproduction or have any selective advantage under natural conditions. It must be stressed that genetical variation both in the host and in the fungus which can affect structure, physiology and biochemistry of the partners and also the ecological background are important in the establishment and maintenance of the symbiosis.

It would seem that future work on specificity in mycorrhizal relationships should follow two important lines. First, the symbiotic effectiveness of different combinations of fungus and host requires investigation. Comparisons should be made at the level of strains or genotypes of both partners as well as at higher taxonomic levels. They should be carried out under controlled and standardized conditions so that the rates of defined processes can be compared. Such processes might be the uptake of a nutrient per unit weight or length of infected root, or perhaps eventually more complex comparisons such as the uptake per unit carbohydrate consumption might be made. Secondly, the total process of infection and the development of mycorrhizal organs might be described in detail from the primary meeting of the root with the infective propagule or hypha of the fungus. The exact sequence of the stages gone through and the exact points at which development ceases in different incompatible or less effective combinations should be determined. By these means a knowledge of the properties needed by both host and fungus for the successful passage through each stage of mycorrhiza formation would be obtained, and a much clearer appreciation of what is meant by and involved in mycorrhizal specificity and effectiveness would emerge. Both the materials and the necessary methods to gain this information are already available but would be improved if more knowledge of the genetics of mycorrhizal fungi were also obtained.

## Recognition

A consideration of specificity of the symbionts to one another leads on to the related problem of how the organisms recognize each other. This is a very important question which has received little attention in mycorrhizal research. Work with fungi has been concerned with pathogenic associations and with sexual reproduction. Parallels may be drawn and contrasts made between pathogenic and mutualistic associations including mycorrhiza, which might shed light on the relationship of both types of association. As a preliminary to the discussion of recognition, some definition of the term or at any rate some appreciation of what is meant by the concept is necessary. We may ask first: Is it a once-and-for-all process? The answer to this for mycorrhizal symbiosis cannot be other than in the negative. There can be no unique moment or process when a mycorrhizal fungal hypha and a potential host root "recognize" that a structurally complex and functionally elaborate mycorrhizal organ may be built by them, although there must be a first stage through which the process must always pass. There are indeed many processes of considerable complexity involved. In vesicular-arbuscular mycorrhiza the attempts at modelling the process of infection by S. E. Smith and Walker (1981), J. G. Buwalda *et al.* (1982) and F. E. Sanders (personal communication) are based on realization of this.

In order to expose the subject and lay bare areas of ignorance, it is necessary first to rehearse the process of infection of the host and the building of mycorrhizal organs and to consider what "recognition" factors are likely to be necessary at any stage.

## Recognition in ectomycorrhiza

We know from the work of Theodorou and Bowen (1971b) that some ectomycorrhizal fungi at least can colonize and spread along the roots of non-host plants. We have a rationale for this behaviour. First, the essential carbohydrates and the accessories such as vitamins and special amino acids needed by ectomycorrhizal fungi occur in the root region. Secondly, root exudates contain substances like the so-called M-factor of Melin which stimulate fungal growth. Thirdly, spores of ectomycorrhizal fungi germinate more readily in the root region than in other sites. None of these (except sometimes the last) is specific and mycorrhizal fungi hold in common with many other soil-borne fungi the ability to spread vegetatively on the root surface but not in the soil in general. The only appearance of specificity is the recent observation of Birraux and Fries (1981) that the spores of *Thelephora terrestris* are stimulated by the roots of mycorrhizal hosts but not by those of others. The building up of an infection potential by ramifying on the root-surface in this way is not associated with the production of antifungal substances which might inhibit competitors, for their production is not a universal or general property of ectomycorrhizal fungi.

The next phase following the colonization of the surface of the root involves two changes in fungal activity: penetration between the cells of the epidermis and cortex of the host (Hartig net production) and the formation of a pseudoparenchymatous fungal tissue—the sheath or mantle. These two fungal activities may be simultaneous or one may precede the other. In any event the hyphal occupation of the root surface may give rise to a patchy or sparse sheath tissue while on the one hand the penetration is proceeding, or on the other a considerable development of sheath may occur before Hartig net penetration is observable.

We have already discussed the factors or stimuli that may be connected with sheath development. It seems possible that they are similar to those concerned in the formation of fruit bodies, hyphal strands and rhizomorphs. First there is the nutritional aspect. The aggregation of hyphae together to form strands and "ectotrophic" sheets of fungal tissue is explained by Day (see Watkinson 1979) and Garrett (see 1970) as depending upon the exudation of nutrients from the main hyphae, which encourages the lateral hyphae to adhere and bind to it. This occurs in an environment low in nutrients or of low carbon:nitrogen ratio, when the leading hyphae are attached to a distant food base and maintained by translocation from it. Such a mechanism, if operating in sheath formation, might explain why the full development of the sheath often follows the formation of the Hartig net as Nylund (1981) emphasized. His sheath-forming factor might then be carbon nutrient supply. Other environmental factors such as oxygen supply are no doubt also important, as Read and Armstrong (1972) showed. At the same time it seems probable some hormonal influence is necessary in strand formation because the hyphae differentiate into main, cortical, and binding hyphae and a degree of differentiation also occurs in all ectomycorrhizal sheaths. The main argument against this view would seem to be that sheaths form on long roots in which no Hartig net develops. However in such cases the fungus is indeed in close contact and forming full mycorrhizas elsewhere on the root system and it may be spreading "ectotrophically" in Garrett's sense in a relatively nutrient-deficient area. Nevertheless the fact that the formation of a sheath occurs on roots that can form ectomycorrhizas and not on all roots would seem to argue for a recognition factor. However the formation of a sheath on silicone tubes in Read and Armstrong's experiments argues against that. Comparative investigation is clearly needed of the early stages of association of fungus and autobiont in compatible pairs, in pairs in which incomplete joint organs are formed, in incompatible pairs in which both symbionts are potentially ectomycorrhizal, and in incompatible pairs where the autobiont does not form ectomycorrhizas.

The formation of the Hartig net clearly requires separate consideration for the hyphae emanating from those of the sheath penetrate between the cell walls of the host tissues. This development does not occur uniformly over the whole primary root surface as sheath formation may. It is restricted to a zone, called the "mycorrhizal infection zone" by Marks and Foster (1973); that is a zone, lying behind the dividing meristem, where cell maturation is occurring. In attempts to initiate mycorrhiza in regions well behind the apex a Hartig net was never formed, although considerable spread of hyphae over the surface often occurred (Atkinson, 1975). Clowes (1951) states of *Fagus* "the fungus never seems to form a Hartig net between cells that have elongated normally but it does so sometimes where cells are intermediate in length between the elongated state and the extreme mycorrhizal state". Chilvers and Pryor (1965) say of *Eucalyptus* "radial elongation of the epidermal cells or associated changes in cell physiology and wall structure are requisite for hyphal entry between cells". Both these observations fundamentally agree with the view of Warren-Wilson that there is a change in the growth of the root-apices before infection takes place. The rate of differentiation of cells tends to overtake the rate of formation. It does not matter in the present context what causes these changes; whether they be generated by hormones from the fungus or whether they arise from internal changes within the plant, as Warren-Wilson's work indicates; it is in this zone of cell-wall building and cell maturation that the Hartig net is formed. In this zone the carbohydrate precursors of the walls of the host cells are being released through the plasmalemma of the cortical cells. The initial hyphae of the Hartig net grow into a region richer in carbohydrate than the young sheath zone and tend to penetrate the walls. We have therefore to suggest a mechanism whereby the Hartig net hyphae continue to penetrate along the walls as they form. The hypothesis that endears itself most to us is that the force is generated by hydrostatic pressure within the hyphae working against the incipient sheath which acts as a firm base or appressorium; the proteins of the enzymes at the hyphal apex or on the hyphal walls might inhibit or complex with the host enzymes which build wall polymers in the manner envisaged by Vanderplank (1978). By this means the polymerizing enzymes would be inhibited, wall formation altered, and soluble carbohydrates destined for them made available to the fungus. Such a process is one of recognition.

If this mechanism is accepted it inevitably implies that a very large number of fungi of diverse taxa possess enzymes or proteins on their surfaces capable of reacting with and deactivating the enzymes of the walls of a great variety of hosts. The alternative hypothesis would be that all ectomycorrhizal fungi possess enzymes on their walls capable of hydrolysing the wall material of their hosts, although no evidence has been obtained that these enzymes are formed by many, let alone all, of the fungi in culture.

Such a process of Hartig net formation would agree with the alteration of the walls to form a "zone of apposition", a "contact zone" or "involving layer" and the extensive branching and fanning out of the Hartig net in the wall region, as well as the movement of carbon compounds from the host to the fungus. Whether or not there might be specific inhibitors preventing colonization of the total depth of the cortex by the fungus other than by a suberized layer which is sometimes present is still in doubt. The specificity of the association which distinguishes mycorrhizal from nonmycorrhizal fungi would lie in the reaction between the enzymes or proteins of the walls of fungus and host. The hypothesis may be rejected as too naïve but it at least has testable points. Any other would require the operation of specific factors or substances in addition to enzymes capable of dissolving cell-wall material, although we know the latter are not consistently formed by ectomycorrhizal fungi in culture.

*Recognition of mycorrhizas with intracellular penetration*

The infection of roots by mycorrhizal fungi which penetrate host cells must also be considered in terms of the way the fungi spread outside the root, make contact with root cells and grow, both inter- and intracellularly, within the cortex. Some growth of hyphae on the surface of the root precedes penetration of the epidermal cells by ericoid, vesicular-arbuscular and orchid mycorrhizal fungi, although this extramatrical mycelium becomes more extensive after penetration of the tissues has taken place. This feature is analogous to the full development of the sheath after Hartig net formation in ectomycorrhizas and the differences deserve to be experimentally exploited. We have already seen how some development of fungal hyphae on the surface of the roots may occur in ericoid infections. In vesicular-arbuscular mycorrhizas appressoria are formed by surface hyphae on epidermal cells, or occasionally on root hairs, and these can perhaps be compared with incipient sheath development. Formation of appressoria certainly provides a visually well-defined stage (like the sheath) at which cell-to-cell contact has occurred, and also indicates that some degree of mutual "recognition" has taken place between the two organisms. Appressorium formation is followed by the development of an infection peg, penetration of the cell wall and the formation of typical arbuscules or intracellular hyphal coils. In vesicular-arbuscular and orchid mycorrhizas (but to a much lesser extent in ericoid mycorrhizas), infection proceeds by longitudinal growth within the roots as well as by infection from external hyphae. The extent of this growth within the root might then possibly depend upon compatibility between the symbionts in this regard and might reflect the degree to which they "recognize" each other, and so proceed to develop the integrated anatomy and physiology required for successful symbiotic existence. Growth between the cells and into the cells of the hosts of vesicular-arbuscular mycorrhiza might seem to require that the fungus forms enzymes capable of hydrolysing the components of the walls of the cells of its host. The evidence is against this. An alternative would be that surface proteins on the hyphae attach themselves to, react or copolymerize with the cell-wall enzymes of the host and inhibit them. Such a process is essentially one of recognition between the two organisms.

Growth of the fungus into the cells of orchid mycorrhiza would seem to be more readily explicable because most of them form enzymes hydrolysing complex carbon polymers. However, the control of these enzymes so that destruction of the host does not take place beyond small limits also must involve a recognition which results in enzyme control. It seems unlikely that end-product inhibition or catabolite repression could control such destructive parasites as *Armillaria mellea* or *Rhizoctonia solani*.

Tinker (1975a) presented an equation in which rate of development of infected root length in vesicular-arbuscular mycorrhiza is proportional to a constant ($S$) which he thought could be a "specificity" or "susceptibility" parameter. He did not however show that the value of $S$ varied in different combinations of host and fungus or in different environmental conditions. S. E. Smith and Walker (1981) have subsequently developed a model for the early stages of infection using two differential equations, from which it is possible to calculate both the frequency of infection of the root system from the soil (via appressoria and infection pegs) and the rate of linear extension of the fungus within the root. This model has so far only been applied to infection of *Trifolium subterraneum* under a limited range of soil conditions, but should be applicable to comparison of infection processes in different hosts and by different species or strains of fungi. It might then give valuable information about the compatibility of fungus and autobiont and their mutual recognition. Integration of the infection of roots by mycorrhizal fungi with physiological aspects of the symbiosis (e.g. phosphate uptake and growth) can also be approached using modelling techniques. Sanders (personal communication), making assumptions of the rate at which the fungus infects the host from the soil and spreads within the root, has successfully modelled data for phosphate uptake and growth in mycorrhizal and non-mycorrhizal onion plants. All these modelling attempts are in their infancy and it is clear that, although they complement each other to some extent, all may produce information about recognition.

One interesting feature is that models of the infection process (e.g. Smith and Walker, 1981) can be used to predict the expected distance between the root apex and the most apical infection point and therefore help to define the mycorrhizal infection zone of the root. By assuming that new infections occur with constant probability on any uninfected part of the root, the frequency of infection $(m^{-1}d^{-1})$ as an average for the root system as a whole can be calculated if the rate of root growth in length is measured. Experimental determination of the frequency of occurrence of the most apical infection point at different distances behind the apex indicates that the root tip may be considerably more, up to ten-fold, infectible than this average. The results need verification on a variety of host plants under different environmental conditions, but the finding is of interest as it can be directly compared with the observation that Hartig net formation in ectomycorrhizas also occurs in the root tip region behind the apical meristem.

Intracellular development of hyphal coils or arbuscules provides additional surfaces of contact between the symbionts of mycorrhizas in which the fungus is intracellular. An apoplast, composed of the fungal wall and interfacial matrix, is present between the plasmalemma of the cells of the root and the fungus and may be a resultant of the effect of hyphal proteins on enzymes forming wall materials of the host. Altered physiology of the autobiont is manifest in the increased area of the plasmalemma, increased volume of the cytoplasm and endoreduplication of nuclear material. Redistribution of membrane-bound ATPases also occurs. These changes must take place as a consequence of a sequence of recognition stages of fungal infection and must be important in the successful functioning of a mycorrhizal root. We can also infer that there must be delicate mechanisms of turgor regulation in both organisms which operate as intracellular penetration of root cells by fungal hyphae takes place. Indeed the invagination of the host plasmalemma may put it under tension which might act as a trigger for biochemical or physiological changes.

## Specificity and recognition in other symbioses

Although there is no close specificity in mycorrhizal symbioses, a comparative account of the recognition process in other symbioses may be valuable. Intracellular fungal growth in mycorrhizas can be compared with development of haustoria by some fungal pathogens. Studies of the latter have been much more extensive that those of mycorrhizas and indicate that changes in the cytoplasm of the cells of the autobiont may occur before the cell wall is breached by an infection peg (see Bracker and Littlefield, 1973; Ingram *et al.*, 1976). This is further evidence that the two organisms are "aware of" and "respond to" each other's presence during the formation of typical infection structures. Apart from mentioning the protein co-polymerization hypothesis put forward by Vander-plank (1976, 1978) we have so far made no attempt to consider how mycorrhizal fungi and host roots interact at the molecular level during the process of recognition. It is clear from the anatomical and cytological investigations just discussed that recognition must be a surface phenomenon and must involve molecules located in or on the walls or membranes of the organisms concerned. No experimental work on mycorrhizas has been carried out in this regard, so we must again fall back on a consideration of information derived from other systems (such as host/parasite, *Rhizobium*/legume and pollen/stigma interactions and gamete fusion) in which cell-to-cell contact and recognition are important. The involvement of lectins in these processes has been thoroughly discussed by Callow (1975, 1977). The following provides a brief summary only.

Lectins are proteins or glycoproteins which have the ability to bind carbohydrate-containing molecules. There is some evidence that lectins in soybean roots may be important in binding *Rhizobium* cells to the root hairs, thus perhaps initiating infection. How far the lectins are involved in the *specific* binding of *Rhizobium* strains capable of nodulating particular host biotypes, as suggested by the work of Bohlool and Schmidt (1974), has recently been questioned (Pueppke *et al.*, 1981), but there is no doubt that the lectin molecules can carry a considerable amount of variability and might certainly exert specific binding properties in some circumstances. Callow (1975) has suggested that lectin action involves positive recognition and binding of compatible cells in the case of *Rhizobium*/legume interactions and he further suggests that a similar system might operate in the infection of roots by mycorrhizal fungi. There is no

doubt that firm binding of fungus and plant surfaces precedes infection in many
associations, especially where appressoria are formed, and might be important in
vesicular-arbuscular mycorrhiza. As we have seen, a high degree of specificity is
not required in the recognition between the organism in either mycorrhizal or
rhizobial symbioses. The relatively unspecific binding by lectins might be the
first recognition step in them. Mutation and selection of the genes coding for
lectins, or for the carbohydrates to which they bind, might superimpose further
specificity upon this generalized system to produce the more precise, for
example, gene-for-gene, pollen/stigma and gamete fusion systems.

   Some aspects of Vanderplank's hypothesis are similar to the lectin hypothesis.
Once again proteins associated with the cell surfaces are envisaged as associating
together (co-polymerizing) in *compatible* fungus/host combinations. The proteins
provide molecules which store a great deal of variability and provide, where
required and selected for, a basis for specific recognition and cell-to-cell binding.
Common to both hypotheses is the concept of binding in compatible combinations of cells and the important feature that such **binding could bring about**
alterations in enzyme activity and perhaps membrane properties. These changes
can be envisaged as having a central role in determining the physiology and
biochemistry of a symbiosis, as distinct from those of the two separate organisms.
It does not seem necessary to accept Vanderplank's further idea that protein co-
polymerization and change in enzyme activity is coupled with continued and
possibly increased protein synthesis and results in a continuing supply of protein
"food" for the fungal partner. All the evidence in mutualistic systems is that
carbohydrates, rather than proteins, are the major organic molecules transferred
from autotroph to heterotroph.

   The ideas presented here contrast with those widely held by some plant
pathologists who envisage that positive binding between organisms is a prerequisite for the resistant, rather than the compatible, response.

## Conclusion

As with the study of specificity, advance in the understanding of recognition in
mycorrhizal symbiosis must depend on a much more detailed description of the
sequence of infection and the development of the composite mycorrhizal organ.
It also requires a clear decision as to where infection is initiated and upon the
condition of the cells and cell walls of the host in the regions of the root shown to
be susceptible to infection. Plants of very diverse taxa readily recognize a single
strain of mycorrhizal fungus and those of a few taxa recognize none or do not
recognize them except under special conditions. Information such as this may
well make the search for cell-wall components, which seems likely to be involved,
somewhat simpler. With such considerations in mind, the methods already
available for the study of glycoproteins, lectins and surface enzymes may yield
results.

## Chapter 19

# Ecological Aspects of Mycorrhizal Symbiosis

## The biology of the root region

We have been mainly concerned so far with the initiation, development, structure and physiological functioning of mycorrhizal organs and of isolated mycorrhizal plants in controlled conditions. In this essay we ask questions about the importance and the success of the functioning of mycorrhizal infection in ecosystems whether natural or man-made. Mycorrhizas are active living components of the soil population having some properties like those of roots and some like those of microorganisms. It is important to note that they are not only widespread in their distribution amongst the taxa of land plants, but also that together the different kinds of mycorrhiza are more common as absorbing organs than uninfected roots. From an evolutionary point of view, symbiotic associations of higher plants with root-inhabiting mycorrhizal fungi would therefore appear to have some selective value, both for the host and for the fungus, in ecological conditions. In all the ecological aspects of mycorrhiza to be considered there are deficiencies of knowledge, but in this essay we often carry the findings of experiments on mycorrhizas to their logical conclusions in the ecological context. If these sometimes seem to be against the normal dogma of many mycorrhizasts we hope that they will act as a spur to investigation.

The soil used to be considered to be fairly uniformly inhabited by a teeming population of microorganisms. It has however become increasingly obvious that this view was a misconception, based mainly upon inadequate methods of investigation (see Gray and Williams, 1971; and Harley, 1971). For instance, the non-selective methods of soil dilution estimate, in addition to vegetatively active organisms, resting stages and dormant spores. Active organisms are localized on particular sites, whereas in the main body of the soil they are few. There, the population consists particularly of resting spores lying in wait for suitable substrates, including easily available carbon compounds upon which they can grow rapidly and complete their life cycles. This is even true of the humic layers of forest soils where the relatively few active organisms are specialized to use lignin, cellulose and other resistant carbon substrates. Any soluble materials released by them in excess of uptake, leached from new detritus or from faecal material of animals of many kinds, are rapidly snatched up by germinating ephemeral microorganisms.

In contrast to this, the surfaces of the roots of photosynthesizing plants release carbon compounds continually because they secrete mucigel and slough off dead or senescing cells as well as being leaky. Their surfaces and immediate environs, the rhizosphere, are sites of metabolically active and numerous microorganisms. In the rhizosphere and on the surface of the root the competition is intense, and disease organisms are considerably affected in their ability to establish themselves. Moreover the whole population may affect the absorption of nutrients by the plant (see Scott Russell, 1977). Some of the species of these commensal populations are also necrotrophic pathogens or symbionts at other stages of their lives. Mycorrhizal fungi and rhizobia as well as disease organisms may spread and compete in the rhizosphere before setting up their more complicated relationship with the root. Spreading on the root-surface against or in spite of other organisms is an early stage of all mycorrhizal infections. Whereas some explanations, like the availability of nutrients, particular amino acids, vitamins, the M-factor, germination factors etc., have been put forward to explain their success at this stage, none is quite satisfactory. We have to admit that none of them seems specific, or even adequately selective, for mycorrhizal fungi of one sort or another to be successful against the general crowd of rhizosphere organisms. As has been pointed out (see Marx, 1973), some mycorrhizal fungi produce substances antagonistic to pathogens of plants and to various bacteria and fungi, but this trait is not consistent enough amongst them to afford a general explanation of their apparent competitive ability in the root region. Indeed Bowen and Theodorou (1979) found that the presence of some bacteria reduced ectomycorrhizal colonization of the rhizosphere and that species of fungi varied in their response to the presence of particular species of bacteria. The situation is clearly complex, for Christie *et al.* (1978) showed that while rhizosphere populations of fungi and bacteria, as well as internal mycorrhizal infection, could be influenced by the presence of different species of host plant in experimental plots, mycorrhizal infection was altered less than the activities of other microorganisms. Of course in natural plant communities, as opposed to man-made sites or derelict land, primary mycorrhizal infection may take place from root to root rather than by germination of spores. In that case the infecting hypha is attached to a food base in a living root and may therefore be in a posture to compete more successfully than a germinating spore which is dependent on competition for carbon compounds in the rhizosphere. The conclusion has been reached that the mycorrhizal fungi of chlorophyllous plants absorb carbon compounds from their hosts, but there is a considerable disagreement as to the quantities involved. The work upon the exudation of compounds from the roots of young plants has shown that quantities of up to about one-tenth of the carbon dioxide fixed may often be released as carbon compounds from the roots. The lower dry weight increment of experimentally inoculated plants in some conditions might seem to indicate that the mycorrhizal fungi of vesicular-arbuscular, ericoid and ectomycorrhizal plants all make greater demands on their hosts for carbon than the quantities than are naturally released by them into the soil. As yet no actual measurements of the quantities of carbon compounds used by the fungus as a proportion of total photosynthesis have been made, but some estimates are available for some ectomycorrhizal plants and some ecosystems.

## Circulation of carbon in ecosystems

Figure 39 shows in outline the paths of circulation of carbon compounds in an ecosystem. As pointed out by Harley (see 1971, 1973), the estimates of carbon dioxide released from the soil (so-called soil respiration) are usually much greater than can be explained by the breakdown of the amount of detritus falling upon them. Quantities between one-third and two-thirds of the $CO_2$ released cannot be accounted for. Indeed Vogt and her colleagues (1980) showed that the rate of release of $CO_2$ was not related to the rate of litter breakdown. The discrepancy seems too large to be explained solely by the respiration and decay of roots although they may account for a very significant part of it. It is extremely probable that much of the $CO_2$ originates fairly directly from the consumption of photosynthetic products by mycorrhizal hyphae and mycorrhizal structures. Table 84 gives two estimates of the amounts of carbon which might be involved in ectomycorrhizas. This kind of mycorrhiza has received some study because of their bulky fungal sheaths and fruit bodies. The first estimate is based on the measured weight of sporophores produced in a spruce forest. The amount of carbohydrate required for this, assuming an efficiency of 40% (high for fungal growth), is 400 kg ha$^{-1}$ year$^{-1}$. This seems large and some of the later estimates of production of fruit bodies by mycorrhizal species suggest that it might be an upper limit. Fogel and Trappe (1978) give values of dry epigeous sporophore weight of 3 kg to 180 kg ha$^{-1}$ year$^{-1}$ for a number of forest types in the northern hemisphere. Fogel and Hunt (1979) only found 24 kg of hypogeous and 41 kg ha$^{-1}$ year$^{-1}$ of epigeous fruit bodies in Douglas fir forest of which some may have been non-mycorrhizal. These would require a maximum of about 160 kg ha$^{-1}$ year$^{-1}$ of carbohydrate to produce. However, Vogt et al. (1982) measured 30 kg ha$^{-1}$ year$^{-1}$ epigeous and 380 kg ha$^{-1}$ year$^{-1}$ hypogeous fruit bodies in a 180-year-old *Abies amabilis* forest which would require about 450 kg ha$^{-1}$ year$^{-1}$ of carbohydrate from photosynthesis to produce. In addition sclerotia of

*Cenococcum* were produced in quantity in both Douglas fir and *Abies* forests (2158 and 2700 kg h$^{-1}$ year$^{-1}$ respectively). These would require 5098 and 6750 kg ha$^{-1}$ year$^{-1}$ of photosynthate for formation. A more direct estimate is illustrated in Table 85 from the work of Tranquillini (1964) with *Pinus cembra*. Even allowing the possibility of some error in the measurement of photosynthesis and respiration, the amount of unaccounted photosynthate illustrates that the demands of the mycorrhizal fungus must be considerable.

Both Fogel and Hunt (1979) and Vogt *et al.* (1982) have made estimates of the turnover rates of mycorrhiza in forest ecosystems. The former calculated that out of a throughput of 30 324 kg ha$^{-1}$ year$^{-1}$ of plant material in a 35–50 year *Pseudotsuga douglasii* forest, 6104 kg was due to the mycorrhizal sheath (20%), 2323 kg to sclerotia and sporophores (7·6%), and 6991 kg to hyphae (23%). That is a total of 15 418 or 50·8% due to turnover of the fungal component. In this case all this turnover cannot be ascribed to the fungal component of

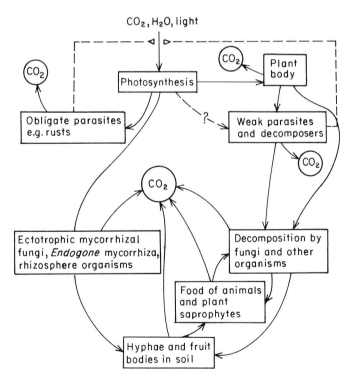

Fɪɢ. 39. Diagram showing the broad lines of carbon cycling in an ecosystem to emphasize the part played by fungi. On the left is the symbiotic cycle directly using photosynthetic products and on the right the decomposition cycle; – – –◁, feed-back effects on photosynthesis.

TABLE 84

Estimates of the consumption of photosynthetic carbon by the fungal component of ectomycorrhizas (see Harley, 1973)

A. Based on sporophore production (data of Rommell, 1939)

*Spruce forest*

| | |
|---|---|
| Dry weight of *Boletus bovinus* | 180 kg ha$^{-1}$ year$^{-1}$ |
| Carbohydrate required | 400 kg ha$^{-1}$ year$^{-1}$ |
| Equivalent volume of timber | 1 m$^3$ ha$^{-1}$ year$^{-1}$ |
| Equivalent in medium quality stand | 10% potential timber production |

B. Based on respiration of fungal sheath (data of Harley, 1969; Möller *et al.* 1954)

*Beech forest*

| | |
|---|---|
| Dry weight of fungal sheath | 40% of weight of mycorrhizas |
| CO$_2$ emission of sheath | 50% (or more) of CO$_2$ production of mycorrhizas |
| Mycorrhizal roots | 10% of root mass |
| Fungal sheath | 4% of root mass |
| CO$_2$ from sheath | 20–25% of total root CO$_2$ |
| CO$_2$ emission per year by beech root systems | 500–900 kg ha$^{-1}$ year$^{-1}$ |
| CO$_2$ emission per year by sheath | 100–250 kg CO$_2$ ha$^{-1}$ year$^{-1}$ |

TABLE 85

Use of photosynthate by mycorrhizal *Pinus cembra* (values are mg CO$_2$ g$^{-1}$ dry weight of leaves) (data of Tranquillini, 1964)

| | |
|---|---|
| Gross photosynthesis | 7830 mg |
| Respiration in growing season | |
|     Leaves day | 1551 mg |
|       night | 556 mg |
|     Roots | 447 mg |
|       when snow-covered | 439 mg |
| Total CO$_2$ emission per g leaf | 2993 mg |
| Net synthesis | 4837 mg |
| Equivalent dry weight gain expected | 2·2 g g$^{-1}$ leaves |
| Actual dry weight increment | 0·65 g g$^{-1}$ leaves |
| Equivalent dry weight unaccounted | 1·55 g g$^{-1}$ leaves |

mycorrhiza alone because it was not verified that all the sclerotia and hyphae were mycorrhizal. In the work of Vogt *et al.* estimates were made of the hyphae, sporophores and sclerotia which were actually mycorrhizal. As Table 86 shows, this estimate, although large—15% of net primary production being due to mycorrhizal turnover—is lower than those of Fogel and Hunt. It should be noted however that it is for net primary production and the gross amount of photosynthate needed to form these fungal structures is more than twice their mass. These observers have estimated the uptake and accumulation of various soil-derived nutrients in the mycorrhizas of forest ecosystems. Fogel (1980) provides an annual nitrogen budget in Douglas fir forest. He estimates that of the annual through-put of nitrogen, 43% is due to the mycorrhizal fungi. Such results agree in principle with the views derived from the physiological work on ectomycorrhizas that the fungal sheath represents an important accumulating and storage tissue which takes on much of the storage function carried out by the cortex in an uninfected root (see Harley, 1969b).

There is, however, very much less information about the carbon metabolism of carbon autotrophic plants which have other kinds of endomycorrhizas. The fact that mycorrhizal infection in all of them may cause, in certain circumstances, decrease in growth rate, strongly suggests that the carbon demands made on them by the fungi are very significant. In temperate forest ecosystems the total biomass of shrubs and herbs may not exceed more than 8–10% of the total biomass and the through-put of carbon and nutrients is proportionately smaller compared with the total in the ecosystem. Although the frequency of vesicular-arbuscular infection in most tropical forests is extremely high and their functioning seems to be similar to that of herbs commonly used in experiments (Janos, 1975, 1980), it is difficult to believe that the demands of their mycorrhizal fungi on the photosynthetic products of their individual plants are proportionately as high as those of ectomycorrhizal fungi. It seems not improbable that ectomycorrhizal infection with the formation of a sheath would have no

TABLE 86

Net primary production of mycorrhizas as a proportion of total (kg ha$^{-1}$ year$^{-1}$ of dry mass) in two *Abies amabilis* forests of different ages (data of Vogt *et al.*, 1982)

|  | kg | % | kg | % |
|---|---|---|---|---|
| Total net primary production | 23 520 |  | 24 870 |  |
| Net primary production of mycorrhizal fungus: |  |  |  |  |
| Mycorrhizal sheath | 530 | 2·3 | 630 | 2·5 |
| Sclerotia and sporocarps | 2 730 | 11·6 | 3 110 | 12·5 |
| Total mycorrhizal fungus | 3 260 | 13·9 | 3 740 | 15·0 |

selective advantage either in an equable climate or for an herbaceous plant with a small carbon budget. This may provide one rationale for the distribution of ecto- and ectendomycorrhizas on woody perennial plants and endomycorrhizas of various sorts on herbaceous plants as well as woody ones. Read (1982) has stressed the abilities of plants of the Ericaceae to accumulate inorganic nutrients in their tissues and there is evidence of high phosphate concentrations in the tissues of vesicular-arbuscular mycorrhizal plants. These might allow for changes in nutrient demand as rate of growth or production of reproductive organs change. In seasonal climates the production of a considerable fungal sheath allows for storage of both soil-derived and carbonaceous nutrients. The first are essential for the periodic activity of the host and the latter for periodic growth of hyphae and production of fruit bodies by the fungi.

## Nutrient accumulation in mycorrhizal fungi

Not only the absorption of inorganic nutrients by mycorrhizal hyphae, their transport and transfer to the hosts, but also their accumulation in fungal structures may be expected to have important ecological consequences. As Harley (see 1952) pointed out, mycorrhizas of all sorts have a relatively short life and on their death their content of nutrients is released by senescence and breakdown into the immediate rooting region of the plant. Hence mycorrhizas may have the effect of concentrating nutrients in the rooting region, not only by their activities in efficient uptake, translocation and storage during the active stage of their life, but also by later local release as they senesce. This last effect has been more strongly emphasized by recent work because the penetration of the fungal hyphae actually into senescing cells of sloughing tissues has been observed (Atkinson, 1975; Nylund, 1981). The ability of the hyphae of vesicular-arbuscular mycorrhizas to remove phosphate and translocate it from a decapitated senescing host to a living one has been demonstrated by Heap and Newman (1980b). More recently Whittingham and Read (1982) have demonstrated that phosphate transport between pairs of healthy grassland plants is much increased if both individuals are mycorrhizal. They used two techniques in their experiments. In short-term labelling experiments they found that $^{32}$P applied to the "source" plant could be detected in the "sink" plant within 24 hours. This is considerably sooner than the 7 days observed by Heap and Newman when the donor plant was senescent. In longer-term experiments, Whittingham and Read found that the growth of the "sink" plant was increased if the "source" plant was fed with nutrient solutions. Although some transport of phosphate between plants occurred when they were not mycorrhizal, these results suggest that one route of transport may well be via common mycorrhizal hyphae, as exudation of phosphate from the roots would be minimal in healthy plants.

It has also been suggested that carbon compounds may be transported between plants via mycorrhizal hyphae (see Hirrell and Gerdemann, 1979). These organic molecules could be amino compounds as well as carbohydrates and their movement could well have significance if one of the host plants was simultaneously symbiotic with nitrogen-fixing microorganisms and mycorrhizal fungi.

There is in consequence a strong tendency to retain the absorbed nutrient in the active biomass and to recycle it in that closed system. A second consequence of absorption and accumulation was demonstrated and emphasized first by Gadgil and Gadgil (1975). The rate of litter and humus decomposition is often dependent on its content of inorganic nutrients. Hence the removal into mycorrhizal plants of the major nutrients, nitrogen and phosphorus in particular, by a mechanism that does not use the carbon of the litter or humus, will directly reduce the rate of litter breakdown. One effect of mycorrhizal infection may therefore be to encourage, as Gadgil and Gadgil showed for *Pinus radiata*, an accumulation of litter and humus at the soil surface. As Singer and Araujo (1979) state: "In the white-sand podsol campinarana type of forests the dominant trees are obligately ectotrophically mycorrhizal: litter is accumulated as raw humus as a consequence of ectotrophic dominance." The magnitude of such an effect will however depend much upon the kind of mineral soil but it will occur with most kinds of mycorrhiza. It will be most obvious in the nutrient-deficient soils such as Singer and Araujo studied.

The factors that may offset the "Gadgil" effect are those which result in the rapid breakdown and decay of the highly nutrient-rich fungal material and its dissemination. Insects and small animals have been shown to be important mycophagists. Not only are the mycelia ingested by a variety of small animals such as Collembola (Warnock *et al.*, 1982) and nematodes (Sutherland and Fortin, 1968), but their fruit bodies and sclerotia are important food for larger animals including many small mammals. Table 87 shows the frequency with which fragments or spores of various kinds of fruiting bodies were identified in digestive tracts of small mammals. It is conspicuous that a large number of these were of mycorrhizal fungi. The Endogonaceae which were identified included species of *Endogone* (*sensu stricto*), *Glomus* and *Sclerocystis*. These might constitute in total less than 1% or up to 10% of the diet by volume (Fogel and Trappe, 1978). The larger fruiting bodies and the spores of ectomycorrhizal Basidiomycetes and Ascomycetes were much more abundant in the digestive tracts of small mammals and might also form the food of some insects (Fogel and Peck, 1975), and fragments of them might form the bulk of the ingested food of a variety of small mammals at certain times of the year. Endogonaceous spores also form part of the diet of ants, worms, wasps and birds (McIlveen and Cole, 1976). Although these activities of animals as mycophagists have been usually thought to contribute only to the dissemination of the spores of the fungi, of hypogeous fungi especially, they must play a further part in the ecosystem. They break the efficient capture and close cycling of nutrients within the mycorrhizal system, and redistribute the nutrient-rich material, as urine, faeces and cadavers, in the soil layers so offsetting the "Gadgil" effect.

TABLE 87

The frequency of fungal taxa in the digestive tracts of small mammals in coniferous forests in Oregon (data of Maser *et al.*, 1978)

| Group | % of all sporocarps |
| --- | --- |
| Basidiomycetes | 61 |
| Ascomycetes | 23 |
| Endogonaceae | 13 |
| Others (lichens) | 3 |
| Hypogeous | 88 |
| Epigeous | 9 |
| Lichen | 3 |
| Ectomycorrhizal | 79 |
| Vesicular-arbuscular | 10 |
| Others | 11 |

## Carbon availability and the achlorophyllous habit

The implications of the discussion in previous sections of this chapter are that the supply of easily available carbon compounds in soil horizons is limited, and that specialized organisms capable of degrading resistant carbon polymers are the only important denizens of the soil which at the same time obtain all their nutrients from the soil and also have permanent (persistent, perennial) vegetative mycelium. Ephemeral fungi flare into activity when easily absorbed carbon becomes available, while mycorrhizal fungi are dependent upon the photosynthesis of their host for carbohydrate. We therefore now ask: How and where do achlorophyllous higher plants fit into the system? There are more than 400 totally achlorophyllous species of angiosperm (excluding direct parasites) and very many more that are partially achlorophyllous, either having in-adequate photosynthetic equipment or a seedling stage which is totally dependent on carbon compounds absorbed from the substrate (like all orchid seedlings). There are also some dozens of species of achlorophyllous (or partially colourless) prothalli amongst the Pteridophyta, such as those of *Lycopodium*, *Ophioglossum* and the Psilotaceae. There is one known achlorophyllous hepatic, *Cryptothallus mirabilis*. The mycorrhizal infection of all these has not been studied

in great detail but it is known to vary. Some, like the Orchidaceae, Monotropaceae and Gentianaceae, are infected by septate hyphae of basidiomycetous or possibly ascomycetous affinities. Others, like Burmanniaceae and the
pteridophyte prothalli and *Cryptothallus* are usually infected by aseptate
vesicular-arbuscular hyphae. There appear to be three modes of carbon
nutrition. First, a dependence of the association, fungus and host, upon dead
plant material, usually of a resistant sort like cellulosic and lignified tissues.
Secondly, a dependence on living autotrophic (chlorophyllous) plants de-
structively parasitized by the fungus as in the case of many orchids. Thirdly, a
dependence upon living green plants, the mycorrhizal fungus of which is
common to the achlorophyllous partner and passes carbon compounds to it,
there being three active living organisms in the mycorrhizal system, as in the case
of *Monotropa*. References to this matter are given in reviews by Furman and
Trappe (1971) and Harley (1973), but it is clear that it needs further
investigation. In any event, "saprophytism" is a specialized habit. The fungi are
specialized in one of three ways and the normal decomposition and decay of soil
organic material is either shorted out of the circuit completely or closely short-
cycled.

## Ecological importance of non-specificity

The physiological work on all kinds of mycorrhizal symbiosis leads to the
conclusion that when the host plants are grown *in isolation* with their fungi they
develop and increase their dry weight more rapidly than without the fungi,
especially in conditions of low or deficient nutrient supply. A second and equally
important conclusion from laboratory and field research is that a single fungal
strain may be capable of being symbiotic with many individual host plants of the
same or different species and vice versa. We also know that a single host may
form mycorrhizas *simultaneously* with many different fungi. This has been clearly
shown by many observers, e.g. by Marks and Foster (1967) with ectomycorrhizas, and by Molina *et al.* (1978) with vesicular-arbuscular mycorrhizas to give
two examples only. It would appear from these results that in any ecosystem the
root systems of the individual plants of diverse species would be enmeshed
together in a complicated network of hyphal systems—many individuals of
many species of both host and fungus contributing to each network. There might
be separate and different kinds of network—ectomycorrhizal, vesicular-
arbuscular mycorrhizal, ericoid mycorrhizal and so on—in any ecosystem.
Indeed, many years ago Magrou (1936) and Stahl (1949) and others showed
that connexions between hosts of different species by hyphae of vesicular-
arbuscular mycorrhiza were possible and this has been confirmed to some extent
by later observers such as Heap and Newman (1980a) and Whittingham and
Read (1982). It has also been observed by Reid and Woods (1969), Reid (1971)

and Woods and Brock (1964) in ectomycorrhizas. The ecological consequences of the possible formation of what can be called a "social complex of organisms" require consideration. At first sight one might be attracted to conclude that this kind of arrangement, if it exists, would have impacts upon establishment of seedlings in shaded situations beneath a canopy. Indeed it might. The main carbon drain upon photosynthesis by the fungal mycelium would be borne by the adults or dominants of the photosynthesizing hosts and would be, as it were, smoothed out between all participants. There would however be a competition between individual hosts for nutrients derived from the fungal complex. Hence competition would operate not between plant and plant absorbing nutrients from the soil but between plant and plant absorbing nutrients from a hyphal complex which was itself absorbing nutrients from the soil. Success in this competition would be dependent on efficiency of the relationship between particular individual plants or different species of plant with that fungal system. The various ectomycorrhizal host plants of a temperate forest community, for instance, would with minor exceptions be in one kind of social complex and the vesicular-arbuscular and ericoid host plants each in others.

In any ecosystem not only are most species and individuals of plants mycorrhizal, but most are extensively infected in situations of nutrient deficiency. Hence, although an uninfected plant would be at a disadvantage and heavily selected against, mycorrhizal infection, as such, would be likely to have no important impact on the competition between mycorrhizal plants infected in the same way, even if they were not connected in a social complex by the hyphae of a symbiotic fungus. We have seen, however, that the root systems are very likely to be enmeshed in hyphae in association with other plants of the same and different species and competition between hosts is likely to reside in the processes of extracting most from and giving least to the fungal system. Fitter (1977) has shown that this might well be the case in competition between species. In his experiments he grew *Lolium perenne* and *Holcus lanatus* together with a single endogonaceous strain, E3 (probably *Glomus fasciculatus*). Separately these grasses are each responsive to infection by this mycorrhizal fungus, but together the proportion of their combined weight due to *L. perenne* decreased with increasing infection. At the same time the proportion of total phosphate uptake which was into the *Lolium* plants was decreased. It is possible that with a different species or strain of fungus there might be a different effect, or perhaps a reversal of the effect. Indeed other work with other paired species has suggested that the growth of plants and development of mycorrhizas on individuals of one species may also be influenced by the presence of another species of host plant in a somewhat different way. Hall (1978b) found that rye grass (*Lolium perenne*) was not responsive (in terms of dry weight increase of tops) to application of phosphate fertilizer or to inoculation with vesicular-arbuscular mycorrhizal fungi either when grown alone or with white clover in his experimental conditions. In contrast, clover was responsive to both treatments. When grown alone at low levels of phosphate fertilization, clover showed a five-fold increase in growth following inoculation. When grown with ryegrass the growth of clover was reduced in absolute terms, but the response to mycorrhizal infection was forty-fold, indicating a considerable increase in competitive ability following development of mycorrhizas. Clearly the possibility of the formation of social complexes of organisms with common mycorrhizal fungi opens up new lines of speculation about root competition and about the causation of ecological specificity in the fungi, which it is best not to pursue pending further experimental investigation of mycorrhiza in plant communities.

The corollary to these considerations is that the greatest ecological importance of mycorrhizal infection lies especially in the establishment of individuals or species in those habitats where the main conditions to be overcome arise from the environment rather than from competition with other plants. Whereas in the competition of a closed grassland community with mixed herbs, vesicular-arbuscular mycorrhiza, as such, might not confer any clear advantage to any one species over any other because they all possess it, the colonization of a bare area or landslip in the same region would be more certain if the individuals became mycorrhizal. By the same token, the colonization of grassland areas by ectomycorrhizal trees and shrubs might be materially aided by their having different mycorrhizal equipment from the grassland herbs. A further point needs emphasis: it is important that each individual rapidly becomes mycorrhizal as it develops and that the inoculum potential in any soil is such as to make infection highly probable.

Read et al. (1976) showed that infection of seedlings in grassland communities occurred at a very early stage and they considered that it was probably essential for survival of seedlings under conditions where availability of both water and nutrients could be limiting. They went so far as to suggest that intense infection of adult plants might merely reflect the general susceptibility of roots under conditions of low nutrients. The spore dispersal methods by animals which were mentioned earlier, and the wind dispersal mechanisms of epigeous Ascomycetes and Basidiomycetes, must be important in colonization. It is possible that the very ancient vesicular-arbuscular infection is so widespread, and the resting spores so efficient, that dispersal of the fungus is no longer necessary on the same grand scale as the Ascomycetes and Basidiomycetes of ectomycorrhiza require. Nevertheless some habitats (indeed those in which mycorrhizal infection of individual plants may be most advantageous) do have generally lower numbers of mycorrhizal propagules; these are disturbed and eroded soils (see, for example, Hall and Armstrong, 1979; Reeves et al., 1979; Powell, 1980).

It is worth noting, however, that the extensive dissemination of spores by the Basidiomycetes of ectomycorrhiza does not necessarily lead to an inoculum being present and maintained in all regions of the north temperate or coniferous zones.

From the early experiments (Hatch, 1937; White, 1941; McComb, 1938; Mikola, 1953; and see Lobanov, 1960), prairie and steppe soils have been shown to be deficient in ectomycorrhizal inocula. This seems likely to arise partly from the limitation of dispersal, but much more from the short term of persistence of the spores. If it is true that some are stimulated to grow in non-host rhizospheres these will soon be eliminated. In such a case spores, like those of *Thelephora terrestris* described by Birraux and Fries (1981), that are stimulated only by potential-host roots would have a selective advantage in the colonization of new environments. On the other hand, the distance of dispersal required by vesicular-arbuscular fungi may be minimal because of the universality of this type of infection. The ericoid mycorrhizas are special to particular habitats and a faculty for very widespread dispersal in them might not be expected. However, their propagules are, as described earlier, extremely widespread.

## Mycorrhizal infection and the structure of ecosystems

In the previous discussion the ecosystem has been viewed as an approximately uniform area of interacting plants and animals in a uniform habitat. This is of course a crude generalization based on statistics. There will be an area over which any climax system may be considered approximately stable but it will consist of different units making up a pattern. In a simple sense this pattern arrangement arises, as Watt (1947) showed, from the fact that each plant, particularly each dominant plant in a community, has a limited life span. Hence the area which it occupies will change not only as it grows but more suddenly and completely when it dies. Its area of influence will then become occupied by other more mobile species which in time will give place to successors until the area or the unit of the pattern returns to its former structure. The ecosystem whether developing and changing, or stable in a statistical sense, therefore provides a great variety of niches available for colonization and development. These allow for endless repetitions of mycorrhizal infection and reinfection of plants as they colonize.

Superimposed on this pattern are the changes brought about by rarer events than these cyclic changes. Repetitive calamities, such as avalanches in montane forest, cyclones in temperate and tropical forest or fire in *Eucalyptus* forest, are examples of frequent occurrence compared with the life period of the dominant life-form (Whitmore, 1974). The rebuilding of the bare areas caused by these abrupt changes is by colonization from propagules of the host plants and often by those of the symbiotic fungi. Mycorrhizal infection by one or another kind is of selective advantage in this process, and has been found to be present in colonizing plants.

Examples of colonization of glacial areas, sand dunes, and other bare areas were discussed by Harley (1970). He pointed out that many of the primary colonizers combined both the ability to fix atmospheric nitrogen, arising from symbiosis with bacteria or actinomycetes, with a second symbiosis with mycorrhizal fungi. Amongst the colonizing plants some had vesicular-arbuscular mycorrhiza and bacterial nodules such as *Hedysarum*, *Oxytropis* and *Astragalus* from the Leguminoseae, some actinomycetes and vesicular-arbuscular mycorrhizas such as *Casuarina*, *Shepherdia* and *Hippophaë* and some actinomycete nodules and ectomycorrhiza such as *Alnus* and *Dryas*. The list could be much extended and it is of interest that in natural communities of New Zealand where native *Dryas*, *Alnus*, *Shepherdia* and *Hippophaë* do not occur and there are no herbaceous legumes, *Coriaria*, *Discaria* and woody legumes like *Carmichaelia*, all vesicular-arbuscular mycorrhizal but also with nitrogen-fixing nodules, occur (together with *Gunnera* spp. symbiotic with nitrogen-fixing algae) in similar places.

Mycorrhizal infection may have an important rôle in stabilization of the soil in bare and disturbed sites. Aggregation of sand grains by hyphae of mycorrhizal fungi in dune systems has now been amply confirmed (Clough and Sutton, 1978; Koske, 1975; Koske *et al.*, 1975; Forster and Nicolson, 1979; and see Sutton and Sheppard, 1976). It is perhaps not surprising that mycorrhizal, as opposed to saprophytic fungi, are important in habitats where organic carbon in the soil is low and hyphal growth must be dependent upon the products of very recent photosynthesis. However, mycorrhizas may also be important in influencing structure in other soils where they are important in stabilizing soil aggregates (see Tisdall and Oades, 1979).

Of course superimposed on these relatively short-term and intermittent changes are long-term climatic changes and also man-made changes. The long-term climatic changes such as those of the Quaternary period and the climatic and vegetation oscillations of the Flandrian are presumably still proceeding, but the ten thousand years or so which have passed since the last retreat of the ice in the northern hemisphere have seen considerable vegetation migrations in which there would have been, and still is, an active selection for mycorrhizal combinations of variable physiological and ecological potential rather than uniformity.

## Types of mycorrhiza and ecological conditions

The very ancient vesicular-arbuscular mycorrhiza is present almost universally in all ecosystems of the world, whereas the other kinds of mycorrhiza appear to be restricted to and characteristic of particular kinds of habitat. The ericoid mycorrhizas for instance, with their hair-roots and extramatrical hyphal networks, seem to be especially adapted to acid soils very poor in nutrients, in both the northern and southern hemispheres. The ectomycorrhizas, although also world-wide, are adapted to habitats with marked seasonal changes especially in the flushes of nutrients in the soil. These are of course broad generalizations and little has been done except in a relatively minor way to describe or explain the finer points of their distribution. For instance, the "fine endophyte" *Rhizophagus* (*Glomus*) *tenuis*, although very widely distributed, appears to be important in the highest altitudes and most rigorous or pioneer conditions (Crush, 1973b), and the propensity of *Cenococcum geophilum* (*graniforme*) to form ectomycorrhizas in nutrient-poor, dry sites with low light intensity has been widely recorded although it is very common in almost all habitats.

Read, in an enlightening paper (1982), discussed in more detail the relationship of some of the different kinds of mycorrhizal infection to different environments. He constructed a series from soils of high mineral nitrogen availability with low polyphenol content in low latitudes and at low altitudes, through intermediate soils with seasonal nitrogen and phosphate availability, to those with extremely low contents of inorganic nitrogen and phosphorus with high or very high polyphenol content at high altitudes and in high latitudes. At the first end of this series were plants characterized by vesicular-arbuscular mycorrhizas. In the intermediate environments were the ectotrophic mycorrhizal plants. The ericoid mycorrhizal plants were characteristic of the deficient, high polyphenol, high altitude environments. These generalizations bring some order to the mass of physiological facts and the broad distribution of species as at present known, but they must not be pressed too far. Not only are the vesicular-arbuscular mycorrhizal plants universally distributed, but some species occur at the highest altitudes and in the most acid mor and bog soils. On the other hand they do find their greatest development in moist temperate and tropical lowland soils where vesicular-arbuscular mycorrhizal shrubs, herbs and trees are dominant in the vegetation. Similarly, as Read himself pointed out, ectomycorrhizal trees vary. Some are associated on the one hand with vesicular-arbuscular mycorrhizal understory plants, both herbs and shrubs; others are associated with an ericoid and arbutoid understory. Arbutoid mycorrhizal plants are by no means all dwarf creeping plants, or small shrubs of peat and mor soils such as *Arctostaphylos uva-ursi*. Many species of *Arbutus* and *Arctostaphylos* are shrubs and small trees characteristic, as their fungal symbionts would lead us to expect, of drier not necessarily acid soils, bearing scrub, chaparral or forest, both deciduous and coniferous. These arbutoid mycorrhizal plants probably have more in common with ectomycorrhizal plants (including the same fungi) than with ericoid mycorrhizal ones. It seems probable that an important physiological and ecological distinction between vesicular-arbuscular mycorrhizas and those of ectomycorrhiza, arbutoid and ericoid mycorrhiza may be found in their more extensive development in mineral soils, rather than in the highly organic and surface layers of soils which are the habitats *par excellence* of the others.

Read expressed the opinion that ericoid mycorrhizas were especially important in resistance to heavy metal toxicity in the soil. Ions of aluminium, copper and zinc are much more soluble in acid than in neutral or alkaline soils and there they may often achieve high toxic proportions. Bradley *et al.* (1981, 1982) have shown that Ericaceae seem to resist this toxicity by a mechanism which prevents the accumulation of toxic metals in their shoot systems. Read has suggested that this property may reside in an interfacial matrix between the fungal hyphae and host plasmalemma. On the other hand, the simultaneous death of fungal hyphae and host cells, and the short life and frequent replacement of the hair-roots, may also be a contributory feature to the exclusion mechanism.

These suggestions, made by Read, of the rôle of the interfacial matrix in ericoid mycorrhizas, raise the possibility that it may function in a similar way in other mycorrhizas. The "contact" or "involving" layer of ectomycorrhizas might also have a similar role. A case in point is the resistance of some ectomycorrhizal plants to highly calcareous substrates. Piou (1979) classified a number of species of tree according to their degree of sensitivity to chalk and showed that mycorrhizal infection was essential to the sensitive ones. Clement *et al.* (1977) examined the strains of *Pinus nigra* capable of growing on calcareous soils and showed that this was not a genetic matter but depended on mycorrhizal infection. Again, the observations on the colonization of mine spoil, often toxic, by mycorrhizal rather than non-mycorrhizal plants also raises the question whether their success is solely due to the nutritional results of mycorrhizas and nitrogen-fixing nodules. It is possible that an ability to resist or escape potentially toxic substances or conditions might also be involved. In this regard, mycorrhizal fungi may vary, for Marx and Artman (1979) observed that *Pisolithus tinctorius* encouraged the growth of pine on coal spoils more than did *Thelephora terrestris*. It also prevented excessive accumulation of Fe, Al, Mn and S in the leaves.

## Mycorrhizal infection and colonization of derelict areas

Schramm (1966) showed that amongst the early plants colonizing anthracite waste in USA were ectomycorrhizal trees including seedlings of *Pinus rigida*, *P. virginiana*, *Betula lutea*, *B. populifolia*, *Quercus rubra*, *Q. velutina* and *Populus tremuloides*. Fruit bodies of several well-known ectomycorrhizal fungal species were associated with them. These included *Scleroderma aurantium*, *Amanita rubescens*, *Pisolithus tinctorius*, *Thelephora terrestris* and *Inocybe lacera*. Of these *P. tinctorius* has proved to be widespread in America and in Europe on such sites. Emanating from these ectomycorrhizas considerable growths of hyphae, rhizomorphs and hyphal strands are often observed. These may bind the whole soil matrix together and run considerable distances from plant to plant. This subject has been reviewed periodically by D. H. Marx (see 1971, 1980) and he and his colleagues have demonstrated the efficacy of ectomycorrhiza, especially that formed by *P. tinctorius*, in the forestation of such sites.

Daft and his colleagues (Daft and Nicolson, 1974; Daft *et al.*, 1975; Daft and Hacskaylo, 1976, 1977) have shown that vesicular-arbuscular infection of herbaceous plants is almost universal in the colonization of coal spoils both in Britain and USA. The only plants lacking infection were from families which are usually uninfected such as *Reseda* and *Carex*. Some species such as *Elaeagnus*, *Robinia* and *Alnus* had double infection of a symbiotic nitrogen fixer and a fungus. It is of interest that they record a species of *Alnus* (*A. alnus*), a genus usually ectomycorrhizal, as having 63% vesicular-arbuscular infection (presumably estimated by the root-slide method) by *Gigaspora gigantea* and a *Glomus* sp. Several species of *Glomus* including *G. fasciculatus*, *G. macrocarpus* var. *geosporus* and *G. calospora* were commonly present in the vegetation of the spoil. However, in the pioneer stages the fine endophyte *Rhizophagus* (*?Glomus*) *tenuis* was the most conspicuous endophyte.

Daft and his colleagues were able to show that host plants such as maize, alfalfa and strawberry grew successfully on coal spoil only in a mycorrhizal state. These findings have been confirmed by Lindsey *et al.* (1977) growing *Chrysothamnus nauseosus* and *Atriplex canescens* on coal spoil. They followed the work of Aldon (1975) who had found that *A. canescens* grew and survived on strip mine spoils in New Mexico if infected by *G. mosseae*. In their experiments, however, *A. canescens* was not improved in growth by *G. fasciculatus*. This might be an interesting case of ecological specificity worth following up. We are led to conclude from these researches that vesicular-arbuscular mycorrhizal or ectomycorrhizal infections are essential to the colonization of nutrient-deficient spoil heaps, as they are for the revegetation of sand dunes and other bare areas of low nutrient availability. It is not yet clear whether there is always a considerable protection from excessive supply of metals or other toxic ions or compounds such as has been shown for ericoid mycorrhiza.

There are many kinds of derelict areas other than the difficult and often toxic spoil heaps left after mining. These include roadside verges, old building sites, quarry areas etc. In such places the population of propagules or living mycorrhizal hyphae may be small. According to Menge (1982) the production of mycorrhizal plants for the revegetation of these, as well as of mine spoil, is becoming a reality. In some parts of the world nurserymen are well aware of the importance of mycorrhizal plants for this purpose. What is needed is first the provision of suitable inocula for widespread use, pending research aimed at deciding how and when it is practical to produce inocula for special purposes.

## Mycorrhiza and crop production

The establishment of a crop has much in common with the colonization of an unvegetated area. This is obviously so with most agricultural crops which are sown or planted into fields cleared of weeds, for such clearing may result in a reduction in numbers of propagules of vesicular-arbuscular mycorrhizal fungi in the soil (see Kruckelmann, 1975; T. F. Smith, 1978). Forest crops too are commonly established first by seeding in nursery beds and then outplanting young plants into slits or small cleared pits in the open. In both these systems efforts could be, and sometimes are, made to equip the young plants with endophytes suitable to the host, soil and other features of the environment. The establishment of a monoculture such as a field of maize or wheat, or the seeding of a mixed pasture of clover and grass in a suitable infected condition does not present any *theoretical* difficulty. However, the practical difficulties are considerable and after establishment the crop would gradually develop mixed infection by the advent of new endophytes or by the spread of indigenous endophytes if they had not been completely eliminated. Similar considerations and difficulties arise with an outplanted tree crop. Although the individual might be inoculated with a preferred strain of fungus in the nursery, it will, after outplanting, form mycorrhizal associations with other fungi instead of, or in addition to, the preferred one. It has to be admitted therefore that although we know that some combinations of host and fungus are more efficient than others it is practical only to aim to equip young plants with symbionts suited to encourage rapid early growth in the site chosen. Later, the mycorrhizas on that crop will diversify and change. From a practical point of view it has to be decided first whether the effort is economically worthwhile. If it is worthwhile in principle there is still room for a great deal more research to perfect the methods used. Over the years inoculation has been shown to be essential in the establishment of ectomycorrhizal forest trees outside their usual habitats. It is also extremely important in the rapid establishment of conifers and broadleaf trees in reforestation and in new plantations even on old forest sites. As D. H. Marx has shown, certain broad-host-range fungi such as *Pisolithus tinctorius* can be used successfully as inoculum for many species of tree. It is of course more economically viable to produce inoculum of a single sort on a commercial scale than to attempt to produce many, each adapted to special conditions; but future research will determine whether the inocula at present recommended are actually the best for general purposes.

There are commercial practices that include the use of fungicides applied to the soil. Many of these substances not only eliminate disease organisms but also greatly diminish the whole fungal population of the soil, as Menge (1982) described. The crops grown on soils treated with fungicide include, in addition to forest tree crops, *Citrus*, avocado, cotton, strawberries, tomatoes and tobacco in some localities, as well as greenhouse and nursery crops. Many of the fungicides eliminate or greatly reduce the vesicular-arbuscular fungi and their propagules, and unless these are replaced by inoculation, the crop usually shows signs of deficiency of phosphate and trace elements. Menge (1982) discussed the subject enumerating the types of fungicide used. He pointed out that the absence of the mycorrhizal fungi could often be compensated by the application of nutrients, especially phosphate, to the crop. He showed how some crops, even cultivars of one genus, e.g. *Citrus*, varied in their ability to extract phosphate from soils of deficient phosphate status, some being more mycorrhiza-dependent than others. He pointed out that the cost of mycorrhizal inoculation which would allow a crop to extract phosphate from a soil of low supplying power was about equal to that of the amount of phosphate needed to produce similar uptake by uninfected plants. In such a case, the profitability of using mycorrhizal inoculum depends upon refining and reducing the cost of the production of the inoculum and increasing its efficiency. On the other hand, when sufficient knowledge is available of the mycorrhizal inoculum to use, the application of fungicides as a preliminary to inoculation may be advantageous. Mycorrhizal fungi may also be adversely affected by agricultural chemicals other than fungicides, including weedkillers, nematicides and the nitrification inhibitor, N-serve. The investigations on the effects of these chemicals have been largely empirical, but they may have significance in increasing awareness of the side-effects of such chemicals on "non-target" organisms.

The future use of mycorrhizal inoculation for general agricultural and forestry purposes lies in the production of fungal inocula whose properties are known and predictable. They must be genetically stable and competitive as symbionts so that they persist in the crop. At the moment we know too little of the genetic variation and variability of mycorrhizal fungi, and almost nothing of their breeding systems and genetics. Nor do we know clearly what properties are required for efficient symbiotic relationship between any host and its fungus. There are those who believe that there is minimal gene flow within the species of most mycorrhizal fungi because sexual pairing, nuclear fusion and reduction appear to be rare. If this were so, and it has not been demonstrated, any selected strain of fungus suited for some chosen conditions or host would be stable, and once recognized and isolated could be grown in quantity and used with confidence in stated conditions without great fear of genetic change. On the other hand, many of the fungi, those of ectomycorrhiza in particular, are clearly very variable physiologically in many ways. If their breeding systems were like those of other related fungi, the production and selection of new strains would be possible. Competitive ability might be combined with efficiency of nutrient absorption, translocation and low economic coefficient. In any event, for both these purposes the definition of what makes a particular fungal strain efficient as a symbiont in cooperation with a particular host needs study. The characters required may vary from physiological traits to those that concern the structure of the mycorrhizal organ and the extent of extramatrical hyphal development. It would seem to be essential that the increasing researches which compare the effects of different strains of fungus on a given host should give as complete descriptions of the structure of the mycorrhizas as possible, as well as the cultural and physiological characteristics of the fungi. Only in this way can escape be made from the totally empirical investigations of this kind of mycorrhizal research.

## Mycorrhizas and diseases of roots

The subject of biological antagonism in the rhizosphere and root region is of long standing. Garrett wrote a review in which he considered the effects of the population of the root region on root disease as long ago as 1934. There have been conferences, of more recent years, given up to related problems (see Baker and Snyder, 1965; see Bruehl, 1975). The mycorrhizal fungi are members of the microbial population of the root region and they can compete with other members of that population sufficiently to have a unique niche and sometimes, as with ectomycorrhizal sheath formation, dominate it. Hence they might be expected to play some part in the antagonism, prevention or exclusion of disease organisms. In 1964, Zak emphasized the probable importance of ectomycorrhizal fungi in this regard and it was subsequently investigated by D. H. Marx to whom we owe the credit for much of the valuable work on the subject. He emphasized (1975a) that ectomycorrhizal fungi might affect the activities of root pathogens (a) by competition, that is the absorption and use of essential nutrients in the rhizosphere and on the root surface; (b) by the presence, in the form of the fungal sheath, of a mechanical barrier against penetration; (c) by the encouragement of a rhizosphere and sheath-surface population of organisms which might compete with the pathogens; and (d) the formation and secretion of toxic or inhibitory compounds.

Marx (1969) showed that some mycorrhizal fungi such as *Leucopaxillus cerealis* var. *piceina* which produced active antibiotics not only inhibited the attack of *Phytophthora cinnamomi* on the actual mycorrhizas of pines but also prevented pathogenic attack on neighbouring uninfected roots. On the other hand, mycorrhizal fungi like *Pisolithus tinctorius* which produced no substance toxic to *P. cinnamomi* protected the actual mycorrhizas by the sheath which was a barrier to the pathogen, but did not protect other roots. There have been considerable researches on the so-called mycorrhizosphere population which was reviewed by Rambelli (1973). In spite of the large amount of work done on this subject it is difficult to reach any very crisp conclusion about its importance to disease escape. We are no clearer about the importance of mycorrhizosphere organisms than about that of the rhizosphere populations of "uninfected" roots. Indeed many of the "rhizosphere" populations examined from natural soils were essentially "mycorrhizosphere populations" of vesicular-arbuscular mycorrhizas. The general biologically antagonistic effect of the mycorrhizosphere populations arises partly from competition for essential substances in the way that Marx envisages the mycorrhizal fungi operating.

The interactions between vesicular-arbuscular mycorrhizal infections and plant disease have also received experimental consideration. Interactions are clearly complex and may involve (a) competition for actual sites of infection in the root; (b) changes in the nutrition of the host plant; and (c) increase of tolerance of the host plant to infection when the mycorrhizal fungi compensate for the damage to roots caused by the pathogen. Effects are not confined to reductions in disease incidence or severity but may involve increases.

Among root infecting fungi, *Phytophthora* species have received considerable attention. In one investigation (Mataré and Hattingh, 1978), pre-establishment of *Glomus fasciculatus* on roots of Avocado had no effect on the course of disease. Subsequent work (Davis *et al.*, 1978; Davis and Menge, 1981) suggests that under some conditions disease may be more severe in mycorrhizal plants and that mycorrhizal infection itself may be reduced in the presence of *Phytophthora*. The magnitude of the effects depended in the experiments mentioned upon the number of chlamydospores of *Phytophthora* used as inoculum and upon the species of mycorrhizal fungus. Bärtschi *et al.* (1981) found that establishment of a mixed population of vesicular-arbuscular mycorrhizal fungi in the roots of *Chamaecyparis lawsoniana* six months before inoculation with *Phytophthora cinnamomi* gave almost complete protection against disease for 120 days, but when both microorganisms were inoculated together the mycorrhizal fungi conferred no advantage. *Glomus mosseae* was less effective than mixed inoculum, even when established in the roots 8 months before inoculation with the pathogen. Interactions between mycorrhizal infection, phosphate nutrition and Take-all disease of wheat have been investigated by Graham and Menge (1982). Phosphate deficient wheat was more susceptible to infection by *Gäumannomyces graminis* than phosphate sufficient plants, whether or not phosphate nutrition was improved by fertilizer application or by mycorrhizal infection. Increase in severity of disease in mycorrhizal plants has also been observed with both root and shoot pathogens: *Helminthosporium sativum* and *Erysiphe graminis* on barley, *Colletotrichum lindemuthianum* and *Uromyces phaseoli* on bean (Dehne and Schönbeck, 1979a). Development of viruses may also be increased, an effect which is apparently correlated with improved phosphate nutrition of the plants (Daft and Okusanya, 1973b; Schönbeck and Schinzer, 1972).

In contrast, damage to cotton seedlings by *Thielaviopsis basicola* and to *Brassica* seedlings by *Olpidium brassicae* was reduced in mycorrhizal plants (Schönbeck and Dehne, 1972; Dehne and Schönbeck, 1979a). In the former example, increased levels of free amino acids in mycorrhizal plants, together with compensating effects of generally improved plant growth, were suggested as reasons for the decrease in disease severity. Vascular-wilt symptoms are almost always alleviated in mycorrhizal plants and this is probably due in part to increased lignin synthesis and partly to increased chitinase synthesis in them (Dehne and Schönbeck, 1978, 1979b; see also Daft and Okusanya, 1973b; Pegg and Vessey, 1973; Vessey and Pegg, 1973).

Interactions between vesicular-arbuscular mycorrhizal infection and nematode development have also received attention. As with other aspects of mycorrhizal response the results are clearly complicated by the species or strains of organisms used and the inoculum levels of microorganisms, as well as by the nutrient status of the host plants. Schenck *et al.* (1975) found that varieties of soybean susceptible to *Meloidogyne incognita* increased in tolerance to nematode infection if they were also mycorrhizal, but that the magnitude of the effect depended upon the amount of nematode inoculum used. Mycorrhizal infection did not decrease the numbers of nematode juveniles recovered at the end of the experiment, indeed the largest numbers of juveniles were recovered from mycorrhizal plants with the largest root systems. The responses differed somewhat, depending upon the species of mycorrhizal fungus used. More recently Strobel *et al.* (1982) found that *Glomus etunicatus* and *Gigaspora margarita* conferred different degrees of tolerance to *M. incognita* infection in peach. Fertilizer treatments also improved tolerance to disease in this investigation. In one experiment reproduction of *M. incognita* was partially inhibited in the presence of *G. margarita*, but in other experiments the results were similar to those of Schenck *et al.* (1975), using soybean, and Roncadori and Hussey (1977) with cotton, which showed increased reproduction of the nematodes in larger, mycorrhizal plants.

Clearly interactions between mycorrhizal fungi and plant pathogens are not at all simple. In any case they form only a part of the complex pattern of relationships which exists between plants, mycorrhizal fungi, plant pathogens and other mutualistic root-inhabiting symbionts (e.g. $N_2$-fixing bacteria and actinomycetes), together with the whole gamut of soil-inhabiting microorganisms.

# References

ABBOTT, L. K. and ROBSON, A. D. (1978). Growth of subterranean clover in relation to the formation of endomycorrhizas by introduced and indigenous fungi in a field soil. *New Phytol.* **81**, 575–585.

ABBOTT, L. K. and ROBSON, A. D. (1979). A quantitative study of the spores and anatomy of mycorrhizas formed by a species of *Glomus*, with reference to its taxonomy. *Aust. J. Bot.* **27**, 363–375.

AHRENS, J. R. and REID, C. P. P. (1973). Distribution of $^{14}$C-labelled metabolites in mycorrhizal and non-mycorrhizal lodgepole pine seedlings. *Can. J. Bot.* **51**, 1029–1035.

AITCHISON, P. A. and BUTT, V. S. (1973). The relationship between synthesis of inorganic polyphosphate and phosphate uptake by *Chlorella vulgaris*. *J. exp. Bot.* **24**, 197–510.

ALDON, F. F. (1975). Endomycorrhizae enhance survival and growth of fourwing saltbush on coalmine spoils. USDA. For. Service Research Note RM. 294; 2p.

ALDRICH-BLAKE, R. N. (1930). The root system of the Corsican pine. *Oxf. For. Mem.* **12**, 1–64.

ALLEN, M. F. (1982). Influence of vesicular-arbuscular mycorrhiza on water movement through *Bouteloua gracilis* (H.B.K.) Lag ex Steud. *New Phytol.* **91**, 191–196.

ALLEN, M. F., MOORE, T. S. and CHRISTENSEN, M. (1980). Phytohormone changes in *Bouteloua gracilis* infected by vesicular-arbuscular mycorrhizae. I. Cytokinin increases in the host plant. *Can. J. Bot.* **58**, 371–374.

ALLEN, M. F., SMITH, W. K., MOORE, T. S. and CHRISTENSEN, M. (1981). Comparative water relations and photosynthesis of mycorrhizal and non-mycorrhizal *Bouteloua gracilis*. *New Phytol.* **88**, 683–693.

ALEXANDER, I. J. (1981). *Picea sichensis* and *Lactarius rufus* mycorrhizal association and its effects on seedling growth and development. *Trans. Br. mycol. Soc.* **76**, 417–423.

ALEXANDER, I. J. (1982). The significance of ectomycorrhizas in the nitrogen cycle. Brit. Ecol. Soc. Symposium. Nitrogen as an ecological factor. (In press.)

ALEXANDER, I. J. and HARDY, K. (1981). Surface phosphatase activity of Sitka spruce mycorrhizas from a serpentine site. *Soil Biol. Biochem.* **13**, 301–305.

ALVAREZ, M. R. (1968). Quantitative changes in nuclear D.N.A. accompanying post-germination embryonic development in *Vanda* (Orchidaceae). *Am. J. Bot.* **55**, 1036–1041.

ALWIS, D. P. and ABEYBAYAKE, K. (1980). A survey of mycorrhizae in some forest trees of Sri Lanka. *In* "Tropical Mycorrhizal Research" (Ed. P. Mikola), pp. 135–155. Cambridge University Press, Cambridge.

ANIKSTER, Y. and WAHL, I. (1979). Co-evolution of rust fungi on Graminae and Liliaceae and their hosts. *A. Rev. Phytopath.* **17**, 367–403.

ANTIBUS, R. K., TRAPPE, J. M. and LINKINS, A. E. (1980). Cyanide resistant repiration in *Salix nigra* endomycorrhizae. *Can. J. Bot.* **58**, 14–20.

ARDITTI, J. (1967). Factors affecting the germination of orchid seeds. *Bot. Rev.* **33**, 1–97.

ARDITTI, J. (1979). Aspects of the physiology of orchids. *Adv. bot. Res.* **7**, 421–655.

ARDITTI, J., FLICK, B. H., EHMANN, A. and FISCH, M. H. (1975). Orchid Phytoalexins II. Isolation and characterization of possible sterol companions. *Am. J. Bot.* **62**, 738–742.

ARMSTRONG, W. (1979). Aeration in higher plants. *Adv. bot. Res.* **7**, 225–332.

ARMSTRONG, W. and READ, D. J. (1972). Some observations on oxygen transport in conifer seedlings. *New Phytol.* **71**, 55–62.

ARTHUR, J. C. (1897). The movement of protoplasm in coenocytic hyphae. *Ann. Bot. (Lond.)* **11**, 491–507.

ASAI, T. (1943). Die Bedeutung der Mycorrhiza für da Pflanzenleben. *Jap. J. Bot.* **12**, 359–436.

ASAI, T. (1944). Uber die Mycorrhizenbildung der leguminosen Pflanzen. *Jap. J. Bot.* **13**, 463–485.

ASHFORD, A. E. and ALLAWAY, W. G. (1982). A sheathing mycorrhiza on *Pisoris grandis* R.Br. (Nyctaginaceae) with development of transfer cells rather than a Hartig net. *New Phytol.* **90**, 511–519.

ASHFORD, A. E., LING LEE, M. and CHILVERS, G. A. (1975). Polyphosphate in eucalypt mycorrhizas: a cytochemical demonstration. *New Phytol.* **74**, 447–453.

ASHTON, D. H. (1976). Studies on the mycorrhizae of *Eucalyptus regnans* F. Muell. *Aust. J. Bot.* **24**, 723–741.

ATKINSON, M. A. (1975). The fine structure of mycorrhizas. D. Phil. Thesis, Oxford.

AZCON, R., BAREA, J. M. and HAYMAN, D. S. (1976). Utilization of rock phosphate in alkaline oils by plants inoculated with mycorrhizal fungi and phosphate solubilizing bacteria. *Soil Biol. Biochem.* **8**, 135–138.

AZCON, R., MARIN, A. D. and BAREA, J. M. (1978). Comparative role of phosphate in soil or inside the host on the formation and effects of endomycorrhiza. *Pl. Soil* **49**, 561–567.

BAGYARAJ, D. J., MARIJUNATH, A. and PATIL, R. B. (1979). Occurrence of vesicular-arbuscular mycorrhizas in some tropical aquatic plants. *Trans. Br. mycol. Soc.* **72**, 164–167.

BAIN, H. F. (1937). Production of synthetic mycorrhiza in the cultivated cranberry. *J. Agric. Res.* **55**, 811–835.

BAKER, K. F. and SNYDER, W. C. (1965). "*Ecology of Soil-borne Pathogens*" University of California Press, Berkeley, California. 571 pp.

BAKSHI, B. K. (1974). "Mycorrhiza and its Role in Forestry" Forestry Research Institute and College, Dehra Dun.

BALDWIN, J. M., LEINHARD, G. E. and BALDWIN, S. A. (1980). The monosaccharine transport system of the human erythrocyte—orientation upon reconstructions. *Biochim. et Biophys. Acta* **599**, 699–714.

BANNISTER, P. and NORTON, W. M. (1974). The response of mycorrhizal and non-mycorrhizal rooted cuttings of heather (*Calluna vulgaris* (L.) Hull) to variations in nutrient and water regimes. *New Phytol.* **73**, 81–89.

BARBER, D. A. and MARTIN, J. K. (1976). The release of organic substances by cereal roots into soil. *New Phytol.* **76**, 69–80.

BARNARD, E. L. and JORGENSON, J. R. (1977). Respiration of field-grown loblolly pine roots as influenced by temperature and root type. *Can. J. Bot.* **55**, 740–743.

BARNES, R. L. and NAYLOR, A. W. (1959a). Effect of various nitrogen sources on the growth of isolated roots of *Pinus serotina*. *Forest Sci.* **12**, 82–89.

BARNES, R. L. and NAYLOR, A. W. (1959b). In vitro culture of pine roots and the use of *Pinus serotina* in metabolic studies. *Forest Sci.* **12**, 158–162.

BARRETT, J. T. (1947). Observations on the root endophyte, *Rhizophagus*, in culture. *Phytopathology* **37**, 359–360.

BARRETT, J. T. (1958). Synthesis of mycorrhizas with pure cultures of *Rhizophagus*. *Phytopathology* **48**, 391.

BARRETT, J. T. (1962). Isolation, culture and host relation of the phycomycetoid vesicular-arbuscular mycorrhizal endophyte, *Rhizophagus*. *Recent Advances in Botany* **2**, 1725.

BARROW, N. J., MALAJCZUK, N. and SHAW, T. C. (1977). A direct test of the ability of vesicular-arbuscular mycorrhiza to help plants take up fixed soil phosphate. *New Phytol.* **78**, 269–276.

BARTLETT, E. M. and LEWIS, D. H. (1973). Surface phosphatase activity of mycorrhizal roots of beech. *Soil Biol. Biochem.* **5**, 249–257.

BARTOS, P., DYCK, P. L. and SAMBORSKI, D. J. (1969). Adult-plant leaf rust resistance in Thatcher and Marquis wheat: genetic analysis of the host-parasite situation. *Can. J. Bot.* **47**, 267–269.

BÄRTCHI, H., GIANINAZZI-PEARSON, V. and VEGH, I. (1981). Vesicular-arbuscular mycorrhiza formation and root rot disease (*Phytophthora cinnamoni*) development in *Chamaecyparis lawsoniana*. *Phytopath. Z.* **102**, 213–218.

DE BARY, A. (1887). "Comparative Morphology and Biology of the Fungi, Mycetozoa and Bacteria" English translation of 1884 edition. Clarendon Press, Oxford.

BAUMEISTER, W. (1958). Die Aschenstoffe. Encyclopaedia of Plant Physiology IV., 5–36.

BAYLIS, G. T. S. (1961). The significance of mycorrhizas and root nodules in New Zealand vegetation. *Proc. R. Soc. N.Z.* **89**, 45–50.

BAYLIS, G. T. S. (1962). *Rhizophagus*—the catholic symbiont. *Aust. J. Sci.* **25**, 195–200.

BAYLIS, G. T. S. (1967). Experiments on the ecological significance of phycomycetous mycorrhizas. *New Phytol.* **66**, 231–243.

BAYLIS, G. T. S. (1969). Host treatment and spore production by *Endogone*. *N.Z. J. Bot.* **7**, 173–174.

BAYLIS, G. T. S. (1972). Fungi, phosphorus and the evolution of plant roots. *Search* **3**, 257–258.

BAYLIS, G. T. S. (1975). The magnolioid mycorrhiza and mycotrophy in root systems derived from it. *In* "Endomycorrhizas" (Eds F. E. Sanders, B. Mosse and P. B. Tinker), pp. 373–389. Academic Press, London and New York.

BEAU, C. (1920). Sur le rôle trophique des Endophytes d'orchidées. *C. R. Acad. Sci., Paris* **171**, 675–677.

BECKER, W. N. and GERDEMANN, J. W. (1977). Colorimetric quantification of vesicular-arbuscular mycorrhizal infection in onion. *New Phytol.* **78**, 289–295.

BEEVER, R. E. and BURNS, D. J. W. (1980). Phosphorus uptake, storage and utilization by fungi. *Adv. bot. Res.* **8**, 128–219.

BENEDICT, R. C., TYLER, V. E. and BRADY, R. (1965). Studies in spore germination and growth of some mycorrhizal-associated basidiomycetes. *Planta med.* **13**, 319–326.

BERNARD, N. (1909). L'évolution dans la symbiose. Les orchidées et leur champignons commenseaux. *Annls Sci. Nat. (Bot.)*, **9**, 1–196.

BERNARD, N. (1911). Sur la fonction fungicide des bulbes d'ophrydées. *Annls Sci. Nat. (Bot.)* **14**, 221–224.

BERTHAU, Y., GIANINAZZI-PEARSON, V. and GIANINAZZI, S. (1980). Dévelopment et expression d'association endomycorrhizienne chez le Blé. I. Mise en evidence d'un effet variétale. *Annales Amélioration des Plantes*, **30**, 67–68.

BETHENFALVAY, G. J., BROWN, M. S. and PACOVSKY, R. S. (1982). Relationships between host and endophyte development in mycorrhizas of soybeans. *New Phytol.* **90**, 537–543.

BEVAN, E. A. and KAMP, R. F. O. (1958). Stipe regeneration and fruit-body production in *Collybia velutipes* (Curt.) F. *Nature, Lond.* **181**, 1145–1146.

BEVEGE, D. I. (1971). Vesicular-arbuscular mycorrhizas of *Araucaria*: aspects of their ecology, physiology and role in nitrogen fixation. PhD. thesis. University of New England, Armidale, Australia.

BEVEGE, D. I. and BOWEN, G. D. (1975). *Endogone* strain and host plant differences in development of vesicular-arbuscular mycorrhizas. *In* "Endomycorrhizas" (Eds F. E. Sanders, B. Mosse and P. B. Tinker), pp. 77–86. Academic Press, London and New York.

BEVEGE, D. I., BOWEN, G. D. and SKINNER, M. F. (1975). Comparative carbohydrate physiology of ecto- and endo-mycorrhizas. *In* "Endomycorrhizas" (Eds F. E. Sanders, B. Mosse and P. B. Tinker) pp. 149–174. Academic Press, London and New York.

BIERMAN, B. and LINDERMANN, R. G. (1981). Quantifying vesicular-arbuscular mycorrhizae: A proposed method towards standardisation. *New Phytol.* **87**, 63–67.

BIGG, W. L. (1981). Some effects of nitrate, ammonium and mycorrhizal fungi on the growth of Douglas fir and Sitka spruce. PhD. Thesis. Aberdeen University.

BILLE HANSEN, E. (1973). Rhizomorph formation in culture by *Clitocybe geotropa*. *Botanisk. Tidsk.* **68**, 329–332.

BIRRAUX, D. and FRIES, N. (1981). Germination of *Thelephora terrestris* basidiospores. *Can. J. Bot.* **59**, 2062–2064.

BJÖRKMAN, E. (1942). Über die Bedingungen der Mykorrhizabildung bei Keifer und Fichte. Symb. Bot. Upsaliens, VI. 190 pp.

BJÖRKMAN, E. (1956). Über die Natur der Mykorrhizabildung unter besonderer Berüchsichtigung der Waldbäume und die Anwendung in der Forstlichen Praxis. *Forstwiss. Zent Bl.* **75**, 265–286.

BJÖRKMAN, E. (1960). *Monotropa hypopitys*. L. an epiparasite on tree roots. *Physiol. plant.* **13**, 308–327.

BJÖRKMAN, E. (1970). Mycorrhiza and tree nutrition in poor forest soils. *Stud. Forest Suec.* **83**, 1–24.

BLACK, C., MASTENBROEK, C., MILLS, W. E. and PETERSON, L. C. (1953). A proposal for an international nomenclature of races of *Phytophthora infestans* and of genes controlling immunity in *Solanum demissum* derivatives. *Euphytica* **2**, 173–179.

BLACK, R. and TINKER, P. B. (1979). The development of endomycorrhizal root systems. II. Effect of agronomic factors and soil conditions on the development of vesicular-arbuscular mycorrhizal infection in barley and on endophyte spore density. *New Phytol.* **83**, 401–413.

BLACKWELL, E. (1943a). The life history of *Phytophthora cactorum*. *Trans. Br. mycol. Soc.* **26**, 71–89.

BLACKWELL, E. (1943b). On germinating the oospores of *Phytophthora cactorum*. *Trans. Br. mycol. Soc.* **26**, 93–103.

BLACKEMAN, J. P., MOKAHEL, M. A. and HADLEY, G. (1976). The effect of mycorrhizal infection on respiration and activity of some oxidase enzymes of orchid protocorms. *New Phytol.* **77**, 697–704.

BOHLOOL, B. B. and SCHMIDT, E. L. (1974). Lectins as Rhizobial recognition signals. *Science* **185**, 269–271.

BOKHARY, H. A. and COOKE, R. C. (1974). Translocation of [$^{14}$C]-glucose by *Phytophthora cactorum*. *Trans. Br. Mycol. Soc.* **63**, 535–540.

BONFANTE-FASOLO, P. (1973). Nuclear division in the vegetative hyphae of *Tuber* species pluremae. *Mycopath. Mycol. appl.* **49**, 161–167.

BONFANTE-FASOLO, P. (1978). Some ultrastructural features of the vesicular-arbuscular mycorrhiza in the grape vine. *Vitis* **17**, 386–395.

BONFANTE-FASOLO, P. (1980). Occurrence of a basidiomycete in living cells of mycorrhizal hair roots of *Calluna vulgaris*. *Trans. Br. mycol. Soc.* **75**, 320–325.

BONFANTE-FASOLO, P. and BRUNEL, A. (1972). Cytological features of a mycorrhizal fungus: *Tuber melanosporum* Vitt. *Allionia* **18**, 5–11.

BONFANTE-FASOLO, P. and GIANINAZZI-PEARSON, V. (1979). Ultrastructural aspects of endomycorrhiza in the Ericaceae. I. Naturally infected hair roots of *Calluna vulgaris* L. Hull. *New Phytol.* **83**, 739–744.

BONFANTE-FASOLO, P. and SCANNERINI, S. (1977). A cytological study of the vesicular-arbuscular mycorrhiza in *Ornithogalum umbellatum* L. *Allionia* **22**, 5–21.

BONFANTE-FASOLO, P., BERTA, G. and GIANINAZZI-PEARSON, V. (1981a). Ultrastructural aspects of endomycorrhiza in the Ericaceae. II. Host endophyte relationships in *Vaccinium myrtillus*. *New Phytol.* **89**, 219–224.

BONFANTE-FASOLO, P., DEXHEIMER, J., GIANINAZZI, S., GIANINAZZI-PEARSON, V. and SCANNERINI, S. (1981b). Cytochemical modifications in the host-fungus interface during intracellular interactions in a vesicular-arbuscular mycorrhiza. *Pl. Sci. Lett.* **22**, 13–21.

BOOKER, C. E. (1980). Free and bound amino acids in the ectomycorrhizal fungus *Pisolithus tinctorius*. *Mycologia* **72**, 868–881.

BOONE, D. M. AND KEITT, (1957). *Venturia inequalis* (Cke.) Wint. XII. Genes controlling pathogenicity of wild-type lines. *Phytopathology* **47**, 403–409.

BOULLARD, B. (1956). Progrès recents dans l'étude des mycorrhizes endotrophs. *Bull. Eoc. bot. France* **103**, 75–90.

BOULLARD, B. (1957). Premières observations concernant l'influence du photopériodisme sur la formation des mycorrhizas. *Mém. Soc. Sci. Nature et Math. Cherbourg* **48**, 1–12.

BOULLARD, B. (1959). Relations entre la photopériode et l'abondance des mycorrhizes chez l'*Aster tripolium* L. (Composées). *Bull. Soc. Bot. France* **106**, 131–134.

BOULLARD, B. (1960). La lumière et les mycorrhizes. *Annls. Biol. Copenh.* **36**, 231–248.

BOULLARD, B. (1961). Influence du photopériodisme sur la mycorrhization de jeunes conifères. *Bull. Soc. linn. Normandie Ser. 10*, **2**, 30–46.

BOULLARD, B. and DOMINIK, T. (1960). Recherches comparatives entre le mycotrophisme du *Fagetum carpaticum* de Babia Góra et celui d'autres *Fageta* précédement étudies. *Zesz. nauk. wyzsz. Szk. roln. Szczec.* **3**, 3–20.

BOULLARD, B. and DOMINIK, T. (1966). Les associations mykorrhiziennes dans les hêtraies francaises. II. *Bull. Mus. Hist. nat. Marseilles* **26**, 5–19.

BOURSNELL, J. G. (1950). The symbiotic seed-borne fungus in the Cistaceae. *Ann. Bot. (Lond.)* **14**, 217–243.

BOWEN, G. D. (1968). Phosphate uptake by mycorrhizas and uninfected roots of *Pinus radiata* in relation to root distribution. *9th Int. Congr. Soil Sci. Trans.* **2**, 219–228.

BOWEN, G. D. (1973). Mineral nutrition of mycorrhizas. *In* "Ectomycorrhizas" (Eds G. C. Marks and T. T. Kozlowski), pp. 151–201. Academic Press, New York and London.

BOWEN, G. D. and SMITH, S. E. (1981). The effects of mycorrhizas on nitrogen uptake by plants. Terrestrial nitrogen cycles. *Ecol. Bull. Stockholm* **33**, 237–247.

BOWEN, G. D. and THEODOROU, C. (1967). Studies on phosphate uptake by mycorrhizas. *14 IUFRO Cong.* **5**, 116–138.

BOWEN, G. D. and THEODOROU, C. (1973). Growth of ectomycorrhizal fungi around seeds and roots. *In* "Ectomycorrhizae" (Eds G. C. Marks and T. T. Kozlowski), pp. 107–150. Academic Press, New York and London.

BOWEN, G. D. and THEODOROU, C. (1979). Interactions between bacteria and ectomycorrhizal fungi. *Soil Biol. Bioch.* **11**, 119–126.

BOWEN, G. D., SKINNER, M. F. and BEVEGE, D. I. (1974). Zinc uptake by mycorrhizal and uninfected roots of *Pinus radiata* and *Araucaria cunninghamii*. *Soil Biol. Biochem.* **6**, 141–144.

BOWEN, G. D., BEVEGE, D. I. AND MOSSE, B. (1975). Phosphate physiology of vesicular-arbuscular mycorrhizae. *In* "Endomycorrhizas" (Eds F. E. Sanders, B. Mosse and P. B. Tinker), pp. 241–260. Academic Press, London and New York.

BRACKER, C. E. (1967). Ultrastructure of fungi. *A. Rev. Phytopath.* **5**, 343–374.

BRACKER, C. E. and BUTLER, E. E. (1963). Ultrastructure and development of septa in hyphae of *Rhizoctonia solani. Mycologia* **55**, 35–58.

BRACKER, C. E. and LITTLEFIELD, L. J. (1973). Structural concepts of host-pathogen interfaces. *In* "Fungal Pathogenicity and the Plant's Response" (Eds R. J. W. Byrde and C. V. Cutting), pp. 159–317. Academic Press, London and New York.

BRADLEY, R., BURT, A. J. and READ, D. J. (1981). Mycorrhizal infection and resistance and heavy metal toxicity in *Calluna vulgaris. Nature,* Lond. **292**, 335–337.

BRADLEY, R., BURT, A. J. and READ, D. J. (1982). The biology of mycorrhiza in the Ericaceae. VIII. The role of mycorrhizal infection in heavy metal resistance. *New Phytol.* **91**, 197–201.

BREED, R. S., MURRAY, E. G. D. and SMITH, N. R. (1957). *In* "Bergey's Manual of Determinitive Bacteriology" 7th edition. Williams and Wilkins, Baltimore.

BREWSTER, J. L. and TINKER, P. B. (1972). Nutrient flow rates into roots. *Soils Fertil.* **35**, 355–359.

BRIAN, P. W. (1967). Obligate parasitism in fungi. *Proc. R. Soc. Lond. B.* **168**, 101–118.

BRIAN, P. W. (1976). The phenomenon of specificity in plant disease. *In* "Specificity in Plant Diseases" (Eds R. K. S. Wood and A. Graniti). Plenum Press, New York.

BRIERLEY, J. K. (1953). Absorption of salts by mycorrhizal roots of *Fagus sylvatica,* p. 60. D. Phil. thesis. Oxford University, Oxford.

BRIERLEY, J. K. (1955). Seasonal fluctuations of the concentration of oxygen and carbondioxide in the litter layer of beech woods with reference to salt uptake by excised mycorrhizal roots of the beech. *J. Ecol.* **43**, 404–408.

BRIGGS, D. and WALTERS, S. M. (1969). "Plant Variation and Evolution". Weidenfeld and Nicolson, London.

BROOK, J. P. (1952). Mycorrhiza of *Pernettya macrostigma. New Phytol.* **51**, 388–397.

BROWN, R. T. and MIKOLA, P. (1974). The influence of fructicose soil lichens upon the mycorrhizae and seedling growth of forest trees. *Acta for. fenn.* **141**, 23 pp.

BRUCHET, G. (1973). Contribution à l'étude du genre *Hebeloma* (Fr.) Kumnm. Essay taxonomique et ecologique. Thesis University Claude Bernard, Lyon.

BRUEHL, J. W. (1975). Biology and control of soil-borne plant pathogens, Am. Path. Soc., St. Paul, Minnesota. p. 216.

BUDD, K. and HARLEY, J. L. (1962a). The uptake and assimilation of ammonia by *Neocosmospora vasinfecta. New Phytol.* **61**, 138–149.

BUDD, K. and HARLEY, J. L. (1962b). The uptake and assimilation of ammonium by *Neocosmospora vasinfecta.* II. Increases in ammonium level in the mycelium during uptake of ammonium. *New Phytol.* **61**, 244–255.

BULLER, A. H. R. (1933). Translocation of protoplasm through the septate mycelium of certain Pyrenomycetes, Discomycetes and Hymenomycetes. *In* "Researches in Fungi" vol. V, pp. 75–167. Longmans, London.

BULTEL, G. (1926). Les orchidées germinées sans champignons ont des plantes normales. *Revue hort. (Paris),* **98**, 125.

BURDON, J. J. (1978). Mechanisms of disease control in heterogeneous plant populations—an ecologist's view. *In* "Plant Disease Epidemiology" (Eds P. R. Scott and A. Bainbridge), pp. 193–200. Blackwell's Scientific Publications, Oxford.

BURGEFF, H. (1932). "Saprophytismus und Symbiose" Gustav Fisher, Jena.

BURGEFF, H. (1936). "Samenkeimung der Orchideen" Gustav Fisher, Jena.

Burgeff, H. (1959). *In* "The Orchids" (Ed. C. L. Withner). The Ronald Press Company, New York.

Burgeff, H. (1961). "Mikrobiologie des Hochmores" Gustav Fischer Verlag, Stuttgart, 197 pp.

Burges, N. A. (1939). The defensive mechanism in orchid mycorrhiza. *New Phytol.* **38**, 273–283.

Burnett, J. H. (1976). "Fundamentals of Mycology", 2nd Edition. Edward Arnold, London.

Bushnell, W. R. (1972). Physiology of fungal haustoria. *A. Rev. Phytopath.* **10**, 151–176.

Butler, E. J. (1939). The occurrence and systematic position of the vesicular-arbuscular type of mycorrhizal fungi. *Trans. Br. mycol. Soc.* **22**, 274–301.

Buwalda, J. G., Ross, G. J. S., Stribley, D. P. and Tinker, P. B. (1982a). Development of endomycorrhizal root systems III. The mathematical representation of the spread of vesicular-arbuscular mycorrhizal infection in root systems. *New Phytol.* **91**, 669–682.

Buwalda, J. G., Ross, G. J. S., Stribley, D. P. and Tinker, P. B. (1982b). The development of endomycorrhizal root systems. III The mathematical analysis of effects of phophorus on the spread of vesicular-arbuscular mycorrhizal infection in root systems. *New Phytol.* **92**, 391–399.

Calleja, M., Mousain, D., Lecouvreur, B. and d'Auzac, J. (1980). Influence de la carence phosphatée sur les activités phosphatases acides de trois champignons mycorrhiziens: *Hebeloma edurum* Metrod, *Suillus granulatus* (L. and Fr.) O. Kuntze et *Pisolithus tinctorius* (Pers.) Coker and Couch. *Physiol. Vég.* **18**, 489–504.

Callow, J. A. (1975). Plant lectins. *Curr. Adv. Pl. Sci.* **7**, 181–193.

Callow, J. A. (1976). Nucleic acid metabolism in biotrophic infections. *In* "Biochemical Aspects of Plant Parasite Relationships" (Eds J. Friend and D. R. Threlfall), pp. 305–330. Academic Press, London and New York.

Callow, J. A. (1977). Recognition, resistance and the role of plant lectins in host-parasite interactions. *Adv. bot. Res.* **4**, 1–49.

Callow, J. A., Capaccio, L. C. M., Parish, G. and Tinker, P. B. (1978). Detection and estimation of polyphosphate in vesicular-arbuscular mycorrhizas. *New Phytol.* **80**, 125–134.

Campbell, E. O. (1963). *Gastrodia minor* Petrie—an epiparasite on Manuka. *Trans. R. Soc. N.Z. (Bot.)* **2**, 73–81.

Campbell, E. O. (1971). Notes on the fungal associations of two *Monotropa* sp. in Michigan. *Mich. Bot.* **10**, 63–67.

Canny, M. J. P. (1960). The rate of translocation. *Biol. Rev.* **35**, 507–532.

Canny, M. J. P. (1962). The mechanism of translocation. *Ann. Bot. (Lond.)* **26**, 603–617.

Capaccio, L. C. M. and Callow, J. A. (1982). The enzymes of polyphosphate metabolism in vesicular-arbuscular mycorrhizas. *New Phytol.* **91**, 81–91.

Carling, D. E., White, J. A. and Brown, M. F. (1977). The influence of fixation procedure on the ultrastructure of the host-endophyte interface of vesicular-arbuscular mycorrhizae. *Can. J. Bot.* **55**, 48–51.

Carling, D. E., Riehle, W. G., Brown, M. F. and Johnson, D. R. (1978). Effects of a vesicular-arbuscular mycorrhizal fungus on nitrate reductase and nitrogenase activities in nodulating and non-nodulating legumes. *Phytopathology* **68**, 1590–1596.

Carling, D. E., Brown, M. F. and Brown, R. A. (1979). Colonization rates and growth responses of soybean plants infected by vesicular-arbuscular mycorrhizal fungi. *Can. J. Bot.* **57**, 1769–1771.

Carr, D. J. and Burrows, W. J. (1966). Evidence of the presence in xylem sap of substances with kinetin-like activity. *Life Sci.* **5**, 2061–2077.

CARRODUS, B. B. (1965). Absorption and assimilation of nitrogen compounds by micorrhiza. D. Phil. thesis, Oxford University, Oxford.

CARRODUS, B. B. (1966). Absorption of nitrogen by mycorrhizal roots of beech. I. Factors affecting assimilation of nitrogen. *New Phytol.* **65**, 358–371.

CARRODUS, B. B. (1967). Absorption of nitrogen by mycorrhizal roots of beech. 2. Ammonium and nitrate as sources of nitrogen. *New Phytol.* **66**, 1–4.

CARRODUS, B. B. and HARLEY, J. L. (1968). Note on the incorporation of acetate and the TCA cycle in mycorrhizal roots of beech. *New Phytol.* **67**, 557–560.

CATALFOMO, P. and TRAPPE, J. M. (1970). Ectomycorrhizal fungi: A phytochemical survey. *NW. Sci.* **44**, 19–24.

CATONI, G. (1929). La fruttificazione basidiofora di un endofita dell' Orchidee. *Boll. Staz. Patol. veg. Roma* **9**, 66–74.

CHALONER, W. G. (1970). The rise and fall of the first land plants. *Biol. Rev.* **45**, 353–377.

CHAMBERS, C. A., SMITH, S. E. and SMITH, F. A. (1980). Effects of ammonium and nitrate ions on mycorrhizal infection, nodulation and growth of *Trifolium subterraneum*. *New Phytol.* **85**, 47–62.

CHEN, H. K. and THORNTON, H. G. (1940). The structure of ineffective nodules and its influence on nitrogen fixation. *Proc. R. Soc. B* **129**, 208–229.

CHEVALIER, G. MOUSAIN, D. and COUTEAUDIER, Y. (1975). Associations ectomycorrhiziennes entre les Tuberaceae et des Cistacées. *Annls. phytopath. Soc. Jap.* **7**, 335.

CHILVERS, G. A. (1968a). Some distinctive types of eucalypt mycorrhiza. *Aust. J. Bot.* **26**, 49–70.

CHILVERS, G. A. (1968b). Low power electronmicroscopy of the root cap region of eucalypt mycorrhizas. *New Phytol.* **67**, 663–665.

CHILVERS, G. A. (1972). Tree root pattern in a mixed eucalypt forest. *Aust. J. Bot.* **20**, 229–334.

CHILVERS, G. A. (1974). The morphogenesis of eucalypt mycorrhizas. *Aust. Pl. Path. Soc. News Letter* **3**, 33–34.

CHILVERS, G. A. and GUST, L. W. (1982a). The development of mycorrhizal populations on pot-grown seedlings of *Eucalyptus St. Johnii*. R. T. Bak. *New Phytol.* **90**, 667–679.

CHILVERS, G. A. and GUST, L. W. (1982b). Comparisons between growth rates of mycorrhizas, uninfected roots and mycorrhizal fungus of *Eucalyptus St. Johnii* R. T. Bak. *New Phytol.* **91**, 453–456.

CHILVERS, G. A. and HARLEY, J. L. (1980). Visualization of phosphate accumulation in beech mycorrhizas. *New Phytol.* **84**, 319–326.

CHILVERS, G. A. and PRYOR, L. D. (1965). The structure of eucalypt mycorrhizas. *Aust. J. Bot.* **13**, 245–259.

CHRISTIE, P., NEWMAN, E. I. and CAMPBELL, R. (1978). The influence of neighbouring grassland plants on each other's endomycorrhizas and root surface micro-organisms. *Soil Biol. Biochem.* **10**, 521–527.

CHRISTOPH, H. (1921). Untersuchungen über die mykotrophen Verhältnisse der Ericales und die Keimung von Pyrolaceen. *Beih. Bot. Zbl.* **38**, 115–117.

CLEMENT, A., GARBAYE, J. and LE TACON, F. (1977). Importance des ectomycorrhizes dans la résistance au calcaire du Pin noir (*Pinus nigra* Arn. ssp. *nigricans*. Host). *Oecol. Pl.* **12**, 111–131.

CLODE, J. J. E. (1965). As micorrizas na nigração do fosforo estudo com O $^{32}$P. *Publ Serv. Flor. Aquic. Portugal* **23**, 167–205.

CLOUGH, K. S. and SUTTON, J. C. (1978). Direct observation of fungal aggregates in sand-dune soil. *Can. J. Microbiol.* **24**, 326–333.

CLOWES, F. A. L. (1949). The morphology and anatomy of the roots associated with ectotrophic mycorrhiza. D. Phil. thesis, Oxford University, Oxford.

CLOWES, F. A. L. (1950). Root apical meristems of *Fagus sylvatica*. *New Phytol*. **49**, 250–268.

CLOWES, F. A. L. (1951). The structure of mycorrhizal roots of *Fagus sylvatica*. *New Phytol*. **50**, 1–16.

CLOWES, F. A. L. (1954). The root-cap of ectotrophic mycorrhizas. *New Phytol*. **53**, 525–529.

CLOWES, F. A. L. (1981). Cell proliferation in ectotrophic mycorrhizas of *Fagus sylvatica*. L. *New Phytol*. **87**, 547–555.

COCHRANE, V. W. (1958). "Physiology of Fungi". John Wiley, New York; Chapman and Hall, London.

COLEMAN, J. O. D. and HARLEY, J. L. (1976). Mitochondria of mycorrhizal roots of *Fagus sylvatica*. *New Phytol*. **10**, 317–330.

COOKE, R. (1977). "The Biology of Symbiotic Fungi". John Wiley, London, New York, Sydney, Toronto.

COOPER, K. M. (1973). Mycorrhizal associations of New Zealand Ferns. PhD Thesis, University Otago.

COOPER, K. M. (1975). Growth responses to the formation of endotrophic mycorrhizas in *Solanum leptospermum* and New Zealand ferns. *In* "Endomycorrhizas" (Eds F. E. Sanders, B. Mosse and P. B. Tinker), pp. 391–407. Academic Press, London and New York.

COOPER, K. M. (1976). A field survey of mycorrhizas in New Zealand ferns. *N. Z. Jl Bot*. **14**, 169–181.

COOPER, K. M. and LÖSEL, D. (1978). Lipid physiology of vesicular-arbuscular mycorrhiza. I. Composition of lipids in roots of onion, clover and ryegrass infected with *Glomus mosseae*. *New Phytol*. **80**, 143–151.

COOPER, K. M. and TINKER, P. B. (1978). Translocation and transfer of nutrients in vesicular-arbuscular mycorrhizas. II. Uptake and translocation of phosphorus, zinc and sulphur. *New Phytol*. **81**, 43–52.

COOPER, K. M. and TINKER, P. B. (1981). Translocation and transfer of nutrients in vesicular-arbuscular mycorrhizas. IV. Effect of environmental variables on movement of phosphorus. *New Phytol*. **88**, 327–339.

CORNER, E. J. W. (1950). "A Monograph of *Clavaria* and Allied Genera". Oxford University Press, Oxford.

COWAN, M. C., THAIN, J. F. and LEWIS, B. G. (1972). Mechanism of translocation of potassium in sporangiophores of *Phycomyces blakesleeanus* in an aqueous environment. *Trans. Br. mycol. Soc*. **58**, 91–102.

COX, G. and SANDERS, F. E. (1974). Ultrastructure of the host-fungus interface in a vesicular-arbuscular mycorrhiza. *New Phytol*. **73**, 901–912.

COX, G. and TINKER, P. B. (1976). Translocation and transfer of nutrients in vesicular-arbuscular mycorrhizas. I. The arbuscule and phosphorus transfer: a quantitative ultrastructural study. *New Phytol*. **77**, 371–378.

COX, G., SANDERS, F. E., TINKER, P. B. and WILD, J. A. (1975). Ultrastructural evidence relating to host endophyte transfer in a vesicular-arbuscular mycorrhiza. *In* Endomycorrhizas" (Eds F. E. Sanders, B. Mosse and P. B. Tinker), pp. 297–312. Academic Press, London and New York.

COX, G., MORAN, K. J., SANDERS, F., NOCKOLDS, C. and TINKER, P. B. (1980). Translocation and transfer of nutrients in vesicular-arbuscular mycorrhizas. III. Polyphosphate granules and phosphorus translocation. *New Phytol*. **84**, 649–659.

CRAFTS, C. B. and MILLER, C. D. (1974). Detection and identification of cytokynins produced by mycorrhizal fungi. *Pl. Physiol*. **54**, 586–588.

CRESS, W. A., THRONEBERRY, G. O. and LINDSEY, D. L. (1979). Kinetics of phosphorus absorption by mycorrhizal and non-mycorrhizal tomato roots. *Pl. Physiol*. **64**, 484–487.

CROMACK, K., SOLLINS, P., GRANSTEIN, W. C., SPEIDEL, K., TODD, A. W., SPYCHER, G., CHING, Y-LI. and TODD, R. L. (1979). Calcium oxalate accumulation and soil weathering in mats of the hypogeous fungus *Hysterangium crassum*. *Soil Biol. Biochem.* **11**, 463–468.

CROMER, D. A. N. (1935). The significance of the mycorrhiza of *Pinus radiata*. *Bull. Forest Bur. Australia* **16**, 1–19.

CROSSETT, R. N. and LOUGHMAN, B. C. (1966). The absorption and translocation of phosphorus by seedlings of *Hordeum vulgare* L. *New Phytol.* **65**, 459–468.

CRUSH, J. R. (1973a). The effect of *Rhizophagus tenuis* mycorrhizas on rye grass, cocksfoot and sweet vernal. *New Phytol.* **72**, 965–973.

CRUSH, J. R. (1973b). Significance of endomycorrhizas in tussock grassland in Otago, New Zealand. *N. Z. Jl. Bot.* **11**, 645–660.

CRUSH, J. R. (1974). Plant growth responses to vesicular-arbuscular mycorrhiza. VII. Growth and nodulation of some herbage legumes. *New Phytol.* **73**, 743–749.

CRUSH, J. R. (1976). Endomycorrhizas and legume growth in some soils of the Mackenzie Basin, Canterbury, New Zealand. *N. Z. Jl. agric. Res.* **19**, 473–476.

CRUSH, J. and CARADUS, J. R. (1980). Effect of mycorrhizas on the growth of some white clovers. *N. J. Jl. agric. Res.* **23**, 233–237.

CURTIS, J. T. (1937). Non-specificity of orchid mycorrhizal fungi. *Proc. soc. exp Biol. Med.* **36**, 43–44.

CURTIS, J. T. (1939). The relation of specificity of orchid mycorrhizal fungi to the problem of symbiosis. *Am. J. Bot.* **26**, 390–399.

DAFT, M. J. and EL-GIAHMI, A. A. (1974). Effect of *Endogone* mycorrhiza on plant growth. VII. Influence of infection on the growth and nodulation in french bean (*Phaseolus vulgaris*). *New Phytol.* **73**, 1139–1147.

DAFT, M. J. and EL-GIAHMI, A. A. (1975). Effect of *Glomus* infection on three legumes. *In* "Endomycorrhizas" (Eds F. E. Sanders, B. Mosse and P. B. Tinker), pp. 581–592. Academic Press, London and New York.

DAFT, M. J. and EL-GIAHMI, A. A. (1978). Effects of arbuscular mycorrhiza on plant growth. VIII. Effects of defoliation and light on selected hosts. *New Phytol.* **80**, 365–372.

DAFT, M. J. and HACSKAYLO, E. (1976). Arbuscular mycorrhizas in anthracite and bituminous coal wastes of Pennsylvania. *J. appl. Ecol.* **13**, 523–531.

DAFT, M. J. and HACSKAYLO, E. (1977). Growth of endomycorrhizal and non-mycorrhizal red maple seedlings in sand and anthracite spoil. *For. Sci.* **23**, 207–216.

DAFT, M. J. and NICOLSON, T. H. (1966). Effect of *Endogone* mycorrhiza on plant growth. *New Phytol.* **65**, 343–350.

DAFT, M. J. and NICOLSON, T. H. (1969a). Effect of *Endogone* mycorrhiza on plant growth. III. Influence of inoculum concentration on growth and infection in tomato. *New Phytol.* **68**, 953–963.

DAFT, M. J. and NICOLSON, T. H. (1969b). Effect of *Endogone* mycorrhiza on plant growth. II. Influence of soluble phosphate on endophyte and host in maize. *New Phytol.* **68**, 945–952.

DAFT, M. J. and NICOLSON, T. H. (1972). Effect of *Endogone* mycorrhiza on plant growth. IV. Quantitative relationship between the growth of the host and the development of the endophyte in tomato and maize. *New Phytol.* **71**, 287–295.

DAFT, M. J. and NICOLSON, T. H. (1974). Arbuscular mycorrhizas in plants colonizing coal wastes in Scotland. *New Phytol.* **73**, 1129–1137.

DAFT, M. J. and OKUSANYA, B. O. (1973a). Effect of *Endogone* mycorrhiza on plant growth. VI. Influence of infection on the anatomy and reproductive development in four hosts. *New Phytol.* **72**, 1333–1339.

DAFT, M. J. and OKUSANYA, B. O. (1973b). Effect of *Endogone* mycorrhiza on plant growth. V. Influence of infection on the multiplication of viruses in tomato, petunia and strawberry. *New Phytol.* **72**, 975–983.

DAFT, M. J., HACSKAYLO, E. and NICOLSON, T. H. (1975). Arbuscular mycorrhizas in plants colonizing coal spoils in Scotland and Pennsylvania. *In* "Endomycorrhizas" (Eds F. E. Sanders, B. Mosse and P. B. Tinker), pp. 561–580. Academic Press, London and New York.

DAFT, M. J., CHILVERS, M. T. and NICOLSON, T. H. (1980). Mycorrhizas of the Liliflorae. I. Morphogenesis of *Endymion non-scriptus* (L.) Garcke and its mycorrhizas in nature. *New Phytol.* **85**, 181–189.

DANGEARD, P. A. (1896). Une maladie du peuplier dans l'ouest de la France. *Le Botaniste* **5**, 38–43.

DANGEARD, P. A. (1900). Le *Rhizophagus populinis*. *Le Botaniste* **7**, 285–287.

DANIELS, B. A. and GRAHAM, S. O. (1976). Effects of nutrition and soil extracts on germination of *Glomus mosseae* spores. *Mycologia* **68**, 108–116.

DANIELS, B. A., McCOOL, P. M. and MENGE, J. A. (1981). Comparative inoculum potential of spores of six vesicular-arbuscular mycorrhizal fungi. *New Phytol.* **89**, 385–391.

DANIELS, B. A. and MENGE, J. A. (1980). Secondary sporocarp formation by *Glomus epigaeus*, a vesicular-arbuscular mycorrhizal fungus, in long-term storage. *Mycologia* **72**, 1235–1238.

DANIELSON, R. M. (1982). Taxonomic affinities and criteria for identification of the common ectendomycorrhizal symbiont of pine. *Can. J. Res.* **60**, 7–18.

DART, P. J. (1974). The infection process. *In* "The Biology of Nitrogen Fixation" (Ed. A. Quispel), pp. 381–349. North Holland, Amsterdam.

DAVIS, R. M. and MENGE, J. A. (1981). *Phytophthora parasitica* inoculation and intensity of vesicular-arbuscular mycorrhizae on citrus. *New Phytol.* **87**, 705–715.

DAVIS, R. M., MENGE, J. A. and ZENTMYER, G. A. (1978). Influence of vesicular-arbuscular mycorrhizae on *Phytophthora* root rot of three crop plants. *Phytopathology* **68**, 1614–1617.

DAY, P. R. (1956). Race names of *Cladosporium fulvum*. *Tomato genet. Co-op. Rep.* **6**, 13–14.

DAY, P. R. (1960). Variation in phytopathogenic fungi. *A. Rev. Microbiol.* **14**, 1–16.

DAY, P. R. (1974). "Genetics of Host–Parasite Interaction" Freeman, San Francisco.

DEHNE, H. W. (1978). Increased susceptibility of mycorrhizal plants to rust, powdery mildew and anthracuose. Abstracts of 3rd International Congress of Plant Pathology, Munich, p. 260.

DEHNE, H. W. and SCHOENBECK, F. (1978). Untersuchungen zum Einfluss der endotrophen Mykorhiza auf Pflanzen-krankenheiten. II. Chitinase aktivität und ornithinzyklus. *Z. PflKrankh. PflPath. PflSchutz.* **85**, 666–678.

DEHNE, H. W. and SCHOENBECK, F. (1979a). Untersuchungen zum Einfluss der endotrophen Mykorhiza auf Pflanzenkrankenheiten. IV. Pilzliche Sprossparasiten, *Olpidium brassicae*, T.M.V. *Z. PflKrankh. PflPath. PflSchutz.* **186**, 103–112.

DEHNE, H. W. and SCHOENBECK, F. (1979b). Untersuchungen zum Einfluss der endotrophen Mykorhiza auf Pflanzenkrankenheiten. III. Phenol stoffwecksel und Lignifizierung. *Phytopath. Z.* **95**, 210–216.

DEXHEIMER, J., GIANINAZZI, S. and GIANINAZZI-PEARSON, V. (1979). Ultrastructural cytochemistry of the host–fungus interfaces in the endomycorrhizal association *Glomus mosseae/Allium cepa*. *Z. PflPhysiol.* **92**, 191–206.

DIXON, G. R. and PEGG, G. F. (1969). Hyphal lysis and tylose formation in tomato cultivars infected with *Verticillium albo-atrum*. *Trans. B. mycol. soc.* **53**, 109–118.

DOAK, K. D. (1928). The mycorrhizal fungus of *Vaccinium*. *Phytopathology* **18**, 101–108.

DOMINIK, T. (1956). Vorschlag einer neuer Klassifikation der ectotrophen Mykorrhizen auf morphologisch-anatomishen Merkmalen begrundet. *Roczn. Nauk. Leśn.* **14**, 223–245.

DORR, I. and KOLLMAN, R. (1969). Fine structure of mycorrhiza in *Neottia nidus avis*. *Planta* **89**, 372–375.

DOWDING, E. S. (1959). Ecology of *Endogone*. *Trans. Br. mycol. Soc.* **42**, 449–457.

DOWNIE, D. G. (1943). The source of the symbiont of *Goodyera repens*. *Trans. Proc. bot. Soc. Edinb.* **33**, 383–390.

DREW, E. A. and SMITH, D. C. (1967). Studies in the physiology of lichens. VIII. Movement of glucose from alga to fungus during photosynthesis in the thallus of *Peligera polydactyla*. *New Phytol.* **66**, 389–400.

DUDDRIDGE, J. A. (1980). "A Comparative Ultrastructural Analysis of a Range of Mycorrhizal Associations" Ph.D thesis, University of Sheffield, Sheffield.

DUDDRIDGE, J. and READ, D. J. (1982). An ultrastructural analysis of the development of mycorrhizas in *Monotropa hypopitys*. L. *New Phytol.* **92**, 203–214.

DUDDRIDGE, J. A., MALIBARI, A. and READ, D. J. (1980). Structure and function of mycorrhizal rhizomorphs with special reference to their role in water transport. *Nature, Lond.* **287**, 834–836.

DUVIGNEAUD, P. and DENAYER-DE-SMET, S. (1971). Cycle des élements biogènes dans les ecosystèmes forestiers d'Europe. *In* "Productivité des Ecosystèmes Forestiers" (Ed. P. Duvigneaud), pp. 527–542.

EDMONDS, A. S., WILSON, J. M. and HARLEY, J. L. (1976). Factors affecting potassium uptake by beech mycorrhizas. *New Phytol.* **76**, 307–315.

EGER, G. (1965a). Untersuchungen über die Bildung und Regeneration von Fruchtkörpen bei Hutpilzen. I. *Pleurotus florida*. *Arch. Mikrobiol.* **50**, 343–356.

EGER, G. (1965b). Untersuchungen über die Bildung und Regeneration von Fruchtkörpen bei Hutpilzen. II. Weitere regenerations versuche mit *Pleurotus florida*. *Arch. Mikrobiol.* **51**, 85–93.

EGER, G. (1965c). Untersuchungen über die Bildung und Regeneration von Fruchtkörpen bei Hutpilzen. III. *Flammulina velutipes* (Aust. and Fr.) und Agaricus bisporus (Lge.)**52**, 282–290.

ENGLANDER, L. and HULL, R. J. (1980). Reciprocal transfer of nutrients between ericaceous plants and a *Clavaria* sp. *New Phytol.* **84**, 661–667.

ERNST, R., (1967). The effect of carbohydrate selection on the growth rate of freshly germinated *Phalaenopsis* and *Dendrobium* seed. *Am. Orchid Soc. Bull.* **36**, 1068–1073.

ERNST, R., ARDITTI, J. and HEALEY, P. L. (1971). Carbohydrate physiology of orchid seedlings. II. Hydrolysis and effects of oligosaccharides. *Am. J. Bot.* **58**, 827–835.

ESHED, N. and WAHL, I. (1970). Host range and interrelations of *Erysiphe graminis hordeii*, *E. graminis tritici* and *E. graminis avenae*. *Phytopathology* **60**, 628–634.

FASSI, B. (1965). Micorrize ectotrofiche di *Pinus strobus* L. prodotte da un'endogone (*Endogone lactiflua* Berk). *Alliona* **11**, 7–15.

FASSI, B. and FONTANA, A. (1967). Sintesi micorrizica tra *Pinus strobus* e *Tuber maculatum*. I. Micorrize e sviluppo die semenzali nel secondo anno. *Allionia* **13**, 177–186.

FASSI, B. and PALENZONA, M. (1969). Mycorrhizal synthesis between *Pinus strobus*, *Pseudotsuga douglasii* and *Endogone lactiflua*. *Allionia* **15**, 105–114.

FASSI, B. and DE VECCHI, E. (1967). Recherche sulla micorrize ectotrofiche del pino strobo in vivaio. I. Discurzione di alcune forme piu diffuse in Piedmonte. *Allionia* **8**, 133–152.

FASSI, B., FONTANA, A. and TRAPPE, J. M. (1969). Ectomycorrhizae formed by *Endogone lactiflua* with species of *Pinus* and *Pseudotsuga*. *Mycologia* **61**, 412–414.

FAYE, M., RANCILLAC, M. and DAVID, A. (1981). Determination of the mycorrhizogenic root formation in *Pinus pinaster* Sol. *New Phytol.* **87**, 557–565.

FENSOM, D. S. (1975). Work with isolated phloem strands. *In* "Encyclopaedia of Plant Physiology. New Series Vol. 1. Transport in plants. I. Phloem transport" (Eds M. H. Zimmermann and J. A. Milburn), pp. 223–244. Springer-Verlag, Berlin, Heidelberg, New York.

FERRY, B. W. and DAS, N. (1968). Carbon nutrition of some mycorrhizal *Boletus* spp. *Trans Br. mycol. Soc.* **51**, 795–798.

FINN, R. F. (1942). Mycorrhizal inoculation in soil of low fertility. *Black Rock for. Pap.* **1**, 116–117.

FISCH, M. H., FLICK, B. H. and ARDITTI, J. (1973). Structure and antifungal activity of hircinol, loroglossol and orcinol. *Phytochemistry* **12**, 437–441.

FITTER, A. H. (1977). Influence of mycorrhizal infection on competition for phosphorus and potassium by two grasses. *New Phytol.* **79**, 119–125.

FLANGAS, A. L. and DICKSON, J. G. (1969). The genetic control of pathogenicity, serotypes and variability in *Puccinia sorghi*. *Am. J. Bot.* **48**, 275–285.

FLOR, H. H. (1942). Inheritance of pathogenicity in *Melampsora lini*. *J. Agri. Res.* **32**, 653–669.

FLOR, H. H. (1971). Current status of the gene-for-gene hypothesis. *A. Rev. Phytopath.* **9**, 275–296.

FOGEL, R. (1980). Mycorrhizae and nutrient cycling in natural forest ecosystems. *New Phytol.* **86**, 199–212.

FOGEL, R. and HUNT, G. (1979). Fungal and arboreal biomass in a western Oregon Douglas fir ecosystem: distribution patterns and turnover. *Can. J. for. Res.* **9**, 245–256.

FOGEL, R. and PECK, S. B. (1975). Ecological studies of hypogeous fungi. I. Coleoptera associated with sporocarps. *Mycologia* **67**, 741–747.

FOGEL, R. and TRAPPE, J. M. (1978). Fungus consumption (mycophagy) by small animals. *N. W. Sci.* **52**, 1–31.

FONTANA, A. (1977). Ectomycorrhizae in *Polygonum viviparum* L. abstracts 3rd NACOM, Athens, Georgia, p. 53.

FONTANA, A. and CENTRELLA, E. (1967). Ectomicorize prodotte da funghi ipogee. *Allionia* **13**, 149–176.

FONTANA, A. and GIOVANETTI, G. (1978–79). Simbiosi micorrizica fra *Cistus incanus* L. ssp. *icanus* e *Tuber melanosporum* Vitt. *Allionia* **23**, 5–11.

FONTANA, A. and PALENZONA, M. (1969). Sintesi micorrizica di *Tuber albidum* in coltura pura, con *Pinus strobus* pioppo euroamericano. *Allionia* **15**, 99–104.

FORSTER, S. M. and NICOLSON, T. H. (1979). Microbial aggregation of sand in a maritime dune succession. *Soil Biol. Biochem.* **13**, 205–208.

FORTIN, J. A. (1966). Synthesis of mycorrhiza on explants of the root hypocotyl of *Pinus sylvestris* L. *Can. J. Bot.* **44**, 1087–1092.

FORTIN, J. A. (1970). Interaction entre Basidiomycètes mycorrhizateurs et racines de pin en presence d'acide indol-3cyl acetique. *Physiologia Pl.* **23**, 365–371.

FORTIN, J. A. and PICHÉ, Y. (1979). Cultivation of *Pinus strobus* root-hypocotyl explants for synthesis of ectomycorrhizae. *New Phytol.* **83**, 109–119.

FORTIN, J. A., PICHÉ, Y. and LALONDE, M. (1980). Technique for observation of early morphological changes during ectomycorrhiza formation. *Canad. J. Bot.* **58**, 361–365.

FOSTER, J. W. (1949). "Chemical Activities of Fungi". Academic Press, New York and London.

FOSTER, R. C. (1981a). Mycelial strands of *Pinus radiata* D. Don: ultra-structure and histochemistry. *New Phytol.* **88**, 705–712.

FOSTER, R. C. (1981b). The ultrastructure and histochemistry of the rhizosphere. *New Phytol.* **89**, 263–273.

FOSTER, R. C. (1983). Fine structure of the epidermal cell mucelages of roots. *New Phytol.*

FOSTER, R. C. and MARKS, G. C. (1966). The fine structure of the mycorrhizas of *Pinus radiata*, D. Don. *Aust. J. biol. Sci.* **18**, 1027–1038.

FOSTER, R. C. and MARKS, G. C. (1967). Observations on the mycorrhizas of forest trees. II. The rhizomorphs of *Pinus radiata* D. Don. *Aust. J. biol. Sci.* **20**, 915–926.

FRANCE, R. C., CLINE, M. I. and REID, C. P. P. (1979). Recovery of ectomycorrhizal fungi after exposure to sub-freezing temperatures. *Can. J. Bot.* **57**, 1845–1848.

FRANCE, R. C. and REID, C. P. P. (1978). Absorption of ammonium and nitrate by mycorrhizal and non-mycorrhizal roots of pine. *In* "Root Physiology and Symbiosis" (Eds A. Reidacker and M. J. Gagnaire-Michard), pp. 410–424. Proc. IUFRO Symp. Nancy, France.

FRANCE, R. C. and REID, C. P. P. (1979). Amino acid inhibition analysis of pine ectomycorrhizas. Abstract. 4 NACOM. Colorado, USA.

FRANCKE, H. L. (1934). Beiträge zut Kentius der Mykorrhiza von *Monotropa hypopitys* L. Analyse und Synthese der Symbiose. *Flora (Jena)* **129**, 1–52.

FRANK, A. B. (1885). Über die auf Wurzelymbiose beruhende Ernährung gewisser Bäume durch unterirdische Pilze. *Ber. dt. bot. Ges.* **3**, 128–145.

FRANK, A. B. (1887). Ueber neue Mykorrhiza-formen. *Ber. dt. bot. Ges.* **5**, 395.

FRANK, A. B. (1894). Die Bedeutung der Mykorrhiza-pilze fur die gemeine Kiefer. *Forshwiss. Zbl.* **16**, 1852–1890.

FREISLEBEN, R. (1933). Über experimentelle Mycorrhizabildung bei Ericaceen. *Ber. du. bot. Ges.* **60**, 351–356.

FREISLEBEN, R. (1934). Zur Frage der Mykotrophia in der gattung *Vaccinium* L. *Jb. wiss. Bot.* **80**, 421–456.

FREISLEBEN, R. (1936). Wieterer Untersuchungen über die Mykotrophie der Ericaceen. *Jb. wiss. Bot.* **82**, 413–459.

FRIES, N. (1941). Über die Sporenkeimung bei einiger Gasteromyceten und Mykorrhiza-bildenen Hymenomyceten. *Arch. Mikrobiol.* **12**, 266–284.

FRIES, N. (1943). Untersuchungen über Sporenkeimung und Mycelentwicklung boden-bewohnender Hymenomyceten. *Symb. bot. upsaliens.* **VI**.

FRIES, N. (1966). Chemical factors in the germination of spores of Basidiomycetes. *In* "The Fungal Spore" (Ed. M. F. Madelin), pp. 189–199. Butterworth Scientific Publications, London.

FRIES, N. (1976). Spore germination in *Boletus* induced by amino acids. Proc. Koninklÿke Nederlande Acadamie von Wetenschappen C. **79**, 142–146.

FRIES, N. (1977). Germination of *Laccaria laccata* spores in vitro. Mycologia **60**, 848–850.

FRIES, N. (1978). Basidiospore germination in some mycorrhiza-forming Hymenomycetes. *Trans. Br. mycol. Soc.* **70**, 319–324.

FRIES, N. (1979a). The taxon-specific spore germination reaction in *Leccinum*. *Trans. Br. mycol. Soc.* **73**, 337–341.

FRIES, N. (1979b). Germination of spores of *Cantharellus cibarius*. *Mycologia* **71**, 216–219.

FRIES, N. (1981a). Effects of plant roots and growing mycelia on basidiospore germination in mycorrhiza-forming fungi. *In* "Arctic and Alpine Mycology" Proc. 1st Int. Symp. (FISAM) Barrow, Alaska, Aug. 1980. (Eds G. A. Laursen and J. F. Amirati).

FRIES, N. (1981b). Recognition reactions between basidiospores and hyphae in *Leccinum*. *Trans Br. mycol. Soc.* **77**, 9–14.

FRIES, N. and BIRRAUX, D. (1980). Spore germination in *Hebeloma* stimulated by living plant roots. *Experientia* **36**, 1056–1057.

FROIDEVAUX, L. (1973). The ectomycorrhizal association of *Alnus rubra* and *Lactarius obscurantus*. *Can. J. for. Res.* **3**, 601–603.

FURLAN, V. and FORTIN, J. A. (1973). Formation of endomycorrhizae by *Endogone calospora* on *Allium cepa*. *Nat. Can.* **100**, 467–477.

FURLAN, V. and FORTIN, J. A. (1973). Formation of endomycorrhizas by *Endogone calospora* on *Allium cepa* under three temperature regimes. *Naturaliste Can.* **100**, 467–477.

FURLAN, V. and FORTIN, J. A. (1975). A flotation-bubbling method for collecting Endogonaceae spores from sieved soil. *Naturaliste Can.* **102**, 663–667.

FURLAN, V., BARTSCHI, H. and FORTIN, J. A. (1980). Media for density gradient extraction of endomycorrhizal spores. *Trans. Br. mycol. Soc.* **75**, 336–338.

FURMAN, T. E. (1966). Symbiotic relationship of *Monotropa*. *Am. J. Bot.* **53**, 627.

FURMAN, T. E. and TRAPPE, J. M. (1971). Phylogeny and ecology of mycotrophic achlorophyllous angiosperms. *Q. Rev. Biol.* **46**, 214–225.

GADGIL, P. D. (1972). Effect of waterlogging on mycorrhizas of radiata pine and Douglas fir. *N.Z. Jl. for. Sci.* **2**, 222–226.

GADGIL, R. L. and GADGIL, P. D. (1975). Suppression of litter decomposition by mycorrhizal roots of *Pinus radiata*. *N.Z. Jl. for. Sci.* **5**, 35–41.

GARBAYE, J., KABRE, A., LE TACON, F., MOUSAIN, D. and PIOU, D. (1979). Fertilization minéral et fructification des champignons supérieurs en hêtraie. *Annls. Sci. for.* **35**, 151–164.

GARREC, J. P. and GAY, G. (1978). Influence des champignons mycorrhiziens sur l'accumulation des element mineraux dans les racines de pin d'alep. Analyse directe par microsonde electronique. *In* "Root Physiology and Symbiosis" (Eds A. Reidacker and M. J. Gagnaire-Michard), pp. 486–488. Proc. IUFRO Symp. Nancy, France.

GARRETT, S. D. (1934). Factors affecting the pathogenicity of cereal root-rot fungi. *Biol. Rev.* **9**, 351–361.

GARRETT, S. D. (1956). "The Biology of Root-infecting Fungi" Cambridge University Press, Cambridge.

GARRETT, S. D. (1970). "Pathogenic Root-infecting Fungi" Cambridge University Press, Cambridge.

GAÜMANN, E. (1956). Abwehrreactionen bei Pflanzenkrankenheiten. *Experientia* **12**, 411–418.

GAÜMANN, E. (1960). Nouvelles données sur les réactions chimiques de défense chez les orchidées. *C.r. hebd. Séanc. Acad. Sci., Paris* **250**, 1944–1947.

GAÜMANN, E. and HOHL, H. R. (1960). Weiterer Untersuchungen uber die chemischen Abwehrreactionen der Orchideen. *Phytopath. Z.* **38**, 93–104.

GAÜMANN, E. and KERN, H. (1959a). Über die isolierung und den chemischen Nachweis des Orchinols. *Phytopath. Z.* **35**, 347–356.

GAÜMANN, E. and KERN, H. (1959b). Über chemische Abwehrreactionen bei Orchideen. *Phytopath. Z.* **36**, 1–26.

GAÜMANN, E., BRAUN, R. and BAZZIGHER, G. (1950). Uber induzierte Abwehrreactionen bei Orchideen. *Phytopath. Z.* **17**, 36–62.

GAÜMANN, E., MULLER, E., NÜESCH, J. and RIMPAU, R. H. (1961). Über die wurzelpilze von *Loroglossum hircinum* (L.) Rich. *Phytopath. Z.* **41**, 89–96.

GAY, C. and GARREC, J. P. (1980). Premiers essais de microlocalisation de quelques éléments mineraux dans les racines courtes et les micorhizes de *Pinus halepensis* Niell. *C.r. hebd. Séanc. Acad. Sci., Paris* **290 D**, 69–71.

GERDEMANN, J. W. (1955). Relation of a large soil-borne spore to phycomycetous mycorrhizal infections. *Mycologia* **47**, 619–632.

GERDEMANN, J. W. (1964). The effect of mycorrhizas on the growth of maize. *Mycologia* **56**, 342–349.

GERDEMANN, J. W. (1968). Vesicular-arbuscular mycorrhiza and plant growth. *A. Rev. Phytopath.* **6**, 397–418.

GERDEMANN, J. W. (1975). Vesicular-arbuscular mycorrhizae. *In* "The Development and Function of Roots" (Eds J. G. Torrey and D. T. Clarkson), pp. 491–575. Academic Press, London and New York.

GERDEMANN, J. W. and NICOLSON, T. H. (1963). Spores of mycorrhizal *Endogone* species extracted from soil by wet sieving. *Trans. Br. mycol. Soc.* **46**, 234–235.

GERDEMANN, J. W. and TRAPPE, J. M. (1974). The Endogonaceae in the Pacific Northwest. *Mycologia. Memoir 5.*

GERDEMANN, J. W. and TRAPPE, J. M. (1975). Taxonomy of the Endogonaceae. *In* "Endomycorrhizas" (Eds F. E. Sanders, B. Mosse and P. B. Tinker), pp. 35–51. Academic Press, London and New York.

GIANINAZZI-PEARSON, V. and GIANINAZZI, S. (1976). Enzymatic studies on the metabolism of vesicular-arbuscular mycorrhiza. I. Effect of mycorrhiza formation and phosphorus nutrition on soluble phosphatase activities in onion roots. *Physiol. Vég.* **14**, 833–841.

GIANINAZZI-PEARSON, V. and GIANINAZZI, S. (1978). Enzymatic studies on the metabolism of vesicular-arbuscular mycorrhiza. II. Soluble alkaline phosphatase specific to mycorrhizal infection in onion roots. *Physiol. pl. Path.* **12**, 45–53.

GIANINAZZI, S., GIANINAZZI-PEARSON, V. and DEXHEIMER, J. (1979). Enzymatic studies on the metabolism of vesicular-arbuscular mycorrhiza. III. Ultrastructural localisation of acid and alkaline phosphatase in onion roots infected with *Glomus mosseae* (Nicol. & Gerd.) *New Phytol.* **82**, 127–132.

GIANINAZZI-PEARSON, V., TROUVELOT, A., MORANDI, D. and MAROCKE, R. (1980). Ecological variations in endomycorrhizas associated with wild raspberry populations in the Vosges region. *Acta Oecologica, Oecol. Plant* **1**, 111–119.

GIANINAZZI-PEARSON, V., FARDEAU, J. C., ASINNI, S. and GIANINAZZI, S. (1981a). Source of additional phosphorus absorbed from soil by vesicular-arbuscular mycorrhizal soybeans. *Physiol. Vég.* **19**, 33–43.

GIANINAZZI-PEARSON, V., MORANDI, D., DEXHEIMER, J. and GIANINAZZI, S. (1981b). Ultrastructural and ultracytochemical features of a *Glomus tenuis* mycorrhiza. *New Phytol.* **88**, 633–639.

GIL, F. and GAY, J. L. (1977). Ultrastructural and physiological properties of the host interfacial components of haustoria of *Erisiphe pisi in vivo* and *in vitro*. *Physiol. pl. Path.* **10**, 1–12.

GILDON, A. and TINKER, P. B. (1981). A heavy metal tolerant strain of a mycorrhizal fungus. *Trans. Br. mycol. Soc.* **77**, 648–649.

GILMORE, A. E. (1971). The influence of endotrophic mycorrhizae on the growth of peach seedlings. *J. Am. Soc. Hort. Sci.* **96**, 35–38.

GILTRAP, N. J. (1979). Experimental studies on the establishment and stability of ectomycorrhizas. Ph.D. thesis, Sheffield University, Sheffield.

GILTRAP, N. J. (1981). Formation of primordia and immature fruit bodies by ectomycorrhizal fungi in culture. *Trans. Br. mycol. Soc.* **77**, 204–205.

GILTRAP, N. J. (1982). Production of polyphenol oxidases by ectomycorrhizal fungi with special reference to *Lactarius* spp. *Trans. Br. mycol. Soc.* **78**, 75–81.

GILTRAP, N. J. and LEWIS, D. H. (1981). Inhibition of growth of ectomycorrhizal fungi in culture by phosphate. *New Phytol.* **87**, 669–675.

GILTRAP, N. J. and LEWIS, D. H. (1982). Catabolic repression of the synthesis of pectin degrading enzymes of *Suillus luteus* (L. ex Fr) S. F. Gray and *Hebeloma oculatum* Bruchet. *New Phytol.* **90**, 485–493.

GIOVANETTI, M. and MOSSE, B. (1980). An evaluation of techniques for measuring vesicular-arbuscular mycorrhizal infection in roots. *New Phytol.* **84**, 489–500.

GODFREY, R. M. (1957a). Studies on British species of *Endogone*. I. Taxonomy and Morphology. *Trans. Br. mycol. Soc.* **40**, 117–135.

GODFREY, R. M. (1957b). Studies on British species of *Endogone*. II. Fungal parasites. *Trans. Br. mycol. Soc.* **40**, 136–144.

GODFREY, R. M. (1957c). Studies on British species of *Endogone*. III. Germination of spores. *Trans. Br. mycol. Soc.* **40**, 203–210.

GOGALA, N. (1967). Die Wuchsstoffe des Pilzes *Bolteus edulis* var. *pinicolus* Vitt. und ihre Wirkung auf die Keimung Samen der Kiefer, *Pinus sylvestris* L. *Biol. Vestn.* (*Lublin*) **15**, 29.

GOGALA, N. (1970). Einfluss der natürlichen Cytokynine von *Pinus sylvestris* L. und andere Wuchsstoffe auf das mycelwachstum von *Boletus edulis* var. *pinicolus* Vitt. *Ost. bot. Z.* **118**, 321–333.

GOODENOUGH, P. W. and KEMPTON, R. J. (1977). Comparative distribution of soluble sugars in species of *Lycopersicon* which are tolerant or susceptible to infection by *Pyrenochaeta lycopersici*. *Phytopath. Z.* **88**, 312–321.

GOODENOUGH, P. W. and MAW, G. A. (1975). Studies on the root-rotting fungus *Pyrenochaeta lycopersici*: the cellulose complex and regulation of its extracellular appearance. *Physiol. pl. Path.* **6**, 145–157.

GOSS, R. W. (1960). Mycorrhiza of ponderosa pine in Nebraska grassland soils. *Univ. Nebr. agr. exp. Res. Bull.* **192**, 47 pp.

GOSS, M. J. (1977). Effects of mechanical impedence on root growth in Barley (*Hordeum vulgare* L.). *J. exp. Bot.* **28**, 96–111.

GRAHAM, J. H. (1976). Identification and classification of root nodule bacteria. *In* "Symbiotic Nitrogen Fixation in Plants" (Ed. P. Nutman), pp. 99–112. Cambridge University Press, Cambridge.

GRAHAM, J. H. (1982). Effects of citrus root exudates on germination of chlamydospores of the vesicular-arbuscular mycorrhizal fungus *Glomus epigaeus*. *Mycologia* **74**, 831–835.

GRAHAM, J. H. and LINDERMAN, R. G. (1979). Ethylene production by ectomycorrhizal fungi on *Fusarium oxysporum* ssp. *pini in vitro* and by aseptically synthesized mycorrhizal and *Fusarium* infected roots of Douglas fir. Proc. 4 NACOM Fort Collins, Colorado.

GRAHAM, J. H. and MENGE, J. A. (1982). Influence of vesicular-arbuscular mycorrhizae and soil phosphate on Take-all disease of wheat. *Phytopathology* **72**, 95–98.

GRAHAM, J. H., LINDERMAN, R. C. and MENGE, J. A. (1982). Development of external hyphae by different isolates of mycorrhizal *Glomus* spp. in relation to root colonization and growth of Troyer Citrange. *New Phytol.* **91**, 183–189.

GRAND, L. F. (1968). Conifer associates and mycorrhizal synthesis of some Pacific Northwest *Suillus* species. *For. Sci.* **14**, 304–312.

GRAW, D., MOAWAD, M. and REHM, S. (1979). Untersuchungen zur wirts—und wirkungs spezifität der V.A. mykorrhiza. *Z. Acker-u. PflBau.* **148**, 85–98.

GRAY, L. E. and GERDEMANN, J. W. (1969). Uptake of phosphorus[32] by vesicular-arbuscular mycorrhizae. *Pl. Soil* **30**, 415–422.

GRAY, T. R. G. and WILLIAMS, S. T. (1971). Microbial productivity in soil. *In* "Microbes and Biological Productivity". *Symp. Soc. gen. Microbiol.* **25**, 255–286.

GREEN, G. J. (1964). A colour mutation, its inheritance, and the inheritance of pathogenicity in *Puccinia graminis* Pers. *Can. J. Bot.* **42**, 1653–1664.

GREEN, G. J. (1966). Selfing studies with races 10 and 11 of wheat stem rust. *Can. J. Bot.* **44**, 1255–1260.

GREEN, N. E., GRAHAM, S. E. and SCHENK, N. C. (1976). The influence of pH on the germination of vesicular-arbuscular mycorrhizal spores. *Mycologia* **68**, 929–934.

GREEN, T. G. A. and SMITH, D. C. (1974). Lichen Physiology. XIV. Differences between algae in symbiosis and in isolation. *New Phytol.* **73**, 753–766.

GREENALL, J. M. (1963). The mycorrhizal endophytes of *Griselinia littoralis* (Cornaceae). *N. Z. Jl. Bot.* **1**, 389–400.

GUNNING, B. E. S. and ROBARDS, A. W. (1976). "Intercellular Communication in Plants. Studies on Plasmodesmata". Springer-Verlag, Berlin, Heidelberg, New York.

GUNNING, B. E. S., PATE, J. S., MINCHIN, F. R. and MARKS, I. (1974). Quantitative aspects of transfer cell structure and function. *Symp. Soc. exp. Biol.* **28**, 87–126.

GUNZE, C. M. B. and HENNESSY, C. M. R. (1980). Effect of host-applied auxin on development of endomycorrhiza in cowpeas. *Trans. Brit. mycol. Soc.* **74**, 247–251.

HACSKAYLO, E. (1973). Carbohydrate physiology of ectomycorrhizae. *In* "Ectomycorrhizae" (Eds G. C. Marks and T. T. Kozlowski), pp. 207–230. Academic Press, New York and London.

HACSKAYLO, E., PALMER, J. G. and VOZZO, J. A. (1965). Effect of temperature on growth and respiration of ectotrophic mycorrhizal fungi. *Mycologia* **57**, 748–756.

HADLEY, G. (1969). Cellulose as a carbon source for orchid mycorrhiza. *New Phytol.* **68**, 933–939.

HADLEY, G. (1970). Non-specificity of symbiotic infection in orchid mycorrhiza. *New Phytol.* **69**, 1015–1023.

HADLEY, G. (1975). Fine structure of orchid mycorrhiza. *In* "Endomycorrhizas" (Eds F. E. Sanders, B. Mosse and P. B. Tinker), pp. 335–351. Academic Press, London and New York.

HADLEY, G. and ONG, S. H. (1978). Nutritional requirements of orchid endophytes. *New Phytol.* **81**, 561–569.

HADLEY, G. and PURVES, S. (1974). Movement of $^{14}$Carbon from host to fungus in orchid mycorrhiza. *New Phytol.* **73**, 475–482.

HADLEY, G. and WILLIAMSON, B. (1971). Analysis of post-infection growth stimulus in orchid mycorrhiza. *New Phytol.* **70**, 445–455.

HADLEY, G., JOHNSON, R. P. C. and JOHN, D. A. (1971). Fine structure of the host fungus interface in orchid mycorrhiza. *Planta* **100**, 191–199.

HAINES, B. L. and BEST, G. R. (1976). *Glomus mosseae*, endomycorrhizal with *Liquidambar styraciflua* L. seedlings retards $NO_3$, $NO_2$ and $NH_4$ nitrogen loss from a temperate forest soil. *Pl. Soil* **45**, 257–261.

HALKET, A. C. (1930). The rootlets of *Amyelon radicans* Will; their anatomy, apices and endophytic fungus. *Ann. Bot. (Lond.)* **44**, 865–904.

HALL, I. R. (1976). Vesicular mycorrhizas in the orchid *Corybas macranthus*. *Trans. Br. mycol. Soc.* **66**, 160.

HALL, I. R. (1977). Species and mycorrhizal infections of New Zealand Endogonaceae. *Trans. Br. mycol. Soc.* **68**, 341–356.

HALL, I. R. (1978a). Effect of vesicular-arbuscular mycorrhizas on two varieties of maize and one of sweetcorn. *N. Z. Jl. agric. Res.* **21**, 517–519.

HALL, I. R. (1978b). Effects of endomycorrhizas on the competitive ability of white clover. *N. Z. Jl. agric. Res.* **21**, 509–515.

HALL, I. R. and ABBOTT, L. K. (1981). Photographic slide collection illustrating features of the Endogonaceae. Ed. 3, pp. 1–27, plus 400 colour transparencies. Invermay Agricultural Research Centre, and Soil Science Department, University of Western Australia.

HALL, I. R. and ARMSTRONG, P. (1979). Effect of vesicular-arbuscular mycorrhizas on growth of white clover, lotus and ryegrass in some eroded soils. *N. Z. Jl agric. Res.* **22**, 479–484.

HALL, I. R. and FISH, B. J. (1978). A key to the Endogonaceae. *Trans. Br. mycol. Soc.* **73**, 261–270.

HAMADA, M. (1940). Studien über die Mykorrhiza von *Galeola septentrionalis* Reichb. f. Ein. Neuer fall der Mykorrhiza-bildung durch intra-radicale Rhizomorphen. *Jap. J. Bot.* **10**, 151–212.

HAMADA, M. and NAKAMURA, S. I. (1963). Wurzelsymbiose von *Galeola altissima* Reichb. f., einer Chlorophyll-freien orchidee, mit dem hölzstorenden pilze *Hymenochcete crocicreas* Berk. *et Br. Sci. Rep. Tôpoku Univ. Ser. 4* **29**, 227–238.

HANDLEY, W. R. C. (1963). Mycorrhizal associations and *Calluna* heathland afforestation. *Bull. For. Commn, Lond.* **36**, 70 pp.

HANDLEY, W. R. C. and SANDERS, C. J. (1962). The concentration of easily soluble reducing substances in roots and the formation of ectotrophic mycorrhizal associations. A re-examination of Björkman's hypothesis. *Pl. Soil* **16**, 42–61.

HARDIE, K. and LEYTON, L. (1981). The influence of vesicular-arbuscular mycorrhiza on growth and water relations of red clover. I. In phosphate deficient soil. *New Phytol.* **89**, 599–608.

HARLEY, J. L. (1934). Some critical experiments upon culture methods used for fungi. *New Phytol.* **33**, 372–385.

HARLEY, J. L. (1936). Mycorrhiza of *Fagus sylvatica*. D.Phil. thesis, Oxford University, Oxford.

HARLEY, J. L. (1939). Beech mycorrhiza: re-isolation and the effects of root extracts upon *Mycelium radicis-fagi* (Chan.) *New Phytol.* **38**, 352–363.

HARLEY, J. L. (1940). A study of the root system of the beech in woodland soils with especial reference to mycorrhizal infection. *J. Ecol.* **28**, 107–117.

HARLEY, J. L. (1948). Mycorrhiza and soil ecology. *Biol. Rev.* **23**, 127–158.

HARLEY, J. L. (1950). Recent progress in the study of endotrophic mycorrhiza. *New Phytol.* **49**, 213–247.

HARLEY, J. L. (1952). Associations between micro-organisms and higher plants (mycorrhiza). *A. Rev. Microbiol.* **6**, 367–386.

HARLEY, J. L. (1959). "The Biology of Mycorrhiza" Leonard Hill, London.

HARLEY, J. L. (1964). Incorporation of carbon dioxide into excised beech mycorrhizas in the presence and absence of ammonia. *New Phytol.* **63**, 203–208.

HARLEY, J. L. (1968). Fungal Symbiosis. *Trans. Br. mycol. Soc.* **51**, 1–11.

HARLEY, J. L. (1969a). "The Biology of Mycorrhiza" 2nd Edition. Leonard Hill, London.

HARLEY, J. L. (1969b). A physiologist's viewpoint. *In* "Ecological Aspects of Mineral Nutrition of Plants" (Ed. I. Rorison), pp. 437–447. Blackwell, Oxford.

HARLEY, J. L. (1970). The importance of microorganisms in colonizing plants. *Trans. Proc. bot. Soc. Edinb.* **41**, 65–70.

HARLEY, J. L. (1971). Fungi in ecosystems. *J. Ecol.* **59**, 653–668.

HARLEY, J. L. (1973). Symbiosis in the ecosystem. *J. Nat. Sci. Council Sri Lanka* **1**, 31–48.

HARLEY, J. L. (1975). Problems of mycotrophy. *In* "Endomycorrhizas" (Eds F. E. Sanders, B. Mosse and P. B. Tinker), pp. 1–24. Academic Press, London and New York.

HARLEY, J. L. (1978a). Ectomycorrhizas as nutrient absorbing organs. *Proc. R. Soc. Lond. B* **203**, 1–21.

HARLEY, J. L. (1978b). Nutrient absorption by ectomycorrhizas. *Physiol. Vég.* **16**, 543–555.

HARLEY, J. L. (1981). Salt uptake and respiration of excised mycorrhizas. *New Phytol.* **87**, 325–332.

HARLEY, J. L. and BRIERLEY, J. K. (1954). The uptake of phosphate by excised mycorrhizal roots of the beech. VI Active transport of phosphorus from the fungal sheath into the host tissue. *New Phytol.* **53**, 240–252.

HARLEY, J. L. and BRIERLEY, J. K. (1955). The uptake of phosphate by excised mycorrhizal roots of beech. VIII. Active transport of $^{32}$P from fungus to host during uptake. *New Phytol.* **54**, 296–301.

HARLEY, J. L. and JENNINGS, D. H. (1958). The effect of sugars on the respiratory responses of beech mycorrhiza to salts. *Proc. R. Soc. Lond. B* **148**, 403–418.

HARLEY, J. L. and LEWIS, D. H. (1969). The physiology of ectotrophic mycorrhiza. *In* "Advances in Microbial Physiology" (Eds R. H. Rose and J. F. Wilkinson), pp. 50–81. Academic Press, London and New York.

HARLEY, J. L. and LOUGHMAN, B. C. (1963). The uptake of phosphate by excised mycorrhizal roots of the beech. IX. The nature of the phosphate compounds passing to the host. *New Phytol.* **62**, 350–359.

HARLEY, J. L. and LOUGHMAN, B. C. (1966). Phosphohexoisomerase in beech mycorrhiza. *New Phytol.* **65**, 157–160.

HARLEY, J. L. and McCREADY, C. C. (1950). Uptake of phosphate by excised mycorrhizas of beech. I. *New Phytol.* **49**, 388–397.

HARLEY, J. L. and McCREADY, C. C. (1952a). Uptake of phosphate by excised mycorrhiza of the beech. II. Distribution of phosphate between host and fungus. *New Phytol.* **51**, 56–64.

HARLEY, J. L. and McCREADY, C. C. (1952b). Uptake of phosphate by excised mycorrhizas of the beech. III. The effect of the fungal sheath on the availability of phosphate to the core. *New Phytol.* **51**, 343–348.

HARLEY, J. L. and McCREADY, C. C. (1953). A note on the effect of sodium azide on the respiration of beech mycorrhiza. *New Phytol.* **52**, 83–85.

HARLEY, J. L. and McCREADY, C. C. (1981). Phosphate accumulation in *Fagus* mycorrhizas. *New Phytol.* **89**, 75–80.

HARLEY, J. L. and ap REES, T. (1959). Cytochrome oxidase in mycorrhizal and non-infected roots of *Fagus sylvatica*. *New Phytol.* **58**, 364–386.

HARLEY, J. L. and WAID, J. S. (1955). The effect of light upon the roots or beech and its surface population. *Pl. Soil* **7**, 96–112.

HARLEY, J. L. and WILSON, J. M. (1959). The absorption of potassium by beech mycorrhizas. *New Phytol.* **58**, 281–298.

HARLEY, J. L. and WILSON, J. M. (1963). Die Wechselwirkung der Kationen Während der Aufnahme durch Buchenmykorrhizen. International Symposium Weimar (1960), 261–272.

HARLEY, J. L., McCREADY, C. C. and BRIERLEY, J. K. (1953). Uptake of phosphate by excised mycorrhizal roots of beech. IV The effect of oxygen concentrations upon host and fungus. *New Phytol.* **52**, 124–132.

HARLEY, J. L., BRIERLEY, J. K. and McCREADY, C. C. (1954). The uptake of phosphate by excised mycorrhizal roots of the beech. V. The examination of possible sources of misinterpretation of the quantities of phosphorus passing into the host. *New Phytol.* **53**, 92–98.

HARLEY, J. L., McCREADY, C. C., BRIERLEY, J. K. and JENNINGS, D. H. (1956). The salt respiration of excised beech mycorrhizas. II. The relationship between oxygen consumption and phosphate absorption. *New Phytol.* **55**, 1–28.

HARLEY, J. L., McCREADY, C. C. and BRIERLEY, J. K. (1958). The uptake of phosphorus by excised mycorrhizal roots of beech. VII. Translocation of phosphorus in mycorrhizal roots. *New Phytol.* **57**, 353–362.

HARLEY, J. L., McCREADY, C. C. and WEDDING, R. T. (1977). Control of respiration of beech mycorrhizas during ageing. *New Phytol.* **78**, 147–159.

HARLEY, S. E. (1965). The ecology of orchid mycorrhizal fungi. Ph.D. Thesis, Cambridge University.

HAROLD, F. M. (1966). Inorganic polyphosphates in biology: structure, metabolism and function. *Bact. Rev.* **30**, 772–793.

HARVAIS, G. and HADLEY, G. (1967). The relation between host and endophyte in orchid mycorrhiza. *New Phytol.* **66**, 205–215.

HASELWANDTER, K. (1979). Mycorrhizal status of ericaceous plants in alpine and subalpine areas. *New Phytol.* **83**, 427–431.

HATCH, A. B. (1937). The physical basis of mycotrophy in the genus *Pinus*. *Black Rock for. Bull.* **6**, 168 pp.

HATCH, A. B. and DOAK, K. D. (1933). Mycorrhizal and other features of the root system of *Pinus*. *J. Arnold Arbor.* **14**, 85–99.

HATTINGH, M. J. (1975). Uptake of $^{32}$P-labelled phosphate by endomycorrhizal roots in soil chambers. *In* "Endomycorrhizas" (Eds F. E. Sanders, B. Mosse and P. B. Tinker), pp. 289 295. Academic Press, London and New York.

HATTINGH, M. J., GRAY, L. E. and GERDEMANN, J. W. (1973). Uptake and translocation of $^{32}$P-labelled phosphate to onion roots by endomycorrhizal fungi. *Soil Sci.* **116**, 383–387.

HAYMAN, D. S. (1970). *Endogone* spore numbers in soil and vesicular-arbuscular mycorrhiza in wheat as influenced by season and soil treatment. *Trans. Br. mycol. Soc.* **54**, 53–63.

HAYMAN, D. S. (1974). Plant growth responses to vesicular-arbuscular mycorrhiza. VI. Effect of light and temperature. *New Phytol.* **73**, 71–80.

HAYMAN, D. S. and MOSSE, B. (1971). Plant growth responses to vesicular-arbuscular mycorrhiza. I. Growth of *Endogone*-inoculated plants in phosphate deficient soils. *New Phytol.* **70**, 19–27.

HAYMAN, D. S. and MOSSE, B. (1972). Plant growth responses to vesicular-arbuscular mycorrhiza. III. Increased uptake of labile P from soil. *New Phytol.* **71**, 41–47.

HAYMAN, D. S. and STOVOLD, G. E. (1979). Spore populations and infectivity of vesicular-arbuscular mycorrhizal fungi in New South Wales. *Aust. J. Bot.* **27**, 227–233.

HAYMAN, D. S., JOHNSON, A. M. and RUDDLESDIN, I. (1975). The influence of phosphate and crop species on *Endogone* spores and vesicular-arbuscular mycorrhiza under field conditions. *Pl. Soil* **43**, 489–495.

HEAP, A. J. and NEWMAN, E. I. (1980a). Links between roots by hyphae of vesicular-arbuscular mycorrhizas. *New Phytol.* **85**, 169–171.

HEAP, A. J. and NEWMAN, E. I. (1980b). The influence of vesicular-arbuscular mycorrhizas on phosphorus transfer between plants. *New Phytol.* **85**, 173–179.

HEPPER, C. M. (1977). A colorimetric method for estimating vesicular-arbuscular mycorrhizal infection in roots. *Soil Biol. Bioch.* **9**, 15–18.

HEPPER, C. M. (1979). Germination and growth of *Glomus caledonius* spores: the effects of inhibitors and nutrients. *Soil Biol. Bioch.* **11**, 269–277.

HEPPER, C. M. (1981). Techniques for studying the infection of plants by vesicular-arbuscular fungi under axenic conditions. *New Phytol.* **88**, 641–647.

HEPPER, C. M. and MOSSE, B. (1975). Techniques used to study the interaction between

*Endogone* and plant roots. *In* "Endomycorrhizas" (Eds F. E. Sanders, B. Mosse and P. B. Tinker), pp. 65–75. Academic Press, London and New York.

HEPPER, C. M. and SMITH, G. A. (1976). Observations on the germination of *Endogone* spores. *Trans. Br. mycol. Soc.* **66**, 189–194.

HESLOP-HARRISON, J. (1978). "Cellular Recognition Systems in Plants". Edward Arnold, London.

HESSELMAN, H. (1900). Om mykorrhizabilningar hos arktiska växter. *Bih. Svenska Akad. Handl.* **26**, 1–46.

HILL, E. P. (1965). Uptake and translocation. 2. Translocation. *In* "The Fungi" (Eds G. C. Ainsworth and A. S. Sussman), Vol. I, pp. 457–463. Academic Press, New York and London.

HILTNER, L. (1904). Über neuer Erfahrungen und Probleme auf den Gebiet der Bodenbacteriologie und unter besonderer Beruchsichtigung der Grundüngung und Brache. *Arb. dt. LandW. Ges.* **98**, 59–78.

HINTIKKA, V. (1969). Acetic acid tolerance in wood- and litter-decomposing Hymenomycetes. *Karstenia* **10**, 177–183.

HIRCE, G. and FINOCCHIO, A. F. (1973). Stem and root anatomy of *Monotropa uniflora*. *Bull. Torrey Bot. Club* **99**, 89–94.

HIRRELL, M. C. and GERDEMANN, J. W. (1979). Enhanced carbon transfer between onions infected with a vesicular-arbuscular mycorrhizal fungus. *New Phytol.* **83**, 731–738.

HIRRELL, M. C., MEHRAVARAN, H. and GERDEMANN, J. W. (1978). Vesicular-arbuscular mycorrhiza in Chenopodiaceae and Cruciferae, do they occur? *Can. J. Bot.* **56**, 2813–2817.

HO, I. and TRAPPE, J. M. (1973). Translocation of the $^{14}C$ from *Festuca* plants to their endomycorrhizal fungi. *Nature, New Biol.* **244**, 30–31.

HO, I. and TRAPPE, J. M. (1975). Nitrate reducing capacity of two vesicular-arbuscular mycorrhizal fungi. *Mycologia* **67**, 886–888.

HO, I. and TRAPPE, J. M. (1980). Nitrate reductase activity of non-mycorrhizal Douglas fir rootlets and some mycorrhizal fungi. *Pl. Soil* **54**, 395–398.

HO, I. and ZAK, B. (1979). Acid phosphatase activity of six ectomycorrhizal fungi. *Can. J. Bot.* **79**, 1203–1205.

HO, I., ZAK, B. and TRAPPE, J. M. (1977). Nitrate reductase activity of mycorrhizal fungi of Douglas fir. *Abst. 2nd Int. Mycol. Cong. Tanipa.*

HOFSTEN, A. (1969). The ultrastructure of mycorrhiza. I. Ectotrophic and endotrophic mycorrhiza of *Pinus sylvestris*. *Svensk. bot. Tidskr.* **63**, 455–000.

HÖGBERG, P. (1980). Occurrence and ecological importance of ectomycorrhizas and nitrogen-fixing root nodules of trees in the Mcombo woodlands of Tanzania. *Report Umeå* 40 pp.

HOLEVAS, C. D. (1966). The effect of vesicular-arbuscular mycorrhiza on the uptake of phosphorus by Strawberry (*Fragaria* sp. var Cambridge Favourite) *J. hort. sci.* **41**, 57–64.

HOLLEY, J. D. and PETERSON, R. L. (1979). Development of a vesicular-arbuscular mycorrhiza in bean roots. *Can. J. Bot.* **57**, 1960–1978.

HOLLIGAN, P. M., CHEN, C., McGEE, E. E. M. and LEWIS, D. H. (1974). Carbohydrate metabolism in healthy and rusted leaves of coltsfoot. *New Phytol.* **73**, 881–888.

HOOKER, A. L. and RUSSELL, W. A. (1962). Inheritance of resistance to *Puccina sorghi* in six corn inbred lines. *Phytopathology* **52**, 122–128.

HORA, F. B. (1959). Quantitative experiments on toadstool production in woods. *Trans. Br. mycol. Soc.* **42**, 1–14.

HORAK, E. (1964). Die Bildung von IES-Derivation durch ectotrophe Mikorrhizapilze (*Phlegmacium* spp.) von *Picea abies* Karsten. *Phytopath. Z.* **51**, 491–515.

HORTON, J. C. and KEEN, N. T. (1966a). Regulation of induced cellulase synthesis in *Pyrenochaeta terrestris* Gorenz. *et al.* by utilisable carbon compounds. *Can. J. Microbiol.* **12**, 209–220.

HORTON, J. C. and KEEN, N. T. (1966b). Sugar repression of endopolygalaturonase and cellulase synthesis during pathogenesis by *Pyrenochaeta terrestris* as a resistance mechanism in onion pink root. *Phytopathology* **56**, 908–916.

HOWARD, A. J. (1978). Translocation in fungi. *Trans. Br. mycol. Soc.* **70**, 265–269.

HOWARD, H. W. (1968). The relation between resistance genes in potatoes and pathotypes of potato-root eelworm (*Heterodera rostochiensis*) wart disease (*Synchytrium endobioticum*) and potato virus X. Abstracts of 1st International Congress of Plant Pathology (London), p. 92.

HOWELER, R. H., EDWARDS, D. G. and ASHER, C. J. (1979). The effect of soil sterilisation and mycorrhizal inoculation on the growth, nutrient uptake and critical phosphorus concentration of Cassava. Proc. 5th International Tropical Root and Tuber Crops Symposium, Manila, Philippines.

HOWELER, R. H., EDWARDS, D. G. and ASHER, C. J. (1981). Application of the flowing solution culture techniques to studies involving mycorrhizas. *Pl. Soil* **59**, 179–183.

HU HUNG TAO (1977). The influences of differences in soil moisture on survival, growth and histological changes in ectomycorrhizal red pine seedlings. *Bull. exp. For. Nat. Taiwan Univ.* **120**, 13–37.

HUBERMANN, M. A. (1940). Normal growth and development of Southern Pine seedlings in the nursery. *Ecology* **21**, 323–334.

HUNT, R. (1975). Further observations on root-shoot equilibria in perennial ryegrass *Lolium perenne* L. *Ann. Bot. (Lond.)* **39**, 745–755.

HURRELL, M. C. and GERDEMANN, J. W. (1979). Carbon transfer between onions infected with vesicular-arbuscular mycorrhizal fungus. *New Phytol.* **83**, 731–738.

HYPPEL, A. (1968). Antagonistic effects of some soil fungi on *Fomes annosus* in laboratory experiments. *Stud. For. Suecica* **64**, 1–18.

HYPPEL, A. (1968). Effect of *Fomes annosus* on seedlings of *Picea abies* in the presence of *Boletus bovinus. Stud. For. Suecica* **66**, 1–16.

INGRAM, D. S., SARGENT, J. A. and TOMMERUP, I. C. (1976). Structural aspects of infection by biotrophic fungi. *In* "Biochemical Aspects of Plant-Parasite Relationships" (Eds J. Friend and D. R. Threlfall), pp. 43–78. Academic Press, London and New York.

ISAAC, P. K. (1964). Cytoplasmic streaming in filamentous fungi. *Can. J. Bot.* **42**, 787–792.

IQBAL, S. H., YOUSAF, M. and YOUNUS, M. (1981). A field survey of mycorrhizal associations in ferns in Pakistan. *New Phytol.* **87**, 69–79.

JAHN, A. (1934). Über Wachstum, Plasmaströmung und vegetative fusionen bei *Humaria leucoloma* Hedw. *Z. Bot.* **27**, 193–250.

JANOS, D. P. (1975). Effects of vesicular-arbuscular mycorrhizae on lowland tropical rain forest trees. *In* "Endomycorrhizas" (Eds F. E. Sanders, B. Mosse and P. B. Tinker), pp. 437–446. Academic Press, London and New York.

JANOS, D. P. (1980). Vesicular-arbuscular mycorrhizae affect tropical rain forest plant growth. *Ecology* **61**, 151–152.

JANSE, J. M. (1897). Les endophytes radicaux de quelques plantes Javanaise. *Annales du Jardin Botanique Buitenzorg*, **14**, 53–201.

JASPER, D. A., ROBSON, A. D. and ABBOTT, L. K. (1979). Phosphorus and the formation of vesicular-arbuscular mycorrhizas. *Soil Biol. Biochem.* **11**, 501–505.

JAYKO, L. G., BABER, T. T., STUBBERFIELD, R. D. and ANDERSON, R. F. (1962). Nutrition and metabolic products of *Lactarius* species. *Can. J. Microbiol.* **8**, 361–371.

JENNINGS, D. H. (1964a). Changes in the size of orthophosphate pools in mycorrhizal roots of beech with reference to absorption of the ion from the external medium. *New Phytol.* **63**, 181–193.

JENNINGS, D. H. (1964b). The effect of cations on the absorption of phosphate by beech mycorrhizal roots. *New Phytol.* **63**, 348–357.

JENNINGS, D. H., THORNTON, D. J., GALPIN, M. F. J. and COGGINS, C. R. (1974). Translocation in Fungi. *Symp. Soc. exp. Biol.* **28**, 139–156.

JINKATANON, S., EDWARDS, D. G. and ASHER, C. J. (1979). An anomalous high external phosphorus requirement for young Cassava plants in solution culture. Proceedings of the 5th International Tropical Root and Tuber Crops Symposium, Manila, Philippines.

JOHNSON, P. N. (1977). Mycorrhizal Endogonaceae in a New Zealand forest. *New Phytol.* **78**, 161–170.

JOHNSON, C. R., GRAHAM, J. H., LEONARD, R. T. and MENGE, J. A. (1982). Interaction of photoperiod and vesicular-arbuscular mycorrhiza on growth and metabolism of sweet orange. *New Phytol.* **90**, 665–669.

JONSSON, L. and NYLUND, J. E. (1979). *Favolaschia dybowskayana* Singer (Aphyllophorales). A new orchid mycorrhizal fungus from tropical Africa. *New Phytol.* **83**, 121–128.

JOHNSTON, H. W. (1956). Chelation between calcium and organic anions. *N.Z. Jl Sci. Technol.* **37B**, 522–537.

JOHNSTON, H. W. and MILLER, R. B. (1959). The solubilisation of 'insoluble' phosphate. IV. The reaction between organic acids and tricalcium phosphate. *N.Z. Jl Sci.* **2**, 109–120.

KAMIENSKI, F. (1881). Die Vegetationsorgane der *Monotropa hypopitys* L. *Bot. Ztg.* **29**, 458.

KAO, K. N. and KNOTT, D. R. (1969). The inheritance of pathogenicity in races 111 and 29 of wheat stem rust. *Can. J. Genet. Cytol.* **11**, 266–274.

KASPARI, H. (1973). Electronenmikroscopische untersuchung zur Feinstructur der endotrophen Tabak mycorrhiza. *Archs Mikrobiol.* **92**, 201–207.

KASPARI, H. (1975). Fine structure of the host parasite interface in endotrophic mycorrhiza of tobacco. *In* "Endomycorrhizas" (Eds F. E. Sanders, B. Mosse and P. B. Tinker), pp. 323–334. Academic Press, London and New York.

KEELEY, J. E. (1980). Endomycorrhizae influence growth of blackgum seedlings in flooded soils. *Am. J. Bot.* **67**, 6–9.

KELLEY, A. P. (1950). Mycotrophy in plants. *Chronica bot.*, *Waltham, Massachusetts.*

KEEN, N. T. and HORTON, J. C. (1966). Induction and repression of polygalacturonase synthesis by *Pyrenochata terrestris*. *Can. J. Microbiol.* **12**, 443–453.

KENDE, H. (1964). Preservation of chlorophyll in leaf sections by substances obtained from root exudates. *Science* **145**, 1066–1067.

KHAN, A. H. (1974). The occurrence of mycorrhizas in halophytes and xerophytes and of *Endogone* spores in adjacent soils. *J. gen. Microbiol.* **81**, 7–14.

KIANMEHY, H. (1978). The response of *Helianthemum chamaecistis* (Mill.) to mycorrhizal infection in two different types of soil. *Pl. Soil* **50**, 719–722.

KIDSTON, R. and LANG, W. H. (1921). On the old red sandstone plants showing structure from the Rhynie chart bed Aberdeenshire. Part V. The thallophyte occurring in the peat bed; the succession of the plants through a vertical section of the bed and the conditions of accumulation and preservation of the deposit. *Trans. R. Soc. Edinb.* **52**, 855–902.

KINDEN, D. A. and BROWN, M. F. (1975a). Electron microscopy of vesicular-arbuscular mycorrhizae of yellow poplar. I. Characterization of endophytic structures by scanning electron stereoscopy. *Can. J. Microbiol.* **21**, 989–993.

KINDEN, D. A. and BROWN, M. F. (1975b). Electron microscopy of vesicular-arbuscular mycorrhizas of yellow poplar. II. Intracellular hyphae and vesicles. *Can. J. Microbiol.* **21**, 1768–1780.

KINDEN, D. A. and BROWN, M. F. (1975c). Electron microscopy of vesicular-arbuscular mycorrhizae of yellow poplar. III. Host endophyte interactions during arbuscular development. *Can. J. Microbiol.* **21**, 1930–1939.

KINDEN, D. A. and BROWN, M. F. (1976). Electron microscopy of vesicular-arbuscular mycorrhizae of yellow poplar. IV. Host endophyte interactions during arbuscular deterioration. *Can. J. Microbiol.* **22**, 64–75.

KNUDSON, L. (1927). Symbiosis and asymbiosis relative to orchids. *New Phytol.* **26**, 328–336.

KNUDSON, L. (1930). Flower production by orchid grown non-symbiotically. *Bot. Gaz.* **89**, 192–199.

KOCH, H. (1961). Untersuchungen über die Mykorrhiza der Kulturpflanzen unter besonderer Berucksichtigung von *Althaea officinalis* L., *Atropa belladonna*, L., *Helianthus annuus* L., und *Solanum lycopersicum* L. *Gartenbauwiss* **26**, 5–11.

KOSKE, R. E. (1975). *Endogone* spores in Australian sand dunes. *Can. J. Bot.* **53**, 668–672.

KOSKE, R. E. (1981a). Multiple germination of spores of *Gigaspora gigantea*. *Trans. Br. mycol. Soc.* **76**, 328–330.

KOSKE, R. E. (1981b). A preliminary study of interactions between species of vesicular-arbuscular fungi in a sand dune. *Trans. Br. mycol. Soc.* **76**, 411–416.

KOSKE, R. E. (1982). Evidence for a volatile attractant from plant roots affecting germ tubes of a VA fungus. *Trans. Br. mycol. Soc.* **79**, 305–310.

KOSKE, R. E. and HALVORSON, W. L. (1981). Ecological studies of vesicular-arbuscular mycorrhizae in a barrier sand dune. *Can. J. Bot.* **59**, 1413–1422.

KOSKE, R. E., SUTTON, J. C. and SHEPPARD, B. R. (1975). Ecology of *Endogone* in Lake Huron sand dunes. *Can. J. Bot.* **53**, 87–93.

KRAMER, P. J. and WILBUR, K. M. (1949). Absorption of radioactive phosphorus by mycorrhizal roots of pine. *Science* **110**, 8–9.

KRUCKELMANN, H. W. (1975). Effects of fertilizers, soils, soil tillage and plant species on the frequency of *Endogone* chlamydospores and mycorrhizal infection in arable soils. *In* "Endomycorrhizas" (Eds F. E. Sanders, B. Mosse and P. B. Tinker), pp. 511–525. Academic Press, London and New York.

KRUPA, S. and BRÄNSTRÖM, G. (1974). Studies in the nitrogen metabolism of ectomycorrhizae. II. Free and bound amino acids in the mycorrhizal fungus *Boletus variegatus* in the root system of *Pinus sylvestris* and during their association. *Physiologia Pl.* **31**, 279–283.

KRUPA, S. and FRIES, N. (1971). Studies in ectomycorrhizae of pine. I. Production of volatile compounds. *Can. J. Bot.* **49**, 1425–1432.

KRUPA, S., FONTANA, A. and PALENZONA, M. (1973). Studies on the nitrogen metabolism in ectomycorrhizae. I. Status of free and bound amino acids in mycorrhizal and non-mycorrhizal root systems of *Pinus nigra* and *Corylus avellana*. *Physiologia Pl.* **28**, 1–6.

KRUPA, S. and NYLUND, J. E. (1971). Studies on ectomycorrhizae of pine. III. Growth inhibition of two root pathogenic fungi by volatile organic constituents of ectomycorrhizal root systems of *Pinus sylvestris* L. *Physiologia Pl.* **25**, 1–7.

KRYWOLAP, G. (1971). Production of antibiotics by certain mycorrhizal fungi. *In* "Mycorrhizae" (Ed. E. Hacskaylo), pp. 210–222. USDA Forest Service misc. publ. No. 1189. US Government Printing Office, Washington.

KUSANO, S. (1911). *Gastrodia elata* and its symbiotic association with *Armillaria mellea*. *J. Coll. Agric. Univ. Tokyo* **4**, 1–66.

LAIHO, O. (1965). Further studies on the ectendotrophic mycorrhiza. *Acta for. fenn.* **79**, 1–35.

LAIHO, O. (1970). *Paxillus involutus* as a mycorrhizal symbiont of forest trees. *Acta for. fenn.* **106**, 1–65.

LAIHO, O. and MIKOLA, P. (1964). Studies on the effects of some eradicants on mycorrhizal development in forest nurseries. *Acta for. fenn.* **77**, 1–34.

LAING, E. V. (1923). Tree roots: their action and development. *Trans. R. Scott. arboric. Soc.* **37**, 6–21.

LALOVE, M. and HALL, R. H. (1973). Cytokinase in *Rhizopogon roseolus* secretion of N[9-($\beta$-D-Riboperanonyl-9H) purin-6²-yl carbonyl] theonine in the culture medium. *Pl. Physiol.* **51**, 559–562.

LAMB, R. J. (1974). Effect of D-glucose on utilization of simple carbon sources by ectomycorrhizal fungi. *Trans. Br. mycol. Soc.* **63**, 295–306.

LAMB, R. J. and RICHARDS, B. N. (1970). Some mycorrhizal fungi of *Pinus radiata* and *P. elliottii* var. *elliottii* in Australia. *Trans. Br. mycol. Soc.* **54**, 371–378.

LAMB, R. J. and RICHARDS, B. N. (1974). Survival potential of sexual and asexual spores of ectomycorrhizal fungi. *Trans. Br. mycol. Soc.* **62**, 181–191.

LAMBERT, D. H., BAKER, D. F. and COLE, H. (1979). The role of mycorrhizae in the interactions of phosphorus with zinc, copper and other elements. *Soil Science, Society of America, Journal* **43**, 976–980.

LAMBERT, D. H., COLE, H. and BAKER, D. E. (1980). Variation in the response of alfalfa clones and cultivars to mycorrhizae and phosphorus. *Crop Sci.* **20**, 615–618.

LANGE, R. T. (1961). Nodule bacteria associated with the indigenous leguminosae of South Western Australia. *J. gen. Microbiol.* **26**, 351–359.

LANGLOIS, C. G. and FORTIN, J. A. (1978). Absorption of phosphorus (³²P) by excised ectomycorrhizae in balsam fir (*Abies balsamea* (L.) Mill.) from low concentrations of $H_2PO_4$. *Naturaliste Canadienne* **105**, 417–424.

LANOWSKA J. (1966). Influence of different sources of nitrogen on the development of mycorrhiza in *Pisum sativum*. *Pamietnik Pulowski* **21**, 365–386.

LARGENT, D. L., SUGIHARA, N. and WISHNER, G. (1980a). Occurrence of mycorrhizae on ericaceous and pyrolaceous plants in Northern California. *Can. J. Bot.* **58**, 2274–2279.

LARGENT, D. L., SUGIHARA, N. and BRINITZER, A. (1980b). *Amanita gemmatum*, a non host-specific mycorrhizal fungus of *Arctostaphylos manzanita*. *Mycologia* **72**, 435–439.

LEISER, A. T. (1968). A mucilaginous root sheath in Ericaceae. *Am. J. Bot.* **55**, 391–398.

LEPPIK, E. E. (1970). Gene centres of plants as sources of disease resistance. *A. Rev. Phytopath.* **8**, 323–344.

LEVISOHN, I. (1954). Aberrant infections of pine and spruce seedlings. *New Phytol.* **53**, 284–290.

LEVISOHN, I. (1960). Physiological and ecological factors influencing the effect of mycorrhizal inoculation. *New Phytol.* **59**, 42–00.

LEVY, Y. and KRIKUN, J. (1980). Effect of vesicular-arbuscular mycorrhiza on *Citrus jambhiri* water relations. *New Phytol.* **85**, 25–31.

LEWELLEN, R. T., SHARP, E. L. and HEHN, E. R. (1967). Major and minor genes in wheat for resistance to *Puccinia striiformis* and their responses to temperature changes. *Can. J. Bot.* **42**, 2155–2172.

LEWIS, D. H. (1963). Uptake and utilisation of substances by beech mycorrhiza. D.Phil. thesis, Oxford University.

LEWIS, D. H. (1973). Concepts in fungal nutrition and the origin of biotrophy. *Biol. Rev.* **48**, 261–278.

LEWIS, D. H. (1974). Microorganisms and plants: The evolution of parasitism and mutualism. *Sym. Soc. gen. Microbiol.* **24**, 367–392.

LEWIS, D. H. (1975). Comparative aspects of the carbon nutrition of mycorrhizas. *In* "Endomycorrhizas" (Eds F. E. Sanders, B. Mosse and P. B. Tinker), pp. 119–148. Academic Press, London and New York.

LEWIS, D. H. (1976). Interchange of metabolites in biotrophic symbioses between angiosperms and fungi. *In* "Perspectives in Experimental Biology", Vol. 2. Botany (Ed. N. Sutherland), pp. 207–219. Pergamon Press, Oxford.

LEWIS, D. H. and HARLEY, J. L. (1965a). Carbohydrate physiology of mycorrhizal roots of beech. I. Identity of endogenous sugars and utilization of exogenous sugars. *New Phytol.* **64**, 224–237.

LEWIS, D. H. and HARLEY, J. L. (1965b). Carbohydrate physiology of mycorrhizal roots of beech. II. Utilization of exogenous sugars by uninfected and mycorrhizal roots. *New Phytol.* **64**, 238–256.

LEWIS, D. H. and HARLEY, J. L. (1965c). Carbohydrate physiology of mycorrhizal roots of beech. III. Movement of sugars between host and fungus. *New Phytol.* **64**, 256–269.

LI, C. Y., LU, K. C., TRAPPE, J. M. and BOLLEN, W. B. (1972). Nitrate reducing capacity of roots and nodules of *Alnus rubra* and roots of *Pseudotsuga menziesii. Pl. Soil* **37**, 409–414.

LIE, T. A. (1978). Symbiotic specialisation in pea plants: the requirement of specific *Rhizobium* strains for pea plants from Afghanistan. *Ann. appl. Biol.* **88**, 462–465.

LIE, T. A., HILLE, D., LAMBERS, R. and HOUWERS, A. (1976). Symbiotic specialization in pea plants: some environmental effects on nodulation and nitrogen fixation. *In* "Symbiotic Nitrogen Fixation in Plants" (Ed. P. S. Nutman), pp. 319–333. Cambridge University Press, Cambridge.

LIE, T. A., TIMMERMANS, P. C. J. M. and LADIZINSKI, G. (1981). Host controlled nitrogen fixation in *Pisum sativum* ecotype *fulvum. In* "Current Perspectives in Nitrogen Fixation" (Ed. A. H. Gibson and W. E. Newton), p. 419. Australian Academy of Science, Canberra.

LIHNELL, D. (1942). *Cenococcum graniforme* aus Mykorrhizabildner von Waldbäumen. *Symb. bot. Upsaliens.* **5**, 1–8.

LIN, S., LIN, D. C. and FLANAGAN, M. D. (1978). Specificity of the effects of cytochalasin B on transport and mobile processes. *Proc. natn. Acad. Sci.* **75**, 329–333.

LINDEBERG, G. (1944). Über die physiologie lignin-abbauender Boden-Hymenomyceten. *Symb. bot. Upsaliens.* **8**, 1–183.

LINDEBERG, G. (1948). On the occurrence of polyphenol oxidases in soil inhabiting Basidiomycetes. *Physiol. Plantarum* **1**, 196–205.

LINDEBERG, G. and LINDEBERG, M. (1974). Effect of short chain fatty acids on the growth of some mycorrhizal and saprophytic Hymenomycetes. *Arch. Microbiol.* **101**, 109–114.

LINDEBERG, G. and LINDEBERG, M. (1977). Pectinolytic ability of some mycorrhizal and saprophytic Hymenomycetes. *Arch. Microbiol.* **115**, 9–12.

LINDEBERG, G., LINDEBERG, M., LUNDGREW, L., POPOFF, T. and THEANDER, O. (1980). Stimulation of litter-decomposing Basidiomycetes by flavonoids. *Trans. Br. mycol. Soc.* **75**, 455–459.

LINDSEY, D. C., CRESS, W. A. and ALDON, E. F. (1977). Effects of endomycorrhizae on growth of Rabbit brush, four-winged salt bush, and corn in coal mine spoil material. USDA For. Service Research Note, R. M. 343, 6 pp.

LINE, R. F., SHARP, E. L. and POWELSON, R. L. (1970). A system for differentiating races of *Puccinia striiformis* in the United States. *Pl. Dis. Rep.* **54**, 992–994.

LING LEE, M., CHILVERS, G. A. and ASHFORD, A. E. (1975). Polyphosphate granules in three different kinds of tree mycorrhiza. *New Phytol.* **75**, 447.

LING LEE, M., CHILVERS, G. A. and ASHFORD, A. E. (1977a). A histochemical study of phenolic materials in mycorrhizal and uninfected roots of *Eucalyptus fastigiata* Deane & Maiden. *New Phytol.* **78**, 313–318.

LING LEE, M., ASHFORD, A. F. and CHILVERS, G. A. (1977b). A histochemical study of polysaccharide distribution in Eucalypt mycorrhizas. *New Phytol.* **78**, 329–335.

LINNEMAN, G. (1955). Untersuchung über die Mykorrhiza von *Pseudtsuga taxifolia* Britt. *Zbl. Bact.* **108**, 398–410.

LINNEMANN, G. (1969). Erfahrungen bei Synthese-Versuchen, isobesondere mit *Pseudotsuga menziesii* (Mirbel.) Franco I. *Zentbl. Bakt. ParasitKde* **123**, 453–463.

LINNEMANN, G. (1971). Erfahrungen bei Synthese-Versuchen mit *Pseudotsuga menziesii* (Mirbel.) Franco II. *Zentbl. Bakt. ParasitKde* **126**, 229–241.

LITTLEFIELD, L. J. (1967). Phosphorus-32 accumulation in *Rhizoctonia solani* sclerotia. *Phytopathology* **57**, 1053–1055.

LITTLEFIELD, L. J. and BRACKER, C. E. (1972). Ultrastructural specialization at the host-pathogen interface in rust-infected flax. *Protoplasma* **74**, 271–305.

LITTLEFIELD, I. J., WILCOXSON, R. D. and SUDIA, T. W. (1965a). Translocation in sporophores of *Lentinus tigrinis*. *Am. J. Bot.* **52**, 599–605.

LITTLEFIELD, L. J., WILCOXSON, R. D. and SUDIA, T. W. (1965b). Translocation of phosphorus-32 in *Rhizoctonia solani*. *Phytopathology* **55**, 536–542.

LOBANOV, N. W. (1960). "Mykotrophie der Holzpflanzen". V.E.B. Deutscher Verlag der Wissenschaften, Berlin. 352 pp.

LODE. A. and PEDERSEN, T. A. (1970). Fatty acid-induced leaking of organic compounds from *Boletus variegatus*. *Physiol. Plant* **23**, 715–727.

LOEGERING, W. Q. and POWERS, H. R. (1962). Inheritance of pathogenicity in a cross of physiological races 111 and 36 of *Puccinia graminis* f.sp. *tritici*. *Phytopathology* **52**, 547–554.

LONG, D. E., FUNG, A. K., McGEE, E. E. M. COOKE, R. C. and LEWIS, D. H. (1975). The activity of invertase and its relevance to the accumulation of storage polysaccharides in leaves affected by biotrophic fungi. *New Phytol.* **74**, 173–182.

LÖSEL, D. M. (1964). The stimulation of spore germination of *Agaricus bisporus* by living mycelium. *Ann. Bot. (Lond.)* **28**, 541–554.

LUCAS, R. L. (1960). Transport of phosphorus by fungal mycelium. *Nature, Lond.* **188**, 763–764.

LUCAS, R. L. (1970). Migration des substances soluble et croissance chez les champignons. *Physiol. Vég.* **8**, 387–394.

LÜCK, R. (1940). Zur Biologie der heimischen *Pyrola* arten. *Schr. Phys.—öken Ges. Königsberg* **71**, 300–334.

LÜCK, R. (1941). Zur Keimung der heimischen *Pyrola* arten. *Flora* (Jena) **135**, 1–5.

LUEDDERS, V. D., CARLING, D. E. and BROWN, M. F. (1979). Effect of soybean plant growth on spore production by *Glomus mosseae*. *Pl. Soil* **53**, 393–397.

LUIG, N. H. and WATSON, I. A. (1961). A study of inheritance of pathogenicity in *Puccinia graminis* var. *tritici*. *Proc. Linn. Soc. N.S.W.* **86**, 217–229.

LUNDEBERG, G. (1970). Utilization of various nitrogen sources, in particular bound soil nitrogen, by mycorrhizal fungi. *Stud. For. Suec.* **79**, 1–95.

LUPPI, A. M. and FONTANA, A. (1967). Sulla utilizzazione dell' azoto ammoniacale e dell' azoto nitrico da paste del *Boletus luteus* L. in coltura. *Alliona* **13**, 195–200.

LUPPI, A. M. and GAUTERO, C. (1967). Recerche sulle micorize di *Quercus robur*, *Q. petraea* R. and *Q. pubescens* in Piedmonte. *Allionia* **13**, 129–148.

LUTZ, R. W. and SJOLUND, R. D. (1973). *Monotropa uniflora*: ultrastructural details of its mycorrhizal habit. *Am. J. Bot.* **60**, 339–345.

LYNCH, J. M. (1972). Identification of substrate and isolation of microorganisms responsible for ethylene production in the soil. *Nature, Lond.* **240**, 45–46.

LYNCH, J. M. and HARPER, S. H. T. (1974). Formation of ethylene by a soil fungus. *J. gen. Microbiol.* **80**, 187–195.

LYON, A. J. E. and LUCAS, R. L. (1969a). The effect of temperature on the translocation of phosphorus by *Rhizopus stolonifer*. *New Phytol.* **68**, 963–969.

LYON, A. J. E. and LUCAS, R. L. (1969b). Phosphorus metabolism of *Rhizopus stolonifer* and *Chaetomium* sp. with respect to phosphorus translocation. *New Phytol.* **68**, 971–976.

LYR, H. (1963). Zur Frage des Streuabbauer durch ectotrophe Mykorrhiza pilze. Mycorrhiza International Symp. Jena, pp. 123–145.

McCOMB, A. L. (1938). The relation between mycorrhizae and the development and nutrient absorption of pine seedlings in a prairie nursery. *J. For.* **36**, 1148–1154.

McCOMB, A. L. (1943). Mycorrhiza and phosphorus nutrition of pine seedlings. *Bull. Iowa Exp. Stn* **314**, 582–612.

McCOMB, A. L. and GRIFFITH, J. E. (1946). Growth stimulation and phosphorus absorption of mycorrhizal and non-mycorrhizal White Pine and Douglas Fir seedlings in relation to fertilizer treatments. *Pl. Physiol.* **21**, 11–17.

MacDONALD, R. M. (1981). Routine production of axenic vesicular-arbuscular mycorrhizas. *New Phytol.* **89**, 87–93.

MacDONALD, R. M. and CHANDLER, M. S. (1981). Bacterium-like organelles in the vesicular-arbuscular mycorrhiza fungus *Glomus caledonius*. *New Phytol.* **89**, 241–246.

MacDONALD, R. M. and LEWIS, M. (1978). The occurrence of some acid phosphatases and dehydrogenases in the vesicular-arbuscular mycorrhizal fungus *Glomus mosseae*. *New Phytol.* **80**, 135–141.

MacDONALD, R. M., CHANDLER, M. R. and MOSSE, B. (1982). The occurrence of bacterium-like organelles in vesicular-arbuscular mycorrhizal fungi. *New Phytol.* **90**, 659–663.

McILVEEN, W. D. and COLE, H. (1976). Spore dispersal of Endogonaceae by worms, ants, wasps and birds. *Can. J. Bot.* **54**, 1486–1489.

McILVEEN, W. D. and COLE, H. (1979). Influence of zinc on development of the endomycorrhizal fungus *Glomus mosseae* and its mediation of phosphorus uptake by *Glycine max* "Amsoy 71". *Agriculture and Environment* **4**, 245–256.

McNABB, R. F. R. (1961). Mycorrhiza in the New Zealand Ericales. *Aust. J. Bot.* **9**, 57–61.

McQUEEN, D. R. (1968). The quantitative distribution of absorbing roots of *Pinus sylvestris* and *Fagus sylvatica* in a forest succession. *Oecol. Planta Gauthier-Villas* **3**, 83–99.

MAGROU, J. (1924). À propos du pouvoir fungicide des tubercles d'Ophrydées. *Annls Sci. nat. Bot.* **6**, 265–270.

MAGROU, J. (1936). Culture et inoculation du champignon symbiotique di l'*Arum maculatum*. *C.r. hebd. Séanc. Acad. Sci., Paris.* **203**, 887–888.

MALABARI, A. A. (1979). Biology of ectomycorrhizas with special reference of their possible role in plant water relations. Ph.D. Thesis, Sheffield University.

MALAJCZULK, N., MOLINA R. and TRAPPE, J. M. (1982). Ectomycorrhiza formation in *Eucalyptus*. I. Pure culture synthesis, host specificity and mycorrhizal compatability in *Pinus radiata*. *New Phytol.* **91**, 467–482.

MALLOCH, D. W., PIROZYNSKI, K. A. and RAVEN, P. H. (1980). Ecological and evolutionary significance of mycorrhizal symbioses in vasicular plants (a review). *Proc. natn. Acad. Sci. U.S.A.* **77**, 2113–2118.

MANGIN, L. (1910). Introduction a l'étude des mycorhizes des arbres forestières. *Nouv. Archs Mus. Hist. nat. Paris, Ser. 5*, **2**, 245–260.

MARCHANT, H. J. (1976). Plasmodesmata in algae and fungi. *In* "Intercellular Communication in Plants. Studies on Plasmodesmata" (Eds B. E. S. Gunning and A. W. Robards), pp. 59–80. Springer-Verlag, Berlin, Heidelberg, New York.

MARKS, G. C. and FOSTER, R. C. (1967). Succession of mycorrhizal associations on individual roots of radiata pine. *Australian Forestry* **31**, 194–201.

MARKS, G. C. and FOSTER, R. C. (1973). Structure, morphogenesis and ultrastructure of ectomycorrhizae. *In* "Ectomycorrhizae" (Eds G. C. Marks and T. T. Kozlowski), pp. 1–41. Academic Press, New York and London.

MARKS, G. C., DITCHBURNE, N. and FOSTER, R. C. (1967). A technique for making quantitative estimates of mycorrhiza populations in *P. radiata* forests. Proc. 14 IUFRO Congress, Munchen, **24**, 67–70.

MARSH, B. A'B (1971). Measurement of length in a random arrangement of lines. *J. appl. Ecol.* **8**, 265–267.

MARTENS, J. W., McKENZIE, R. I. H. and GREEN, G. J. (1970). Gene-for-gene relationships in the *Avena: Puccinia graminis* host-parasite system in Canada. *Can. J. Bot.* **48**, 969–975.

MARX, C., DEXHEIMER, J., GIANINAZZI-PEARSON, V. and GIANINAZZI, S. (1982). Enzymatic studies on the metabolism of vesicular-arbuscular mycorrhiza. IV. Ultracytoenzymological evidence (ATPase) for active transfer processes in the host arbuscule interface. *New Phytol.* **90**, 37–43.

MARX, D. H. (1967). Ectotrophic mycorrhizae as biological deterrents to pathogenic root infections by *Phytophthora cinnamonii*. Proc. 14 IUFRO Congress, Munchen, **24**, 172–181.

MARX, D. H. (1969–70). The influence of ectotrophic mycorrhizal fungi on the resistance of pine roots to pathogenic infection. *Phytopathology* **59**, 153–163; 411–417; 549–558; 559–565; **60**, 1472–1473.

MARX, D. H. (1972). Ectomycorrhizae as biological deterrents to pathogenic root infections. *A. Rev. Phytopath.* **10**, 429–434.

MARX, D. H. (1973). Mycorrhizae and feeder root disease. *In* "Ectomycorrhizae" (Eds G. C. Marks and T. T. Kozlowski), pp. 351–382. Academic Press, New York and London.

MARX, D. H. (1975). Role of ectomycorrhizae in the protection of pine from root infection by *Phytophthora cinnamoni*. *In* "Biology and Control of Soil-borne Plant Pathogens" (Ed. J. W. Bruehl). American Phytopathology Society.

MARX, D. H. (1975). Mycorrhiza and establishment of trees on strip-mined land. *Ohio J. Sci.* **75**, 288–297.

MARX, D. H. (1979a). Synthesis of *Pisolithus* ectomycorrhizae on White Oak seedlings in fumigated nursery soil. For. Service Research Note, USDA, 3E 280.

MARX, D. H. (1979b). Synthesis of ectomycorrhizae by different fungi in Northern Red Oak seedlings. For. Service Research Note, USDA, SE 282.

MARX, D. H. (1980). Role of mycorrhizae in forestation of surface mines. Proc. symp. Trees for Reclamation Interstate mining, Lexington, Kentucky.

MARX, D. H. and ALTMAN, J. D. (1979). *Pisolithus tinctorius* ectomycorrhiza improve survival and growth of pine seedlings on acid coal spoil in Kentucky and Virginia. *Reclam. Rev.* **2**, 23–37.

MARX, D. H. and BRYAN, W. C. (1971). Formation of ectomycorrhizae on half-sib progenies of slash pine in aseptic culture. *For. Sci.* **17**, 488–492.

MARX, D. H. and ROSS, E. W. (1970). Aseptic synthesis of ectomycorrhizae on *Pinus taeda* by basidiospores of *Thelephora terrestris*. *Can. J. Bot.* **48**, 197–198.

MARX, D. H. and ZAK, B. (1965). Effect of pH on mycorrhizal formation of slash pine in aseptic culture. *For. Sci.* **11**, 66–75.

MARX, D. H., HATCH, A. B. and MENDICINO, J. F. (1977). High fertility decreases sucrose content and susceptibility of Loblolly pine roots to ectomycorrhizal infection by *Pisolithus tinctorius*. *Can. J. Bot.* **55**, 1569–1574.

MASER, C., TRAPPE, J. M. and NUSSBAUM, R. A. (1978). Fungal-small mammal inter-relationships with emphasis on Oregon coniferous forests. *Ecology* **59**, 799–809.

MASON, P. A. (1975). The functions of mycorrhizal associations between *Amanita muscaria* and *Betula verrucosa*. *In* "Development and Function of Roots" (Eds J. G. Torrey and D. T. Clarkson), pp. 567–574. Academic Press, London and New York.

MASUI, K. (1926). A study of mycorrhiza of *Abies firma* S & K with special reference to its mycorrhizal fungus *Cantharellus floccosus* Schu. *Mem. Coll. Sci. Kyoto Imp. Univ. B.* **2**, 85–92.

MATARÉ, R. and HATTINGH, M. J. (1978). Effect of mycorrhizal status of Avocado seedlings on root rot caused by *Phytophthora cinnamoni*. *Pl. Soil* **49**, 433–435.

MEJSTŘÍK, V. K. (1965). Study on the development of endotrophic mycorrhiza in the association of *Cladietum marisci*. *In* "Plant Microbes Relationships" Symposium on Relationships between Soil micro-organisms and Plant roots. Prague. p. 193.

MEJSTŘÍK, V. K. (1970). The uptake of $^{32}$P by different kinds of mycorrhizas of *Pinus*. *New Phytol.* **69**, 295–298.

MEJSTŘÍK, V. K. (1972). Vesicular-arbuscular mycorrhizas of the species of a *Molinietum coeruleae* L.1. association: the ecology. *New Phytol.* **71**, 883–890.

MEJSTŘÍK, V. K. and HADAČ, E. (1975). Mycorrhizas of *Arctostaphylos uva-ursi*. *Pedobiologia* **15**, 336–342.

MEJSTŘÍK, V. K. and KRAUSE, H. H. (1973). Uptake of $^{32}$P by *Pinus radiata* inoculated with *Suillus luteus* and *Cenococcum graniforme* from different sources of available phosphate. *New Phytol.* **72**, 137–140.

MELHUISH, J. H. and HACSKAYLO, E. (1980a). Fatty acids of selected *Athelia* spp. *Mycologia* **702**, 251–258.

MELHUISH, J. H. and HACSKAYLO, E. (1980b). Fatty acid content of *Pisolithus tinctorius* in response to changing ratios of nitrogen and carbon source. *Mycologia* **22**, 1041–1044.

MELHUISH, J. H., HACSKAYLO, E. and BEAN, G. A. (1975). Fatty acid composition of ectomycorrhizal fungi. *Mycologia* **67**, 952–960.

MELIN, E. (1917). Studier över de norrlandska myrmarkernas vegetation med Särskil-dhänsyn till deras Skogsvegetation eftertorrläggning Akad. Avhandl. Uppsala 1–426.

MELIN, E. (1923). Experimentelle Untersuchungen über die Konstitution and Ökologie der Mykorrhizen von *Pinus sylvestris* und *Picea abies*. *Mykol. Unters. Ber.* **2**, 72–331.

MELIN, E. (1925). Untersuchungen über die Bedeutung der Baummykorriza. G. Fischer, Jena, 152 pp.

MELIN, E. (1927). Studier över barrträdsplatans utveckling i rahumus. II. Mykorrhizans utbildning hos tallplantan i olika rahumus former. *Meddeln St. SkogsforskInst.* **23**, 433–494.

MELIN, E. (1946). Die Einfluss von Waldstreuextrakten auf das Wachstum von Bodenpilzen mit besonderer Beruckssichtigung der Wurzelpilze von Baümen. *Symb. Bot. Upsaliens.* **8**, 1–116.

MELIN, E. (1959). "Mycorrhiza. Encyclopedia of Plant Physiology" (Ed. W. Ruhland), vol. XI, pp. 605–638. Springer Verlag, Berlin.

MELIN, E. (1962). Physiological aspects of mycorrhizae of forest trees. *In* "Tree Growth" (Eds T. T. Kozlowski), pp. 247–263. Ronald Press, New York.

MELIN, E. (1963). Some effects of forest tree roots on mycorrhizal Basidiomycetes. *In* "Symbiotic Associations" (Eds B. Mosse and P. S. Nutman), pp. 124–145. Cambridge University Press, Cambridge.

MELIN, E. and DAS, V. S. R. (1954). The influence of root-metabolites on the growth of tree mycorrhizal fungi. *Pl. Physiol.* **7**, 851–858.

MELIN, E. and KRUPA, S. (1971). Studies on the ectomycorrhizae of pine. II. Growth inhibition of mycorrhizal fungi by volatile organic constituents of *Pinus sylvestris* L. (Scots pine) roots. *Physiologia Pl.* **25**, 337–340.

MELIN, E. and NILSSON, H. (1950). Transfer of radioactive phosphorus to pine seedlings by means of mycorrhizal hyphae. *Physiologia Pl.* **3**, 88–92.

MELIN, E. and NILSSON, H. (1952). Transport of labelled nitrogen from an ammonium source to pine seedlings through mycorrhizal mycelium. *Svensk bot. Tidskr.* **46**, 281–285.

MELIN, E. and NILSSON, H. (1953a). Transfer of labelled nitrogen from glutamic acid to pine seedlings through the mycelium of *Boletus variegatus* (S.W.) Fr. *Nature, Lond.* **171**, 434.

MELIN, E. and NILSSON, H. (1953b). Transport of labelled phosphorus to pine seedlings through the mycelium of *Cortinarius glaucopus* (Shaeff ex. Fr.) Fr. *Svensk bot. Tidskr.* **48**, 555–558.

MELIN, E. and NILSSON, H. (1955). $^{45}$Ca used as indicator of transport of cations to pine seedlings by means of mycorrhizal mycelium. *Svensk bot. Tidskr.* **49**, 119–121.

MELIN, E. and NILSSON, H. (1957). Transport of C$^{14}$-labelled photosynthate to the fungal associate of pine mycorrhiza. *Svensk bot. Tidskr.* **51**, 166–186.

MELIN, E. and NILSSON, H. (1958). Translocation of nutritive elements through mycorrhizal mycelia to pine seedlings. *Bot. Notiser* **111**, 251–256.

MELIN, E. and NORKRANS, B. (1948). Amino acids and the growth of *Lactarius deliciosus*. *Physiologia Pl.* **1**, 178–184.

MELIN, E. and NYMAN, P. (1940). Weitere untersuchungen über die Wirkung von Aneurin und Biotin auf das Wachstum von Wurzel pilze. *Arch. Mikrobiol.* **11**, 318–328.

MELIN, E. and NYMAN, B. (1941). Über das Wuchstoffbedürfnis von *Boletus granulatus* (L.). *Arch. Mikrobiol.* **12**, 254–259.

MELIN, E., NILSSON, H. and HACSKAYLO, E. (1958). Translocation of cations to seedlings of *Pinus virginiana* through mycorrhizal mycelium *Bot. Gaz.* **119**, 243–246.

MENGE, J. A. (1982). Current and potential uses of vesicular-arbuscular fungi. NACOM, Quebec. *Can. J. Bot.* (In press.)

MENGE, J. A., STEIRLE, D., BAGARAJ, D. J., JOHNSON, E. L. V. and LEONARD, R. T. (1978a). Phosphorus concentration in plant responsible for inhibition of mycorrhizal infection. *New Phytol.* **80**, 575–578.

MENGE, J. A., JOHNSON, E. L. V. and PLATT, R. G. (1978b). Mycorrhizal dependency of several citrus cultivars under three nutrient regimes. *New Phytol.* **81**, 553–559.

MENGE, J. A., LABANAUSKAS, C. K., JOHNSON, E. L. V. and PLATT, R. G. (1980). Partial substitution of mycorrhizal fungi for phosphorus fertilization in the greenhouse culture of citrus. *Soil Sci. Soc. Am. J.* **42**, 926–930.

METZGER, R. J. and TRIONE, E. J. (1962). Application of the gene-for-gene relationship hypothesis to the *Triticum Tilletia* system. *Phytopathology* **52**, 363.

MEXAL, J. and REID, C. P. P. (1973). The growth of selected mycorrhizal fungi in response to induced water stress. *Can. J. Bot.* **51**, 1579–1588.

MEYER, F. H. (1962). Die Buchen und Fichten mykorrhiza in vershiedenen Bodentypen, ihr Beeinflussing durch Mineraldüngung Sowie für die Mykorrhizabildung wichtige Factoren. *Mitt. Bund Forsch Anst Forst-u. Holzw.* **54**, 73 pp.

MEYER, F. H. (1964). The role of the fungus *Cenococcum graniforme* (Sow.) Ferd. et Winge in the formation of mor. *In* "Soil Micromorphology" (Ed. A. Jongerius), pp. 23–31. Elsevier, Amsterdam.

MEYER, F. H. (1973). Distribution of ectomycorrhizae in native and man-made forests. *In* "Ectomycorrhizae" (Eds G. C. Marks and T. T. Kozlowski), pp. 79–105. Academic Press, New York and London.

MEYER, F. H. and GÖTTSCHE, D. (1971). Distribution of root tips and tender roots of beech. *In* "Ecological Studies: Analysis and Synthesis" (Ed. H. Ellenberg), pp. 48–52. Springer-Verlag, Berlin, Heidelberg, New York.

MIAH, M. A. J. and SACKSTON, W. E. (1970). Genetics of pathogenicity in sunflower rust. *Phytoprotection* **51**, 17–35.

MIKOLA, P. (1948). On the physiology and ecology of *Cenococcum graniforme*. *Communs Inst. for. fenn.* **36**, 1–104.

MIKOLA, P. (1953). An experiment on the invasion of mycorrhizal fungi into prairie soil. *Karstenia* **2**, 33–34.

MIKOLA, P. (1965). Studies on ectendotrophic mycorrhiza of pine. *Acta for. fenn.* **79**, 1–56.

MIKOLA, P. and LAIHO, O. (1962). Mycorrhizal relations in the raw humus layer of northern spruce forests. *Communs Inst. for. fenn.* **55**, 1–13.

MILLER, C. O. (1971). Cytokinin production by mycorrhizal fungi. *In* "Mycorrhizae" (Ed. E. Hacskaylo). Proc. 1st NACOM (1969).

MILNE, L. and COOKE, R. C. (1969). Translocation of $^{14}$C glucose by *Rhizoctonia solani*. *Trans. Br. mycol. Soc.* **53**, 279–289.

MINCHIN, F. R. and PATE, J. S. (1973). The carbon balance of a legume and the functional economy of its nodules. *J. exp. Bot.* **24**, 259–271.

MITCHELL, D. T. and READ, D. J. (1981). Utilization of inorganic and organic phosphates by the mycorrhizal endophytes of *Vaccinium macrocarpon* and *Rhododendron ponticum*. *Trans. Br. mycol. Soc.* **76**, 255–260.

MITCHELL, H. L., FINN, R. F. and ROSENDAHL, R. O. (1937). The growth and nutrition of white pine (*Pinus strobus* L.) seedlings in culture. *Black Rock For. Paper 1.* 58–73.

MIURA, G. and HALL, R. H. (1973). *Trans*-ribosylzeatin, its biosynthesis in *Zea mais* endosperm and the mycorrhizal fungus *Rhizopogon roseolus*. *pl. Physiol.* **51**, 563–569.

MODESS, O. (1941). Zur Kentniss der Mykorrhizabildner von Kiefer und Fichte. *Symb. Bot. Upsaliens* **5**, 1–147.

MOLINA, R. (1979). Pure culture synthesis and host specificity of Red Alder mycorrhiza. *Can. J. Bot.* **57**, 1223–1225.

MOLINA, R. J. (1980). Patterns of ectomycorrhizal host-fungus specificity in the Pacific Northwest. PhD Thesis Oregon State University. 299 pp.

MOLINA, R. (1981). Ectomycorrhizal specificity in the genus *Alnus*. *Can. J. Bot.* **59**, 325–334.

MOLINA, R. and TRAPPE, J. M. (1982a). Lack of mycorrhizal specificity by the ericaceous hosts *Arbutus menziesii* and *Arctostaphylos uva-ursi*. *New Phytol.* **90**, 495–509.

MOLINA, R. J. and TRAPPE, J. M. (1982b). Patterns of ectomycorrhizal host specificity and potential amongst Pacific Northwest conifers and fungi. *Forest Sci.* **28**, 423–457.

MOLINA, R. J., TRAPPE, J. M. and STRICKLER, G. S. (1978). Mycorrhizal fungi associated with *Festuca* in western United States and Canada. *Can. J. Bot.* **56**, 1691–1695.

MÖLLER, C. (1947). Mycorrhizae and nitrogen assimilation. *Forstt. Forsöksr. Danm.* **19**, 105–208.

MÖLLER, C. M., MÜLLER, D. and NIELSEN, J. (1954). The dry matter production of european beech. *Forstlige Forsogsvaesen, Danmark* **21**, 253–335.

MOLLISON, J. E. (1943). *Goodyera repens* and its Endophyte. *Trans. Bot. Soc. Edinb.* **33**, 391–403.

MONSON, A. M. and SUDIA, T. W. (1963). Translocation in *Rhizoctonia solani*. *Bot. Gaz.* **124**, 440–443.

MOORMAN, T., REEVES, F. B. and GOOD, C. W. (1979). The role of endomycorrhiza in revegetation practices in the semi-arid West. II. A bioassay to determine the effect of land disturbance on endomycorrhiza populations. *Am. J. Bot.* **66**, 14–18.

MORIMOTO, M., IAWI, M. and LUKUMOTO, J. (1954). Antibiotic substances from mycorrhizal fungi. I. Isolation of antibiotic strains. *Kagaku to Kagyo* **28**, 111–116.

MORLEY, C. D. and MOSSE, B. (1976). Abnormal vesicular-arbuscular mycorrhizal infections in white clover induced by lupin. *Trans. Br. mycol. Soc.* **67**, 510–513.

MORRISON, T. M. (1956). Mycorrhiza of silver beech. *N.Z. Jl For.* **7**, 47–52.

MORRISON, T. M. (1957a). Mycorrhiza and phosphorus uptake. *Nature, Lond.* **179**, 907.

MORRISON, T. M. (1957b). Host-endophyte relationship in mycorrhiza of *Pernettya macrostigma*. *New Phytol.* **56**, 247–257.

MORRISON, T. M. (1962a). Absorption of phosphorus from soils by mycorrhizal plants. *New Phytol.* **61**, 10–20.

MORRISON, T. M. (1962b). Uptake of sulphur by mycorrhizal plants. *New Phytol.* **61**, 21–27.

MORRISON, T. M. (1963). Uptake of sulphur by excised beech mycorrhizas. *New Phytol.* **62**, 44–49.

MOSCA, A. M. L. and FONTANA, A. (1965). Sulle colture di miceli isolati da micorrhize. I. *Boletus luteus* L. *Allionia* **11**, 63–71.

MOSCA, A. M. L. and FONTANA, A. (1975). Sull utilizzazione dell azoto proteico da paste del micelio di *Boletus luteus* L. *Allionia* **20**, 47–52.

MOSEMAN, J. G. (1957). Host parasite interactions between culture 12A1 of the powdery mildew fungus and the Mlk and MLg genes in barley. *Phytopathology* **47**, 453.

MOSEMAN, J. G. (1959). Host-pathogen interaction of the genes for resistance in *Hordeum vulgare* and for pathogenicity in *Erysiphe* f. sp. *hordei*. *Phytopathology* **49**, 469–472.

MOSER, M. (1958). Der einfluss liefer temperaturen auf des Wachstum und Lebenstädig-heit höherer Pilze mit spezieller Beruchsichtigung von Mykorrhizapilzen. *Sydowia* **12**, 386–399.

MOSER, M. (1959). Beiträge zur Kentnis der wuchstoffbezieungen in bereich ectotrophen Mykorrhizen. *Arch. Mikrobiol.* **34**, 251–264.

MOSSE, B. (1953). Fructifications associated with mycorrhizal strawberry roots. *Nature, Lond.* **171**, 974.

MOSSE, B. (1956). Fructifications of an *Endogone* species causing endotrophic mycorrhiza in fruit plants. *Ann. Bot. (Lond.)* **20**, 349–362.

MOSSE, B. (1957). Growth and chemical composition of mycorrhizal and non-mycorrhizal apples. *Nature, Lond.* **179**, 922–924.

MOSSE, B. (1959a). The regular germination of resting spores and some observations on the growth requirements of an *Endogone* sp. causing vesicular-arbuscular mycorrhiza. *Trans. Br. mycol. Soc.* **42**, 273–286.

MOSSE, B. (1959b). Observations on the extramatrical mycelium of a vesicular-arbuscular endophyte. *Trans. Br. mycol. Soc.* **42**, 439–448.

MOSSE, B. (1961). Experimental techniques for obtaining pure inoculum of *Endogone* sp. and some observations on vesicular-arbuscular infestations caused by it and other fungi. *Recent Advances in Botany*, **2**, 1728–1732.

MOSSE, B. (1962). The establishment of vesicular-arbuscular mycorrhiza under aseptic conditions. *J. gen. Microbiol.* **27**, 509–520.

MOSSE, B. (1963). Vesicular-arbuscular mycorrhiza: an extreme form of fungal adaptation. *In* "Symbiotic Associations" (Eds P. S. Nutman and B. Mosse), pp. 146–170. Cambridge University Press, Cambridge.

MOSSE, B. (1972a). Effect of different *Endogone* strains on the growth of *Paspalum notatum* (Batatai) in two Brazilian soils. *Nature, Lond.* **239**, 221–223.

MOSSE, B. (1972b). The influence of soil type and *Endogone* strain on the growth of mycorrhizal plants in phosphate-deficient soil. *Rev. Écol. Biol. Sol.* **9**, 529–537.

MOSSE, B. (1973). Plant growth responses to vesicular-arbuscular mycorrhiza IV. In soil given additional phosphate. *New Phytol.* **72**, 127–136.

Mosse, B. (1975). Specificity of VA Mycorrhizas. *In* "Endomycorrhizas" (Eds F. E. Sanders, B. Mosse and P. B. Tinker), pp. 469–484. Academic Press, London and New York.

Mosse, B. (1978). Mycorrhiza and Plant Growth. *In* "Structure and Functioning of Plant Populations" (Eds A. H. J. Freysen and J. W. Waldendorp), pp. 269–297. North Holland Publishing Co., Amsterdam.

Mosse, B. and Bowen, G. D. (1968a). A key to the recognition of some *Endogone* spore types. *Trans. Br. mycol. Soc.* **51**, 469–483.

Mosse, B. and Bowen, G. D. (1968b). The distribution of *Endogone* spores in some Australian and New Zealand soils and in experimental field soil at Rothamsted. *Trans. Br. mycol. Soc.* **51**, 485–492.

Mosse, B. and Hayman, D. S. (1971). Plant growth responses to vesicular-arbuscular mycorrhiza. II. In unsterilised field soils. *New Phytol.* **70**, 29–34.

Mosse, B. and Hepper, C. M. (1975). Vesicular-arbuscular mycorrhizal infections in root organ cultures. *Physiol. Pl. Path.* **5**, 215–223.

Mosse, B. and Phillips, J. M. (1971). The influence of phosphate and other nutrients on the development of vesicular-arbuscular mycorrhiza in culture. *J. gen. Microbiol.* **69**, 157–166.

Mosse, B., Hayman, D. S. and Arnold, D. J. (1973). Plant growth responses to vesicular-arbuscular mycorrhiza. V. Phosphate uptake by three plant species from P-deficient soils labelled with $^{32}$P. *New Phytol.* **72**, 809–815.

Mosse, B., Powell, C. L. and Hayman, D. S. (1976). Plant growth responses to vesicular-arbuscular mycorrhiza. IX. Interactions between V-A mycorrhiza, rock phosphate and symbiotic nitrogen fixation. *New Phytol.* **76**, 331–342.

Mousain, D. (1971). Essai d'analyse de la symbiose ectomycorrhizienne chez le pin maritime. Dr. Ing. Thesis Acadamie de Montpelier. 236 pp.

Murdoch, C. L. Jakobs, J. A. and Gerdemann, J. W. (1967). Utilization of phosphorus sources of different availability by mycorrhizal and non-mycorrhizal maize. *Pl. Soil* **27**, 329–334.

Nagy, S., Nordby, H. E. and Nemec, S. (1980). Composition of lipids in roots in six citrus cultivars infected with the vesicular-arbuscular mycorrhizal fungus *Glomus mosseae*. *New Phytol.* **85**, 377–384.

Nakamura, S. J. (1982). Nutritional conditions required for non-symbiotic culture of an achlorophyllous orchid, *Galeola septentrionalis*. *New Phytol.* **90**, 701–715.

Napoli, C., Sanders, R., Carlson, R. and Albersheim, P. (1980). Host Symbiont Interactions: Recognising *Rhizobium*. *In* "Nitrogen Fixation Vol. II" (Eds W. E. Newton and W. H. Orme-Johnson), pp. 189–203. University Park Press, Baltimore.

Neal, J. L., Lu, K. C., Bollen, W. B. and Trappe, J. M. (1968). Some ectomycorrhizae of *Alnus rubra*. *In* "Biology of Alder" (Eds J. M. Trappe, J. L. Franklin, A. F. Tanant and G. M. Hanson), pp. 179–192. Pacific Northwest Range Exp. Sta. USDA.

Nelson, C. D. (1964). The production and translocation of photosynthate $C^{14}$ in conifers. *In* "Formation of Wood in Forest Trees" (Ed. M. H. Zimmerman), pp. 235–257. Maria Mons Cabot Foundation, New York.

Newman, E. I. (1966). A method of estimating the total length of root in a sample. *J. appl. Ecol.* **3**, 139–145.

Ng, P. P., Cole, A. L. J., Jameson, P. E. and McWha, J. A. (1982). Cytokinin production by ectomycorrhizal fungi. *New Phytol.* **91**, 57–62.

Nicolson, T. H. (1955). The mycotrophic habit in grasses. Thesis, Nottingham University.

Nicolson, T. H. (1959). Mycorrhiza in the Graminae. I. Vesicular-arbuscular endophytes with special reference to the external phase. *Trans. Br. mycol. Soc.* **42**, 421–438.

NICOLSON, T. H. (1960). Mycorrhiza in the Graminae. II. Development in different habitats particularly sand dunes. *Trans. Br. mycol. Soc.* **43**, 132–145.

NICOLSON, T. H. (1967). Vesicular-arbuscular mycorrhiza—a universal plant symbiosis. *Sci. Prog., Oxford* **55**, 561–581.

NICOLSON, T. H. (1975). Evolution of vesicular-arbuscular mycorrhizas. *In* "Endomycorrhizas" (Eds F. E. Sanders, B. Mosse and P. B. Tinker), pp. 25–34. Academic Press, New York and London.

NIELSEN, N. (1930). Untersuchungen über einen neuen Wachstumregulierenden Stoff: Rizopin. *Jb. wiss. Bot.* **73**, 125–191.

NIEUWDORP, P. J. (1969). Some investigations on the mycorrhiza of *Calluna, Erica* and *Vaccinium. Acta bot. neerl.* **18**, 180–196.

NIEUWDORP, P. J. (1972). Some observations with light and electron microscope on the endotrophic mycorrhiza of orchids. *Acta bot. neerl.* **21**, 128–144.

NIEWIECZERZALOWNA, B. (1932). Recherches morphologiques sur la mycorrhize des orchidées indigenes. *C.r. Soc. Sci., Varsovie* **25**, 86–115.

NOBÉCOURT, P. (1923). Sur la production d'anticorps par les tubercles des Ophrydées. *C.r. hebd. Séanc. Acad. Sci., Paris* **177**, 1055–1057.

NORKRANS, B. (1950). Studies in growth and cellulolytic enzymes of *Tricholoma. Sym. Bot. Upsaliens* **11**, 126 pp.

NORKRANS, B. (1953). The effect of glutamic acid, aspartic acid and related compounds on the growth of certain *Tricholoma* species. *Physiologia Pl.* **6**, 584–593.

NORKRANS, B. and ASCHAN, K. (1953). A study of the cellulolytic variation of wild types and mutants of *Collybia velutipes*. II. Relations to different media. *Physiologia Pl.* **6**, 829–836.

NORONHA-WAGNER, M. and BETTENCOURT, A. J. (1967). Genetic study of the resistance of *Coffea* spp. to leaf rust. I. Identification and behaviour of four factors conditioning disease reaction in *Coffea arabica* to twelve physiologic races of *Hemileia vastatrix. Can. J. Bot.* **45**, 2021–2031.

NUESCH, J. (1963). Defence reactions in orchid bulbs. *Symp. Soc. gen. Microbiol.* **13**, 335–343.

NUTMAN, P. S. (1969). Genetics of symbiosis and nitrogen fixation in legumes. *Proc. R. Soc. Lond. B.* **172**, 417–437.

NYE, P. H. and TINKER, P. B. (1977). 'Solute Movement in the Soil-Root System". Blackwell Scientific Publications, Oxford.

NYLUND, J. E. (1980). Symplastic continuity during Hartig net formation in Norway Spruce ectomycorrhizae. *New Phytol.* **86**, 373–378.

NYLUND, J. E. (1981). The formation of ectomycorrhiza in conifers: structural and physiological studies with special reference to the microbiont *Piloderma croceum* Erikss. & Agorts J. Ph.D. thesis, Faculty of Science, University of Uppsala.

NYLUND, J. E. and UNISTAM, T. (1982). Structure and physiology of ectomycorrhizae I. The process of mycorrhiza formation in Norway spruce *in vitro. New Phytol.* **91**, 63–79.

OADES, J. M. (1978). Mucilages at the root surface. *J. Soil Sci.* **29**, 1–16.

O'BANNON, J. H., EVANS, D. W. and PEADEN, R. N. (1980). Alfalfa varietal response to seven isolates of vesicular-arbuscular mycorrhizal fungi. *Can. J. Pl. Sci.* **60**, 859–863.

OCAMPO, J. A. (1980). Effect of crop rotations involving host and non-host plants on vesicular-arbuscular mycorrhizal infection of host plants. *Pl. Soil* **56**, 283–291.

OCAMPO, J. A., MARTIN, J. and HAYMAN, D. S. (1980). Influence of plant interactions on vesicular-arbuscular mycorrhizal infections. I. Host and non-host plants grown together. *New Phytol.* **84**, 27–35.

OLD, K. M. and NICOLSON, T. H. (1975). Electron microscopical studies of the mycorrhiza of roots of sand dune grasses. *New Phytol.* **74**, 51–58.

OLIVER, A. J., SMITH, S. E., NICHOLAS, D. J. D., WALLACE, W. and SMITH, F. A. (1983). Nitrate reductase activity in *Trifolium subterraneum*: effects of mycorrhizal infection and phosphate nutrition. (In press, *New Phytol.*)

OLSEN, R. A. (1971). Glycosides as inhibitors of fungal growth and metabolism. *Physiologia Pl.* **24**, 534–543; **25**, 204–212; **25**, 503–508.

OLSEN, R. A., ODHAM, G. and LINDEBERG, G. (1971). Aromatic substances and leaves of *Populus tremula* as inhibitors of mycorrhizal fungi. *Physiologia Pl.* **25**, 122–129.

OORT, A. J. P. (1963). A gene-for-gene relationship in the *Triticum-Ustilago* system and some remarks on host-parasite combinations in general. *Netherlands Journal of Plant Pathology* **69**, 104–109.

ORLOV, A. (1957). Observations on absorbing roots of spruce (*Picea excelsa* Link.) in natural conditions. *Bot. Zh. SSSR* **45**, 888–896.

ORLOV, A. Y. (1960). Further observations on the growth of absorbing roots of spruce (*Picea excelsa* Link.) in natural conditions. *Bot. Zh. SSSR* **45**, 888–896.

OSBORNE, T. G. B. (1909). Lateral roots of *Amyelon radicans* and their mycorrhiza. *Ann. Bot. (Lond.)* **23**, 603–610.

OSMOND, C. B. and SMITH, F. A. (1976). Symplastic transport of metabolites during $C_4$ photosynthesis. *In* "Intercellular Communication in Plants: Studies on Plasmodsmata" (Eds B. E. S. Gunning and A. W. Robards), pp. 229–241. Springer-Verlag, Berlin, Heidelberg, New York.

PACHLEWSKI, R. and PACHLEWSKA, J. (1974). Studies on symbiotic properties of mycorrhizal fungi of pine *Pinus sylvestris* L. with the aid of methods of mycorrhizal synthesis in pure culture in agar. Forest Res. Inst. Warsaw.

PAIRUNAN, A. K., ROBSON, A. D. and ABBOTT, L. K. (1980). The effectiveness of vesicular-arbuscular mycorrhizas in increasing growth and phosphorus uptake of subterranean clover from phosphorus sources of different solubilities. *New Phytol.* **84**, 327–338.

PALENZONA, M. (1969). Sintesi micorrizica tra *Tuber aestivum* Vitt. e semenzali di *Corylus avellana* L. *Allionia* **15**, 121–131.

PALMER, J. G. and HACSKAYLO, E. (1970). Ectomycorrhizal fungi in pure culture. I. Growth on single carbon sources. *Physiologia Pl.* **23**, 1187–1197.

PANG, P. C. and PAUL, E. A. (1980). Effects of vesicular-arbuscular mycorrhiza on $^{14}C$ and $^{15}N$ distribution in nodulated fababeans. *Can. J. Soil Sci.* **60**, 241–250.

PEARSON, V. and READ, D. J. (1973a). The biology of mycorrhiza in the Ericaceae. I. The isolation of the endophyte and synthesis of mycorrhizas in aseptic culture. *New Phytol.* **72**, 371–379.

PEARSON, V. and READ, D. J. (1973b). The biology of mycorrhiza in the Ericaceae. II. The transport of carbon and phosphorus by the endophyte and the mycorrhiza. *New Phytol.* **72**, 1325–1331.

PEARSON, V. and READ, D. J. (1975). The physiology of the mycorrhizal endophyte of *Calluna vulgaris*. *Trans. Br. mycol. Soc.* **64**, 1–7.

PEARSON, V. and TINKER, P. B. (1975). Measurement of phosphorus fluxes in the external hyphae of endomycorrhizas. *In* "Endomycorrhizas" (Eds F. E. Sanders, B. Mosse and P. B. Tinker), pp. 277–287. Academic Press, London and New York.

PEDERSEN, T. A. (1980). Effect of fatty acids and methyl octanoate on resting mycelium of *Boletus variegatus*. *Physiologia Pl.* **23**, 654–666.

PEDERSEN, T. A. and LODE, A. (1971). Interaction of 8-anilino-1-naphthalene sulfonate with fatty acid treated mycelium. *Arch. Mikrobiol.* **77**, 118–126.

PEGG, G. F. and VESSEY, J. C. (1973). Chitinase activity in *Lycopersicon esculentum* and its relationship to the *in vivo* lysis of *Verticillium alboatrum* mycelium. *Physiol. pl. Path.* **3**, 207–222.

PEGLER, D. N. and FIARD, J. P. (1979). Taxonomy and ecology of *Lactarius* (Agaricales) in the Lesser Antilles. *Kew Bull.* **33**, 601–628.

PÉROMBELON, M. and HADLEY, G. (1965). Production of pectic enzymes by pathogenic and symbiotic *Rhizoctonia* strains. *New Phytol.* **64**, 144–151.

PETERSON, T. A., MUELLER, W. C. and ENGLANDER, L. (1980). Anatomy and ultrastructure of *Rhododendron* root-fungus association. *Can. J. Bot.* **58**, 244–253.

PEUSS, H. (1958). Untersuchungen zür Ökologie und Bedeutung der Tabakmycorrhiza. *Arch. Mikrobiol.* **29**, 112–142.

PEYRONEL, B. (1923). Fructification de l'endophyte à arbuscules et à vesicules des mycorhizes endotrophes. *Bull. Soc. Mycol. France.* **39**, 119–126.

PEYRONEL, B. (1924). Prime ricerche sulle micorize endotrofiche e sulla microflora radicicola normale delle famerogame *Memorie R. Staz. Patol. Veg.* 1–61.

PICHÉ, Y. and FORTIN, J. A. (1982). Development of mycorrhizae, extramatrical mycelium and sclerotia on *Pinus strobus* seedlings. *New Phytol.* **91**, 211–220.

PICHÉ, Y., FORTIN, J. A. and LAFONTAINE, J. G. (1981). Cytoplasmic phenols and polysaccarides in ectomycorrhizal and non-mycorrhizal short roots of pine. *New Phytol.* **88**, 695–703.

PICHOT, J. and BINH, T. (1976). Action des endomycorrhizes sur la croissance et la nutrition phosphatée de l'*Agrostis* en vase de vegetation et sur le phosphore isotopiquement diluable du sol. *Agron. trop. Nogent* **4**, 375–378.

PIOU, D. (1979). Importance de mycorrhization dans la résistance au calcaire de diverses espèces forestières. *Revue for. fr.* **31**, 116–125.

PLUNKETT, B. E. (1956). Translocation and pileus formation in *Polyporus brumalis*. *Ann. Bot. (Lond.)* **20**, 563–585.

PORTER, W. M. (1979). The "most probable number" method for enumerating infective propagules of vesicular-arbuscular mycorrhizal fungi in soil. *Aust. J. Soil Res.* **17**, 515–519.

PORTER, W. M., ABBOTT, L. K. and ROBSON, A. D. (1978). Effect of rates of application of superphosphate on populations of vesicular-arbuscular endophytes. *Aust. J. exp. agric. anim. Husbandry* **18**, 573–578.

POSSINGHAM, J. V. and GROOT-OBBINK, J. (1971). Endotrophic mycorrhiza and the nutrition of grape vines. *Vitis* **10**, 120–130.

POWELL, C. L. (1975a). Plant growth responses to vesicular-arbuscular mycorrhiza. VIII. Uptake of P by onion and clover infected with different *Endogone* spore types in $^{32}$P-labelled soils. *New Phytol.* **75**, 563–566.

POWELL, C. L. (1975b). Potassium uptake by endotrophic mycorrhizas. *In* "Endomycorrhizas" (Eds F. E. Sanders, B. Mosse and P. B. Tinker), pp. 460–468. Academic Press, London and New York.

POWELL, C. L. (1976). Development of mycorrhizal infections from *Endogone* spores and infected root segments. *Trans. Br. mycol. Soc.* **66**, 439–445.

POWELL, C. L. (1977). Mycorrhizas in hill country soils. I. Spore-bearing mycorrhizal fungi in thirty-seven soils. *N.Z. Jl agric. Res.* **20**, 53–57.

POWELL, C. L. (1979). Spread of mycorrhizal fungi through soil. *N.Z. Jl Agric. Res.* **22**, 335–339.

POWELL, C. L. (1980). Mycorrhizal infectivity in eroded soils. *Soil Biol. Biochem.* **12**, 247–250.

POWELL, C. L. and DANIEL, J. (1978). Mycorrhizal fungi stimulate uptake of soluble and insoluble phosphate fertilizer from a phosphate-deficient soil. *New Phytol.* **80**, 351–358.

Powers, H. R. and Sando, W. J. (1957). Genetics of host-parasite relationship in powdery mildew of wheat. *Phytopathology* **47**, 453.

Pueppke, S. G., Freund, T. H., Schulz, B. C. and Friedman, H. P. (1981). Interaction of soybean lectin with Rhizobia from peanut (*Arachis hypogaea* L.) *In* "Current Perspectives in Nitrogen Fixation" (Eds A. H. Gibson and W. E. Newton), pp. 423. Australian Academy of Science, Canberra.

Purves, S. and Hadley, G. (1975). Movement of carbon compounds between partners in orchid mycorrhiza. *In* "Endomycorrhizas" (Eds F. E. Sanders, B. Mosse and P. B. Tinker), pp. 173–194. Academic Press, London and New York.

Purves, S. and Hadley, G. (1976). The physiology of symbiosis in *Goodyera repens*. *New Phytol.* **77**, 689–696.

Pyrozinski, K. A. and Malloch, D. W. (1975). The origin of land plants: a matter of mycotrophy. *Biosystems* **6**, 153–164.

Quednow, K. G. (1930). Beiträge zur Frage der Aufnahme gelöster Kohlenstoffverbindungen durch Orchideen und andere Pflanzen. *Bot. Arch.* **30**, 51–108.

Rambelli, A. (1973). The rhizosphere of mycorrhizae. *In* "Ectomycorrhizae" (Eds G. C. Marks and T. T. Kozlowski), pp. 299–349. Academic Press, New York and London.

Ramsbottom, J. (1922). Orchid Mycorrhiza. *Trans. Br. mycol. Soc.* **8**, 28–66.

Rast, D. and Stäuble, E. J. (1970). On the mode of action of isovaleric acid in stimulating the germination of *Agarius bisporus* spores. *New Phytol.* **69**, 557–566.

Ratnayake, M., Leonard, R. T. and Menge, J. A. (1978). Root exudation in relation to supply of phosphorus and its possible relevance to mycorrhizal formation. *New Phytol.* **81**, 543–552.

Raven, J. A. (1974). Phosphate transport in *Hydrodictyon africanum*. *New Phytol.* **73**, 421–432.

Raven, J. A. and Smith, F. A. (1976). Nitrogen assimilation and transport in vascular land plants in relation to intracellular pH regulation. *New Phytol.* **76**, 415–431.

Raven, J. A., Smith, S. E. and Smith, F. A. (1978). Ammonium assimilation and the role of mycorrhizas in climax communities in Scotland. *Trans. Proc. bot. Soc. Edinb.* **43**, 27–35.

Rawald, W. (1962). Zur Abhängigkeit des Mycelwachstums höheren Pilze von Versongung mit Kohlenhydraten. *Z. allg. Mikrobiol.* **2**, 303–313.

Rawald, W. (1963). Untersuchungen zur Stickstoffernährung der höheren Pilze. *In* "Mykorrhiza" (Eds W. Rawald and H. Lyr), pp. 67–83. Internal Mykorrhiza Symposium Weimar, 1960. G. Fisher, Jena.

Rayner, M. C. (1927). Mycorrhiza. New Phytol. Reprint 15. 246 pp.

Rayner, M. C. and Levisohn, I. (1940). Production of synthetic mycorrhiza in cultivated cranberry. *Nature, Lond.* **145**, 461.

Rayner, M. C. and Nielson-Jones, W. (1944) "Problems in tree nutrition". Faber and Faber. London.

Read, D. J. (1974). *Pezizella ericae* sp. nov. the perfect state of a typical mycorrhizal endophyte of Ericaceae. *Trans. Br. mycol. Soc.* **65**, 381–383.

Read, D. J. (1978). Biology of mycorrhiza in heathland ecosystems with special reference to the nitrogenous nitrition of Ericaceae. *In* "Microbial Ecology" (Eds M. W. Loutit and J. A. R. Miles), pp. 324–328. Springer-Verlag, Berlin, Heidelberg, New York.

Read, D. J. (1982). The biology of mycorrhiza in the Ericales. Proc. 5th NACOM. *Can. J. Bot.* (In press.)

Read, D. J. and Armstrong, W. (1972). A relationship between oxygen transport and the formation of the ectotrophic mycorrhizal sheath in conifer seedlings. *New Phytol.* **71**, 49–53.

READ, D. J. and HASELWANDTNER, K. (1981). Observations on the mycorrhizal status of some alpine plant communities. *New Phytol.* **88**, 341–352.

READ, D. J., KIANMEHR, H. and MALIBARI, A. (1977). The biology of mycorrhiza in *Helianthemum* Mill. *New Phytol.* **78**, 305–312.

READ, D. J., KOUCHEKI, H. K. and HODGSON, J. (1976). Vesicular-arbuscular mycorrhiza in natural vegetation systems. I. The occurrence of infection. *New Phytol.* **77**, 641–653.

READ, D. J. and MALIBARI, A. (1979). Water transport through mycelial strands of ectomycorrhizal roots of pine. *In* "Root Physiology and Symbiosis" (Eds A. Riedacher and M. J. Gagnaire-Michard), pp. 410–423. IUFRO Symposium, Nancy, France.

READ, D. J. and STRIBLEY, D. P. (1973). Effect of mycorrhizal infection on nitrogen and phosphorus nutrition of ericaceous plants. *Nature, Lond.* **244**, 81.

READ, D. J. and STRIBLEY, D. P. (1975). Some mycological aspects of the biology of mycorrhiza in the Ericaceae. *In* "Endomycorrhizae" (Eds F. E. Sanders, B. Mosse and P. B. Tinker), pp. 105–117. Academic Press, London and New York.

REEVES, F. B., WAGNER, D., MOORMAN, T. and KIEL, J. (1979). The role of endomycorrhizae in revegetation practices in the semi-arid west. I. A comparison of incidence of mycorrhizae in severely disturbed vs. natural environments. *Am. J. Bot.* **66**, 6–13.

REID, C. P. P. (1971). Transport of $^{14}$C-labelled substances in mycelial strands of *Thelephora terrestris. In* "Mycorrhizae" Proc. 1st NACOM Washington, DC. pp. 227–2

REID, C. P. P. (1979). Mycorrhizae and water stress. *In* "Root Physiology and Symbiosis" (Eds A. Riedacker and M. J. Gagnaire-Michard), pp. 392–408. IUFRO Symposium, Nancy, France.

REID, C. P. P. and BOWEN, G. D. (1978). Effect of soil moisture on V-A mycorrhiza formation and root development in *Medicago. In* "The Root-Soil Interface" (Eds J. L. Harley and R. Scott Russell), pp. 211–219. Academic Press, London and New York.

REID, C. P. P. and BOWEN, G. D. (1979). Effect of water stress on phosphorus uptake by mycorrhizas of *Pinus radiata.* New Phytol. **83**, 103–107.

REID, C. P. P. and WOODS, F. W. (1969). Translocation of $^{14}$C-labelled compounds in mycorrhiza and its implications in interpreting nutrient cycling. *Ecology* **50**, 179–181.

REISSEK, S. (1847). Endophyten der Pflanzenzelle. *Naturw. Abh.* (Ed. W. Haidinger), **1**, 31.

RENNIE, P. J. (1955). The uptake of nutrients by mature forest growth. *Pl. Soil* **7**, 49–95.

RHODES, L. H. and GERDEMANN, J. W. (1975). Phosphate uptake zones of mycorrhizal and non-mycorrhizal onions. *New Phytol.* **75**, 555–561.

RHODES, L. H. and GERDEMANN, J. W. (1978a). Translocation of calcium and phosphate by external hyphae of vesicular-arbuscular mycorrhizae. *Soil Sci.* **126**, 125–126.

RHODES, L. H. and GERDEMANN, J. W. (1978b). Hyphal translocation and uptake of sulphur by vesicular-arbuscular mycorrhizae of onions. *Soil Biol. Biochem.* **10**, 355–360.

RHODES, L. H. and GERDEMANN, J. W. (1978c). Influence of phosphorus nutrition on sulphur uptake by vesicular-arbuscular mycorrhizae of onions. *Soil Biol. Biochem.* **10**, 361–364.

RHODES, L. H. and GERDEMANN, J. W. (1980). Nutrient translocation in vesicular-arbuscular mycorrhizae. *In* "Cellular Interactions in Symbiosis and Parasitism" (Eds C. B. Cooks, P. W. Pappas and E. D. Rudolph), pp. 173–195. Ohio State University Press, Columbus, Ohio, U.S.A.

RICH, J. R. and BIRD, G. W. (1974). Association of early season vesicular-arbuscular mycorrhizae with increased growth and development of cotton. *Phytopathology* **64**, 1421–1425.

RICHARDS, B. N. (1961). Soil pH and mycorrhiza development in *Pinus. Nature, Lond.* **190**, 105–106.

RICHARDS, B. N. (1965). Mycorrhiza development of loblolly pine seedlings in relation to soil reaction and supply of nitrate. *Pl. Soil* **22**, 187–199.

RICHARDS, B. N. and WILSON, G. I. (1963). Nutrient supply and mycorrhiza development in Caribbean pine. *For. Sci.* **9**, 40–52.

RICHARD, C., FORTIN, J. A. and FORTIN, A. (1971). Protective effect of ectomycorrhizal fungus against the root-pathogen *Mycelium radicis altovirens*. *Can. J. for. Res.* **1**, 246–251.

RICHARDS, C. and FORTIN, J. A. (1973). The identification of *Mycelium radicis atrovirens* (*Phialocephala dimorphospora*). *Can. J. Bot.* **51**, 2247–2248.

RITTER, G. (1964). Vergleichende untersuchungen über die Bildung von Ectoenzymen durch Mykorrhizenpilze. *Z. allg. Mikrobiol.* **4**, 295–312.

RIVETT, M. (1924). The root tubercles in *Arbutus unedo*. *Ann. Bot.* (*Lond.*) **38**, 661–677.

ROBBINS, W. J., HARVEY, A., DAVIDSON, R. W., HA, R. and ROBBINS, W. C. (1945). A survey of wood-destroying and other fungi for antibacterial activity. *Bull. Torrey bot. Club* **72**, 165–190.

ROBERTS, O. (1950). Translocation in the fungi. Proceedings of the 7th International Botanical Congress, Stockholm, pp. 453–454.

ROBERTSON, D. C. and ROBERTSON, J. A. (1983). Ultrastructure of *Pterospora andromeda* and *Sarcodes sanguinea* mycorrhizas. *New Phytol.* (In press.)

ROBERTSON, N. F. (1954). Studies on the mycorrhiza of *Pinus sylvestris*. *New Phytol.* **53**, 253–283.

ROMMELL, L. G. (1938). A trenching experiment in spruce forest and its bearing on the problems of mycotrophy. *Svensk Bot. Tidskr.* **32**, 89–99.

ROMMELL, L. G. (1939a). The ecological problem of mycotrophy. *Ecology* **20**, 163–167.

ROMMELL, L. G. (1939b). Barrskogens marksvampar och deras roll i skogens liv. *Svensk fören. Tidskr.* **37**, 348–375.

RONCADORI, R. W. and HUSSEY, R. S. (1977). Interaction of the endomycorrhizal fungus *Gigaspora magarita* and root-knot nematode on cotton. *Phytopathology* **67**, 1507–1511.

ROSE, S. and YOUNGBERG, C. T. (1981). Tripartite associations in snow-brush (*Ceanothus velutinus*): effect of vesicular-arbuscular mycorrhizae on growth, nodulation and nitrogen fixation. *Can. J. Bot.* **59**, 34–39.

ROSS, J. P. (1971). Effect of phosphate fertilization on yield of mycorrhizal and non-mycorrhizal soybeans. *Phytopathology* **61**, 1400–1403.

ROVIRA, A. D. (1965). Plant root exudates and their influence upon soil micro-organisms. *In* "Ecology of Soil-borne Plant Pathogens" (Eds K. F. Baker and W. C. Snyder), pp. 170–186. University of California Press, Berkeley.

RUINEN, J. (1953). Epiphytosis. A second view of Epiphytism. *Ann. bogor.* **I**, 101–157.

SACKSTON, W. E. (1962). Studies on sunflower Rust. III. Occurrence, distribution and significance of races of *Puccinia helianthi* Schw. *Can. J. Bot.* **40**, 1449–1458.

SAFIR, G. R., BOYER, J. S. and GERDEMANN, J. W. (1971). Mycorrhizal enhancement of water transport in soybean. *Science* **172**, 581–583.

SAFIR, G. R., BOYER, J. S. and GERDEMANN, J. W. (1972). Nutrient status and mycorrhizal enhancement of water transport in soybean. *Pl. Physiol.* **49**, 700–703.

SAIF, S. R. (1977). The influence of stage of root development on vesicular-arbuscular mycorrhizae and Endogonaceous spore population in field-grown vegetable crops. I. Summer grown crops. *New Phytol.* **79**, 341–348.

SAIF, S. R. (1981). The influence of soil aeration on the efficiency of vesicular-arbuscular mycorrhizae. I. Effect of soil oxygen on the growth and mineral uptake of *Eupatorium odoratum* L. inoculated with *Glomus macrocarpus*. *New Phytol.* **88**, 649–659.

SAMBORSKI, D. J. and DYCK, P. L. (1968). Inheritance of virulence in wheat leaf rust on the standard differential wheat varieties. *Can. J. Genet. Cytol.* **10**, 24–32.

SANDERS, F. E. (1975b). The effect of foliar-applied phosphate on the mycorrhizal infections of onion roots. *In* "Endomycorrhizas" (Eds F. E. Sanders, B. Mosse and P. B. Tinker), pp. 261–276. Academic Press, London and New York.

SANDERS, F. E. and TINKER, P. B. (1971). Mechanism of absorption of phosphate from soil by *Endogone* mycorrhizas. *Nature, Lond.* **33**, 278–279.

SANDERS, F. E. and TINKER, P. B. (1973). Phosphate flow into mycorrhizal roots. *Pestic. Sci.* **4**, 385–395.

SANDERS, F. E., TINKER, P. B., BLACK, R. L. and PALMERLEY, S. M. (1977). The development of endomycorrhizal root systems. I. Spread of infection and growth promoting effects with four species of vesicular-arbuscular mycorrhizae. *New Phytol.* **78**, 257–268.

SANDS, R., REID, C. P. P. and FISCUS, E. C. (1981). Hydraulic conductivity of mycorrhizal pine roots. 5 NACOM abstracts, p. 33.

SANDS, R. and THEODOROU, C. (1978). Water uptake of mycorrhizal roots of radiata pine seedlings. *Aust. J. pl. Phys.* **5**, 301–309.

SANTORO, T. and CASSIDA, L. E. (1962). Elaboration of antibiotics by *Boletus luteus* and certain other fungi. *Can. J. Microbiol.* **8**, 43.

SARGENT, J. A., TOMMERUP, I. C. and INGRAM, D. S. (1973). The penetration of a susceptible lettuce variety by the down mildew fungus *Bremia lactucae* Regl. *Phys. pl. Path.* **3**, 231–240.

SCANNERINI, S. (1968). Sull' ultrastruttura delle ectomicorrize. II. Ultrastruttura di una micorriza di ascomycete: *Tuber albidum X Pinus strobus. Allionia* **14**, 77–95.

SCANNERINI, S. (1972). Ultrastruttura delle endomicorrize di *Ornithogalum umbellatum* L. all'inizio dell' attivita vegetativa. *Allionia* **18**, 129–150.

SCANNERINI, S. (1975). Le ultrastructure delle micorize. *G. Bot. Ital.* **109**, 109–144.

SCANNERINI, S. and BELLANDO, M. (1967). Some ultrastructural features of endotrophic mycorrhiza in *Ornithogalum umbellatum. G. Bot. Ital.* **101**, 313–324.

SCANNERINI, S. and BONFANTE-FASOLO, P. (1975). Dati preliminari sull' ultrastruttura di vesicole intracellulari nell' endomicorriza di *Ornithogalum umbellatum* L. *Atti. Accad. Sci. Torenc* **109**, 619–621.

SCANNERINI, S. and BONFANTE-FASOLO, P. (1976). Ultrastructural features of a vesicular-arbuscular mycorrhiza. Proc. VI European Congress on Electron Microscopy, Vol. 2, pp. 492–494.

SCANNERINI, S. and BONFANTE-FASOLO, P. (1979). Ultrastructural cytochemical demonstration of polysaccharides and proteins within the host-arbuscule interfacial matrix in endomycorrhiza. *New Phytol.* **83**, 87–94.

SCANNERINI, S. and BONFANTE-FASOLO, P. (1982). Analysis of mycorrhizal associations. Proc. 5th NACOM, *Canadian J. Bot.* (In press.)

SCANNERINI, S. and PALENZONA, M. (1967). Recherche sulle ectomicorrize di *Pinus strobus* in vivaio. III. Micorrize di *Tuber albidum* Pico. *Allionia* **13**, 187–194.

SCANNERINI, S., BONFANTE-FASOLO, P. and FONTANA, A. (1975). An ultra-structural model for the host-symbiont interaction in the endotrophic mycorrhizae of *Ornithogalum umbellatum* L. *In* "Endomycorrhizas" (Eds F. E. Sanders, B. Mosse and P. B. Tinker, pp. 313–324. Academic Press, London and New York.

SCHAFFERSTEIN, G. (1938). Untersuchungen über die Avitaminose der Orchideen Keimlinge. *Jb. wiss. Bot.* **86**, 720–752.

SCHAFFERSTEIN, G. (1941). Die Avitaminose der Orchideen Keimlinge. *Jb. wiss. Bot.* **90**, 141–198.

SCHENCK, N. C. and SCHROEDER, V. N. (1974). Temperature response of *Endogone* mycorrhiza on soybean roots. *Mycologia* **66**, 600–605.

SCHENCK, N. C., GRAHAM, S. O. and GREEN, N. E. (1975a). Temperature and light effect on contamination and spore germination of vesicular-arbuscular mycorrhizal fungi. *Mycologia* **67**, 1189–1192.

SCHENCK, N. C., KINLOCH, R. A. and DICKSON, D. W. (1975b). Interaction of endomycorrhizal fungi and root-knot nematode on soybean. *In* "Endomycorrhizas" (Eds F. E. Sanders, B. Mosse and P. B. Tinker), pp. 605–617. Academic Press, London and New York.

SCHÖNBECK, F. and DEHNE, H. W. (1972). Damage to mycorrhizal and non-mycorrhizal cotton seedlings by *Thielaviopsis basicola*. *Pl. Dis. Reptr* **61**, 266–267.

SCHOENBECK, F. and DEHNE, H. W. (1979). The influence of endotrophic mycorrhiza on plant diseases. IV. Fungal parasitism on aerial plant parts. *Z. PflKrankh. PflPath. PflSchutz* **86**, 103–112.

SCHÖNBECK, F. and SCHINZER, U. (1972). Untersuchungen über den Einfluss der endotrophen Mycorrhiza auf die TMV läsionenbildung in *Nicotiana tabaccum* L. var. *Xanthi*—nc. *Phytopath. Z.* **73**, 78–80.

SCHRADER, R. (1958). Untersuchungen zur Biologie der Erbsenmycorrhiza. *Arch. Mikrobiol.* **32**, 81–114.

SCHÜTTE, K. H. (1956). Translocation in the fungi. *New Phytol.* **55**, 164–182.

SCHWEERS, W. and MEYER, F. H. (1970). Einfluss der Mykorrhiza auf den transport von assimilaten in die Wurzel. *Ber. dt. bot. Ges.* **83**, 109–119.

SCOTT, G. D. (1969). "Plant Symbiosis" Edward Arnold, London.

SCOTT RUSSELL, R. (1977). "Plant Root Systems" McGraw Hill Book Company (UK), Maidenhead. 298 pp.

SCHRAMM, J. E. (1966). Plant colonization studies on black wastes from anthracite mining in Pennsylvania. *Am. Phil. Soc.* **56**, 1–94.

SEVIOUR, R. J., WILLING, R. R. and CHILVERS, G. A. (1973). Basidiocarps associated with ericoid mycorrhizas. *New Phytol.* **72**, 381–385.

SEVIOUR, R. J., CHILVERS, G. A. and CROW, W. D. (1974). Characterization of eucalypt mycorrhizas by pyrolysis gas chromatography. *New Phytol.* **73**, 321–323.

SEVIOUR, R. J., HAMILTON, D. and CHILVERS, D. A. (1978). Scanning electron microscopy of surface features of Eucalypt mycorrhizas. *New Phytol.* **80**, 153–156.

SHEMAKHANOVA, N. M. (1967). "Mycotrophy of Wood Plants" USDA & NSF Washington, DC. Israel Program for Scientific Translations. 329 pp.

SHIROYA, T., LISTER, G. R., SLANKIS, V., KROTKOV, G. and NELSON, C. D. (1962). Translocation of products of photosynthesis to roots of pine seedlings. *Can. J. Bot.* **40**, 1125–1136.

SIDHU, G. and PERSON, C. (1971). Genetic control of virulence in *Ustilago hordei*. II. Segregations for higher levels of virulence. *Can. J. Genet. Cytol.* **14**, 209–213.

SIDHU, G. and PERSON, C. (1972). Genetic control of virulence in *Ustilago hordei*. Identification of genes for host resistance and demonstration of gene-for-gene relations. *Can. J. Genet. Cytol.* **14**, 209–213.

SIHANONTH, P. and TODD, R. L. (1977). Transfer of nutrients from ectomycorrhizal fungi to plant roots. *In* "Soil organisms as components of ecosystems". *Ecol. Bull. Stockholm*, **25**, 392–397.

SILSBURY, J. H., SMITH, S. E. and OLIVER, A. J. (1983). A comparison of growth efficiency and specific rate of dark respiration of uninfected and vesicular-arbuscular mycorrhizal plants of *Trifolium subterraneum* L. *New Phytol.* (In press.)

DE SILVA, R. L. and WOOD, R. K. S. (1964). Infection of plants by *Corticium solani* and *C. praticola*—effect of plant exudates. *Trans. Br. mycol. Soc.* **47**, 15–24.

SINGER, R. and MORELLO, J. (1960). Ectotrophic tree mycorrhiza and forest communities. *Ecology* **41**, 549–551.

SINGER, R., ARAUJO, L. DE J. and DA, S. (1979). Litter decomposition and ectomycorrhiza in Amazonian forests. I. A comparisons of litter decomposing and ectomycorrhizal Basidiomycetes in latosol-terra-firme rain forest and white podsol Campinarana. *Acta Amazonica* **9**, 25–41.

SINGH, K. G. (1964). Fungi associated with roots and rhizosphere of Ericaceae. PhD. thesis, Durham University.

SINGH, K. G. (1966). Ectotrophic mycorrhiza in equatorial rain forest. *Malay. Forester* **39**, 13–19.

SINGH, K. G. (1974). Mycorrhiza in the Ericaceae with special reference to *Calluna vulgaris*. *Svensk bot. Tidskr.* **68**, 1–16.

SKINNER, M. F. and BOWEN, G. D. (1974a). The uptake and translocation of phosphate by mycelial strands of pine mycorrhizas. *Soil Biol. Biochem.* **6**, 53–56.

SKINNER, M. F. and BOWEN, G. D. (1974b). The penetration of soil by mycelial strands of ectomycorrhizal fungi. *Soil Biol. Biochem.* **6**, 57–81.

SLANKIS, V. (1948). Einfluss von exudaten von *Boletus variegatus* auf die dichotomische Verzweigung isolieste Kiefernwurzehn. *Physiologia Plant.* **1**, 390–400.

SLANKIS, V. (1949). Wirkung von $\beta$-Indolylessigsäure auf die dichotomische verzweigung isolierter Wurzeln von *Pinus sylvestris* L. *Svensk bot. Tidsk.* **43**, 603–607.

SLANKIS, V. (1950). Effect of $\alpha$-naphthalene acetic acid on dichotomous branching of isolated roots of *Pinus sylvestris*. *Pl. Physiol.* **3**, 40–44.

SLANKIS, V. (1951). Über den Einfluss von $\beta$-Indolylessigsäure und anderen Wuchstoffen auf das Wachstum von Kiefernwurzeln. *Symb. Bot. Upsaliens* **11**, 1–63.

SLANKIS, V. (1958). The role of auxin acid and other exudates in mycorrhizal symbiosis of forest trees. *In* "Physiology of Forest Trees" (Ed. K. V. Thimann), pp. 427–443. Ronald Press, New York.

SLANKIS, V. (1960). Der gegenwärtige Stand unseres Wissens von der Bildung der ektotrophen Mykorrhiza bei Waldbäumen. *In* "Mykorrhiza" (Eds W. Rawald and H. Lyr), pp. 175–183. Gustav Fisher, Jena.

SLANKIS, V. (1963). Der gegenwärtige Stand unseres Wissens von der Bildung der ectotrophen Mykorrhizen bei Waldbaumen. *In* "Mykorrhiza" (Eds W. Rawald and H. G. Lyr), pp. 173–183, Fischer, Jena.

SLANKIS, V. (1967). Renewed growth of ectotrophic mycorrhizae as an indication of an unsuitable symbiotic relationship. Proc. 14 IUFRO Congress **24**, 84–89.

SLANKIS, V. (1971). Formation of ectomycorrhizae of forest trees in relation to light, Carbohydrates and auxins. *In* "Mycorrhizae" (Ed. E. Hacskayo). Proc. 1st NACOM. US Govt. Printing Office, Washington, DC.

SLANKIS, V. (1973). Hormonal relationship in mycorrhizal development. *In* "Ectomycorrhizae" (Eds G. C. Marks and T. T. Kozlowski). Academic Press, New York and London.

SLAYMAN, C. L. (1977). Energetics and control of transport in *Neurospora*. *In* "Water Relations in Membrane Transport in Plants and Animals" (Eds A. M. Jungreis, T. K. Hodges, A. Kleinzeller and S. G. Schultz). pp. 69–86. Academic Press, New York and London.

SMITH, D. C. (1974). Transport from symbiotic algae and symbiotic chloroplast to host cells. *Symp. Soc. exp. Biol.* **28**, 487–508.

SMITH, D. C. (1975). Symbiosis and the biology of lichenised fungi. *Symp. Soc. exp. Biol.* **29**, 373–405.

SMITH, D. C. (1978). Symbiosis in the microbial world. *In* "Essays in Microbiology" (Eds J. R. Norris and M. H. Richmond), pp. 15/1–15/32. John Wiley & Sons, New York.

SMITH, D. C., MUSCATINE, L. and LEWIS, D. H. (1969). Carbohydrate movement from autotrophs to heterotrophs in parasitic and mutualistic symbiosis. *Biol. Rev.* **44**, 17–90.

SMITH, F. A. (1966). Active phosphate uptake by *Nitella translucens*. *Biochim. biophys. Acta* **126**, 94–99.

SMITH, F. A. (1972). A comparison of the uptake of nitrate, chloride, and phosphate by excised beech mycorrhizas. *New Phytol.* **71**, 875–882.

SMITH, F. A. and SMITH, S. E. (1981). Mycorrhizal infection and growth of *Trifolium subterraneum*: comparison of natural and artificial inocula. *New Phytol.* **88**, 311–325.

SMITH, G. W. and SKIPPER, H. D. (1979). Comparison of methods to extract spores of vesicular-arbuscular mycorrhizal fungi. *J. Soil Soc. Am.* **43**, 722–725.

SMITH, J. E. and BERRY, D. R. (1977). "*The Filamentous Fungi*", Vol. 3. "Developmental Mycology". Edward Arnold, London.

SMITH, S. E. (1966). Physiology and ecology of orchid mycorrhizal fungi with reference to seedling nutrition. *New Phytol.* **65**, 488–499.

SMITH, S. E. (1967). Carbohydrate translocation in orchid mycorrhizal fungi. *New Phytol.* **66**, 371–378.

SMITH, S. E. (1973). Asymbiotic germination of orchid seeds on carbohydrates of fungal origin. *New Phytol.* **72**, 497–499.

SMITH, S. E. (1974). Mycorrhizal fungi. *Crit. Rev. Microbiol.* **3**, 275–313.

SMITH, S. E. (1980). Mycorrhizas of autotrophic higher plants. *Biol. Rev.* **55**, 475–510.

SMITH, S. E. (1982). Inflow of phosphate into mycorrhizal and non-mycorrhizal plants of *Trifolium subterraneum* at different levels of soil phosphate. *New Phytol.* **90**, 293–303.

SMITH, S. E. and BOWEN, G. D. (1979). Soil temperature, mycorrhizal infection and nodulation of *Medicago truncatula* and *Trifolium subterraneum*. *Soil Biol. Bioch.* **11**, 469–473.

SMITH, S. E. and DAFT, M. J. (1977). Interaction between growth, phosphate content and $N_2$ fixation in mycorrhizal and non-mycorrhizal *Medicago sativa*. *Aust. J. pl. Physiol.* **4**, 403–413.

SMITH, S. E. and DAFT, M. J. (1978). The effect of mycorrhizas on the phosphate content, nitrogen fixation and growth of *Medicago sativa*. *In* "Microbial Ecology" (Eds M. W. Loutit and J. A. R. Miles), pp. 314–319. Springer-Verlag, Berlin, Heidelberg, New York.

SMITH, S. E. and SMITH, F. A. (1973). Uptake of glucose, trehalose and mannitol by leaf slices of the orchid *Bletilla hyacinthina*. *New Phytol.* **72**, 957–964.

SMITH, S. E. and WALKER, N. A. (1981). A quantitative study of mycorrhizal infection in *Trifolium*: separate determination of the rates of infection and of mycelial growth. *New Phytol.* **89**, 225–240.

SMITH, S. E., NICHOLAS, D. J. D. and SMITH, F. A. (1979). The effect of early mycorrhizal infection on nodulation and nitrogen fixation in *Trifolium subterraneum* L. *Aust. J. Pl. Physiol.* **6**, 305–311.

SMITH, S. E., SMITH, F. A. and NICHOLAS, D. J. D. (1981). Effects of endomycorrhizal infection on phosphate and cation uptake by *Trifolium subterraneum*. *Pl. Soil* **63**, 57–64.

SMITH, T. F. (1978). A note on the effect of soil tillage on the frequency and vertical distribution of spores of vesicular-arbuscular endophytes. *Aust. J. Soil Res.* **16**, 359–361.

SNELLGROVE, R. C., SPLITSTOESSER, W. E., STRIBLEY, D. P. and TINKER, P. B. (1982). The carbon distribution and the demand of the fungal symbiont in leek plants with vesicular-arbuscular mycorrhizas. *New Phytol.* **92**, 75–81.

SOHN, R. F. (1981). *Pisolithus tinctorius* forms long ectomycorrhizae and alters root development in seedlings of *Pinus resinosa*. *Can. J. Bot.* **59**, 2129–2134.

SØNDERGAARD, M. and LAEGAARD, S. (1977). Vesicular-arbuscular mycorrhiza in some aquatic plants. *Nature, Lond.* **268**, 232–233.

SORTKJAER, O. and ALLERMANN, K. (1972). Rhizomorph formation in fungi. I. Stimulation by ethanol and acetate and inhibition by disulphuram of growth and rhizomorph formation in *Armillaria mellea*. *Physiologia Plant.* **26**, 376–380.

SPARLING, G. P. and TINKER, P. B. (1975). Mycorrhizas in Pennine grasslands. *In* "Endomycorrhizas" (Eds F. E. Sanders, B. Mosse and P. B. Tinker), pp. 545–560. Academic Press, London and New York.

SPARLING, G. P. and TINKER, P. B. (1978). Mycorrhizal infection in Pennine grassland. I. Levels of infection in the field, II. Effects of mycorrhizal infection on the growth of some upland grasses on $\gamma$-irradiated soils. III. Effects of mycorrhizal infection on the growth of white clover. *J. Appl. Ecol.* **15**, 943–950; 951–958; 959–964.

SPANSWICK, R. M. (1976). Symplastic transport in tissues. *In* "Transport in Plants. II. Part B. Tissues and Organs" (Eds U. Lüttge and M. G. Pitman). *Encycl. Plant Physiol.* NS **1**, 35–53. Springer-Verlag, Berlin, Heidelberg, New York.

SPENCER-PHILLIPS, P. T. N. and GAY, J. L. (1981). Domains of ATPase in plasmamembranes and transport through infected plant cells. *New Phytol.* **89**, 393–400.

SPORNE, K. R. (1980). A re-investigation of character correlations among dicotyledons. *New Phytol.* **85**, 419–449.

SPRENT, J. I. (1979). "The Biology of Nitrogen Fixing Organisms". McGraw Hill Book Company (UK), Ltd, Maidenhead.

SPRENT, J. I. (1980). Root nodule anatomy, type of export product and evolutionary origin in some Leguminosae. *Plant Cell and Environment* **3**, 35–43.

STACK, R. W., SINCLAIR, W. A. and LARSEN, A. O. (1975). Preservation of basidiospores of *Laccaria laccata* for use as mycorrhizal inoculum. *Mycologia* **67**, 167–170.

STAHL, E. (1900). Die Sinn der Mycorrhizenbildung. *Jb. wiss. Bot.* **34**, 534–668.

STAHL, M. (1949). Die Mykorrhiza de Lebermoose mit besonderer Beruchsichtigung der thallosen Formen. *Planta* **37**, 103–148.

ST. JOHN, T. V., HAYS, R. I. and REID, C. P. P. (1981). A new method for producing pure vesicular-arbuscular mycorrhiza-host cultures without specialised media. *New Phytol.* **89**, 81–86.

STELZ, THÉRÈSE (1968). Mycorrhizes et végétation des pelouses calcaires. Doctoral thesis, University of Rouen.

STONE, E. L. (1950). Some effects of mycorrhizae on the phosphorus nutrition of Monterey pine seedlings. *Proc. Soil Sci. Am.* **14**, 340–345.

STREET, H. E. (1967). The ageing of meristems. *In* "Aspects of the Biology of Ageing". *Symp. Soc. exp. Biol.* **21**, 517–542.

STREET, H. E. (1968). Factors influencing the initiation of activity of meristems of roots. Proc. 15 Easter School in Agric. Sci., Nottingham (Ed. W. J. Whittington). Butterworths, London.

STRIBLEY, D. P. and READ, D. J. (1974a). The biology of mycorrhiza in Ericaceae. III. Movement of Carbon-14 from host to fungus. *New Phytol.* **73**, 731–741.

STRIBLEY, D. P. and READ, D. J. (1974b). The biology of mycorrhiza in the Ericaceae. IV. The effect of mycorrhizal infection on uptake of $^{15}N$ from labelled soil by *Vaccinium macrocarpon* Ait. *New Phytol.* **73**, 1449–1155.

STRIBLEY, D. P. and READ, D. J. (1975). Some nutritional aspects of the biology of ericaceous mycorrhizas. *In* "Endomycorrhizas" (Eds F. E. Sanders, B. Mosse and P. B. Tinker), pp. 195–207. Academic Press, London and New York.

STRIBLEY, D. P. and READ, D. J. (1976). The biology of mycorrhiza in the Ericaceae. VI. The effects of mycorrhizal infection and concentration of ammonium nitrogen on growth of Cranberry (*Vaccinium macrocarpon* Ait.) in sand culture. *New Phytol.* **77**, 63–72.

STRIBLEY, D. P. and READ, D. J. (1980). The biology of mycorrhiza in the Ericaceae. VII. The relationship between mycorrhizal infection and the capacity to utilize simple and complex organic nitrogen sources. *New Phytol.* **86**, 365–371.

STRIBLEY, D. P., REID, D. J. and HUNT, R. (1975). The biology of mycorrhiza in the Ericaceae. V. The effect of mycorrhizal infection, soil type and partial soil sterilization (by gamma irradiation) on growth of Cranberry (*Vaccinium macrocarpon* Ait.). *New Phytol.* **75**, 119–130.

STRIBLEY, D. P., TINKER, P. B. and RAYNER, J. (1980). Relation of internal phosphorus concentration and plant weight in plants infected by vesicular-arbuscular mycorrhizae. *New Phytol.* **86**, 261–266.

STROBEL, N. E., HUSSEY, R. S. and RONCADORI, R. W. (1982). Interactions of vesicular-arbuscular mycorrhizal fungi, *Meloidogyne incognita* and soil fertility on peach. *Phytopathology* **72**, 690–694.

STRULLU, D. G. (1973). Étude des mycorrhizes ectotrophe de *Pinus brutia* Ten. en microscopie électronique à balayage et à transmission. *C.r. hebd. Séanc. Sci., Paris* **t.277**, 1757–1760.

STRULLU, D. G. (1974). Étude ultrastructurale du réseau de Hartig d'une ectomycorrhize à ascomycetes de *Pseudotsuga menziesii. C.r. hebd. Séanc. Acad. Sci., Paris* **278**, ser.D., 2139–2142.

STRULLU, D. G. (1976a). Recherches de biologie et de microbiologie forestières. Étude des relations nutrition, developpement et cytologie des mycorrhizes chez le douglas (*Pseudotsuga menziesii* Mirb.) et les abietacées. Thèse. Univ. Rennes, 1976.

STRULLU, D. G. (1976b). Contribution a l'étude ultrastructuale des ectomycorrhizes à basidiomycetes de *Pseudotsuga menziesii* (Mirb.) *Bull. Soc. Bot. Fr.* **123**, 5–16.

STRULLU, D. G. (1977). Étude de ectomycorrhizes à Basidomycetes et à Ascomycetes du *Betula pubescens* (Ehrh.) en microscopie electronique. *C.r. hebd. Séanc. Acad. Sci., Paris* **284**, 2243–2246.

STRULLU, D. G. (1978). Histologie et cytologie des endomycorrhizas. *Physiol. Vég.* **16**, 657–669.

STRULLU, D. G. (1979). The ultrastructure and spatial representation of the fungal mantle of ectomycorrhizae. *Can. J. Bot.* **57**, 2319–2324.

STRULLU, D. G. and GERAULT, A. (1977). Études des ectomycorrhizes à Basidiomycètes et à ascomycètes du *Betula pubescens* (Ehr.) en microscopie electronique. *C. R. Acad. Sci. Paris* **284**, Série D, 2243–2246.

STRULLU, D. G. and GOURRET, J. P. (1974). Ultrastructure et évolution du champignon symbiotique des racines de *Dactylorchis maculata. J. Micros., Paris* **20**, 285–294.

STRULLU, D. G., GOURRET, J. P., GARREC, J. P. and FOUREY, A. (1981a). Ultrastructure and electron probe analysis of the metachromatic vacuolar granules occurring in *Taxus* mycorrhizas. *New Phytol.* **87**, 537–545.

STRULLU, D. G., GOURRET, J. P. and GARREC, J. P. (1981b). Microanalyse des granules vacuolaires des ectomycorrhizes, endomycorrhizes et endomycothalles. *Physiol. Vég.* **19**, 367–378.

STRULLU, D. G., HARLEY, J. L., GOURRET, J. P. and GARREC, J. P. (1982). Ultrastructure and microanalysis of polyphosphate granules of the ectomycorrhiza of *Fagus sylvatica. New Phytol.* **92**, 417–423.

SUBBARAYADU, S., WILCOXSON, R. D. and SUDIA, T. W. (1966). Translocation of phosphorus-32 into sclerotia of *Phymatotrichum omnivorum. Phytopathology* **56**, 903.

SUMMERHAYES, V. S. (1951). Wild Orchids of Britain. Collins, London.

SUTTON, J. C. (1973). Development of vesicular-arbuscular mycorrhiza in crop plants. *Can. J. Bot.* **51**, 2487–2493.

SUTTON, J. C. and BARRON, C. L. (1972). Population dynamics of *Endogone* spores in soil. *Can. J. Bot.* **50**, 1909–1914.

SUTTON, J. C. and SHEPPARD, B. R. (1976). Aggregation of sand dune soil by endomycorrhizal fungi. *Can. J. Bot.* **54**, 326–333.

SWAMINATHAN, V. (1979). Nature of the inorganic fraction of soil phosphate fed on by vesicular-arbuscular mycorrhizae of potatoes. *Proc. Indian Acad. Sci.* **88B**, 423–433.

SWARD, R. J. (1981). The structure of the spores of *Gigaspora margarita*. II. Changes accompanying germination. *New Phytol.* **88**, 661–666.

SWARD, R. J., HALLAM, N. D. and HOLLAND, A. A. (1978). *Endogone* spores in a heathland area of South Eastern Australia. *Aust. J. Bot.* **26**, 29–43.

TANDY, P. A. (1975a). Studies of sporocarpic Endogonaceae in Australia. PhD thesis, Adelaide University.

TANDY, P. A. (1975b). Sporocarpic species of Endogonaceae in Australia. *Aust. J. Bot.* **23**, 849–866.

TANENBAUM, S. W. (1978). "Cytochalasins: Biochemical and Cell Biological Aspects". Elsevier, North Holland, Amsterdam.

TAYLOR, D. L. (1978). Nutrition of algal-invertebrate symbiosis. II. Effects of exogenous nitrogen sources on growth, photosynthesis and the rate of excretion by symbionts *in vivo* and *in vitro*. *Proc. R. Soc., B.* **201**, 401–412.

TENNANT, D. (1975). A test of a modified line intersect method of estimating root length. *J. Ecol.* **63**, 995–1001.

TERNETZ, C. (1907). Über die Assimilation des atmosphärischen Stickstoffes durch Pilze. *Jb. wiss. Bot.* **44**, 353–408.

THARPAR, H. S., SINGH, B. and BAKSHI, B. K. (1967). Mycorrhiza in *Eucalyptus*. *Indian Forestry* **93**, 756–759.

THAXTER, R. (1922). A revision of the Endogonaceae. *Proc. Am. Acad. Arts Sci.* **57**, 291–351.

THEODOROU, C. (1968). Inositol phosphate in needles of *Pinus radiata* D. Don and the phytase activity of mycorrhizal fungi. *Proc. Congr. Soil Sci. Adelaide* **3**, 483–493.

THEODOROU, C. (1971). The phytase activity of the mycorrhiza fungus *Rhizopogon roseolus*. *Soil Biol. Biochem.* **3**, 89–90.

THEODOROU, C. (1978). Soil moisture and the mycorrhizal association of *Pinus radiata* D. Don. *Soil Biol. Biochem.* **10**, 33–37.

THEODOROU, C. and BOWEN, G. D. (1969). Influence of pH and nitrate on mycorrhizal association of *Pinus radiata* D. Don. *Aust. J. Bot.* **17**, 54.

THEODOROU, C. and BOWEN, G. D. (1970). Mycorrhizal responses of radiata pine in experiments with different fungi. *Aust. For.* **34**, 183–191.

THEODOROU, C. and BOWEN, G. D. (1971a). Influence of temperature on the mycorrhizal association of *Pinus radiata* D. Don. *Aust. J. Bot.* **19**, 13–20.

THEODOROU, C. and BOWEN, G. D. (1971b). Effects of non-host plants on growth of mycorrhizal fungi of radiata pine. *Aust. For.* **35**, 17–22.

THEODOROU, C. and BOWEN, G. D. (1973). Inoculation of seeds and soil with basidiospores of mycorrhizal fungi. *Soil Biol. Biochem.* **5**, 765–771.

THIMANN, K. V. (1935). Growth substances in plants. *A. Rev. Biochem.* **4**, 545–568.

THOMAS, G. W. and JACKSON, R. M. (1979). Sheathing mycorrhizas of nursery-grown *Picea sichensis*. *Trans. Br. mycol. Soc.* **73**, 117–125.

THROWER, L. B. and THROWER, S. L. (1968). Movement of nutrients in fungi. I. The mycelium. *Aust. J. Bot.* **16**, 71–80.

THROWER, S. L. and THROWER, L. B. (1961). Transport of carbon in fungal mycelium. *Nature, Lond.* **190**, 82–00.

TIMMER, L. W. and LEYDEN, R. F. (1980). The relationship of mycorrhizal infection to phosphorus-induced copper deficiency in sour orange seedlings. *New Phytol.* **85**, 15–23.

TINKER, P. B. (1975a). Effects of vesicular-arbuscular mycorrhizas on higher plants. *In Symp. Soc. Exp. Biol.* **29**, 325–349.

TINKER, P. B. (1975b). Soil chemistry of phosphorus and mycorrhizal effects on plant growth. *In* "Endomycorrhizas" (Eds F. E. Sanders, B. Mosse and P. B. Tinker), pp. 353–371. Academic Press, London and New York.

TINKER, P. B. (1978). Effects of vesicular-arbuscular mycorrhizas on plant nutrition and plant growth. *Physiol. Vég.* **16**, 743–751.

TISDALL, J. M. and OADES, J. M. (1979). Stabilization of soil aggregates by the root systems of ryegrass. *Aust. J. Soil Res.* **17**, 429–441.

TOMASKEWSKI, M. and WOJCIECHOWSKA, B. (1973). The role of growth regulators released by fungi in pine mycorrhizae. Plant Growth Substances (8th Int. Conf. on Plant Growth Substances, Tokyo), pp. 217–227.

TOMMERUP, I. C. and ABBOTT, L. K. (1981). Prolonged survival and viability of V-A mycorrhizal hyphae after root death. *Soil Biol. Biochem.* **13**, 431–433.

TOMMERUP, I. C. and KIDBY, D. K. (1980). Production of aseptic spores of vesicular-arbuscular endophytes and their viability after chemical and physical stress. *Appl. environ. Microbiol.* **39**, 1111–1119.

TOXOPEUS, H. J. (1956). Reflections on the new physiologic races of *Phytophthora* and the breeding of resistance in potatoes. *Euphytica* **5**, 221–237.

TRANQUILLINI, W. (1964). Photosynthesis and dry matter production of trees at high altitudes. *In* "Formation of Wood in Forest Trees" Ed. M. H. Zimmermann), pp. 505–518. Academic Press, New York and London.

TRAPPE, J. M. (1962). Fungus associates of ectotrophic mycorrhizae. *Bot. Rev.* **28**, 538–606.

TRAPPE, J. M. (1964). Mycorrhizal hosts and distribution of *Cenococcum graniforme. Lloydia* **27**, 100–106.

TRAPPE, J. M. (1967). Pure culture synthesis of Douglas fir mycorrhizae with species of *Hebeloma, Suillus, Rhizopogon* and *Astraeus. For. Sci.* **13**, 121–130.

TRAPPE, J. M. (1971). Mycorrhiza-forming Ascomycetes. *In* "Mycorrhizae". Proc. 1 NACOM, pp. 19–37. US Government Printing Office, Washington, D. C.

TRESCHOW, C. (1944). The nutrition of the cultivated mushroom. *Dansk bot. Ark.* **11**, 1–180.

TRINICK, M. J. (1977). Vesicular-arbuscular infection and soil phosphorus utilization in *Lupinus* spp. *New Phytol.* **78**, 297–304.

TU, J. C. (1979). Alterations in the membranes of bacteriodal cells in soybean root nodules as revealed by freeze fracturing. *Physiol. pl. Path.* **15**, 35–41.

TYREE, M. T. (1970). The symplast concept. A general theory of symplastic transport according to the thermodynamics of irreversible processes. *J. theor. Biol.* **26**, 181–214.

UHLIG, S. K. (1972). Untersuchungen zur Trockenresistenz Mykorrizabildner Pilz. *Z. Bact. Parasit. Infections. Krankheiten und Hygiene* **127**, 124–133.

ULRICH, J. M. (1960a). Auxin production by mycorrhizal fungi. *Physiologia Pl.* **13**, 429–443.

ULRICH, J. M. (1960b). Effect of mycorrhizal fungi and auxins on root development of sugar pine seedlings (*Pinus lambertiana* Dougl.). *Physiologia Pl.* **13**, 493–000.

VANDERPLANK, J. E. (1976). Four essays. *A. Rev. Phytopath.* **14**, 1–10.

VANDERPLANK, J. E. (1978). "Genetic and Molecular Basis of Plant Pathogenesis". Springer-Verlag, Berlin, Heidelberg, New York.

VANDERPLOEG, J. F., LIGHTY, R. W. and SASSER, M. (1974). Mycorrhizal association between *Lilium* taxa and *Edogone. Hort. Sci.* **9**, 383–384.

VAVILOV, N. I. (1949). "The Origin, Variation, Immunity and Breeding of Cultivated

Plants". Translated by K. Starr Chester. Chronica Botanica, Waltham, Massachusetts.

VEGH, I., FAHRE, E. and GIANINAZZI-PEARSON, V. (1979). Présence en France de *Pezizella ericae* Read. On endomycorrhizal fungus of horticultural Ericaceae. *Photopath. Z.* **96**, 231–243.

VERMEULEN, P. (1946). "Studies on *Dactylorchis*". Utrecht. 180 pp.

VESSEY, J. C. and PEGG, G. F. (1973). Autolysis and chitinase production in cultures of *Verticillium albo-atrum*. *Trans. Br. mycol. Soc.* **60**, 133–143.

VOGT, K. A., EDMUNDS, R. L., AUTOS, G. C. and VOGT, D. J. (1980). Relationship between $CO_2$ evolution, ATP concentration, and decomposition in four forest ecosystems in Western Washington. *Oikos* **35**, 72–79.

VOGT, K. A., EDMUNDS, R. L. and GRIER, C. C. (1981). Biomass and nutrient concentrations of sporocarp produced by mycorrhizal and decomposer fungi in *Abies amabilis* stands. *Oecologia (Bot.)*, **50**, 170–175.

VOGT, K. A., GRIER, C. C., MEIER, C. E. and EDMUNDS, R. L. (1982). Mycorrhizal role in net primary production and nutrient cycling in *Abies amabilis* (Dougl.) Forbes ecosystems in Western Washington. *Ecology.* **63**, 370–380.

VOZZO, J. H. and HACSKAYLO, E. (1971). Inoculation of *Pinus caribaea* with ectomycorrhizal fungi in Puerto Rico. *For. Sci.* **17**, 239–245.

WAGNER, G. (1979). Actomycin as a basic mechanism of movement in plants and animals. *In* "Physiology of Movement" (Eds W. Haupt and M. E. Feinleib), pp. 114–126. Springer-Verlag, Berlin, Heidelberg, London.

WALKER, C. (1979). *Complexipes moniliformis*: a new genus and species tentatively placed in Endogonaceae. *Mycotaxon* **10**, 99–104.

WALKER, N. A. (1976). Transport of solutes through the plasmodesmata of *Chara* nodes. *In* "Intercellular Communication in Plants. Studies on Plasmodesmata" (Eds B. E. S. Gunning and A. W. Robards), pp. 165–179. Springer-Verlag, Berlin, Heidelberg, New York.

WALKER, N. A., BEILBY, M. J. and SMITH, F. A. (1979). Amine uniport at the plasmalemma of charophyte cells. I. Current-voltage curves, saturation kinetics, and effects of unstirred layers. *J. membrane Biol.* **49**, 21–55.

WALKER, N. A. and SMITH, S. E. Quantitative studies of mycorrhizal infection in *Trifolium*: asymptotic values of percentage infection and the effect of propagule density in soil and early mycorrhizal development. (In preparation.)

WARCUP, J. H. (1971). Specificity of mycorrhizal associations in some Australian orchids. *New Phytol.* **70**, 41–46.

WARCUP, J. H. (1973). Symbiotic germination of some Australian terrestrial orchids. *New Phytol.* **72**, 387–392.

WARCUP, J. H. (1975a). A culturable *Endogone* associated with Eucalypts. *In* "Endomycorrhizas" (Eds F. E. Sanders, B. Mosse and P. B. Tinker), pp. 53–63. Academic Press, London and New York.

WARCUP, J. H. (1975b). Factors affecting symbiotic germination of orchid seeds. *In* "Endomycorrhizas" (Eds F. E. Sanders, B. Mosse and P. B. Tinker), pp. 87–104. Academic Press, London and New York.

WARCUP, J. H. (1980). Ectomycorrhizal associations of Australian indigenous plants. *New Phytol.* **85**, 531–535.

WARCUP, J. H. (1981). The mycorrhizal relationship of Australian orchids. *New Phytol.* **87**, 371–387.

WARCUP, J. H. and TALBOT, P. H. B. (1962). Ecology and identity of mycelia isolated from soil. *Trans. Br. mycol. Soc.* **45**, 495–518.

WARCUP, J. H. and TALBOT, P. H. B. (1965a). Ecology and identity of mycelia isolated from soil. II. *Trans. Br. mycol. Soc.* **46**, 465–472.

WARCUP, J. H. and TALBOT, P. H. B. (1965b). Ecology and identity of mycelia isolated from soil. III. *Trans. Br. mycol. Soc.* **49**, 427–435.

WARCUP, J. H. and TALBOT, P. H. B. (1966). Perfect states of some Rhizoctonias. *Trans. Br. Mycol. Soc.* **49**, 427–435.

WARCUP, J. H. and TALBOT, P. H. B. (1967). Perfect states of Rhizoctonias associated with orchids. *New Phytol.* **66**, 631–641.

WARCUP, J. and TALBOT, P. H. B. (1970). Perfect states of Rhizoctonias associated with orchids. II. *New Phytol.* **70**, 35–40.

WARNOCK, A. J., FITTER, A. H. and ASHER, M. B. (1982). The influence of a springtail, *Folsomia candida* (Insecta Collembola) on the mycorrhizal association of leek, *Allium porrum* and the vesicular-arbuscular endiphyte, *Glomus fasciculatus. New Phytol.* **90**, 283–292.

WARREN WILSON, J. (1951). Micro-organisms in the rhizosphere of Beech. D.Phil. thesis, Oxford University.

WARREN WILSON, J. and HARLEY, J. L. (in preparation). The development of mycorrhizas in seedlings of *Fagus sylvatica.*

WATKINSON, S. C. (1971a). The mechanism of mycelial strand induction in *Serpula lacrymans*: a possible effect of nutrient distribution. *New Phytol.* **70**, 1079–1086.

WATKINSON, S. C. (1971b). Phosphorus translocation in stranded and unstranded mycelia of *Serpula lacrymans. Trans. Br. mycol. Soc.* **57**, 535–539.

WATKINSON, S. C. (1975). The relation between nitrogen nutrition and formation of mycelial strands in *Serpula lacrymans. Trans. Br. mycol. Soc.* **64**, 195–200.

WATKINSON, S. C. (1979). Growth of rhizomorphs, mycelial strands, coremia and sclerotia. *In* "Fungal Walls and Hyphal Growth" (Eds J. H. Burnett and A. P. J. Trinci), pp. 93–113. Cambridge University Press, Cambridge.

WATRUD, L. S., HEITHAUS, J. J. and JAWORSKI, E. G. (1978). Evidence for production of inhibitor by the vesicular-arbuscular mycorrhizal fungus *Gigaspora marganita. Mycologia* **70**, 821–818.

WATT, A. S. (1947). Pattern and process in the plant community. *J. Ecol.* **35**, 1–22.

WEBSTER, J. (1976). *Pezizella ericae* is homothallic. *Trans. Br. mycol. Soc.* **66**, 173.

WEDDING, R. T. and HARLEY, J. L. (1976). Fungal polyol metabolites in the control of carbohydrate metabolism of mycorrhizal roots of Beech. *New Phytol.* **77**, 675–688.

WEINHOLD, A. R. (1963). Rhizomorph production by *Armillaria mellea* induced by ethanol and related compounds. *Science* **142**, 1065–1066.

WELLS, T. C. E. (1967). Changes in a population of *Spiranthes spiralis* (L.) Chevall. at Knocking Hoe National Nature Reserve, Bedfordshire, 1962–1965. *J. Ecol.* **55**, 83–89.

WERLICH, I. and LYR, H. (1957). Über die Mykorriza ausbilding von Kiefer (*Pinus sylvestris* L.) und Buche (*Fagus sylvatica* L.) auf verschiedenen standorten. *Arch. Forstw.* **6**, 1–23.

WHITE, D. P. (1941). Prairie soil as a medium for tree growth. *Ecology* **22**, 398–407.

WHITE, J. A. and BROWN, M. F. (1979). Ultrastructure and X-ray analysis of phosphorus granules in a vesicular-arbuscular mycorrhizal fungus. *Can. J. Bot.* **57**, 2812–2818.

WHITMORE, T. C. (1974). Change with time and role of cyclones in tropical rain forest in Kolombangara in the Solomon Islands. Inst. Pap. Commonwealth For. Inst.

WHITTINGHAM, J. and READ, D. J. (1982). Vesicular-arbuscular mycorrhiza in natural vegetation systems. III. Nutrient transfer between plants with mycorrhizal interconnections. *New Phytol.* **90**, 277–284.

WILCOX, H. E. (1964). Xylem in roots of *Pinus resinosa* Ait. in relation to heterorhizy and growth activity. *In* "Formation of Wood in Forest Trees" (Ed. M. Zimmerman), pp. 459–478.

WILCOX, H. E. (1967). Seasonal patterns of root initiation and mycorrhizal development in *Pinus resinosa* Air. *Proc. 14 IUFRO Kongress*, **5**, 29–30.

WILCOX, H. E. (1968a). Morphological studies of the root of red pine, *Pinus resinosa*. I. Growth characteristics and branching patterns. *Am. J. Bot.* **55**, 247–254.

WILCOX, H. E. (1968b). Morphological studies of the roots of red pine, *Pinus resinosa*. II. Fungal colonization of roots and development of mycorrhizae. *Am. J. Bot.* **55**, 686–700.

WILCOX, H. E. (1978). Growth and development of *Pinus resinosa* with selected mycorrhizal fungi. *In* "Root Physiology and Symbiosis" (Eds A. Reidacker and M. J. Gagnaire-Michard), pp. 427–434. Proc. IUFRO Symp., Nancy, France.

WILCOX, H. E. (1971). Morphology of ectendomycorrhizae in *Pinus resinosa*. *In* "Mycorrhizae" (Ed. E. Hacskaylo), pp. 54–68. Proc. 1st NACOM. US Government Printing Office, Washington, DC.

WILCOX, H. E. and GANMORE-NEUMANN, R. (1974). Ectendomycorrhizae: *Pinus resinosa* seedlings. I. Characteristics of mycorrhiza produced by a black imperfect fungus. *Can. J. Bot.* **52**, 2145–2155.

WILCOX, H. E. and GANMORE-NEUMANN, R. (1975). Effects of temperature on root morphology. *Can. J. for. Res.* **5**, 171–175.

WILCOX, H. E., GANMORE-NEUMANN, R. and WANG, C. J. C. (1975). Ectendomycorrhizas in *Pinus resinosa* seedlings. II. Characteristics of two fungi producing ectendomycorrhizae in *Pinus resinosa*. *Can. J. Bot.* **52**, 2279–2282.

WILCOXSON, R. D. and SUBBARAYADU, S. (1968). Translocation to and accumulation of phosphorus-32 in sclerotia of *Sclerotium rolfsii*. *Can. J. Bot.* **46**, 85–88.

WILCOXSON, R. D. and SUDIA, T. W. (1968). Translocation in Fungi. *Bot. Rev.* **34**, 32–50.

WILKINS, W. H. (1946). Investigations of the production of antibiotic substances by fungi. *Ann. appl. Biol.* **33**, 188–190.

WILKINS, W. H. and HARRIS, G. C. M. (1944). Investigations into the production of bacteriostatic substances by fungi. VI. Examination of the larger basidiomycetes. *Ann. appl. Biol.* **31**, 261–270.

WILLIAMS, N. D., GOUGH, F. J. and RONDON, M. R. (1966). Interaction of pathogenicity genes in *Puccinia graminis* f. sp. *tritici* and reaction genes in *Triticum aestivum* spp. *vulgare* "Marquis" and "Reliance". *Crop Sci.* **6**, 245–248.

WILLIAMSON, B. (1970). Induced DNA synthesis in orchid mycorrhiza. *Planta* **92**, 347–354.

WILLIAMSON, B. (1973). Acid phosphatases and esterase activity in orchid mycorrhiza. *Planta* **112**, 149–158.

WILLIAMSON, B. and ALEXANDER, I. (1975). Acid phosphatases localized in the sheath of beech mycorrhiza. *Soil Biol. Bioch.* **7**, 195–198.

WILLIAMSON, B. and HADLEY, G. (1969). DNA content of nucleii in orchid protocorms symbiotically infected with *Rhizoctonia*. *Nature* **222**, 582–583.

WILLIAMSON, B. and HADLEY, G. (1970). Penetration and infection of orchid protocorms by *Thanetephorus cucumeris* and other *Rhizoctonia* isolates. *Phytopathology* **60**, 1092–1096.

WILLIAMSON, R. E. (1976). Actin and Motility in Plant Cells. *In* "Contractile Systems in Non-muscle Tissues" (Eds S. V. Perry, A. Margreth and R. S. Adelstein), pp. 91–101. Elsevier, Amsterdam, Oxford, New York.

WILSON, J. M. (1957). A study of the factors affecting the uptake of potassium by mycorrhiza of beech. D.Phil. thesis, Oxford University.

WINTER, A. G. and PEUSS-SCHÖNBECK, H. (1963). Zur Bedeutung der endotrophen Mycorrhiza für die Entwicklung von Kulturpflanzen. *In* "Mycorrhiza" (Eds W.

Rawald and H. Lyr), pp. 367–375. International Mycorrhizasymposium, Weimar. Gustav Fischer, Jena.

WOOD, R. K. S. (1967). "Physiological Plant Pathology". Blackwell Scientific Publications, Oxford.

WOOD, R. K. S. and GRANITI, I. (1976). "Specificity in Plant Diseases". Plenum Press, New York.

WOODS, F. W. and BROCK, K. (1964). Interspecific transfer of $Ca^{45}$ and $P^{32}$ by root systems. *Ecology* **45**, 886–889.

WORLEY, J. F. and HACSKAYLO, E. (1959). The effect of available soil moisture on the mycorrhizal association of Virginian pine. *For. Sci.* **5**, 267–268.

WOOLHOUSE, H. W. (1969). Differences in the properties of the acid phosphatases of plant roots and their significance in the evolution of adaptive ecotypes. *In* "Ecological Aspects of Mineral Nutrition of Plants" (Ed. I. H. Robinson). *Symp. Br. ecol. Soc.* **9**, 357–380.

WOOLHOUSE, H. W. (1975). Transport problems in endomycorrhizas. *In* "Endomycorrhizas" (Eds F. E. Sanders, B. Mosse and P. B. Tinker), pp. 209–239. Academic Press, London and New York.

ZADOKS, J. C. (1961). Yellow rust on wheat: studies in epidemiology and physiologic specialization. *Tijdschr. PlZiekt.* **67**, 69–256.

ZAK, B. (1964). Role of mycorrhizae in root disease. *A. Rev. Phytopath.* **2**, 377–392.

ZAK, B. (1971). Characterization and identification of Douglas fir mycorrhizae. *In* "Mycorrhizae" (Ed. E. Hacskaylo) Proc. 1st NACOM, 38–53. US Government Printing Office. Washington, DC.

ZAK, B. (1973). Classification of Ectomycorrhizae. *In* "Ectomycorrhizae" (Eds G. C. Marks and T. T. Kozlowski), pp. 43–78. Academic Press, New York and London.

ZAK, B. (1974). Ectendomycorrhiza of Pacific madrone (*Arbutus menziesii*). *Trans. Br. mycol. Soc.* **62**, 202–204.

ZAK, B. (1976a). Pure culture synthesis of bearberry mycorrhizae. *Can. J. Bot.* **54**, 1297–1305.

ZAK, B. (1976b). Pure culture synthesis of Pacific madrone ectendomycorrhizae. *Mycologia* **68**, 362–369.

# Index

# H

# I